T0213199

Lecture Notes in Bioinformatics 11466

Subseries of Lecture Notes in Computer Science

More information about this series at http://www.springer.com/series/5381

Ignacio Rojas · Olga Valenzuela ·
Fernando Rojas · Francisco Ortuño (Eds.)

Bioinformatics and Biomedical Engineering

7th International Work-Conference, IWBBIO 2019
Granada, Spain, May 8–10, 2019
Proceedings, Part II

 Springer

Editors
Ignacio Rojas
Department of Computer Architecture
and Computer Technology Higher Technical
School of Information Technology
and Telecommunications Engineering
CITIC-UGR
Granada, Spain

Fernando Rojas
CITIC-UGR
University of Granada
Granada, Spain

Olga Valenzuela
ETSIIT
University of Granada
Granada, Spain

Francisco Ortuño
Fundacion Progreso y Salud
Granada, Spain

University of Chicago
Chicago, IL, USA

ISSN 0302-9743   ISSN 1611-3349 (electronic)
Lecture Notes in Bioinformatics
ISBN 978-3-030-17934-2   ISBN 978-3-030-17935-9 (eBook)
https://doi.org/10.1007/978-3-030-17935-9

LNCS Sublibrary: SL8 – Bioinformatics

This Springer imprint is published by the registered company Springer Nature Switzerland AG
The registered company address is: Gewerbestrasse 11, 6330 Cham, Switzerland

Preface

We are proud to present the set of final accepted full papers for the 7th edition of the IWBBIO conference—International Work-Conference on Bioinformatics and Biomedical Engineering—held in Granada (Spain) during May 8–10, 2019.

IWBBIO 2019 sought to provide a discussion forum for scientists, engineers, educators, and students about the latest ideas and realizations in the foundations, theory, models, and applications for interdisciplinary and multidisciplinary research encompassing disciplines of computer science, mathematics, statistics, biology, bioinformatics, and biomedicine.

The aims of IWBBIO are to create a friendly environment that could lead to the establishment or strengthening of scientific collaborations and exchanges among attendees, and therefore IWBBIO 2019 solicited high-quality original research papers (including significant work-in-progress) on any aspect of bioinformatics, biomedicine, and biomedical engineering.

New computational techniques and methods in machine learning; data mining; text analysis; pattern recognition; data integration; genomics and evolution; next-generation sequencing data; protein and RNA structure; protein function and proteomics; medical informatics and translational bioinformatics; computational systems biology; modeling and simulation and their application in the life science domain, biomedicine, and biomedical engineering were especially encouraged. The list of topics in the successive Call for Papers has also evolved, resulting in the following list for the present edition:

1. **Computational proteomics**. Analysis of protein–protein interactions; protein structure modeling; analysis of protein functionality; quantitative proteomics and PTMs; clinical proteomics; protein annotation; data mining in proteomics.
2. **Next-generation sequencing and sequence analysis**. De novo sequencing, re-sequencing and assembly; expression estimation; alternative splicing discovery; pathway analysis; Chip-seq and RNA-Seq analysis; metagenomics; SNPs prediction.
3. **High performance in bioinformatics**. Parallelization for biomedical analysis; biomedical and biological databases; data mining and biological text processing; large-scale biomedical data integration; biological and medical ontologies; novel architecture and technologies (GPU, P2P, Grid etc.) for bioinformatics.
4. **Biomedicine**. Biomedical computing; personalized medicine; nanomedicine; medical education; collaborative medicine; biomedical signal analysis; biomedicine in industry and society; electrotherapy and radiotherapy.
5. **Biomedical engineering**. Computer-assisted surgery; therapeutic engineering; interactive 3D modeling; clinical engineering; telemedicine; biosensors and data acquisition; intelligent instrumentation; patient monitoring; biomedical robotics; bio-nanotechnology; genetic engineering.
6. **Computational systems for modeling biological processes**. Inference of biological networks; machine learning in bioinformatics; classification for

biomedical data; microarray data analysis; simulation and visualization of biological systems; molecular evolution and phylogenetic modeling.

7. **Health care and diseases**. Computational support for clinical decisions; image visualization and signal analysis; disease control and diagnosis; genome–phenome analysis; biomarker identification; drug design; computational immunology.

8. **E-health**. E-health technology and devices; e-Health information processing; telemedicine/e-health application and services; medical image processing; video techniques for medical images; integration of classical medicine and e-health.

After a careful peer review and evaluation process (each submission was reviewed by at least two, and on average 3.2, Program Committee members or additional reviewer), 97 papers were accepted for oral, poster, or virtual presentation, according to the recommendations of reviewers and the authors' preferences, and to be included in the LNBI proceedings.

During IWBBIO 2019 several special sessions were held. Special sessions are a very useful tool to complement the regular program with new and emerging topics of particular interest for the participating community. Special sessions that emphasize multi-disciplinary and transversal aspects, as well as cutting-edge topics, are especially encouraged and welcome, and in this edition of IWBBIO they were the following:

– **SS1. High-Throughput Genomics: Bioinformatic Tools and Medical Applications**

Genomics is concerned with the sequencing and analysis of an organism's genome. It is involved in the understanding of how every single gene can affect the entire genome. This goal is mainly afforded using the current, cost-effective, high-throughput sequencing technologies. These technologies produce a huge amount of data that usually require high-performance computing solutions and opens new ways for the study of genomics, but also transcriptomics, gene expression, and systems biology, among others. The continuous improvements and broader applications on sequencing technologies is generating a continuous new demand of improved high-throughput bioinformatics tools. Genomics is concerned with the sequencing and analysis of an organism genome taking advantage of the current, cost-effective, high-throughput sequencing technologies. Continuous improvement of genomics is in turn leading to a continuous new demand of enhanced high-throughput bioinformatics tools. In this context, the generation, integration, and interpretation of genetic and genomic data are driving a new era of health-care and patient management. Medical genomics (or genomic medicine) is this emerging discipline that involves the use of genomic information about a patient as part of the clinical care with diagnostic or therapeutic purposes to improve the health outcomes. Moreover, it can be considered a subset of precision medicine that is having an impact in the fields of oncology, pharmacology, rare and undiagnosed diseases, and infectious diseases. The aim of this special session is to bring together researchers in medicine, genomics, and bioinformatics to translate medical genomics research into new diagnostic, therapeutic, and preventive medical approaches. Therefore, we invite authors to submit original research, new tools or pipelines, or their update, and review articles on relevant topics, such as (but not limited to):

- Tools for data pre-processing (quality control and filtering)
- Tools for sequence mapping
- Tools for the comparison of two read libraries without an external reference
- Tools for genomic variants (such as variant calling or variant annotation)
- Tools for functional annotation: identification of domains, orthologs, genetic markers, controlled vocabulary (GO, KEGG, InterPro, etc.)
- Tools for gene expression studies
- Tools for Chip-Seq data
- Integrative workflows and pipelines

Organizers: **Prof. M. Gonzalo Claros**, *Department of Molecular Biology and Biochemistry, University of Málaga, Spain*

Dr. Javier Pérez Florido, *Bioinformatics Research Area, Fundación Progreso y Salud, Seville, Spain*

Dr. Francisco M. Ortuño, *Bioinformatics Research Area, Fundación Progreso y Salud, Seville, Spain*

– **SS2. Omics Data Acquisition, Processing, and Analysis**
Automation and intelligent measurement devices produce multiparametric and structured huge datasets. The incorporation of the multivariate data analysis, artificial intelligence, neural networks, and agent-based modeling exceeds the experiences of classic straightforward evaluation and reveals emergent attributes, dependences, or relations. For the wide spectrum of techniques, genomics, transcriptomics, metabolomics, proteomics, lipidomics, aquaphotomics, etc., the superposition of expert knowledge from bioinformatics, biophysics, and biocybernetics is required. The series of systematic experiments have to also deal with the data pipelines, databases, sharing, and proper description. The integrated concepts offer robust evaluation, verification, and comparison.
In this special section a discussion on novel approaches in measurement, algorithms, methods, software, and data management focused on the omic sciences is provided. The topic covers practical examples, strong results, and future visions.

Organizer: **Dipl-Ing. Jan Urban**, *PhD, Head of laboratory of signal and image processing. University of South Bohemia in Ceské Budejovice, Faculty of Fisheries and Protection of Waters, South Bohemian Research Center of Aquaculture and Biodiversity of Hydrocenoses, Institute of Complex Systems, Czech Republic.*

Websites:
www.frov.jcu.cz/en/institute-complex-systems/lab-signal-image-processing

– **SS3. Remote Access, Internet of Things, and Cloud Solutions for Bioinformatics and Biomonitoring**
The current process of the 4th industrial revolution also affects bioinformatic data acquisition, evaluation, and availability. The novel cyberphysical measuring

devices are smart, autonomous, and controlled online. Cloud computing covers data storage and processing, using artificial intelligence methods, thanks to massive computational power. Laboratory and medical practice should be on the apex of developing, implementing, and testing the novel bioinformatic approaches, techniques, and methods, so as to produce excellent research results and increase our knowledge in the field.

In this special section, results, concepts, and ongoing research with novel approaches to bioinformatics, using the Internet of Things (IoT) devices is presented.

Organizer: **Antonin Barta**, *Antonin Barta, Faculty of Fishery and Waters Protection, Czech Republic*

– **SS4: Bioinformatics Approaches for Analyzing Cancer Sequencing Data**
In recent years, next-generation sequencing has enabled us to interrogate entire genomes, exomes, and transcriptomes of tumor samples and to obtain high-resolution landscapes of genetic changes at the single-nucleotide level. More and more novel methods are proposed for efficient and effective analyses of cancer sequencing data. One of the most important questions in cancer genomics is to differentiate the patterns of the somatic mutational events. Somatic mutations, especially the somatic driver events, are considered to govern the dynamics of clone birth, evolution, and proliferation. Recent studies based on cancer sequencing data, across a diversity of solid and hematological disorders, have reported that tumor samples are usually both spatially and temporally heterogeneous and frequently comprise one or multiple founding clone(s) and a couple of sub-clones. However, there are still several open problems in cancer clonality research, which include (1) the identification of clonality-related genetic alterations, (2) discerning clonal architecture, (3) understanding their phylogenetic relationships, and (4) modeling the mathematical and physical mechanisms. Strictly speaking, none of these issues is completely solved, and these issues remain in the active areas of research, where powerful and efficient bioinformatics tools are urgently demanded for better analysis of rapidly accumulating data. This special issue aims to publish the novel mathematical and computational approaches and data processing pipelines for cancer sequencing data, with a focus on those for tumor micro-environment and clonal architecture.

Organizers: **Jiayin Wang, PhD, Professor,** *Jiayin Wang, PhD, Professor, Department of Computer Science and Technology, Xian Jiaotong University, China*

Xuanping Zhang, PhD, Associate Professor, *Xuanping Zhang, PhD, Associate Professor, Department of Computer Science and Technology, Xian Jiaotong University, China.*

Zhongmeng Zhao, PhD, Professor, *Zhongmeng Zhao, PhD, Professor, Department of Computer Science and Technology, Xian Jiaotong University, China*

– **SS5. Telemedicine for Smart Homes and Remote Monitoring**

Telemedicine in smart homes and remote monitoring is implementing a core research to link up devices and technologies from medicine and informatics. A person's vital data can be collected in a smart home environment and transferred to medical databases and the professionals. Most often different from clinical approaches, key instruments are specifically tailored devices, multidevices, or even wearable devices respecting always individual preferences and non-intrusive paradigms. The proposed session focused on leading research approaches, prototypes, and implemented hardware/software co-designed systems with a clear networking applicability in smart homes with unsupervised scenarios.

*Organizers: **Prof. Dr. Juan Antonio Ortega**. Director of the Centre of Computer Scientific in Andalusia, Spain, Head of Research Group IDINFOR (TIC223), University of Seville, ETS Ingeniería Informática, Spain*

***Prof. Dr. Natividad Martínez Madrid**. Head of the Internet of Things Laboratory and Director of the AAL-Living Lab at Reutlingen University, Department of Computer Science, Reutlingen, Germany*

***Prof. Dr. Ralf Seepold**. Head of the Ubiquitous Computing Lab at HTWG Konstanz, Department of Computer Science, Konstanz, Germany*

– **SS6. Clustering and Analysis of Biological Sequences with Optimization Algorithms**

The analysis of DNA sequences is a crucial application area in computational biology. Finding similarity between genes and DNA subsequences provides very important knowledge of their structures and their functions. Clustering as a widely used data mining approach has been carried out to discover similarity between biological sequences. For example, by clustering genes, their functions can be predicted according to the known functions of other genes in the similar clusters. The problem of clustering sequential data can be solved by several standard pattern recognition techniques such as k-means, k-nearest neighbors, and the neural networks. However, these algorithms become very complex when observations are sequences with variable lengths, like genes. New optimization algorithms have shown that they can be successfully utilized for biological sequence clustering.

*Organizers: **Prof. Dr. Mohammad Soruri** Faculty of Electrical and Computer Engineering, University of Birjand, Birjand, Iran. Ferdows Faculty of Engineering, University of Birjand, Birjand, Iran*

– **SS7. Computational Approaches for Drug Repurposing and Personalized Medicine**

With continuous advancements of biomedical instruments and the associated ability to collect diverse types of valuable biological data, numerous recent research studies have been focusing on how to best extract useful information from the 'big biomedical data' currently available. While drug design has been one of the most essential areas of biomedical research, the drug design process for the most part has not fully benefited from the recent explosion in the growth of biological data and bioinformatics algorithms. With the incredible overhead associated with the traditional drug design process in terms of time and cost, new alternative methods, possibly based on computational approaches, are very much needed to propose innovative ways for effective drugs and new treatment options. As a result, drug repositioning or repurposing has gained significant attention from biomedical researchers and pharmaceutical companies as an exciting new alternative for drug discovery that benefits from the computational approaches. This new development also promises to transform health care to focus more on individualized treatments, precision medicine, and lower risks of harmful side effects. Other alternative drug design approaches that are based on analytical tools include the use of medicinal natural plants and herbs as well as using genetic data for developing multi-target drugs.

Organizer: **Dr. Hesham H. Ali**, *UNO Bioinformatics Core Facility College of Information Science and Technology University of Nebraska at Omaha, USA*

It is important to note, that for the sake of consistency and readability of the book, the presented papers are classified under 14 chapters. The organization of the papers is in two volumes arranged basically following the topics list included in the call for papers. The first volume (LNBI 11465), entitled "Bioinformatics and Biomedical Engineering. Part I," is divided into eight main parts and includes contributions on:

1. High-throughput genomics: bioinformatic tools and medical applications
2. Omics data acquisition, processing, and analysis
3. Bioinformatics approaches for analyzing cancer sequencing data
4. Next-generation sequencing and sequence analysis
5. Structural bioinformatics and function
6. Telemedicine for smart homes and remote monitoring
7. Clustering and analysis of biological sequences with optimization algorithms
8. Computational approaches for drug repurposing and personalized medicine

The second volume (LNBI 11466), entitled "Bioinformatics and Biomedical Engineering. Part II," is divided into six main parts and includes contributions on:

1. Bioinformatics for health care and diseases
2. Computational genomics/proteomics
3. Computational systems for modeling biological processes
4. Biomedical engineering
5. Biomedical image analysis
6. Biomedicine and e-health

This seventh edition of IWBBIO was organized by the Universidad de Granada. We wish to thank to our main sponsor and the institutions, the Faculty of Science, Department of Computer Architecture and Computer Technology, and CITIC-UGR from the University of Granada for their support and grants. We wish also to thank to the editors of different international journals for their interest in editing special issues from the best papers of IWBBIO.

We would also like to express our gratitude to the members of the various committees for their support, collaboration, and good work. We especially thank the Organizing Committee, Program Committee, the reviewers and special session organizers. We also want to express our gratitude to the EasyChair platform. Finally, we want to thank Springer, and especially Alfred Hofmann and Anna Kramer for their continuous support and cooperation.

May 2019

Ignacio Rojas
Olga Valenzuela
Fernando Rojas
Francisco Ortuño

Organization

Steering Committee

Miguel A. Andrade	University of Mainz, Germany
Hesham H. Ali	University of Nebraska, EEUU
Oresti Baños	University of Twente, The Netherlands
Alfredo Benso	Politecnico di Torino, Italy
Giorgio Buttazzo	Superior School Sant'Anna, Italy
Gabriel Caffarena	University San Pablo CEU, Spain
Mario Cannataro	Magna Graecia University of Catanzaro, Italy
Jose María Carazo	Spanish National Center for Biotechnology (CNB), Spain
Jose M. Cecilia	Universidad Católica San Antonio de Murcia (UCAM), Spain
M. Gonzalo Claros	University of Malaga, Spain
Joaquin Dopazo	Research Center Principe Felipe (CIPF), Spain
Werner Dubitzky	University of Ulster, UK
Afshin Fassihi	Universidad Católica San Antonio de Murcia (UCAM), Spain
Jean-Fred Fontaine	University of Mainz, Germany
Humberto Gonzalez	University of the Basque Country (UPV/EHU), Spain
Concettina Guerra	College of Computing, Georgia Tech, USA
Roderic Guigo	Center for Genomic Regulation, Pompeu Fabra University, Spain
Andy Jenkinson	Karolinska Institute, Sweden
Craig E. Kapfer	Reutlingen University, Germany
Narsis Aftab Kiani	European Bioinformatics Institute (EBI), UK
Natividad Martinez	Reutlingen University, Germany
Marco Masseroli	Polytechnic University of Milan, Italy
Federico Moran	Complutense University of Madrid, Spain
Cristian R. Munteanu	University of Coruña, Spain
Jorge A. Naranjo	New York University (NYU), Abu Dhabi
Michael Ng	Hong Kong Baptist University, SAR China
Jose L. Oliver	University of Granada, Spain
Juan Antonio Ortega	University of Seville, Spain
Julio Ortega	University of Granada, Spain
Alejandro Pazos	University of Coruña, Spain
Javier Perez Florido	Genomics and Bioinformatics Platform of Andalusia, Spain
Violeta I. Pérez Nueno	Inria Nancy Grand Est (LORIA), France

Horacio Pérez-Sánchez	Universidad Católica San Antonio de Murcia (UCAM), Spain
Alberto Policriti	University of Udine, Italy
Omer F. Rana	Cardiff University, UK
M. Francesca Romano	Superior School Sant'Anna, Italy
Yvan Saeys	VIB, Ghent University, Belgium
Vicky Schneider	The Genome Analysis Centre (TGAC), UK
Ralf Seepold	HTWG Konstanz, Germany
Mohammad Soruri	University of Birjand, Iran
Yoshiyuki Suzuki	Tokyo Metropolitan Institute of Medical Science, Japan
Oswaldo Trelles	University of Malaga, Spain
Shusaku Tsumoto	Shimane University, Japan
Renato Umeton	CytoSolve Inc., USA
Jan Urban	University of South Bohemia, Czech Republic
Alfredo Vellido	Polytechnic University of Catalonia, Spain
Wolfgang Wurst	GSF National Research Center of Environment and Health, Germany

Program Committee and Additional Reviewers

Hisham Al-Mubaid	University of Houston, USA
Hesham Ali	University of Nebraska Omaha, USA
Rui Alves	Universitat de Lleida, Spain
Georgios Anagnostopoulos	Florida Institute of Technology, USA
Miguel Andrade	Johannes Gutenberg University of Mainz, Germany
Saul Ares	Centro Nacional de Biotecnología (CNB-CSIC), Spain
Hazem Bahig	Ain Sham, Egypt
Oresti Banos	University of Twente, The Netherlands
Ugo Bastolla	Centro de Biologia Molecular Severo Ochoa, Spain
Alfredo Benso	Politecnico di Torino, Italy
Paola Bonizzoni	Università di Milano-Bicocca, Italy
Larbi Boubchir	University of Paris 8, France
David Breen	Drexel University, USA
Jeremy Buhler	Washington University in Saint Louis, USA
Gabriel Caffarena	CEU San Pablo University, Spain
Mario Cannataro	Magna Graecia University of Catanzaro, Italy
Rita Casadio	University of Bologna, Italy
Francisco Cavas-Martínez	Technical University of Cartagena, Spain
José M. Cecilia	Catholic University of Murcia, Spain
Keith C. C. Chan	The Hong Kong Polytechnic University, SAR China
Ting-Fung Chan	The Chinese University of Hong Kong, SAR China
Nagasuma Chandra	Indian Institute of Science, India
Bolin Chen	University of Saskatchewan, Canada
Chuming Chen	University of Delaware, USA
Jeonghyeon Choi	Georgia Regents University, USA
M. Gonzalo Claros	Universidad de Málaga, Spain

Darrell Conklin	University of the Basque Country, Spain
Bhaskar Dasgupta	University of Illinois at Chicago, USA
Alexandre G. De Brevern	INSERM UMR-S 665, Université Paris Diderot—Paris 7, France
Fei Deng	University of California, Davis, USA
Marie-Dominique Devignes	LORIA-CNRS, France
Joaquin Dopazo	Fundacion Progreso y Salud, Spain
Beatrice Duval	LERIA, France
Christian Esposito	University of Naples Federico II, Italy
Jose Jesus Fernandez	Consejo Superior de Investigaciones Cientificas (CSIC), Spain
Gionata Fragomeni	Magna Graecia University of Catanzaro, Italy
Pugalenthi Ganesan	Bharathidasan University, India
Razvan Ghinea	University of Granada, Spain
Oguzhan Gunduz	Marmara University, Turkey
Eduardo Gusmao	University of Cologne, Germany
Christophe Guyeux	University of Franche-Comté, France
Juan M. Gálvez	University of Granada, Spain
Michael Hackenberg	University of Granada, Spain
Nurit Haspel	University of Massachusetts Boston, USA
Morihiro Hayashida	National Institute of Technology, Matsue College, Japan
Luis Herrera	University of Granada, Spain
Ralf Hofestaedt	Bielefeld University, Germany
Vasant Honavar	The Pennsylvania State University, USA
Narsis Kiani	Karolinska Institute, Sweden
Dongchul Kim	The University of Texas Rio Grande Valley, USA
Tomas Koutny	University of West Bohemia, Czech Republic
Istvan Ladunga	University of Nebraska-Lincoln, USA
Dominique Lavenier	CNRS/IRISA, France
José L. Lavín	CIC bioGUNE, Spain
Kwong-Sak Leung	The Chinese University of Hong Kong, SAR China
Chen Li	Monash University, Australia
Shuai Cheng Li	City University of Hong Kong, SAR China
Li Liao	University of Delaware, USA
Hongfei Lin	Dalian University of Technology, China
Zhi-Ping Liu	Shandong University, China
Feng Luo	Clemson University, USA
Qin Ma	Ohio State University, USA
Malika Mahoui	Eli Lilly, USA
Natividad Martinez Madrid	Reutlingen University, Germany
Marco Masseroli	Politecnico di Milano, Italy
Tatiana Maximova	George Mason University, USA
Roderick Melnik	Wilfrid Laurier University, Canada
Enrique Muro	Johannes Gutenberg University/Institute of Molecular Biology, Germany

Kenta Nakai	Institute of Medical Science, University of Tokyo, Japan
Isabel Nepomuceno	University of Seville, Spain
Dang Ngoc Hoang Thanh	Hue College of Industry, Vietnam
Anja Nohe	University of Delaware, USA
José Luis Oliveira	University of Aveiro, Portugal
Juan Antonio Ortega	University of Seville, Spain
Francisco Ortuño	Clinical Bioinformatics Aea, Fundación Progreso y Salud, Spain
Motonori Ota	Nagoya University, Japan
Joel P. Arrais	University of Coimbra, Portugal
Paolo Paradisi	ISTI-CNR, Italy
Javier Perez Florido	Genomics and Bioinformatics Platform of Andalusia (GBPA), Spain
Antonio Pinti	I3MTO Orléans, Italy
Hector Pomares	University of Granada, Spain
María M. Pérez	University of Granada, Spain
Jairo Rocha	University of the Balearic Islands, Spain
Fernando Rojas	University of Granada, Spain
Ignacio Rojas	University of Granada, Spain
Jianhua Ruan	Utsa, USA
Gregorio Rubio	Universitat Politècnica de València, Spain
Irena Rusu	LINA, UMR CNRS 6241, University of Nantes, France
Michael Sadovsky	Institute of Computational Modelling of SB RAS, Russia
Jean-Marc Schwartz	The University of Manchester, UK
Russell Schwartz	Carnegie Mellon University, USA
Ralf Seepold	HTWG Konstanz, Germany
Xuequn Shang	Northwestern Polytechnical University, China
Wing-Kin Sung	National University of Singapore, Singapore
Prashanth Suravajhala	Birla Institute of Scientific Research, India
Yoshiyuki Suzuki	Tokyo Metropolitan Institute of Medical Science, Japan
Martin Swain	Aberystwyth University, UK
Sing-Hoi Sze	Texas A&M University, USA
Stephen Tsui	The Chinese University of Hong Kong, SAR China
Renato Umeton	Massachusetts Institute of Technology, USA
Jan Urban	Institute of Complex Systems, FFPW, USB, Czech Republic
Lucia Vaira	Set-Lab, Engineering of Innovation, University of Salento, Italy
Olga Valenzuela	University of Granada, Spain
Alfredo Vellido	Universitat Politècnica de Catalunya, Spain
Konstantinos Votis	Information Technologies Institute, Centre for Research and Technology Hellas, Greece
Jianxin Wang	Central South University, China

Jiayin Wang   Xi'an Jiaotong University, China
Junbai Wang   Radium Hospital, Norway
Lusheng Wang   City University of Hong Kong, SAR China
Ka-Chun Wong   City University of Hong Kong, SAR China
Fang Xiang Wu   University of Saskatchewan, Canada
Xuanping Zhang   Xi'an Jiaotong University, China
Zhongmeng Zhao   Xi'an Jiaotong University, China
Zhongming Zhao   University of Texas Health Science Center at Houston,
                       USA

Contents – Part II

Computational Genomics/Proteomics

Computational Systems for Modelling Biological Processes

Biomedical Engineering

Biomedical Image Analysis

Biomedicine and e-Health

Contents – Part I

Omics Data Acquisition, Processing, and Analysis

Bioinformatics Approaches for Analyzing Cancer Sequencing Data

Next Generation Sequencing and Sequence Analysis

Structural Bioinformatics and Function

Telemedicine for Smart Homes and Remote Monitoring

Clustering and Analysis of Biological Sequences with Optimization Algorithms

Computational Approaches for Drug Repurposing and Personalized Medicine

Bioinformatics for Healthcare and Diseases

Developing a DEVS-JAVA Model to Simulate and Pre-test Changes to Emergency Care Delivery in a Safe and Efficient Manner

Shrikant Pawar[1,2(✉)] and Aditya Stanam[3]

[1] Department of Computer Science, Georgia State University,
34 Peachtree Street, Atlanta, GA 30303, USA
spawar2@gsu.edu
[2] Department of Biology, Georgia State University,
34 Peachtree Street, Atlanta, GA 30303, USA
[3] College of Public Health, The University of Iowa, UI Research Park,
#219 IREH, Iowa City, IA 52242-5000, USA

Abstract. Patients' overcrowding in Emergency Department (ED) is a major problem in medical care worldwide, predominantly due to time and resource constraints. Simulation modeling is an economical approach to solve complex healthcare problems. We employed Discrete Event System Specification-JAVA (DEVS-JAVA) based model to simulate and test changes in emergency service conditions with an overall goal to improve ED patient flow. Initially, we developed a system based on ED data from South Carolina hospitals. Later, we ran simulations on four different case scenarios. 1. Optimum number (no.) of doctors and patients needed to reduce average (avg) time of assignment (avg discharge time = 33 min). 2. Optimum no. of patients to reduce avg discharge time (avg wait time = 150 min). 3. Optimum no. of patients to reduce avg directing time to critical care (avg wait time = 58 min). 4. Optimum no. of patients to reduce avg directing time to another hospital (avg wait time = 93 min). Upon execution of above 4 simulations, 4 patients got discharged utilizing 3 doctors; 5 patients could be discharged from ED to home; 2 patients could be transferred from ED to critical care; 3 patients could be transferred from ED to another hospital. Our results suggest that the generated DEVS-JAVA simulation method seems extremely effective to solve time-dependent healthcare problems.

Keywords: DEVS-JAVA · Simulation · Emergency department (ED)

1 Introduction

1.1 Problem Statement

An emergency department (ED), also known as accident & emergency department (A&E), provides acute care for patients who attend hospital without prior appointment. The EDs of most hospitals customarily operate 24 h a day, 7 days a week and over-crowding of patients in EDs is an increasing problem in countries around the world. ED overcrowding has been shown to have many adverse consequences such as increased medical errors, decreased quality of care and subsequently poor patient outcomes,

© Springer Nature Switzerland AG 2019
I. Rojas et al. (Eds.): IWBBIO 2019, LNBI 11466, pp. 3–14, 2019.
https://doi.org/10.1007/978-3-030-17935-9_1

increased workload, ambulance diversions, increased patient dissatisfaction, prolonged patient waiting times and increased cost of care [1, 2]. There are different types of computer simulations techniques utilized to address these types of issues in the field of medicine and health. Some of such tools include discrete event simulation (DES), system dynamics (SD) and agent-based simulations (ABS). DES is identified to be one of the best method to address this problem due to its capability of replicate the behavior of complex healthcare systems over time. DEVS stands for Discrete Event System Specification which is a time extended finite state machine [3]. Even more suitable method of modeling for this problem is needed due to the fact that ED system is more sensitive for constraints and limitations imposed by time. We believe that the closest technique to replicate this process could be achieved through DEVS-JAVA [4], in the current article we have modelled the system in ED using DEVS-JAVA to explore the possibility of using the developed model to simulate changes of services conditions and understand the outputs so that a better service to its patients can be provide with the ED.

1.2 Modeling and Simulation Goals

To address these concerns, our specific objectives are to conduct patient flow simulation through ED in terms of key assumptions, systems requirements, and input and output data, and to identify the usefulness of this simulation method for service redesign and evaluating the likely impact of changes related to the delivery of emergency care. We compared our simulation model with current ED flow data from known existing database, which will report on differences in conclusions about ED performance with our simulation model and existing ER practice to solve the problem of ED overcrowding. The simulation model was implemented in DEVS-JAVA. There can be different targets of executing simulations based on this system. But for the purpose of simplification, we considered the following objectives as our goals of this project.

1. Identifying how many doctors and ambulance are optimal for decreasing average time of assignment.
2. Understand how many patients are optimal for decreasing average time of discharge.
3. Identifying how many patients are optimal for decreasing average time of directing to critical care treatment.
4. Studying how many patients are optimal for decreasing average time of directing patient to another hospital.

These types of simulations are extremely useful for hospitals to make decisions on its resource allocations. The simulations importance and contribution is even more significant as the results can be observed without practically implementing a case. This approach can be utilized to reduce risks in any real world system with trustworthy estimations.

2 Materials and Methods

The main focus of this simulation process is to identify the possible modifications of the existing system in order to improve the overall efficiency of the system. One of the best approach to solve this problem is to first develop a system with current statistics and then incorporate modifications (changing the parameters) to the system, to derive the conclusions. Furthermore, the experimental conclusions are needed to be compared to existing ED data, for which we went to ED timings in the South Carolina hospitals (Table 1). The South Carolina hospital data was collected between April 2016 and April 2017 from Centers for Medicare and Medicaid Services (CMS) [5]. The data was collected throughout the day for all the hospitals without any distinct day timings. CMS established standard definitions and a common metric to accurately compare different hospitals. In most cases someone must manually write down the time a patient was seen, so the times are not always precise. To combat this, emergency rooms outfit doctors and nurses with electronic badges now wirelessly record exact times. According to CMS, hospitals have 30 days to review their data before submitting it to the government. The agency places most of the responsibility on hospitals for making sure their data is correct before making it public. Average values of timings are strong enough to represent the whole population and comparing to our simulation results. Given the data in Table 1, we built DEVS-JAVA models for each of these four cases/columns for simulating respective timings. We ran different simulations on each of these cases to improve through put within the above stipulated time ranges. For example, the question for case 1 was, what will be the optimal number of doctors to be utilized in a situation so as to maximize the patients discharged within the average waiting time of 33 min.

Table 1. Waiting times of different stages of South Carolina hospitals (Waiting times (minutes) for a patient attended by a doctor). The first column refers to the waiting time taken by each hospital for the process of admitting a patient from the accident cite until the patient is attended by a physician. Second column represents the time taken to the process of going through emergency front desk to a doctor for prescription and then the time taken send the prescribed patient to his home. The third column is the time associated with a patient getting a critical care treatment. The fourth column is a transfer time, which is the total time required by a hospital to redirect a patient to a different hospital in a condition that it cannot be treated due to lack of human or physical resources.

Hospital	Waiting time	Time until sent home	Critical case time	Transfer time
Abbeville Area Medical Center	27	96	64	38
Aiken Regional Medical Center	47	189	60	149
Anmed Health	75	204	87	95
Baptist Easley Hospital	54	142	55	53
Beaufort County Memorial Hospital	19	152	34	120
Bon Secours-St Francis Xavier Hospital	35	170	50	68
Cannon Memorial Hospital	34	94	82	68
Carolina Pines Regional Medical Center	25	118	61	80
Carolinas Hospital System	22	163	60	89
Carolinas Hospital System Marion	19	134	26	65
Chester Regional Medical Center	24	109	52	51
Coastal Carolina Hospital	34	207	64	146

(*continued*)

Table 1. (*continued*)

Hospital	Waiting time	Time until sent home	Critical case time	Transfer time
Colleton Medical Center	6	103	28	70
Conway Medical Center	24	131	47	118
East Cooper Medical Center	26	116	36	56
Ghs Greenville Memorial Hospital	51	221	34	135
Ghs Greer Memorial Hospital	31	173	62	110
Ghs Hillcrest Memorial Hospital	30	149	45	101
Ghs Laurens County Memorial Hospital	36	141	60	104
Ghs Oconee Memorial Hospital	46	176	80	61
Grand Strand Regional Medical Center	8	122	21	90
Hampton Regional Medical Center	37	125	53	64
Hilton Head Regional Medical Center	12	167	66	125
Kershawhealth	33	116	48	70
Lake City Community Hospital	45	108	88	65
Lexington Medical Center	43	202	49	176
Mary Black Health System Gaffney	27	146	56	105
Mary Black Health System Spartanburg	18	150	71	110
Mcleod Health Cheraw	20	117	48	42
Mcleod Health Clarendon	36	150	108	63
Mcleod Loris Hospital	34	137	50	43
Mcleod Medical Center - Dillon	35	168	52	36
Mcleod Regional Medical Center-Pee Dee	70	224	101	104
Mount Pleasant Hospital	5	96	26	41
Musc Medical Center	24	162	47	99
Newberry County Memorial Hospital	32	138	54	68
Palmetto Health Baptist	36	193	59	174
Palmetto Health Baptist Parkridge	35	164	58	144
Palmetto Health Richland	28	205	78	207
Palmetto Health Tuomey Hospital	70	190	70	124
Pelham Medical Center	39	150	70	81
Piedmont Medical Center	38	176	72	143
Providence Health	32	152	65	152
Roper Hospital	10	103	31	61
Self-Regional Healthcare	49	154	79	50
Spartanburg Medical Center	68	218	80	145
Springs Memorial Hospital	22	126	58	75
St Francis-Downtown	44	172	75	100
Tidelands Health	20	122	40	93
Tidelands Waccamaw Community Hospital	28	138	44	89
Trident Medical Center	8	115	24	100
Trmc Of Orangeburg & Calhoun	47	172	64	93
Union Medical Center	31	130	86	66
Associated average time (Minutes)	33	150	58	93

2.1 Models Developed

1. **Simulation case 1 (Simulation for admitting a patient to ED with doctor allocation and performing initial treatments referred as "door to doctor time"):** Compilation of all the java codes was conducted on a computer with a 3.07 GHz quad-core CPU and 24 G memory supported by a NVIDIA GTX 580 GPU card with 3G memory on NetBeans integrated development environment (IDE). All the java classes and compilation codes are submitted on GitHub repository which can be found on the corresponding authors account (https://github.com/spawar2/Devs_Java_Patient_Flow_SimulationinEMS).   The   DEVS-JAVA provides a collection of two methods of execution. One is using the *SimView*, which is a graphical representation and the other being running a program directly through a java class to run a simulation for given time using *genDevs.simulation.coordinator*. *SimView* approach is more suitable to graphically understand the flow of each activity, while the second approach is useful for getting associated observations for a specified time of simulation. The main setup for the project is shown in Fig. 1(a). This case simulation is shown in Fig. 1(b). The simulation is based on the assumptions that ambulances will be available every 10 min and a doctor becomes available every 5 min to the queue. Also, we assume that each patient is taking an average time for 10 min to undergo a treatment process. We have utilized the ambulance and doctor generator models which are namely *Accident Site* and *Doctor Assigned*. Also, we are using a simulator for ED that we name as *Emergency Dept. Operator* here is basically a transducer that observes number of patients and doctors flowing through EMS. Running this simulation provides an output for one iteration of this case (Fig. 1(f)). The logging of the data is important to make sure that simulation is working properly and also to identify how many resources utilized for the given time range and what is the corresponding throughput (essentially the number of patients discharged). With the significant number of iterations with for-loops, identifying patters in simulations is possible. The final goal of this case simulation is to identify the fact *"How many doctors and ambulances are optimal for decreasing average time of assignment (average wait time = 33 min)?"*.

2. **Simulation case 2 (Simulation of the patient being sent home after treatment referred as "door to discharge time"):** This case simulation is shown in Fig. 1(c). This simulation will identify the time required to do the process of a patient discharged to home. It will have a *Front Desk* generator that generates the patients to the process of medication prescribed which is further redirected to the generator *Sent Home*. An *Operator* is seen sitting beside and working as a transducer that works on the count of patients flowing in this flow. We have utilized a range of simulation cycles to answer this case question *"How many patients are optimal for decreasing average time of discharge (Average wait time = 150 min)?"*.

3. **Simulation case 3 (Simulation of a patient being transferred to the critical care):** This case simulation is shown in Fig. 1(d). This simulation identifies the time constraints associated with a patient who is assigned to critical care. The transferred to critical care criteria according to CMS only applies to the "initiation of a critical

care level of intervention" for the current data. This simulation is having the generators for *Emergency Front Desk* and *Critical Care*. Also, it is having a transducer called *Operator* which keeps the track of all the ongoing information. The question of interest in this simulation is "How many patients are optimal for decreasing average time of directing them to critical care treatment (Average wait time = 58 min)?".

4. **Simulation case 4 (Simulation of a patient being transferred to another hospital):** This case simulation is shown in Fig. 1(e). The task of this simulation is to clearly identify the timings associated with a patient being transferred to another hospital. The setup will have an *Emergency Front Desk* generator, *Another Hospital* generator and the *Operator* as an observer. The question of interest here is "How many patients are optimal for decreasing average time of directing a patient to another hospital (Average wait time = 93 min)?".

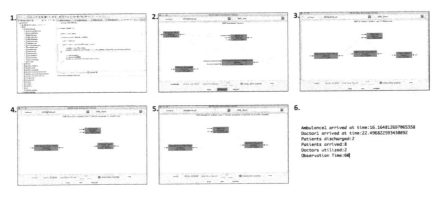

Fig. 1. (a) The main setup for the project. (b)–(e): Different simulation cases. (f): The command line execution output for one of the test simulation. Different generators that we developed for these simulations are *ambulanceGenr*: This model is a ViewableAtomic that is used to generate and simulate patients taken form accident site to the hospital though ambulances. We have set 10 min as the default time for the frequency of ambulance availability. *doctorGenr*: This is also one of the main important generators that will represent one of the main resources of the simulated hospital system which is the doctor. We will be assuming that on average the doctors will be available in each 5 min, while we develop this model. *FrontDeskGenr*: This is also a generator model that will simulate the patient flow thought the Emergency Front Desk. *HomeGenr*: This referees to the generator model that refers to the patients being set to home. *MedGenr*: This is a DEVS model that will represent the process of medications being prescribed by a doctor. *patientEntity*: This is the most important entity that we utilized during these simulations to represent the patient. It represents the behavior of a patient and hence it is utilized in different stages of medical treatments to measure the time for each step. This is one of the most important measurements we used in the discussion of our observations. *Operators*: We are utilizing transducers that we name as operator for each of these simulations. So, we have 4 different transducers that we name as operator0 to operator4.

3 Results

1. *"How many doctors and patients are optimal for decreasing average door to doctor time (average wait time of 1 patient discharge = 33 min)?"*

Results are obtained using execution of simulation using *genDevs.simulation.coordinator* for a range of iterations and observing and summarizing the results against the simulation goal questions. Table 2 states the simulation results of case 1. Using these results and selecting optimized value highlighted in bold, we can calculate the optimal patients discharged by a simple linear Eq. 1 as follows (average wait time = 33 min):

$$Patients\,Discharged = (No\,of\,patients\,discharged/Observed.\,time)$$
$$* \,Current\,average\,comparison\,time. \tag{1}$$

When applied ((8/60) * 33 = 4.4 $\sim$ 4), it gives 4 patients discharged (current time is 1 patient discharge) utilizing 3 doctors as an optimal solution for this situation. We followed the same approach for the rest of cases with a range of simulations with following results.

2. *"How many patients are optimal for decreasing average door to discharge time (Average wait time = 150 min)?"*.

Our optimal experimental results states that at most 5 patients can be discharged (current time is 1 patient discharge) from ED to home, for the average time range of 150 min.

3. *"How many patients are optimal for decreasing average time of directing them to critical care treatment (Average wait time = 58 min)?"*

Our experimental results states that at most 2 patients can be transferred (current time is 1 patient transferred) from ED to critical care, for the average time range of 58 min.

4. *"How many patients are optimal for decreasing average time of directing a patient to another hospital (Average wait time = 93 min)?"*.

Our experimental results states that at most 3 patients can be transferred from transferred (current time is 1 patient transferred) from ED to another hospital, for the average time range of 93 min.

Table 2. Simulation results of case 1.

Observed time (minutes)	No. of doctors	Finish time of last doctor (minutes)	Patients discharged
60	1	9.7	3
60	2	10.4	5
60	**3**	**8.7**	**8**
60	4	10.6	6

4 Discussion and Conclusions

Based on our simulation results, the patient flow from ED can be significantly improved with suggested numbers of doctors and patient arrival timings. There is no mechanism in place to control the number of patients who arrive in the ED for evaluation, and an ideal ED input number cannot be derived. However, the current model can be adopted to the input of patients known to occur at certain day and time. A preliminary survey with average number of input patients arriving in specific ED is needed for the model to operate. This can be easily done in any hospital and a custom parameters can be applied to models evaluation. There are also some other types of computer simulation techniques utilized to address similar issues in the field of medicine and health. Some of such tools include Discrete Event Simulation (DES), System Dynamics (SD) and Agent-based Simulations (ABS). DES is identified to be one of the best method to address this problem due to its capability of replicating the behavior of complex healthcare systems over time. Future validation of our simulated data can be performed using this technique.

Acknowledgments. 1. Support from the Georgia State University Information Technology Department (GSU IT) for server space is gratefully acknowledged.

2. Support from the Artificial Intelligence and Simulation Research Group, Department of Electrical and Computer Engineering, The University of Arizona for providing DEVS-JAVA architecture is gratefully acknowledged.

Supporting Information. No external funding has been utilized for this analysis.

Appendix - Selected Pseudo Code Segments

This section contains selected pseudo code segments that are related to the simulation of case 1 discussed in this article. This could help the readers to reproduce and improve our simulation results.

1. ***Generator: ambulanceGenr***

```
package DEVSJAVALab;
import simView.*;
import java.lang.*;
import genDevs.modeling.*;
import genDevs.simulation.*;
import GenCol.*;
import util.*;
import statistics.*;
public class ambulanceGenr extends ViewableAtomic{
 protected double int_gen_time;
 protected int count;
 protected rand r;
 public ambulanceGenr() {this("ambulanceGenr", 7);}
 public ambulanceGenr(String name,double period){
   super(name);
   addInport("in");
   addOutport("out");
   int_gen_time = period ;
 }
 public void initialize(){
   holdIn("active", int_gen_time);
   r = new rand(123987979);
   count = 0;
 }
 public void deltext(double e,message x)
 {
 Continue(e);
   for (int i=0; i< x.getLength();i++){
     if (messageOnPort(x, "in", i)) { //the stop message from tranducer
       passivate();
     }
   }
 }
 public void deltint( )
 {
 if(phaseIs("active")){
   count = count +1;
// holdIn("active",int_gen_time);
   holdIn("active",6+r.uniform(int_gen_time));
 }
 else passivate();
 }
 public message out( )
 {
 //System.out.println(name+" out count "+count);
   message m = new message();
// content con = makeContent("out", new entity("car" + count));
   content con = makeContent("out", new patientEntity("Ambulance" + count, 5+r.uniform(20), 50+r.uniform(100), 1));
   m.add(con);
   return m;
 }}
```

2.   *Generator: doctorGenr*

```
package DEVSJAVALab;
import simView.*;
import java.lang.*;
import genDevs.modeling.*;
import genDevs.simulation.*;
import GenCol.*;
import util.*;
import statistics.*;
public class doctorGenr extends ViewableAtomic{
  protected double int_gen_time;
  protected int count;
  protected rand r;
  public doctorGenr() {this("doctorGenr", 30);}
public doctorGenr(String name,double period){
   super(name);
   addInport("in");
   addOutport("out");
   int_gen_time = period ;
}
public void initialize(){
   holdIn("active", 2);
   r = new rand(2);
   count = 0;
}
public void deltext(double e,message x)
{
Continue(e);
}
public void deltint( )
{
if(phaseIs("active")){
   count = count +1;
   holdIn("active",20+r.uniform(int_gen_time));
}
else passivate();
}
public message out( )
{
//System.out.println(name+" out count "+count);
   message m = new message();
//  content con = makeContent("out", new entity("truck" + count));
   content con = makeContent("out", new patientEntity("Doctor" + count, 10+r.uniform(30), 100+r.uniform(100), 1));
   m.add(con); return m;
}}
```

3. Generator: *patientEntity*

```
package DEVSJAVALab;
import GenCol.*;
public class patientEntity extends entity{
 protected double processingTime;
 protected double price;
 protected int priority;
 public patientEntity(){
          this("patient", 10, 10, 1);
 }
 public patientEntity(String name, double _procTime, double _price, int _priority){
          super(name);
          processingTime = _procTime;
          price = _price;
          priority = _priority;
 }
 public double getProcessingTime(){
          return processingTime;
 }
 public double getPrice(){
          return price;
 }
 public int getPriority(){
          return priority;
 }
 public String toString(){
          return name+"_"+(double)((int)(processingTime*100))/100;
          //return name+"_"+((double)((int)(processingTime*100)))/100;
}}
```

4. Generator: *EMS_Sim*

```
package DEVSJAVALab;
import simView.*;
import java.awt.*;
import java.io.*;
import genDevs.modeling.*;
import genDevs.simulation.*;
import GenCol.*;
public class EMS_Sim extends ViewableDigraph{
public EMS_Sim(){
  this("EMS Simulation System");
}
public EMS_Sim(String nm){
  super(nm);
  emsConstruct();
}
public void emsConstruct(){
  this.addOutport("out");
  ViewableAtomic amb_genr = new ambulanceGenr("Accident Site",10);
  ViewableAtomic doc_genr = new doctorGenr("Doctor assigned",1);
  ViewableAtomic ems_dpt = new EMSDept("Emergency Dept/Critical care");
  ViewableAtomic operator = new operator ();
  add(amb_genr);
  add(ems_dpt);
  add(doc_genr);
  add(operator);
  addCoupling(amb_genr,"out",ems_dpt,"AmbulanceReached");
  addCoupling(amb_genr,"out",operator,"ariv");
  addCoupling(doc_genr,"out",operator,"ariv");
  addCoupling(ems_dpt,"out",operator,"Assigned");
  addCoupling(doc_genr,"out",ems_dpt,"DoctorAssigned");
  addCoupling(ems_dpt,"out",this,"out");}
```

References

1. Mohiuddin, S., Busby, J., Savović, J.: Patient flow within UK emergency departments: a systematic review of the use of computer simulation modelling methods. BMJ Open (2017)
2. Institute of Medicine (US) Committee on Assuring the Health of the Public in the 21st Century: The Future of the Public's Health in the 21st Century, vol. 5. National Academies Press, Washington (DC) (2002). The Health Care Delivery System. https://www.ncbi.nlm.nih.gov/books/NBK221227/
3. Zeigler, B., Hessam, S.: Introduction to DEVS Modeling and Simulation with JAVA: Developing Component-Based Simulation Models. http://www.cs.gsu.edu/xhu/CSC8350/DEVSJAVA_Manuscript.pdf
4. Artificial Intelligence and Simulation Research Group: DEVS-Java Reference Guide (1997). https://grid.cs.gsu.edu/ ~ fbai1/paper/4560.pdf
5. Hospital Compare datasets: Data.medicare.gov (2018). https://data.medicare.gov/data/hospital-compare/

Concept Bag: A New Method for Computing Concept Similarity in Biomedical Data

Richard L. Bradshaw, Ramkiran Gouripeddi, and Julio C. Facelli[✉]

Department of Biomedical Informatics,
University of Utah, Salt Lake City, UT, USA
{rick.bradshaw,ram.gouripeddi,julio.facelli}@utah.edu

Abstract. Biomedical data are a rich source of information and knowledge, not only for direct patient care, but also for secondary use in population health, clinical research, and translational research. Biomedical data are typically scattered across multiple systems and syntactic and semantic data integration is necessary to fully utilize the data's potential. This paper introduces new algorithms that were devised to support automatic and semi-automatic integration of semantically heterogeneous biomedical data. The new algorithms incorporate both data mining and biomedical informatics methods to create "concept bags" in the same way that "word bags" are used in data mining and text retrieval. The methods are highly configurable and were tested in five different ways on different types of biomedical data. The new methods performed well in computing similarity between medical terms and data elements - both critical for semi/automatic data integration operations.

Keywords: Concept bag · Semantic integration · Biomedical data · Data federation

1 Introduction

Biomedical data are a potentially rich source for information and knowledge discovery. Biomedical data collected for patient care or specific research protocols are valuable for reuse to address broader clinical, translational, comparative effectiveness (CER), population health, or public health research questions. However, reusing biomedical data for these purposes is not trivial. Data from multiple biomedical data systems are typically required to answer research questions and integrating these data is complex. Biomedical data are modeled and stored using various formats, syntaxes, and value sets to represent clinical or biomedical observations or facts about patients, research subjects or other artifacts. For instance, to answer a translational research study question, one would likely need demographic data from one system, diagnostic data from another system, and data from bioinformatics pipelines [1, 2].

Biomedical data integration generally requires homogenization of semantically and syntactically heterogeneous data. Semantic heterogeneity occurs when the same domain is modeled differently and data elements and values have different meanings [3], whereas syntactic heterogeneity occurs when the structural formats between data sets are different. OpenFurther [4], and i2b2/SHRINE (Informatics for Integrating Biology and

© Springer Nature Switzerland AG 2019
I. Rojas et al. (Eds.): IWBBIO 2019, LNBI 11466, pp. 15–23, 2019.
https://doi.org/10.1007/978-3-030-17935-9_2

the Bed-side/Shared Health Research Information Network) [5] are example state-of-the-art biomedical data integration tools. Each employ different integration strategies, but both require experts to perform the semantic and syntactic integration. Between the two forms of heterogeneity, semantic integration is the more challenging and costly aspect of biomedical data integration [6]. Expensive terminologists and/or highly trained knowledge engineers are needed to perform the work [7, 8]. At the rate and scale that biomedical data are created, there are simply not enough humans with these skills to massively scale integration efforts. Moreover much of biomedical data are available as free text and in semi-structured formats that are often not leveraged in semantic integration efforts that may affect results from translation research studies [9]. Semiautomatic, and ultimately automatic semantic integration tools, are required to achieve massive biomedical data integration.

Automatic data integration techniques are based on computing semantic "alignments" between data sets, but none of the existing methods achieve the level of performance needed for biomedical investigations [10]. Here we present new algorithms that incorporate both data mining and biomedical informatics methods to improve on and add to existing biomedical data integration methods. The new methods are based on "concept bags" that are similar to "word bags" used for data mining and text retrieval.

2   Methods

The concept bag (CB) method is based on the n-gram method [11]; however, instead of comparing textual elements, the CB method operates by comparing concept codes. The CB method is defined by two essential steps: (1) Construction of the CB - convert textual or named entities to a representative set of concept codes, and (2) Similarity Measurement – systematically compare CBs using a set or vector-based analysis method. This is a very general definition that leaves a great deal of flexibility in its implementation. Any reasonable selection of a text-to-concept code method or set/vector n-gram-like analysis method could be considered for implementation. This stepwise process describes how the CBs were constructed for the studies that follow:

1. Assign each textual element a unique identifier.
2. Process textual elements and extract representative concept codes. We used the Name-Entity Recognition (NER) software [12] named MetaMap [13] that identifies Unified Medical Language System (UMLS, https://www.nlm.nih.gov/research/umls/) Concept Unique Identifiers (CUI) in the text. MetaMap is a fully integrated NER software which includes, all necessary natural language processes including spell checking for misprints.
3. Associate the distinct set of concept codes from each textual element from step 2 with its unique identifier from step 1.
4. A CB for each textual element is constructed using associations created in step 3. All concept codes associated with each textual identifier comprise the CB for that textual element.

The CB method was originally designed to resolve semantic similarity of strings such as "SBP" and "systolic BP" to the same concept recognizing synonymy between the two strings. However, the method does not consider similarity between words such as "abortion" and "miscarriage" that are not exactly synonyms, but are semantically related [3]. Observing the flexibility of the CB and the potential for ontological relatedness of textual elements, we enhanced CBs by adding UMLS CUIs from the underlying ontological relationships. CUIs from SNOMED's "is-a" hierarchy were extracted and added for each original CB CUI. We named this strategy the Hierarchical Concept Bag (HCB) method. Further details of the method are given in Bradshaw's Dissertation [14].

In order to measure similarities between CBs, we selected the Jaccard similarity algorithm [15]. The Jaccard formula computes a decimal value between 0 and 1, where 0 represents no conceptual similarity, and 1 represents a perfect match, by calculating the ratio between the cardinal numbers of the intersection and union of the CBs of the two entities under comparison. Using the CB to compute the similarity between "abortion" and "miscarriage" literally returned 0.0 similarity. With HCBs, the ontological CUI sets representing "abortion" and "miscarriage" share several CUIs in common. After adding the HCB codes, the similarity between "abortion" and "miscarriage" was upgraded from the CB's 0.0 to 0.89, see Fig. 1 for details.

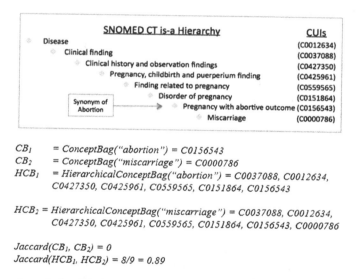

CB_1 = ConceptBag("abortion") = C0156543
CB_2 = ConceptBag("miscarriage") = C0000786
HCB_1 = HierarchicalConceptBag("abortion") = C0037088, C0012634, C0427350, C0425961, C0559565, C0151864, C0156543

HCB_2 = HierarchicalConceptBag("miscarriage") = C0037088, C0012634, C0427350, C0425961, C0559565, C0151864, C0156543, C0000786

Jaccard(CB_1, CB_2) = 0
Jaccard(HCB_1, HCB_2) = 8/9 = 0.89

Fig. 1. Examples of the CB and HCB with hierarchical concept codes (CUIs) from the SNOMED CT is-a hierarchy comparing "miscarriage" and "abortion." The "Parents" figure shows the hierarchies (SNOMED CT is-a hierarchy is poly-hierarchical) and all of the UMLS clinical concepts for both "miscarriage" and "abortion." The Jaccard similarity scores illustrate the differences between the two methods.

To test the performance of the new methods, we considered three test cases that measured the alignment between, (1) heterogeneous data elements (DE) from a controlled vocabulary, (2) DE from an uncontrolled vocabulary, and (3) medical terms. In the following, we present a brief description of the data sets and how they have been used in these test-cases.

- DEs from a controlled vocabulary: 17 DEs from three domains of the UMLS were selected for this study, seven from demographics, five from vital signs, and five echocardiogram measures.
- DEs from an uncontrolled vocabulary: 899,649 DEs extracted from REDCap [16] from 5 sites engaged in clinical and translational research.
- Medical terms: A benchmark containing a set of 30 medical term pairs that have been curated and judged by physicians and terminologists [17].

The textual features of the data sets were evaluated [14] and each had different textual characteristics are depicted in Table 1. The text size averages varied from 13.1 to 74.9 characters per element, while the mean number of concepts per word varied from 1.5 to 3.2. A more detailed description can be found in [14].

Table 1. Descriptive statistics for the studied data sets, UMLS-selected DEs, REDCap DEs, and the medical terms reference.

	Datasets		
	UMLS DEs	REDCap DEs	Medical terms
Element counts	315	899649	60
Mean characters/Element	24 ± 13.1	43.1 ± 74.9	13.4 ± 6.0
Mean words/Element	3.1 ± 1.5	7.1 ± 11.8	1.6 ± 0.7
Concepts/Data set	380	4187	133
Hierarchical concepts/Data set	753	6929	740
Mean Concepts/Element	9.8 ± 6.7	10.4 ± 10.0	2.6 ± 1.9
Mean concepts/Word	3.2 ± 1.8	2 ± 1.4	1.5 ± 0.6
Mean hierarchical concepts/Word	15.6 ± 11.0	12.8 ± 14.5	20.0 ± 14.3

2.1   DE Alignment Measurement

We measured the alignments between the DEs using similarity measurement algorithms. These algorithms returned real values between 0.0 and 1.0 where 1.0 is perfect similarity and 0.0 is no similarity. Alignment decisions were determined by alignment measurement cutoff values. Similarity values do not represent the percentage of semantic alignment or statistical p-values where one chooses a standard cutoff point. Cutoff values that ultimately define alignment decisions are algorithm-specific and needed to be determined for each of the tested algorithms; therefore, cutoff values for each algorithm were calculated using the Youden method [18], which select cutoff values that maximized the specificity and sensitivity of the decision for each algorithm [14].

All DEs were compared to each other to examine all possible alignments. This process required $n(n - 1)/2$ comparisons for each set, or approximately 50,000 comparisons for the UMLS controlled vocabulary concepts for test case 1, and half a trillion comparisons of REDCap DE alignments for test case 2. For the REDCap data set preliminary exploration of the computed CB similarity scores indicated that match candidates were very infrequent, ~ 3 per 1,000 pairs, indicating that an unreasonably large random sample would have been required for human review while maintaining both an accurate sample distribution and conclusive confidence interval. Therefore, a stratified random sample was assembled with 12 buckets based on the computed CB similarity scores, from which a set of 1,200 DE pairs were then randomly sorted and manually reviewed for semantic matches by two professional clinical data architects. Disagreements were reviewed by the two architects to reach consensus. Details of the construction of the data sets used in these comparisons are given in Ref. [14].

Four configurations of the CB and HCB were tested for measuring the alignment between Medical Terms benchmark.

1. CBs created using concepts produced by MetaMap (with default settings) and Jaccard similarity algorithm. This configuration was completely unsupervised.
2. Same as step 1 except the CBs were augmented to HCBs using the SNOMED CT "is-a" hierarchy. This configuration was also completely unsupervised.
3. Same as step 2 but restricting MetaMap to SNOMED CT, and retaining only the single highest-ranking concept for each term. The authors resolved rank ties, which should be considered a minor supervision.
4. Same as step 3 except HCBs were converted to vectors and compared using cosine similarity.

Each of the four CB configurations produced similarity values for each of the 30 pairs from the benchmark data, which were scaled to the same range used by the experts judging the original data set. Correlation between the four CB algorithm configurations given above, seven other published algorithms, and benchmark experts were calculated using Pearson's correlation.

3   Results and Discussion

Receiver operator characteristic (ROC) curves were used to examine the alignment compliance of the DEs from the controlled (UMLS) vocabulary and uncontrolled (REDCap) environment for each data set and each algorithm, see Figs. 2 and 3. The CB and HCB performed very well at the task of aligning UMLS DE as indicated by the AUC = 0.92/0.89, F-measure = 0.79/67 and ROC curves. The performance numbers are slightly lower for the REDCap DE alignment performance AUC = 0.92/0.91, F-measure = 0.55/0.53. This was not surprising due to the nature of REDCap uncontrolled environment where arbitrary abbreviations and local jargon are allowed and supported. None-the-less, even with the added complexities of REDCap DE, the CB and HCB still had much lower combined false positive and false negative rates than the other algorithms considered here.

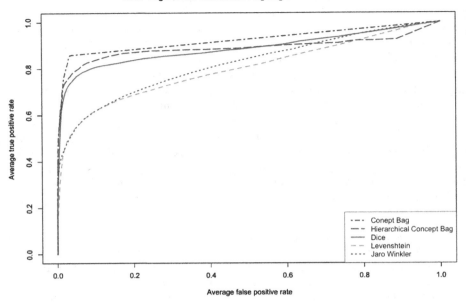

Fig. 2. ROC curve of the UMLS DE alignment algorithm performance. Curves closest to the top left corner are the best performers.

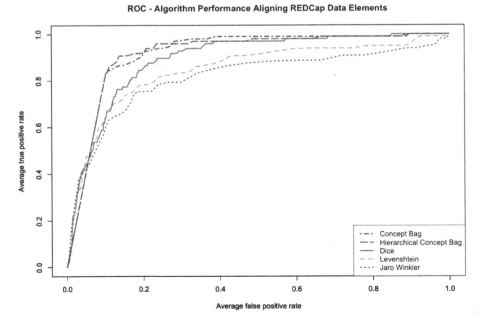

Fig. 3. ROC curve of the REDCap DE alignment algorithm performance. Curves closest to the top left corner are the best performers.

When the goal is automatic DE alignment for semantic integration, alignment decisions are binary, the algorithm either aligns or does not align. In this binary case, degrees of similarity are not tested; therefore, to test the algorithms' ability to assess degree of similarity, we used a medical term similarity benchmark. The CB and HCB algorithms were tested using the medical term similarity benchmark. Table 2 contains the results of the medical term similarity evaluation.

Overall, the HCB correlation scores matched or exceeded 31 other published algorithms [17], for which the top seven are included in Table 2. Of the four CB algorithms tested, the HCB using SNOMED CT with the "is-a" hierarchy measured using Jaccard similarity, tied with the highest correlation with medical terminologists, 0.76, and was the second highest correlation with physicians, 0.72. The tie for the highest correlation with terminologists was with Personalized PageRank algorithm [19]. The highest correlation with physicians, 0.84, was attributed to Pedersen's Context Vector [3] that was successfully augmented with physician-based information content (IC) from a large physician-created corpus. The methods that used terminologist-created SNOMED is-a hierarchy correlated highest with terminologists' similarity judgements, and the method that used physician-created IC correlated highest with physicians.

Table 2. Correlation of the similarity scores obtained with Concept Bag (CB), Hierarchical Concept Bag (HCB), Leacock and Chadorow (LC) [20], Wu and Palmer (WP) [21], Personalized PageRank (PPR) [19], and Context Vector [3].

Method	Physicians	Terminologists
SNOMED HCB	0.72	0.76
SNOMED HCB-Vector	0.65	0.67
UMLS CB	0.46	0.59
UMLS HCB	0.46	0.57
SNOMED LC	0.50	0.66
UMLS LC	0.60	0.65
SNOMED WP	0.54	0.66
UMLS WP	0.66	0.74
SNOMED PPR	0.49	0.61
UMLS PPR	0.67	0.76
Context vector	0.84	0.75

The CB and HCB methods using UMLS (USABase library) and Jaccard similarity both had correlation values of 0.46 with physicians, while the correlation with terminologists was higher for both, 0.59 and 0.57, respectively. The lower correlations are likely due to the broader concept coverage contained in UMLS for which SNOMED CT is a subset, i.e. UMLS produces larger concept bags than it does when there is a reduced set of source vocabularies resulting in more sensitive and less specific similarity measurements. Adding additional hierarchical concept codes magnifies this effect.

4   Conclusions

Automatic alignment of heterogeneous biomedical data is a challenging problem due to the sophisticated semantics of biomedical data. In this paper, we introduced a new class of methods that introduces the idea of using "concept bags" to represent the semantics of textual data elements, described how they can be used to evaluate semantic similarity, and then demonstrated how the similarity measures were tested to automatically align biomedical data elements and compute medical term similarity. Several configurations of the new similarity algorithms were tested for each application and the new methods performed as well or better than well-established methods. Unlike bag of words and n-gram methods, CB and HCB are capable of measuring semantic similarities between synonymous and non-similarly spelled words. We believe we have established the face validity of the CB and HCB methods and recommend them as viable options for computing semantic similarity as demonstrated.

Acknowledgements. This work was partially supported by NCATS award 1ULTR002538 (JCF) and NIH bioCADDIE award 1U24AI117966-01. The authors acknowledge Bernie LaSalle for securing the REDCap data from the CTSA funded institutions and their informatics leaders for providing the data. Computational resources were provided by the Utah Center for High Performance Computing, which has been partially funded by the NIH Shared Instrumentation Grant 1S10OD021644-01A1.

References

1. Gouripeddi, R., Warner, P., Mo, P.: Federating clinical data from six pediatric hospitals: process and initial results for microbiology from the PHIS+ Consortium. In: AMIA Annual Symposium Proceedings (2012)
2. Narus, S.P., Srivastava, R., Gouripeddi, R., Livne, O.E., Mo, P., Bickel, J.P., et al.: Federating clinical data from six pediatric hospitals: process and initial results from the PHIS + Consortium. In: AMIA Annual Symposium Proceedings, pp. 994–1003 (2011). PubMed PMID: 22195159; PubMed Central PMCID: PMCPMC3243196
3. Pedersen, T., Pakhomov, S.V.S., Patwardhan, S., Chute, C.G.: Measures of semantic similarity and relatedness in the biomedical domain. J. Biomed. Inform. **40**(3), 288–299 (2007). https://doi.org/10.1016/j.jbi.2006.06.004
4. Lasalle, B., Varner, M., Botkin, J., Jackson, M., Stark, L., Cessna, M., et al.: Biobanking informatics infrastructure to support clinical and translational research. AMIA Jt. Summits Transl. Sci. Proc. 132–5 (2013). PubMed PMID: 24303252; PubMed Central PMCID: PMC3845745
5. Murphy, S.N., Weber, G., Mendis, M., Gainer, V., Chueh, H.C., Churchill, S., et al.: Serving the enterprise and beyond with informatics for integrating biology and the bedside (i2b2). J. Am. Med. Inform. Assoc. (JAMIA) **17**(2), 124–130 (2010). https://doi.org/10.1136/jamia.2009.000893. PubMed PMID: 20190053; PubMed Central PMCID: PMC3000779
6. Chute, C.G., Beck, S.A., Fisk, T.B., Mohr, D.N.: The enterprise data trust at Mayo clinic: a semantically integrated warehouse of biomedical data. J. Am. Med. Inform. Assoc. (JAMIA) **17**(2), 131–135 (2010). https://doi.org/10.1136/jamia.2009.002691. PubMed PMID: 20190054; PubMed Central PMCID: PMC3000789

7. Fan, J.W., Friedman, C.: Deriving a probabilistic syntacto-semantic grammar for biomedicine based on domain-specific terminologies. J Biomed. Inform. **44**(5), 805–814 (2011). https://doi.org/10.1016/j.jbi.2011.04.006. PubMed PMID: 21549857; PubMed Central PMCID: PMC3172402

8. Jezek, P., Moucek, R.: Semantic framework for mapping object-oriented model to semantic web languages. Front Neuroinform. **9**, 3 (2015). https://doi.org/10.3389/fninf.2015.00003. PubMed PMID: 25762923; PubMed Central PMCID: PMC4340193

9. Price, S.J., Stapley, S.A., Shephard, E., Barraclough, K., Hamilton, W.T.: Is omission of free text records a possible source of data loss and bias in clinical practice research datalink studies? A case–control study. BMJ open **6**(5) (2016). https://bmjopen.bmj.com/content/6/5/e011664

10. Dhombres, F., Charlet, J.: As ontologies reach maturity, artificial intelligence starts being fully efficient: findings from the section on knowledge representation and management for the yearbook 2018. Yearb. Med. Inform. **27**(1), 140–145 (2018). https://doi.org/10.1055/s-0038-1667078. PubMed PMID: 30157517

11. Baayen, R.H., Hendrix, P., Ramscar, M.: Sidestepping the combinatorial explosion: an explanation of n-gram frequency effects based on naive discriminative learning. Lang. Speech **56**(Pt 3), 329–347 (2013). PubMed PMID: 24416960

12. Patrick, J., Li, M.: High accuracy information extraction of medication information from clinical notes: 2009 i2b2 medication extraction challenge. J. Am. Med. Inform. Assoc. (JAMIA) **17**(5), 524–527 (2010). https://doi.org/10.1136/jamia.2010.003939. PubMed PMID: 20819856; PubMed Central PMCID: PMC2995676

13. Aronson, A.R., Lang, F.M.: An overview of MetaMap: historical perspective and recent advances. J. Am. Med. Inform. Assoc. (JAMIA) **17**(3), 229–236 (2010). https://doi.org/10.1136/jamia.2009.002733. PubMed PMID: 20442139; PubMed Central PMCID: PMCPMC2995713

14. Bradshaw, R.: Concept bag: a new method for computing similarity. University of Utah (2015)

15. Jaccard, P.J.: Similarity and shingling. Data mining (2015). http://www.cs.utah.edu/~jeffp/teaching/cs5955/L4-Jaccard+Shingle.pdf

16. Harris, P.A., Taylor, R., Thielke, R., Payne, J., Gonzalez, N., Conde, J.G.: Research electronic data capture (REDCap)–a metadata-driven methodology and workflow process for providing translational research informatics support. J. Biomed. Inform. **42**(2), 377–381 (2009). https://doi.org/10.1016/j.jbi.2008.08.010. PubMed PMID: 18929686; PubMed Central PMCID: PMC2700030

17. Garla, V.N., Brandt, C.: Semantic similarity in the biomedical domain: an evaluation across knowledge sources. BMC Bioinform. **13**, 261 (2012). https://doi.org/10.1186/1471-2105-13-261. PubMed PMID: 23046094; PubMed Central PMCID: PMCPMC3533586

18. Lopez-Raton, M., Rodriguez-Alvarez, M., Cadarso-Suarez, C., Gude-Sampedro, F.: Optimal cutpoints: an R package for selecting optimal cutpoints in diagnostic tests. J. Stat. Softw. **61**(8), 1–36 (2015)

19. Aquire, E., cuadros, M., Rigua, G., Soroa, A. (eds.): Exploring knowledge bases for similarity. In: Proceedings of the Seventh International Conference on Language Resources and Evaluation. European Language Resources Association, Valleta (2010)

20. Leacock, C., Chodorow, M.: Using corpus statistics and wordnet relations for sense identification. In: Fellbaum, C. (ed.) Wordnet: An Electronic Lexical Database, pp. 265–283. MIT Press, Cambridge (1998)

21. Wu, Z., Palmer, M.: Verbs semantics and lexical selection. In: Proceedings of the 32nd Annual Meeting on Association for Computational Linguistics. Association for Computational Linguistics, Las Cruces (1994)

A Reliable Method to Remove Batch Effects Maintaining Group Differences in Lymphoma Methylation Case Study

Giulia Pontali[1,2]([✉]), Luciano Cascione[3]([✉]), Alberto J. Arribas[3]([✉]),
Andrea Rinaldi[3]([✉]), Francesco Bertoni[3]([✉]), and Rosalba Giugno[4]([✉])

[1] Department CIBIO, University of Trento, via Sommarive 9, 38123 Povo, TN, Italy
giulia.pontali@unitn.it
[2] Institute for Biomedicine, EURAC Research, Via Galvani 31, 39100 Bolzano, Italy
giulia.pontali@eurac.edu
[3] Institute of Oncology Research (IOR), Università della Svizzera italiana,
Via Vincenzo Vela 6, Bellinzona, Switzerland
{luciano.cascione,alberto.arribas,andrea.rinaldi,
francesco.bertoni}@ior.usi.ch
[4] Department of Computer Science, University of Verona,
Strada le Grazie 15, 3134 Verona, Italy
rosalba.giugno@univr.it

Abstract. The amount of biological data is increasing and their analysis is becoming one of the most challenging topics in the information sciences. Before starting the analysis it is important to remove unwanted variability due to some factors such as: year of sequencing, laboratory conditions and use of different protocols. This is a crucial step because if the variability is not evaluated before starting the analysis of interest, the results may be undesirable and the conclusion can not be true. The literature suggests to use some valid mathematical models, but experience shows that applying these to high-throughput data with a non-uniform study design is not straightforward and in many cases it may introduce a false signal. Therefore it is necessary to develop models that allow to remove the effects that can negatively influence the study preserving biological meaning. In this paper we report a new case study related lymphoma methylation data and we propose a suitable pipeline for its analysis.

Keywords: Biological data · Variability · Lymphoma · Methylation · Batch effects removal

1 Introduction

The biological data are increasing and scientists have to overcome many challenges to deal with them: how to handle the complexity of information, how

G. Pontali and L. Cascione—Equal contribution.

© Springer Nature Switzerland AG 2019
I. Rojas et al. (Eds.): IWBBIO 2019, LNBI 11466, pp. 24–32, 2019.
https://doi.org/10.1007/978-3-030-17935-9_3

to integrate the data from heterogeneous resources, what kind of principles or standards must be adopted processing data [20]. In this study we faced with two critical concepts that are connected to each other: variability of data sources and accuracy of data quality. One of the most common variations in data derives from the presence of batch effects due to experimental artifacts occurring during data management such as the reagents used or the technicians who prepare the experiments [7].

When batch effects are present in the data it is fundamental to find and remove them for avoiding spurious results. Errors can affect negatively the analysis of the data reducing statistical power [3,8,12]. Biological variations also occur when the data represent different study groups [13]. In this case, a reliable procedure must eliminate the batch effects by maintaining the biological difference between the groups, which in fact the study aims to characterize and quantify [9].

As is commonly the case, with groups of different sizes, batch effects imply false differences between groups due to the confounder introduced. There is currently no universal solution for managing batch effects while maintaining a significant group difference [8]. The well accepted strategy is to filter the data in order to obtain a balanced study project, to detect and remove the batch effects and to perform statistical analyzes on the modified data without further considerations on the batch effects [12]. However, getting a balanced study project is not always possible and can discard significant biological samples [14].

One way is to normalize subtracting the mean of the values affected by the batch. However, this procedure is not sufficient when combining datasets containing large batch-to-batch variations or comparing unbalanced groups [7].

Alternatively, variation removal can be incorporated into the statistical analysis using the two-way analysis of variance, in which the variances between groups is estimated considering the variance in each group derived from batches [13]. If the batches introduced in the group difference estimates are lower than the variance within each group, partially the effects of the batch have been resolved. On the other hand, if the group differences estimates introduce more batches than the variance within the groups, this can lead to excessive estimates of group differences obtaining false signals [13].

There are several R packages to handle batch effects like *Limma* [16] which is based on a two-way variance detection and *sva* based on *ComBat* [7] that uses an empirical Bayes approach to regulate potentials batch effects.

The application of both of them in some high-throughput data studies can introduce false signals [13].

To overcome this, we propose a new framework, called *Remby*, implemented in R/Bioconductor[1], which appears reliable to analyze methylation profiling data without missing relevant information and maintaining the biological significance.

---

[1] Bioconductor repository provides tools for analysis and comprehension of high-throughput genomic data. It has 1560 software packages. The current release of Bioconductor is version 3.7.

2   Case Study

We focused on the analysis of methylation profiling in marginal zone lymphoma (*MZL*). Methylation is an epigenetic mechanism, i.e. an heritable changes in gene function that occur in the absence of alteration in DNA sequence. It takes place in mammalian genome and represents a covalent modification of DNA that does not change the DNA sequence but has an influence on a gene activity.

Errors in methylation could give rise dangerous consequences including cancer such as lymphoma. Methylation can be measured with the Illumina Human-Methylation450 (450k) array that contains 485 512 individual CpG sites, spread across many genes. For each CpG there is a methylated and unmethylated intensity probes[2], and their values determine the proportion of methylation at each CpG site.

In this study we take into account only CpG islands. The WHO (World Health Organization) [19] categorizes MZL into three distinct types based on their site of impact: Splenic Marginal Zone Lymphoma (SMZL); Extranodal Marginal Zone Lymphoma (also called mucosa-associated lymphoid tissue) (EMZL); Nodal Marginal Zone Lymphoma (NMZL). Individual MZL share some common features but are different in their biology and behavior [4,15]. SMZL involving in the spleen, bone marrow, and blood; EMZL arises in a number of extranodal tissues, most commonly from stomach, salivary gland, lung, small bowel, thyroid, ocular adenoxena skin; NMZL has similar morphologic and immunophenotipic similarities with SMZL and MALT.

Our dataset is composed by 44 samples. 22 samples are EMZL: 10 samples have been profiled in 2011 and 12 in 2014; 12 samples NMZL and have been profiled in 2014; 10 samples are SMZL and have been analyzed in 2011. SMZL are from spleen; 7 EMZL are from ocular tissue, 10 from lung and 5 from other tissues. In brief, samples were been collected in 2011 and others in 2014, obtained from different anatomical sites. This means that data can suffer from multiple effects and the unwanted noise/hidden artifacts could dramatically reduce the accuracy of analysis [18].

Our aim was to adjust expression data for both effects maintaining the biological differences between MZL subtypes.

3   *Remby*: Batch Effects Removal Preserving Group Differences in Methylation Lymphoma Data

The proposed framework, named *Remby*, aims to analyze methylation data in presence of several batch effects without missing relevant information and maintaining biological significance. In what follows we describe the steps of *Remby* which are also pictured in Figs. 1 and 2.

a. **Groups identification.** We divided the cohort according to the year of hybridization (first possible batch) before starting the analysis (see Fig. 1-A). For each group that compose the dataset do:

---

[2] Single-stranded fragments of DNA that are complementary to a gene.

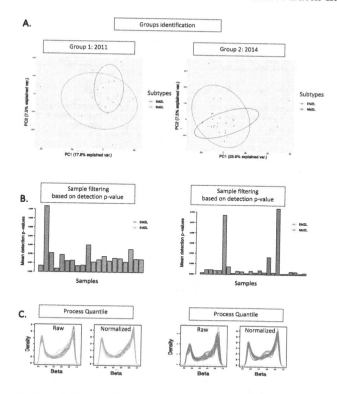

Fig. 1. Description of the proposed framework *Remby* (first part).

- **Samples filtering.** Use detection p-value and remove the poor quality samples from the analysis. Examine the mean of p-values of CpG probes across all samples and keep probes with a mean ≤ 0.05 (see Fig. 1-B). This step uses the method proposed by *Minfi* R package [2]:

  the detection p-value compares the total signal (Methylated + Unmethylated) for each probe to the back-ground signal level. Remove samples based on the number of probes that exceed the detection p-value.

  The result does not depend on the size of the data set but on the quality of the data it contains.

- **Minimize variation within samples.** Apply quantile normalization procedure to the methylated and unmethylated signal intensities separately (see Fig. 1-C) to make the distribution of signal to be the same across samples [6]. Although there is no single normalization method that is universally considered as the best one; a recent study suggests *preprocessQuantile* function [1]. This is the most suited approach for dataset with no expect global differences among samples.

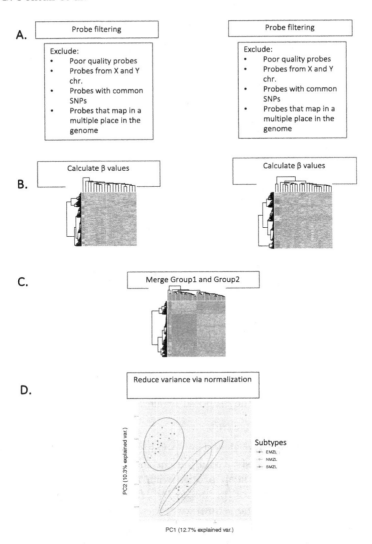

Fig. 2. Description of the proposed framework *Remby* (second part). (Color figure online)

- **Filtering of probes excluding variables**. Remove probes that can have a negative impact on the analysis (see Fig. 2-A) such as:
  - Poor quality probes (detection p-value $>= 0.01$) and probes that have failed in one or more samples.
  - Probes from X and Y chromosomes because autosomal[3]: CpG methylation differs by sex in both of term of individual CpG sites and global autosomal CpG methylation [11];

---

[3] Pertaining to a chromosome that is not a sex chromosome.

- Probes that are known to have common SNPs at the CpG site because are the most important confounding variables [15] and in presence of them the signal is not clear thus the methylation signal can be lost. Use the function called *dropLociWithSnps* in *Minfi* package to remove all probes affected by SNPs.
- Probes that map to multiple places in the genome. In [5], authors published a list of probes that map to multiple places in the genome. Remove all probes that are present both in that list and in your data because redundant.

  - **Calculate β values.** Once data are filtered and normalized, evaluate the methylation level that is estimated based on the intensities of methylated and unmethylated probes (see Fig. 2-B) using *getBeta* function present in *Minfi*. Beta value computes the ratio of the methylated probe intensity and the overall intensity (sum of methylated and unmethylated probes intensities). This value is between 0 and 1: 0 being unmethylated and 1 fully methylated.

b. **Combine the data together and normalize.** Combine data coming from different groups taking into account only the probes that are shared among every groups of interest; apply quantile normalization, i.e. a global adjustment method for making two distributions identical in statistical properties (see Figs. 2-C and D). It is now possible to analyze the whole dataset.

4 Results

Applying *Remby* we divided the dataset into two groups, the first one containing samples sequenced in 2011 and the second one samples sequenced in 2014 (Fig. 1-A). For each group, we evaluated the quality of signal across all the probes in each sample.

After that we check probes quality to remove those that could influence negatively the analysis. We took into account all samples (Fig. 1-B) because each of them had a detection p-value < 0.05. To remove amplitude variations that are present among samples we also performed normalization (Fig. 1-C). We continued filtering out poor performing probes, probes from X and Y chromosomes, probes with common SNPs and probes that map in a multiple place in the genome (Fig. 2-A). We evaluate for each probes the level of methylation by calculating Beta values (Fig. 2-B): yellow means that the probe is not methylated, red that the probe is methylated. Results shows that the majority of probes present a Beta values < 0.3 that is down-methlated. Than, we combined data (Fig. 2-C) taking into account only those shared by both groups, and reduce variance between the groups by applying again the normalization.

After *Remby* application data are clustered heterogeneously compare to the original data (Figs. 2-D and 4). Based on the literature this is what we expected. EMZL clusterized in independent way, instead NMZL overlapped SMZL in agreement with egentic data showing a high overlap of the two entities [17].

This result is crucial because underlines that the method proposed does not lose biological information. This is also shown by evaluating the methylation levels of every MZL subtypes and results showed that unmethylated probes were over represented in CpG in every type of MZL (see Fig. 3). This is in line with the study [10] showing that MZLs exhibit low levels of *de novo* DNA methylation when compared to GC-derived neoplasms. In addition, MZLs subtypes show promoter methylation changes that lead to the inactivation of a series of tumor suppressors but also to the activation of genes that sustain the lymphoma cell survival and proliferation. In more details: 28% of probes were hypomethylated, 57.56% down-methylated, and 16.16% up-methylated in EMZL; 26.90% of probes were hypomethylated, 57.15% down-methylated, 15.95% up-methylated in NMZL; 26.83% of probes were hypomethylated, 57.10% down-methylated, and 16.03% up-methylated in SMZL.

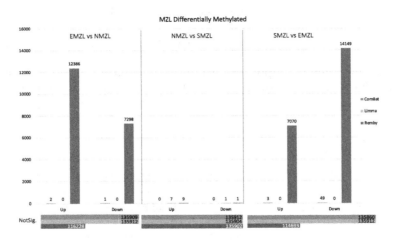

Fig. 3. MZL Differentially Methylated analysis with batch effects removal by *ComBat*, *Limma*, and *Remby*.

Using *ComBat* and *Limma* correction we obtained results that did not reflect the known biology. *ComBat* and *Limma* clusterized the data in a similar way, *Remby* combine them differently from the other two (Fig. 4). *ComBat* and *Limma* appeared to work better than *Remby* since data ended up in one single cluster, but almost all the probes were not methylated (=no Signal) (Fig. 3). This means that if we had used *ComBat* and *Limma* in the MZL data set we could have lost most information on methylation among the subtypes lymphoma.

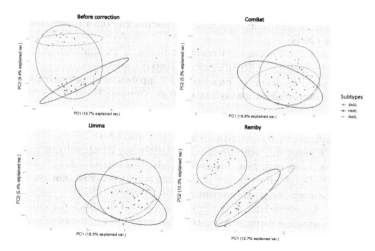

Fig. 4. Principal component analysis (PCA) of the data before correction and after batch effects removal using *ComBat, Limma,* and *Remby.*

5   Conclusion

We proposed a framework, *Remby*, to minimize the variance among samples and among groups of lymphoma methylation data maintaining biological information. The results reflected knowledge on MZL. *Remby* was designed to overcome the limitation of batch effects removal methods in literature which in the current case study were not able to detect them preserving biological subtypes and methylation differentiation.

As we have seen in this paper the analyses of biological data are becoming more and more challenging and biological support is of fundamental importance to really understand if the mathematical models fit the biological data, and can not exist an unique model for all data.

Acknowledgments. This work has been partially supported by the following projects: GNCS-INDAM, Fondo Sociale Europeo, and National Research Council Flagship Projects Interomics. This work has been partially supported by the project of the Italian Ministry of education, Universities and Research (MIUR) "Dipartimenti di Eccellenza 2018-2022".

References

1. https://www.rdocumentation.org/packages/minfi/versions/1.18.4/topics/preprocessQuantile
2. Aryee, M., et al.: Minfi: a flexible and comprehensive bioconductor package for the analysis of infinium DNA methylation microarrays. Biostatistics **30**(10), 1363–1369 (2014)

3. Benito, M., et al.: Adjustment of systematic microarray data biases. Bioinformatics **20**(1), 105–114 (2004)
4. Bertoni, F., Rossi, D., Zucca, E.: Recent advances in understanding the biology of marginal zone lymphoma. F1000Research **7**(406) (2018)
5. Chen, Y., et al.: Discovery of cross-reactive probes and polymorphic CpGs in the illumina infinium humanmethylation450 microarray. Epigenetics **8**(2), 203–209 (2013)
6. Hicks, S., Okrah, K., Paulson, J., Quackenbush, J., Irizarry, R., Bravo, H.: Smooth quantile normalization. Biostatistics **19**(2), 185–198 (2018)
7. Johnson, W., Cheng, L., Rabinovic, A.: Adjusting batch effects in microarray expression data using empirical bayes methods. Biostatistics **8**(1), 118–127 (2007)
8. Lazar, C., et al.: Batch effect removal methods for microarray gene expression data integration: a survey. Brief. Bioinform. **14**(4), 469–490 (2013)
9. Luo, J., et al.: A comparison of batch effect removal methods for enhancement of prediction performance using MAQC-II microarray gene expression data. Pharmacogenomics **10**(4), 278–291 (2010)
10. Martin-Subero, J., et al.: A comprehensive microarray-based DNA methylation study of 367 hematological neoplasms. PLoS One **4**(9), e6986 (2009)
11. McCarthy, N., et al.: Meta-analysis of human methylation data for evidence of sex-specific autosomal patterns. BMC Genomics **15**(1), 981 (2014)
12. Nueda, M.J., Ferrer, A., Conesa, A.: ARSyN: a method for the identification and removal of systematic noise in multifactorial time course microarray experiments. Biostatistics **13**(3), 553–566 (2012)
13. Nygaard, V., Rødland, E., Hovig, E.: Methods that remove batch effects while retaining group differences may lead to exaggerated confidence in downstream analyses. Biostatistics **17**(1), 29–39 (2016)
14. Pourhoseingholi, M., Baghestani, A., Vahedi, M.: How to control confounding effects by statistical analysis. Gastroenterol Hepatol Bed Bench **5**(2), 79–83 (2012)
15. Rinaldi, A., et al.: Genome-wide dna profiling of marginal zone lymphomas identifies subtype-specific lesions with an impact on the clinical outcome. Blood **117**(5), 1595–1604 (2011)
16. Smyth, G., Speed, T.: Normalization of cDNA microarray data. Methods **31**(4), 265–273 (2003)
17. Spina, V., et al.: The genetics of nodal marginal zone lymphoma. Blood **128**(10), 1362–1373 (2016)
18. Sun, Z., et al.: Batch effect correction for genome-wide methylation data with illumina infinium platform. BMC Med. Genomics **4**, 84 (2011)
19. Swerdlow, S., et al.: WHO classification of tumours of haematopoietic and lymphoid tissues. International Agency for Research on Cancer, Lyon (2008)
20. Yixue, L., Luonan, C.: Big biological data: challenges and opportunities. Genomics Proteomics Bioinform. **12**(5), 187–189 (2014)

Predict Breast Tumor Response to Chemotherapy Using a 3D Deep Learning Architecture Applied to DCE-MRI Data

Mohammed El Adoui[1]([⊠]), Stylianos Drisis[2],
and Mohammed Benjelloun[1]

[1] Computer Science Unit, Faculty of Engineering, University of Mons, No. 9,
Rue Houdain, 7000 Mons, Belgium
{Mohammed.Eladoui,Mohammed.Benjelloun}@umons.ac.be
[2] Radiology Department, Jules Bordet Institute, No. 121, Bd Waterloo,
1000 Brussels, Belgium
Stylianos.Drisis@bordet.be

Abstract. Purpose: Many breast cancer patients receiving chemotherapy cannot achieve positive response unlimitedly. The main objective of this study is to predict the intra tumor breast cancer response to neoadjuvant chemotherapy (NAC). This aims to provide an early prediction to avoid unnecessary treatment sessions for no responders' patients.

Method and material: Three-dimensional Dynamic Contrast Enhanced of Magnetic Resonance Images (DCE-MRI) were collected for 42 patients with local breast cancer. This retrospective study is based on a data provided by our collaborating radiology institute in Brussels. According to the pathological complete response (pCR) ground truth, 14 of these patients responded positively to chemotherapy, and 28 were not responsive positively. In this work, a convolutional neural network (CNN) model were used to classify responsive and non-responsive patients. To make this classification, two CNN branches architecture was used. This architecture takes as inputs three views of two aligned DCE-MRI cropped volumes acquired before and after the first chemotherapy. The data was split into 20% for validation and 80% for training. Cross-validation was used to evaluate the proposed CNN model. To assess the model's performance, the area under the receiver operating characteristic curve (AUC) and accuracy were used.

Results: The proposed CNN architecture was able to predict the breast tumor response to chemotherapy with an accuracy of 91.03%. The Area Under the Curve (AUC) was 0.92.

Discussion: Although the number of subjects remains limited, relevant results were obtained by using data augmentation and three-dimensional tumor DCE-MRI.

Conclusion: Deep CNNs models can be exploited to solve breast cancer follow-up related problems. Therefore, the obtained model can be used in future clinical data other than breast images.

Keywords: Breast cancer · Response prediction · DCE-MRI · Deep learning

© Springer Nature Switzerland AG 2019
I. Rojas et al. (Eds.): IWBBIO 2019, LNBI 11466, pp. 33–40, 2019.
https://doi.org/10.1007/978-3-030-17935-9_4

1 Introduction

Over last years, deep Convolutional Neural Networks (CNNs) have been widely applied to many computer vision applications [1]. Many breakthroughs related to healthcare and medical imaging have been achieved, such as skin, lung, prostate and breast cancers detection and response prediction.

The main purpose of this work is to provide an automatic deep learning model to predict breast cancer response to chemotherapy using Dynamic Contrast Enhanced of Magnetic Resonance Image (DCE-MRI) data acquired before and after the first chemotherapy session. This aims to provide an early prediction to avoid unnecessary chemotherapy sessions for no responders' patients. Indeed, Early prediction of neoadjuvant chemotherapy (NAC) treatment in breast cancer is fundamental for clinicians to avoid administration of toxic treatment and select alternative therapeutic strategies to no responders' patients such as surgery. In this study, the pathological complete response (pCR) [2] was used as medical ground truth. In fact, after several sessions, many patients do not respond positively to treatment. However, this negative response cannot be visualized after the first chemotherapy session. Especially, when radiologist use the size change as follow-up metric. Indeed, the size of the tumor remains almost the same in most cases [3]. However, several studies affirm that in one chemotherapy session, intra-tumor changes occur within the breast tumor. Therefore, we propose to train a deep neural network using augmented DCE-MR images acquired before and after the first chemotherapy to predict the breast cancer response to chemotherapy using only two DCE-MRI volumes.

2 Related Work

From last years, breast intra tumor changes during a chemotherapy treatment was used as index to predict the breast cancer response to chemotherapy using medical images [4]. The technical researches monitoring the breast cancer response to chemotherapy using MR images are divided into two categories. The first one is based on classical image processing. In this context, texture analysis, histogram analysis and parametric response mapping (PRM) [5] methods have shown good results. These methods have been reviewed in our previously published paper [6]. However, these methods are based on semi-automatic intermediate steps. This require more processing time and user intervention. The second research category is based on the application of machine learning and deep learning techniques to predict the breast cancer response to chemotherapy. Huynh et al. [7] used CNN architecture to extract the more relevant descriptors from DCE-MRI acquired in three times contrast points of the baseline exam. They used VGGNet [8] architecture to extract features. To classify the responder and no responder patients, the Linear Discriminant Analysis (LDA) classifier [9] based on the pathological response as ground truth was used. In this research, cross validation, with the metric of the Area Under the ROC Curve (AUC) was used to assess prediction performance. The obtained accuracy for this research was 84%. More recently, Ravichandran et al. [10] used an MRI dataset of 166 patients with breast cancer. They achieved an accuracy of 85% using only the baseline DCE-MRI exam propagated on six CNNs blocks. All this research used only the pre-treatment exam as inputs. However, the post treatment exam is crucial to predict if a breast cancer patient can continue to have chemotherapy treatment or not. To the best of our knowledge, the current study is the first using data from the two first exam in a multi view CNN model to predict breast cancer response to NAC.

3 Method and Material

3.1 Data

In this study, we used a clinical dataset of 42 patients with local breast cancer obtained from the Jules Bordet institute in Brussels, Belgium. All the data were used with an agreement confirmed by an ethics committee according to a respected protocol. According to the medical ground truth (pCR) [2], 14 of them were responded to chemotherapy treatment positively and 28 were not responded. For each patient, we possess DCE-MR volumes acquired before and after the first chemotherapy. The MR images were performed with a Siemens $1.5T$ scanner. From DCE-MRI exam, only slices containing a tumor with surrounding regions were extracted (Fig. 1). Consequently, a total of:

- **693** original images acquired before chemotherapy and **693** after chemotherapy using the *transversal* view.
- **721** original images acquired before chemotherapy and **721** after chemotherapy using the *axial* view.
- **696** original images acquired before chemotherapy and **696** after chemotherapy using the *coronal* view.

3.2 Data Preprocessing

Before the training process, the volumes of interest (VOIs) containing tumors were aligned by 3D affine registration. This aims to make correspondences between each slice acquired before chemotherapy to that acquired after. Figure. 1 illustrates the preprocessing steps applied to the DCE-MR images before starting the training process. A part of this work was presented in our previous paper [5].

The first step of image preprocessing was to crop the volume of interest (VOI) before and after chemotherapy. Then, an affine registration was applied to the cropped volumes [11] for matching slices, which enables to compare their respective information. Consequently, the neural network inputs are the two cropped and aligned of DCE-MRI images containing the tumor acquired before and after the first chemotherapy.

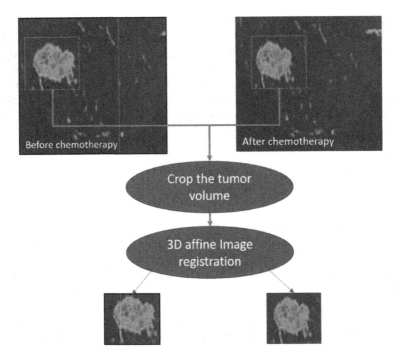

Fig. 1. Image preprocessing steps

3.3   CNN Architecture and Training

The proposed Convolutional Neural Network (CNN) is consisting of two parallel convolutional layers taking as inputs transversal, coronal and axial slices acquired before and after chemotherapy for each patient.

As illustrated in Fig. 2, this architecture contains two similar branches, each one contains 4 blocks of 2D convolution followed by an activation function (ReLU) [12] and a Max pooling layer [13]. In the first and second blocks, 32 kernels were used for each convolutional layer. 64 kernels were used for the third and the fourth blocks. The two branches are concatenated after a fully connected layer of 512 hidden units. Then, a sigmoid function was applied. A mean of the three obtained classifications (responsive or non-responsive) using each tumor view is provided in the output.

The parallel Deep learning architecture was applied for each view using corresponding slices before and after the first chemotherapy. Consequently, this architecture was used three times in parallel, to provide a mean of the three outputs.

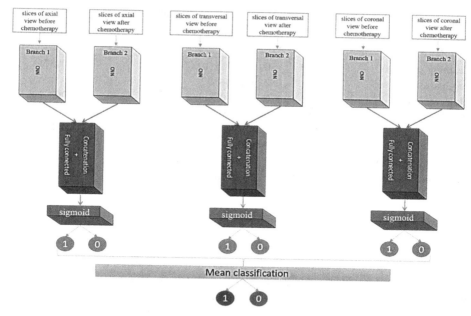

Fig. 2. The proposed CNN architecture to predict breast cancer response to chemotherapy using 3D DCE-MRI

3.4 Data Augmentation and Model Evaluation

To provide a powerful training process, a data augmentation generator function was defined. This function applies and returns random operations of rotations, translations and flips. The random data generator was called for each training epoch. Therefore, within the 150 training epochs, **67680** different samples for *axial* view, **55440** samples for *transversal* view and **55680** samples for the *coronal* view were used.

To define and fix each training parameter, a fine-tuning technique was used. Consequently, we used the Stochastic Gradient Descente (SGD) [14] with a learning rate of *0.0052*. A learning rate decay of $3.46e^{-5}$ [15] was used to schedule a best accuracy. To compile the model, a categorical cross entropy was used as loss function and standard accuracy was used as a metric. Table 1 present the used parameters and their optimal values. To avoid results' bias, a 5-Fold stratified cross validation with AUC as metric was used. The total training time was 9 h and 39 min, using the following hardware's properties:

- **CPU:** 16 cores, 2.10 GHz, 32 GB of RAM memory
- **GPU:** Nvidia GeForce GTX 980, 16 GB of memory

Table 1. Tested and optimal values of the used parameters

Parameter	Tested values	Optimal value
Learning rate (lr)	0.05, 0.005, 0.0005	0.0052
Batch size	2, 4, 8, 16, 32	8
Momentum rate	0.8, 0.9, 0.99	0.99
Weight initialization	Normal, Uniform, Glorot	Normal
Per-parameter adaptive learning rate methods	SGD, RMSprop, Adagrad, Adadelta, Adam	SGD
Learning rate decay	Yes, no	Yes (1e–6)
Activation function	Sigmoid, ReLU, elu	ReLU
Dropout rate	0.1, 0.25, 0.3, 0.5, 0.75	0.25 & 0,30

4 Results

Within 150 epochs with 5-fold stratified cross validation, an accuracy of 90.03 was obtained using 20% of 3D validation data. As shown in Fig. 3, the obtained Area under the ROC Curve (AUC) is 0.92. Using only 2D slices, the obtained accuracies were:

- **88.68%** using only axial views
- **89.88%** using only coronal views
- **91.46%** using only transversal views
- **90.03%** using the three views together

Without using data augmentation for the three views, an accuracy of 75.66 was obtained. An overfitting was observed during training when using only one of the views without data augmentation.

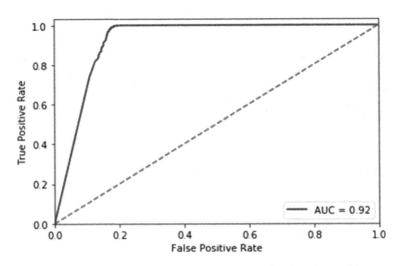

Fig. 3. The area under the curve of the obtained model using 3D views of breast tumors

Even if the accuracy using only transversal view is higher, it is recommended to use the three views together to have more coherence and more extracted features.

Table 2 shows a comparison between the recent results of the methods proposed in the literature using deep neural network applied to DCE-MR images for only pretreatment exam (baseline), and the method proposed in this paper using two exams.

Table 2. Comparison between recent proposed methods results

Authors	Year	3D approach	Data	AUC
Huynh et al. [7]	2017	No	DCE-MRI (one exam)	.84
Ravichandran et al. [10]	2018	No	DCE-MRI (one exam)	.85
Proposed method	**2019**	**Yes**	**DCE-MRI (two exams)**	**.92**

5   Discussion and Conclusion

In this work a parallel architecture was used to predict the breast cancer response to chemotherapy using the three views of a tumor obtained before and after the chemotherapy, during a neoadjuvant treatment. Each of the three architectures takes as inputs one of the views acquired before and after the first chemotherapy. A mean of the three responses is obtained in the output. Clinically, the results were appreciated by the collaborating radiologist. However, the pretreatment step still time consuming. Indeed, a semi-automatic crop of the tumor volume and a 3D affine registration to align each slice before chemotherapy to that after chemotherapy are needed to start the learning process. Therefore, an automatization of this step can be one of our main future works.

The proposed method can be generalized to other types of cancer. Indeed, the method used is based on a deep neural network architecture considering two volumes of interest (tumor) acquired before and after the first chemotherapy. Thus, only the pretreatment steps are required to start the training process and have a prediction of the intra tumor response to chemotherapy of any type of tumor.

Acknowledgments. Our thanks are addressed to Dr. Marc Lemort, the head of the Radiology Department at Jules Bordet Institute in Brussels, for offering us the dataset used to evaluate our proposed method.

References

1. Litjens, G., et al.: A survey on deep learning in medical image analysis. Med. Image Anal. **42**, 60–88 (2017)
2. Cortazar, P., et al.: Pathological complete response and long-term clinical benefit in breast cancer: the CTNeoBC pooled analysis. Lancet **384**(9938), 164–172 (2014)
3. El Adoui, M., Drisis, S., Benjelloun, M.: Analyzing breast tumor heterogeneity to predict the response to chemotherapy using 3D MR images registration. In: Proceedings of the 2017 International Conference on Smart Digital Environment, pp. 56–63. ACM, July 2017

4. Marusyk, A., Almendro, V., Polyak, K.: Intra-tumor heterogeneity: a looking glass for cancer? Nat. Rev. Cancer **12**(5), 323 (2012)
5. El Adoui, M., Drisis, S., Benjelloun, M.: A PRM approach for early prediction of breast cancer response to chemotherapy based on registered MR images. Int. J. Comput. Assist. Radiol. Surg. **13**, 1233–1243 (2018)
6. El Adoui, M., Drisis, S., Larhmam, M.A., Lemort, M., Benjelloun, M.: Breast cancer heterogeneity analysis as index of response to treatment using MRI images: a review. Imaging Med. **9**(4), 109–119 (2017)
7. Huynh, B.Q., Antropova, N., Giger, M.L.: Comparison of breast DCE-MRI contrast time points for predicting response to neoadjuvant chemotherapy using deep convolutional neural network features with transfer learning. In: Medical Imaging 2017: Computer-Aided Diagnosis, vol. 10134, p. 101340U. International Society for Optics and Photonics, March 2017
8. He, K., Zhang, X., Ren, S., Sun, J.: Deep residual learning for image recognition. In: Proceedings of the IEEE Conference on Computer Vision and Pattern Recognition, pp. 770–778 (2016)
9. Bardos, M., Zhu, W.H.: Comparaison de l'analyse discriminante linéaire et des réseaux de neurones. Application à la détection de défaillance d'entreprises. Revue de statistique appliquée **45**(4), 65–92 (1997)
10. Ravichandran, K., Braman, N., Janowczyk, A., Madabhushi, A.: A deep learning classifier for prediction of pathological complete response to neoadjuvant chemotherapy from baseline breast DCE-MRI. In: Medical Imaging 2018: Computer-Aided Diagnosis, vol. 10575, p. 105750C. International Society for Optics and Photonics, February 2018
11. Jenkinson, M., Smith, S.: A global optimization method for robust affine registration of brain images. Med. Image Anal. **5**(2), 143–156 (2001)
12. Radford, A., Metz, L., Chintala, S.: Unsupervised representation learning with deep convolutional generative adversarial networks (2015). arXiv preprint arXiv:1511.06434
13. Krizhevsky, A., Sutskever, I., Hinton, G.E.: Imagenet classification with deep convolutional neural networks. In: Advances in Neural Information Processing Systems, pp. 1097–1105 (2012)
14. Bottou, L.: Large-scale machine learning with stochastic gradient descent. In: Lechevallier, Y., Saporta, G. (eds.) Proceedings of COMPSTAT 2010. Physica-Verlag HD, Heidelberg (2010)
15. Kingma, D.P., Ba, J.: Adam: a method for stochastic optimization (2014). arXiv preprint arXiv:1412.6980

Data Fusion for Improving Sleep Apnoea Detection from Single-Lead ECG Derived Respiration

Ana Jiménez Martín[✉], Alejandro Cuevas Notario,
J. Jesús García Domínguez, Sara García Villa,
and Miguel A. Herrero Ramiro

Electronics Department, Polytechnics School,
University of Alcalá, Barcelona-Madrid Road, 28805 Alcalá de Henares, Spain
ana.jimenez@uah.es

Abstract. This work presents two algorithms for detecting apnoeas from the single-lead electrocardiogram derived respiratory signal (EDR). One of the algorithms is based on the frequency analysis of the EDR amplitude variation applying the Lomb-Scargle periodogram. On the other hand, the sleep apnoeas detection is carried out from the temporal analysis of the EDR amplitude variation. Both algorithms provide accuracies around 90%. However, in order to improve the robustness of the detection process, it is proposed to fuse the results obtained with both techniques through the Dempster-Shafer evidence theory. The fusion of the EDR-based algorithm results indicates that, the 84% of the detected apnoeas have a confidence level over 90%.

Keywords: Central sleep apnea · ECG · EDR · Evidence theory

1 Introduction

It is well known that sleep is fundamental for the optimal physical and mental functioning of the human being, but it has been recently when it has become aware of the consequences that a poor quality of sleep can have on health in the medium and long term. The Spanish Society of Neurology estimates that between 20 and 48% of the adult population suffers, at some point in their lives, difficulty in initiating or maintaining sleep, and at least 10% suffer from some chronic and serious sleep disorder [1]. The most common disorder is insomnia followed by apnoeas. However, it is estimated that 90% of sleep apnoea patients are undiagnosed. Therefore, although there has been a considerable increase in the number of medical consultations for sleep disorders, it is expected that the number will keep growing. One of the main reasons of this growth of clinical cases is the current lifestyle, the increase in life expectancy and, consequently, the elderly population, as the prevalence of these imbalances increases with age.

Sleep apnoea (SA) is a disorder that occurs when a person's breathing is interrupted during sleep from a few seconds to minutes. There are three types depending on whether or not exist respiratory effort while cessations in air flow occur: obstructive apnoea (OSA), when there is respiratory effort but the airways are blocked; central

© Springer Nature Switzerland AG 2019
I. Rojas et al. (Eds.): IWBBIO 2019, LNBI 11466, pp. 41–50, 2019.
https://doi.org/10.1007/978-3-030-17935-9_5

apnoea (CSA) when there is no effort; and mixed, as a combination of both. Polysomnography is the paraclinical gold standard for diagnosis. It is a complex study where multiple physiological signals, such as electroencephalogram (EEG), electrooculogram EOG, electrocardiogram ECG, electromyogram EMG, respiratory effort, airflow and oxygen saturation (SaO2) are recorded during a night. Consequently, the study requires the patient to stay overnight in a controlled environment with multiple sensors connected. This complexity is an inconvenience that makes it an expensive and uncomfortable procedure for the patient, hence the need to look for alternatives that allow to detect these disorders indirectly. The most immediate parameter to evaluate apnoeas could be the respiratory rhythm, which is also a basic physiological parameter in many clinical environments for the identification of anomalies. Respiration can be evaluated through a variety of technologies, however, they can restrict usual breathing and can interrupt sleep due to their position [2]. In the last decade different algorithms have been proposed to estimate respiratory information derived from the electrocardiogram (ECG) and plethysmography (PPG) [3].

The ECG contains a large amount of information including cardiovascular function, breathing, and electrical activity of the heart [4, 5]. Respiratory activity causes variations in lung volume during inspiration and expiration cycles resulting in fluctuation of thoracic electrical impedance and thus in ECG signal morphology. Breathing also results in changes in the position of the cardiac vector relative to the ECG electrodes [3]. Therefore, from the study of the morphology of the cardiac beats an EDR signal can be derived (breathing signal derived from the ECG) that allows the estimation of the respiratory frequency and, therefore, its alterations as the case of the sleep apnoeas [3].

Many studies have proposed different techniques to estimate EDR and some of them also develop algorithms to identify sleep apnoeas. Recently, one of the alternatives proposed in the literature to increase the accuracy and robustness of the respiratory rate BR estimation is the fusion of multiple respiratory signals, either by extracting multiple signals simultaneously, such as ECG and PPG [6], or by using multiple methods [7] from a signal such as ECG [8]. Other authors propose fusion techniques based on the segmentation of a respiratory signal into several windows and treating them as individual signals [9].

This work presents two algorithms for detecting apnoeas based on respiratory signal derived from the single-lead ECG and to improve the accuracy of these algorithms it is proposed to fusion the results of the different alternatives, looking for synergies between them, and thus obtain more reliable results. The novelty of the proposal lies fundamentally in the use of the Dempster-Shafer evidence theory to carry out this fusion.

This paper is organized as follows. Section 2 introduces different methods to obtain the EDR signal, including the database used. Section 3 deals with algorithms for identifying sleep apnoea and Sect. 4 reports the obtained results. Section 5 describes the fusion technique and its results. Finally, Sect. 6 concludes the paper.

2   Techniques for EDR Extraction

In order to evaluate the different detection algorithms, a total of 8 tests were performed on 7 different subjects using a commercial polygraph. The tests were carry out on adults between the ages of 24 and 61, obtaining a total of 68 apnoea episodes and 140 samples of normal breathing. The ECG signal was recorded using a modified lead V2 electrode setup, at a sampling rate of 512 Hz, simultaneously with thoracic and abdominal stress signals and respiratory flow signal, which are the respiratory signal reference. For the evaluation of the fusion process, we use a new data set, with 25 apnoea episodes drawn from two different tests.

2.1   QRS Complex Estimation

The information used to obtain the EDR signal from a single-lead ECG is extracted from the QRS complex, and concretely from R-peaks. These peaks have been detected by using an algorithm developed in our research group [10]. The results obtained are similar to the ones using the well-known Pan-Tompkins algorithm but lower computational load is needed. This QRS detection algorithm has two main stages. The first one is the signal pre-processing to eliminate unwanted interference such as baseline wander noise. Therefore, it is responsible for filtering the ECG signal and then move to the stage of detection of R-peaks. The second stage (detection stage) is based on a finite state machine (FSM) that carries out an adaptive threshold to discriminate between R-peaks or noise. After that, RR interval series are generated by calculating interval time between successive R-peak points.

2.2   EDR Estimation

This work analyses two of the most cited techniques for the EDR signal estimation in the literature [8], both of them based on the analysis of the R-peaks. The ECG signals were recorded simultaneously with the respiratory signal, considering the latter as the reference for the validation of the different techniques to EDR extraction. It is important to highlight the difficulty of working with the ECG signal, since it contains a lot of information, so it is hard to discriminate only the respiration data. It should be noticed that R-peaks based technique provides one sample of the EDR per heartbeat.

A. EDR estimation from the R-peaks amplitude

The first technique measures R-peaks amplitude variation to form the EDR signal, in fact it is formed as R-peaks envelop, as Fig. 1 shows, and it is denoted by EDR_{amp}. This amplitude variation is due to low-frequency modulation caused by the breathing process itself or the breathing movements.

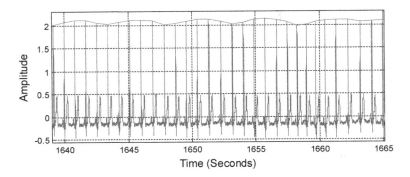

Fig. 1. ECG signal (blue) and the EDR$_{amp}$ signal (red). (Color figure online)

B. EDR estimation from R-peaks area

The second EDR estimation technique is through the analysis of the R-peaks area [11] and it is defined as EDR$_{area}$. In this case the area is calculated from a fix interval of 40 ms around the R-peak, 20 ms before and 20 ms after each R-peak. Every QRS complex area is calculated and used as a point estimate of the EDR signal.

C. Comparison of the EDR methods

The estimated respiratory frequencies from the two EDR methods are compared to that one obtained from a simultaneously recorded respiratory signal (reference signal). Figure 2 shows a typical epoch of normal breathing for the respiratory signal and the two EDR signals for comparison purposes.

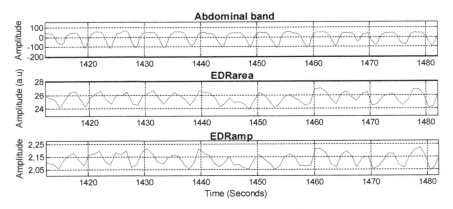

Fig. 2. Respiratory signal obtained with an abdominal band (reference), EDR$_{area}$ and EDR$_{amp}$ estimations.

Comparing the calculated EDR signals with that obtained with a chest band, it has been verified that the EDR_{area} reports the best results. This technique has reported a level of accuracy in the respiratory frequency estimation of 97.5% as opposed to 93.5% obtained with the EDR_{amp}. Therefore, EDR_{area} signal will be used for the development of EDR-based sleep apnoea detection algorithms.

3 Algorithms for Identifying Sleep Apnoea

This section introduces two different algorithms to identify epochs of central sleep apnoeas from the EDR_{area} signal. Our developed CSA detection algorithm is executed on the EDR_{area} signal divided into 30-second segments with an overlap between them of 20 s. Two alternatives are proposed, one in the frequency domain and the other in the time domain, both based on thresholds. The complexity of the ECG signal and its variability for each patient make it difficult the achievement of an efficient algorithm.

Frequency analysis is performed through the Lomb-Scargle periodogram, showing better results than the Fourier transform in resolution and noise, taking into account the aperiodic nature of the EDR signal [11]. Each 30-second epoch is analysed considering the amplitude and position of the predominant spectral peaks. An episode of apnoea is considered to happen if the frequency of the maximum peak is lower than 0.15 Hz and, in addition, the ratio between this maximum peak and the next highest peak below 0.15 Hz is between 1.1 and 1.6. This algorithm is denoted as $F\text{-}EDR_{area}$. A CSA epoch is considered in the reference signal when there is no breathing for more than 10 s.

Figures 3 and 4 show the results of the $F\text{-}EDR_{area}$ algorithm, applying the Lomb-Scargle periodogram, to a normal breathing segment and another during an apnoea episode, respectively. It can be observed how normal breathing is identified at frequencies above 0.15 Hz with a characteristic peak, which decreases significantly when an apnoea episode appears.

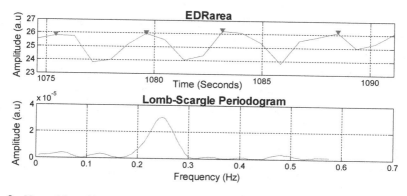

Fig. 3. Normal breathing: EDR signal (top) and its Lomb-Scargle periodo-gram (bottom).

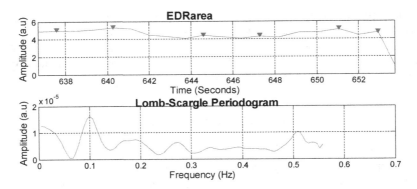

Fig. 4. Apnoea episode: EDR signal (top) and its Lomb-Scargle periodo-gram (bottom).

The second proposed algorithm, called D-EDR$_{area}$, is based on the amplitude variation of the differentiated EDR$_{area}$ signal when an episode of apnoea occurs. In this case, EDR$_{area}$ is sampled peak to peak instead of using windows and it is used an adaptive threshold to estimate CSA epoch. It is consider apnoea when the average of the last 10 analysed areas drops more than 60%. In addition, this drop has to be maintained at least in three consecutive areas.

As an alternative of this second algorithm, the quadratic of the differentiated EDR$_{area}$ signal has been analysed in order to emphasize amplitude changes associated with apnoea episodes, facilitating their detection. This third option is called D2-EDR$_{amp}$. The threshold criterion to consider apnoea in this case is to detect more than 3 consecutive area values lower than 35% of the mean of the last 6 analysed areas.

Figure 5 shows the differentiated EDR$_{area}$ signal (at the top), and its quadratic version (at the bottom). Apnea epochs are identified when the sharp amplitude reduction is close to zero in the quadratic differentiated EDR$_{area}$ signal.

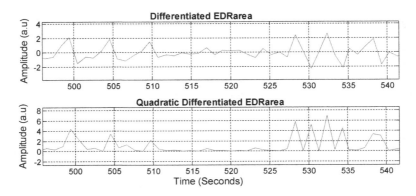

Fig. 5. Differentiated EDR$_{area}$ signal (upper) and its quadratic version (lower) for an apnoea episode.

4   Results of the Algorithms Based on EDR Signal

This section reports the results of EDR-based algorithms and their effectiveness in apnoea detection. The CSA detection algorithms previously explained lead in the classification of apnoea or non-apnoea at a specific time. For each method the statistical parameters of sensitivity (Se), specificity (Sp) and accuracy (Acc) have been calculated. The results of these statistical parameters for the different algorithms are shown in Table 1. The analysis of the differentiated EDR_{area} in the time domain provides the highest sensitivity, while specificity is better in frequency analysis of the EDR_{area}, and the best accuracy with the time domain analysis of the quadratic differentiated EDR_{area}. As each method identifies better different features of the ECG, we could conclude that they are complementary.

Table 1. Se, Sp and Acc parameters for each proposed algorithm for the CSA detection.

	$F\text{-}EDR_{area}$	$D\text{-}EDR_{area}$	$D2\text{-}EDR_{area}$
S_e (%)	81.1	91.7	91.5
S_p (%)	94.0	85.0	90.6
A_{CC} (%)	89.4	86.5	90.9

Accuracy parameters are at the state of the art, around 90% [8, 11, 12]. However, it is very difficult to achieve higher levels with these simple algorithms due to the variability of the ECG signal for each individual. Therefore, we propose to improve the detection by fusing the information provided by each algorithm.

5   Fusion of the CSA Detection Algorithm Results

The aim of this section is to assess automatic apnoea detection by fusing the results of the different described EDR algorithms according to their level of accuracy A_{CC}. We propose to use the Dempster–Shafer evidential theory [13] for performing this objective. This probability-based data fusion classification algorithm is useful when the information sources contributing data cannot associate a 100% probability of certainty to their output decisions. In this theory, each information source associates a declaration or hypothesis with a probability mass, expressing the amount of support or belief directly attributed to the declaration, in other words, the certainty of the declaration. The probability masses m_A for the decisions made by each information source are then combined by using Dempster's rule of combination [14]. The hypothesis favoured by the largest accumulation of evidence from all the information sources is selected as the most likely outcome of the fusion process. The Dempster–Shafer theory estimates how close the evidence is to forcing the truth of a hypothesis, rather than estimating how close the hypothesis is to being true [15].

Each of the three proposed algorithms provides a maximum level of confidence $\alpha_{A(i)}$ (i = 1, 2 and 3), or belief, according to its accuracy, as Table 2 shows. When a recording of the EDR is processed, the probability mass of each algorithm $m_{A(i)}$ is calculated according to the detection parameters of each algorithm, following a weighted function. For example, the probability mass of the first algorithm (F-EDR$_{area}$) is obtained as Fig. 6 shows, considering at which frequency is found the maximum peak.

Table 2. Confidence levels (Belief) for each detection algorithm.

	A_{cc}	Maximum belief, $\alpha_{A(i)}$
F-EDR$_{area}$	89.42%	0.89
D-EDR$_{area}$	86.54%	0.86
D2-EDR$_{area}$	90.87%	0.90

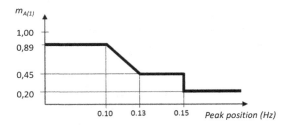

Fig. 6. Calculation of the probability mass of the first algorithm, F-EDR$_{area}$.

The evaluation of the fusion process is performed on a new data set, also acquired with the commercial polygraph at a sampling frequency of 512 Hz. In this case, 25 apnoea epochs are taken from two different tests. These 25 apnoea episodes have been processed by the 3 detection EDR-based algorithms and subsequently the Dempster's rule of combination [14] has been used to combine the results of the respective algorithms, providing a fused value of the level of confidence of the existence of apnoeas. Figure 7 shows the results of the fusion process. The 25 apnoea episodes have been detected reaching 100% detection. According to evidential theory, after fusing the results provided by each algorithm, 84% of the detected apnoeas have a confidence level between 93% and 99%. Therefore we can conclude that not only all the cases have been detected, but most have done with a high level of confidence.

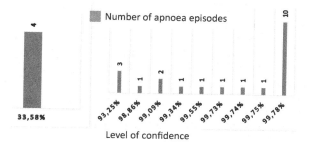

Fig. 7. Number of CSA assessments versus their level of confidence.

6 Conclusions

In this study two algorithms have been developed to detect central sleep apnoeas from the respiratory signal derived from the ECG, the EDR signal. The detection results are similar to the current state of the art in the detection of these disorders, around 90%. The CSA detection provided by the different EDR-based algorithms has been fused to improve performance, by both reducing the errors, and increasing the proportion of segments for which a CSA estimate is detected. This work proposes to use the Dempster-Shafer evidence theory for fusing the information. The data fusion proposal increases the robustness of the detection, indicating that the 84% of the detected apnoeas have a level of certainty above 90%. Although the results are very promising, it is necessary to carry out further studies with more recordings from different patients.

Acknowledgment. This work was supported in part by Junta de Comunidades de Castilla La Mancha (FrailCheck project SBPLY/17/180501/000392) and the Spanish Ministry of Economy and Competitiveness (TARSIUS project, TIN2015-71564-c4-1-R).

References

1. Gaig, C.: Redacción médica (2018). https://www.redaccionmedica.com
2. Moody, G.B., Mark, R.G.: Derivation of respiratory signals from multi-lead ECGs. Comput. Cardiol. **12**, 113–116 (1985)
3. Bailón, R., Sornmo, L., Laguna, P.: A robust method for ECG-based estimation of the respiratory frequency during stress testing. IEEE Trans. Biomed. Eng. **53**(7), 1273–1285 (2006)
4. Malik, M., et al.: Heart rate variability: standards of measurement, physiological interpretation and clinical use. Eur. Heart J. **17**, 354–381 (1996)
5. Correa, L., Laciar, E., Torres, A., Jane, R.: Performance evaluation of three methods for respiratory signal estimation from the electrocardiogram. In: 2008 30th Annual International Conference of the IEEE Engineering in Medicine and Biology Society, EMBS 2008, pp. 4760–4763 (2008)
6. Ahlstrom, C., et al.: A respiration monitor based on electrocardiographic and photoplethys-mographic sensor fusion. In: Conference on Proceeding of Engineering in Medicine and Biology Society, pp. 2311–2314. IEEE (2004)

7. Lakdawala, M.M.: Derivation of the respiratory rate signal from a single lead ECG, MSc. thesis, New Jersey Institute of Technology (2008)
8. Charlton, P.H., et al.: Breathing rate estimation from the electrocardiogram and photoplethysmogram: a review. IEEE Rev. Biomed. Eng. **11**, 2–20 (2017)
9. Bailon, R., Pahlm, O., Sornmo, L., Laguna, P.: Robust electrocardiogram derived respiration from stress test recordings: validation with respiration recordings. In: Conference on Proceedings of CinC, pp. 293–296. IEEE (2004)
10. Gutiérrez-Rivas, R., García, J.J., Marnane, W.P., Hernández, A.: Novel real-time low-complexity QRS complex detector based on adaptive thresholding. IEEE Sensors J. **15**(10), 6036–6043 (2015)
11. Fan, S.H., Chou, C.C., Chen, W.C., Fang, W.C.: Real-time obstructive sleep apnea detection from frequency analysis of EDR and HRV using Lomb Periodogram. In: 2015 Conference of the IEEE Engineering in Medicine and Biology Society, pp. 5989–5992 (2015)
12. Janbakhshi, P., Shamsollahi, M.B.: Sleep apnea detection from single-lead ecg using features based on ECG-Derived Respiration (EDR) signals. IRBM **39**(3), 206–218 (2018)
13. Shafer, G.: A Mathematical Theory of Evidence. Princeton University Press, Princeton (1976)
14. Klein, L.A.: Data and Sensor Fusion: A Tool for Information Assessment and Decision Making. SPIE, Bellingham (2004)
15. Pearl, J.: Probabilistic Reasoning in Intelligent Systems: Networks of Plausible Inference. Morgan Kaufmann, San Mateo (1988)

Instrumented Shoes for 3D GRF Analysis and Characterization of Human Gait

João P. Santos[1], João P. Ferreira[1,2], Manuel Crisóstomo[1(✉)], and A. Paulo Coimbra[1]

[1] Institute of Systems and Robotics, Department of Electrical and Computer Engineering, University of Coimbra, Coimbra, Portugal
mcris@isr.uc.pt
[2] Department of Electrical Engineering,
Superior Institute of Engineering of Coimbra, Coimbra, Portugal

Abstract. The main objective of this work is to develop a computerized system based on instrumented shoes to characterize and analyze the human gait. The system uses instrumented shoes connected to a personal computer to provide the 3D Ground Reaction Forces (GRF) patterns of the human gait. This system will allow a much more objective understanding of the clinical evolution of patients, enabling a more effective functional rehabilitation of a patient's gait. The sample rate of the acquisition system is 100 Hz and the system uses wireless communication.

Keywords: Human gait analysis · Gait pathologies · Ground reaction forces

1 Introduction

The study of the human gait has generated much interest in fields like biomechanics, robotics, and computer animation. It has been studied in medical science [1–3], psychology [4, 5], and biomechanics [6–8] for five decades. Comprehensive measures of gait pathology are useful in clinical practice. They allow stratification of severity, give an indication of the gait quality, and provide an evaluation of treatments outcomes. There are many ways to gauge overall gait pathology. While parent and caregiver assessments are useful and practical, they lack the precision and objectivity provided by three-dimensional quantitative gait data. Gait data can be used to assess pathology in a variety of ways. For example, stride parameters such as walking speed, step length, and cadence provide an overall picture of gait quality. These parameters are especially useful after normalization to account for differences in stature [9]. So, it's important to have systems that provide such kind of information.

To study and analyze the human gait, their patterns must be acquired. During walking, the most common force acting on the body is the GRF that is a three-dimensional force vector. The GRF consists of a vertical component plus two shear components, antero-posterior and medial-lateral directions, acting along the foot support surface. A fourth variable is needed, the location of the center of pressure of this GRF vector.

© Springer Nature Switzerland AG 2019
I. Rojas et al. (Eds.): IWBBIO 2019, LNBI 11466, pp. 51–62, 2019.
https://doi.org/10.1007/978-3-030-17935-9_6

In the particular domain of plantar pressure and GRF, there are a variety of pressure mats or insoles [10] like as F-scan©, Emed© or Zebris© systems. Force platforms mounted in the ground are used in many studies on humans to measure GRF during walking or running [11, 12]. However, force plates are expensive, the number of successive ground contacts is very limited, and they are conditioned to certain types of environment.

The first measurement of dynamic ground reaction force and plantar pressure using instrumented shoes was realized by Marey [13]. In [14] is disclosed a system for detecting abnormal human gait using an inertial measuring unit to measure angular velocity and acceleration of the foot, and four load sensors and a bending sensor installed on the insole of the feet to get force and flexural data respectively. It was specifically designed to detect patterns of normal gait, toe in, toe out, over supination, and heel walking gait abnormalities. The system presented in [15] was designed to measure GRF and CoP using six-degrees-of-freedom force sensors, one under the heel and the other on the forefoot. Because there are only two sensors, the determination of the location of the CoP in every moment of the walking phase as well as the behavior of the reaction force at other points of contact between the foot and the ground are restricted.

Another more complete solution is the combination of the kinematics of the body segments with the GRF, like the system presented in [16, 17]. Human walking patterns data are analyzed, especially in the sagittal plane, based on video camera images and an acquisition system of the CoP and the vertical component of GRF using four force sensors in each shoe sole. This system objective was to obtain gait signatures using computer vision techniques and to extract kinematics features for describing human motion and equilibrium. These results have been used afterwards in gait specification of biped robots. Results show that this approach worked successfully.

For an accurate characterization and analysis of the human gait, it is needed a GRF acquisition system to obtain the three components of the GRF, the vertical force, anterior-posterior shear, medial-lateral shear.

2 Data Acquisition System

The system developed for the 3D ground reaction forces data acquisition consists of a prototype shoes with force sensors, hardware, firmware, and acquisition software.

Figure 1 presents the system architecture showing the data flow from the force sensors located in the shoes up to the computer where the data processing software is hosted. The data can be acquired via an Android application and then sent to the computer to be processed or sent directly form the instrumented shoes to the computer.

2.1 Prototype Shoes

The prototype shoes' soles, where the force sensors are located, were built with ABS 3D-printed pieces that were assembled and glued to the shoes garment. A rubber sole was added to the bottom of the shoe. A prototype of the developed right instrumented shoe is shown in Fig. 2.

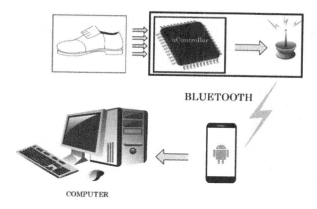

Fig. 1. System architecture.

Fig. 2. Instrumented shoe prototype.

The measures by which the prototype was designed for are shoes sizes from 40 up to 46, according to the European system. The different sizes are obtaining by adjusting the back Velcro strap and sliding the heel forward and backward. The physical characteristics of the shoe are shown in the Table 1 below.

Table 1. Features of the prototype instrumented shoe.

Characteristics	Values
Length	310 (mm)
Height	20 (mm)
Width	100 (mm)
Weight	700 (g)
Sensors	16

The shoe sole consists of two independent parts, the back part that comes into contact with the rear-foot and mid-foot, and the front part that comes into contact with the fore-foot. The upper connection between these parts is made with an insole leather material and a hinge causing the shoe accompanying the dorsiflexion and plantar flexion movements of the foot. Each sole part is made out of a top and bottom platforms. The sensors are placed on the bottom platforms (Fig. 3).

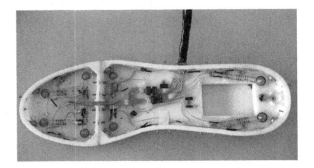

Fig. 3. Interior of the shoe sole prototype. (Color figure online)

In order to centralize the force applied to the sensors and capture 100% of this force, shims (pink semi-spheres) were used between the load and the sensor. Discs with a diameter equal to the diameter of the sensitive area of the sensor and with a thickness of 0.5 mm were also used so that the force is distributed only by the sensitive area of the sensor (Fig. 4).

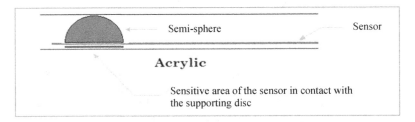

Fig. 4. Semi sphere and disc in contact with the sensor in order to centralize force and distribute the load to the sensitive area of the sensor.

The global system has 32 force sensors, 16 force sensors in each shoe. In each shoe, there are 8 force sensors to measure the vertical component of the GRF, and 8 to measure the horizontal components of the GRF. The sensors are equally distributed by both front and back parts of the bottom platform of the shoe (Fig. 3). The horizontal force sensors at the corners of the soles are each placed at 45°. This way each horizontal force sensor measurement is decomposed in longitudinal (x axis) and lateral (y axis) force components (Fig. 5).

The vertical force sensors are Teskan FlexiForce A201 sensors with a range of 0-440 N, and the horizontal force ones are the same model but with a range of 0-110 N.

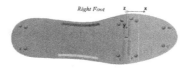

Fig. 5. Sensors position and coordinate system of the right shoe prototype. The left shoe coordinate system is symmetric.

2.2 Hardware

It was chosen to use two commercial boards, one for each shoe, with an embedded Arduino Mega 2560 microcontroller because has a small size and contains 16 analog inputs, meeting the needs of each shoe to read all the 16 sensors. Each board is powered by a battery with 7.4 V and 2250 mAh.

Another two dedicated boards were developed to signal conditioning the force sensors' values. Each of these boards has an RN-42 Bluetooth Module to establish a wireless communication with the Android device or the used computer, operating over the frequency range of 2.402 GHz to 2.48 GHz, and having a signal reception range of about 20 m.

To acquire the values of the sensors, it is used the force-to-voltage circuit of Fig. 6, and the force sensor signals are acquired by 10-bit resolution ADCs integrated in the microcontroller of the Arduino chip board. This circuit uses an inverting operational amplifier to produce an analog output based on the sensor resistance (R_S) and a fixed reference resistance (R_F). The resistance R_F value is 6.81 kΩ for the vertical force sensors and 7.5 kΩ for the horizontal ones.

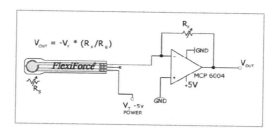

Fig. 6. Force-to-voltage circuit

The force sensors were calibrated. Considering that all sensors of the same type had similar behavior, and since two types of sensors are used in this work, 0-440 N range and 0-110 N range, the calibration process needed to be repeated for both types of sensors. The calibration process consisted in applying different weights with a known mass to a sensor and reading its output voltage. To have more accurate results the sensor should be calibrated in the circuit were it is going to be used. For this reason, the sensor was connected to the Arduino board and the voltage was also read from the same board. With these values, a linear regression was made to obtain an equation that would fit best the behavior of the sensor.

2.3   Firmware

The firmware routines are important parts of the system, allowing the acquisition and transmission of the force values from the shoes to the computer, with error checking.

These routines are hosted and performed by the embedded microcontrollers. The set of operating instructions programmed in the Arduino includes the configuration of the serial communication, reading sensor values from the Analog-Digital Converter (ADC), generating the Cyclic Redundancy Code (CRC), packaging of data to transmit, and sending data via Bluetooth.

There are 4 modes of operation: acquiring the sensor values at a sampling rate of 100 Hz (suitable for the human gait acquisition [18]), acquiring the sensor values at a sampling rate of 1 Hz (for system checking), the tare mode, and idle mode.

The state diagram of the implemented operations (actions) are shown in Fig. 7. In the two acquisition modes, after the interrupt flag is activated, the Arduino draws the Interrupt Service Routine (ISR) to be executed. In the ISR, for each sampling time is determined the current time and read the 16 sensor values. After the acquisition of the sensor values, it is calculated the CRC value for each sensor value and for the time that is later packaged and sent to the Android device. In tare mode, the Arduino reads the sensor values and sets the offsets for each sensor that are sent and stored on the computer which will then be subtracted to the value of the force. In idle mode the system is on standby until it receives a command to start one of the aforementioned modes.

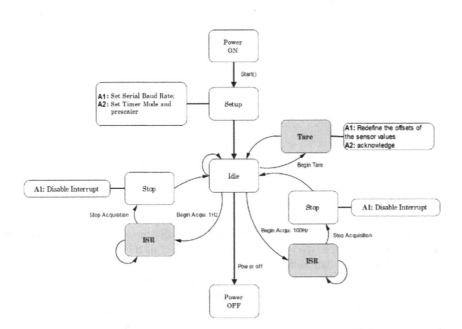

Fig. 7. State diagram of the actions (Ax) implemented in the firmware.

2.4   Acquisition Software

To acquire the human gait, an Android application was developed to obtain the force sensor values. The acquisition is performed at a sampling rate of 100 Hz, and this acquired data will be used to characterize the human gait. Before starting the acquisition, the user needs to fill in data fields related to the person whose gait is going to be analyzed and select the Bluetooth devices corresponding to the two instrumented shoes (Fig. 8). The files produced by the application are stored on the "Downloads" folder of the Android device.

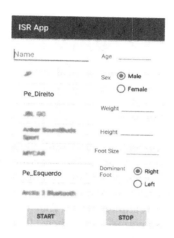

Fig. 8. Layout of the Android application to acquire the human gait.

The processing of the acquired force values is done using a Matlab application. On each foot, the anterior-posterior shear force, corresponding to F_x force, and the medial-lateral shear force, corresponding to F_y force, are calculated using the 8 forces exerted on the horizontal force sensors. The positioning of the horizontal force sensors are at $m \times 45$ degrees around the z axis of the reference system of each foot, where m is an integer. Therefore, the absolute values of the horizontal components of the forces of the sensor n are given by

$$F_{n,x} = F_{n,y} = F_n \times \sin\frac{\pi}{4} = F_n \times \frac{\sqrt{2}}{2} \tag{1}$$

where F_n is the absolute value of the force of sensor n. The anterior-posterior shear force is then calculated by

$$F_x = \left(\sum_{i=1}^{4} F_{i,x} - \sum_{j=1}^{4} F_{j,x} \right) \times \frac{\sqrt{2}}{2} \tag{2}$$

where $F_{i,x}$ is the absolute value of the ith anterior sensor of each shoe part and $F_{j,x}$ is the absolute value of the jth posterior sensor of each shoe part. The medial-lateral shear force is calculated by

$$F_y = \left(\sum\nolimits_{i=1}^{4} F_{i,y} - \sum\nolimits_{j=1}^{4} F_{j,y} \right) \times \frac{\sqrt{2}}{2} \qquad (3)$$

where $F_{i,y}$ is the absolute value of the ith lateral sensor and $F_{j,y}$ is the absolute value of the jth medial sensor.

The values of the vertical sensors are used to calculate the vertical component of the GRF corresponding to the F_z force:

$$F_z = \sum\nolimits_{i=1}^{8} F_{i,z} \qquad (4)$$

where $F_{i,z}$ is the absolute value of the ith vertical sensor.

3   Experiments and Results

The experiments with the instrumented shoes were performed with one 23 years old person, with 76 kg and 1.79 m of height. The experiments consisted in walking at different speeds to acquire the patient's 3D GRF. At the beginning of each experiment the patient had to stand on each feet for a few seconds in order to use the total registered force as the person's weight reference. Then the patient had to walk six times along the corridor starting with a normal speed, then a little faster and then even a faster speed. During the fourth walk the patient should walk at a normal speed again, followed by a slower speed and finally an even slower speed. The experiments were carried out using the shoes without any attachment (friction coefficient of 0.47) and using a 0.25 mm PVC sheet underneath the rubber sole (friction coefficient of 0.27).

In Figs. 9 and 10 it is possible to see the vertical component of the GRF for the five speeds and for the two friction coefficients. All the graphics were normalized for the person's weight. It is noticeable from the graphics that the faster the walking speed was, the higher the first peak corresponding to the contact of the heel on the ground (heel strike). The second peak corresponds to the push-off (toe off). It is also visible that the faster the person walked the shorter is the duration of the step. Comparing the patterns with different friction coefficients, it is possible to see that the vertical component of the GRF is higher for the higher friction coefficient, corresponding to a more confidence walking, as expected.

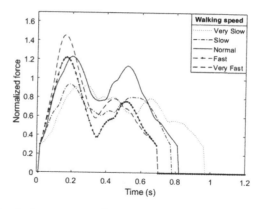

Fig. 9. Vertical component of the GRF with a 0.47 friction coefficient.

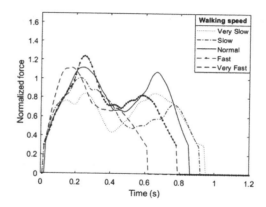

Fig. 10. Vertical component of the GRF with a 0.27 friction coefficient.

In Figs. 11 and 12 are represented the antero-posterior components of the GRF for the two friction coefficients. The amplitude of the forces is about 3 to 4 times smaller than when these forces are measured using horizontal static platforms [7]. This is due to the fact that in the used instrumented shoes, when these antero-posterior forces are higher, the shoe is with its greater inclination to the ground, having its sensors inclined to the horizontal. In other words, these instrumented shoes' sensors coordinate system rotates in the sagittal plane during the gait. These force profiles cannot therefore be compared with the horizontal static platforms results.

In Fig. 11, for the 0.47 friction coefficient, the curve corresponding to the fastest walking speed shows the typical pattern of the antero-posterior component of the GRF, with well-defined heel strike and toe-off forces [7]. As the speed decreases, the amplitude of the antero-posterior component of the GRF also decreases, as expected. At slow speed, the gait is more irregular and the forces absorbed on the heel strike and those created on toe-off are smaller, causing more irregular patterns.

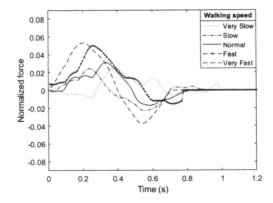

Fig. 11. Antero-posterior component of the GRF with a 0.47 friction coefficient.

In Fig. 12, for the 0.27 friction coefficient, the curves present lower amplitudes and are more irregular than with the 0.47 friction coefficient, showing a lower confidence in the walking. Unlike with the 0.47 friction coefficient, the amplitudes of the antero-posterior components of the GRF are not clearly related with the speed.

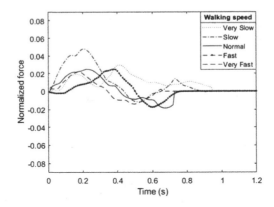

Fig. 12. Antero-posterior component of the GRF with a 0.27 friction coefficient.

The medio-lateral component of the GRF (Figs. 13 and 14) is the most imprecise of all the components because of its low value, about 1–2% of the person weight, similar to the instrumented shoe weight. When the shoe is not fully grounded, the medio-lateral component of the GRF readings suffer from the shoe dynamic forces. These force profiles cannot therefore be compared with horizontal static platforms results. The experiments with the lower friction coefficient, Fig. 14, present lower values of the medio-lateral component of the GRF, as expected.

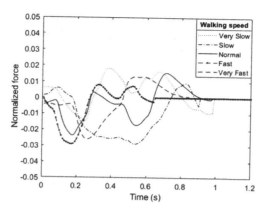

Fig. 13. Medio-lateral component of the GRF with a 0.47 friction coefficient.

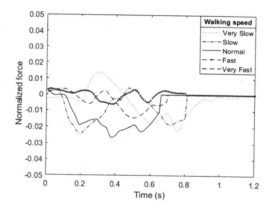

Fig. 14. Medio-lateral component of the GRF with a 0.27 friction coefficient.

4   Conclusions and Future Work

This paper presents an instrumented shoe that measures both medio-lateral and antero-posterior GRF as well as the GRF vertical component. The medio-lateral and antero-posterior GRF are measured in an innovative way by using force sensors placed at 45°.

The measured antero-posterior forces are smaller than when these forces are measured using horizontal static platforms. This is due to the fact that these instrumented shoes' sensors coordinate system rotates in the sagittal plane during the gait. Also, when the shoe is not fully grounded, its weight influences the medio-lateral component of the GRF readings. These force profiles cannot therefore be compared with horizontal static platforms results.

The system can be used in medical areas particularly in diagnosing the degradation of one or more leg joints, by detecting walking disturbances in a patient gait.

In the future, to make a full validation of the prototype and to develop a classification and diagnosis system based on 3D GRF, it is intended to acquire gaits of a large number of people, comparing with results obtained using static force platforms and similar instrumented shoes.

Acknowledgment. The Fundação para a Ciência e Tecnologia (FCT) and COMPETE 2020 program are gratefully acknowledged for funding this work with PTDC/EEI-AUT/5141/2014 (Automatic Adaptation of a Humanoid Robot Gait to Different Floor-Robot Friction Coefficients).

References

1. Whittle, M.W.: Clinical gait analysis: a review. Hum. Mov. Sci. **15**, 369–387 (1996)
2. Murray, M.P.: Gait as a total pattern of movement. Am. J. Phys. Med. **46**(1), 290–333 (1967)
3. Perry, J.: Gait Analysis: Normal and Pathological Function. Slack, Thorofare (1992)
4. Johansson, G.: Visual perception of biological motion and a model for its analysis. Percept. Psychophys. **14**(2), 201–211 (1973)
5. Stevenage, S.V., Nixon, M.S., Vince, K.: Visual analysis of gait as a cue to identity. Appl. Cogn. Psychol. **13**, 513–526 (1999)
6. Inman, V.T., Ralston, H.J., Todd, F.: Human Walking. Williams & Wilkins, Baltimore (1981)
7. Winter, D.A.: The Biomechanics and Motor Control of Human Movement, 2nd edn. Wiley, New York (1990)
8. Aggarwal, J.K., Cai, Q., Liao, W., Sabata, B.: Nonrigid motion analysis: articulated and elastic motion. Comput. Vis. Image Underst. **70**(2), 142–156 (1998)
9. Vaughan, C.L., Langerak, N.G., O'Malley, M.J.: Neuromaturation of human locomotion revealed by non-dimensional scaling. Exp. Brain Res. **153**, 123–127 (2003)
10. Schaff, P.S.: An overview of foot pressure measurement systems. Clin. Podiatr. Med. Surg. **10**, 403–415 (1993)
11. Chao, E.Y., Laughman, R.K., Schneider, E., et al.: Normative data of knee joint motion and ground-reaction forces in adult level walking. J. Biomech. **16**, 219–233 (1983)
12. Bates, B.T., Osternig, L.R., Sawhill, J.A., James, S.L.: An assessment of subject variability, subject-shoe interaction, and the evaluation of running shoes using ground reaction force data. J. Biomech. **16**(3), 181–191 (1983)
13. Marey, E.J.: De la locomotion terrestre chez les bipedes et les quadripedes. J. Anat. Physiol. **9**, 42–80 (1873)
14. Chen, M., Huang, B., Xu, Y.: Intelligent shoes for abnormal gait detection. In: IEEE International Conference on Robotics and Automation, ICRA 2008. IEEE (2008)
15. Veltink, P.H., et al.: Ambulatory measurement of ground reaction forces. IEEE Trans. Neural Syst. Rehabil. Eng. **13**(3), 423–427 (2005)
16. Ferreira, J.P., Crisóstomo, M., Coimbra, A.P.: Human gait acquisition and characterization. IEEE Trans. Instrum. Meas. **58**(9), 2979–2988 (2009)
17. Ferreira, J.P., Crisóstomo, M., Coimbra, A.P., Carnide, D., Marto, A.: A human gait analyzer. In: IEEE International Symposium on Intelligent Signal Processing – WISP 2007, Madrid, 3–5 October 2007
18. Mittlemeier, T.W.F., Morlock, M.: Pressure distribution measurements in gait analysis: dependency on Measurement frequency. In: Abstract presented at 39th annual meeting of the orthopaedic research society (1993)

Novel Four Stages Classification of Breast Cancer Using Infrared Thermal Imaging and a Deep Learning Model

Sebastien Mambou[1], Ondrej Krejcar[1]([⊠]) (ID), Petra Maresova[2] (ID),
Ali Selamat[1,3] (ID), and Kamil Kuca[1] (ID)

[1] Center for Basic and Applied Research,
Faculty of Informatics and Management, University of Hradec Kralove,
Rokitanskeho 62, 500 03 Hradec Kralove, Czech Republic
{jean.mambou, ondrej.krejcar, kamil.kuca}@uhk.cz
[2] Department of Economy, Faculty of Informatics and Management,
University of Hradec Kralove, Rokitanskeho 62, 500 03 Hradec Kralove,
Czech Republic
petra.maresova@uhk.cz
[3] Malaysia Japan International Institute of Technology (MJIIT),
Universiti Teknologi Malaysia, Jalan Sultan Yahya Petra,
Kuala Lumpur, Malaysia
aselamat@utm.my

Abstract. According to a recent study conducted in 2016, 2.8 million women worldwide had already been diagnosed with breast cancer; moreover, the medical care of a patient with breast cancer is costly and, given the cost and value of the preservation of the health of the citizen, the prevention of breast cancer has become a priority in public health. We have seen the apparition of several techniques during the past 60 years, such as mammography, which is frequently used for breast cancer diagnosis. However, false positives of mammography can occur in which the patient is diagnosed positive by another technique. Also, the potential side effects of using mammography may encourage patients and physicians to look for other diagnostic methods. This article, present a Novel technique based on an inceptionV3 couples to k-Nearest Neighbors (InceptionV3-KNN) and a particular module that we named: "StageCancer." These techniques succeed to classify breast cancer in four stages (T1: non-invasive breast cancer, T2: the tumor measures up to 2 cm, T3: the tumor is larger than 5 cm and T4: the full breast is cover by cancer).

Keywords: InceptionV3 · KNN · StageCancer

1 Introduction

The human body naturally manages the creation, growth, and death of cells in its tissues. Once the process begins to function abnormally, the cells do not die as fast as they should. Therefore, there is an increase in the ratio of cell growth to cell death, which is a direct cause of cancer. Likewise, breast cancer appears when cells divide and

I. Rojas et al. (Eds.): IWBBIO 2019, LNBI 11466, pp. 63–74, 2019.
https://doi.org/10.1007/978-3-030-17935-9_7

grow without reasonable control; It is a disease well known all over the world. In the United States, one in eight women is diagnosed with breast cancer in her lifetime, and more than 40,000 die each year in the United States [1]. We can reduce this number by using early detection techniques, awareness campaigns, and better diagnoses, as well as improving treatment options. In this article, we will first explore infrared digital imaging, which assumes that a necessary thermal comparison between a healthy breast and a breast with cancer always shows an increase in thermal activity in precancerous tissues, and the areas surrounding developing breast cancer. This is due to the metabolic activity and vascular circulation that surround the cancer cells. Over the last 20 years, several techniques have been proposed to detect breast cancer much earlier, such as [2], where the author presents a powerful approach to separate healthy to sick (cancerous) breasts. Likewise, our deep convolutional neural networks (DNN) can make this distinction but with a Novelty, which is the classification of the cancerous breast into four stages:

- T1: The tumor in the breast is 20 mm or smaller in size at its widest area.
- T2: The tumor is more significant than 20 mm but not larger than 50 mm.
- T3: The tumor is more significant than 50 mm.
- T4: Cancer has grown into the chest wall.

As per our Knowledge, it is the first time that such classification is done using thermal images of the breasts.

2 Previous Work

The authors in [3] presented an overview of computer-aided design (CAD) systems based on the state-of-the-art techniques developed for mammography and breast histopathology images. They also describe the relationship between histopathology phenotypes and mammography, which takes into account the biological aspects. They propose a computer modeling approach to breast cancer that develops a mapping of the phenotypes/ characteristics between mammographic abnormalities and their histopathological representation. Similarly, the authors in [4] have studied the use of Local Quinary Patterns (LQP) for the classification of mammographic density in mammograms on different neighborhood topologies. They adopted a multiresolution and multi-orientation approach, studied the effects of multiple neighborhood topologies, and selected dominant models to maximize texture information. Nevertheless, they used a Support Vector Machine (SVM) classifier to classify the resulting data.

Despite the preference of mammography for several decades, new techniques have emerged to overcome the limitations of mammography. Similarly, research has shown near-infrared fluorescence (NIRF) as an essential part of the cancer diagnostic process [5], as well as ongoing observation of the disease and its treatment. It is vital that the image processing produces a powerful NIRF light signal so that the image taken contains a lot of information very close to the actual state of the breast. As we know, the earlier the tumor is detected and the sooner the treatment is started, the better the chances of success. Other researchers have discussed the difficulty of obtaining tumor parameters such as metabolic heat, tumor depth and thermogram diameter [6]. Another article [7] mentions the limitations of computed tomography (CT) and magnetic resonance imaging (MRI), which have low sensitivity for lesions less than one centimeter because of their limited spatial resolution. Some research has shown other negative points of successive mammograms for ten years [8]. According to the study [8], the rate of false positive diagnoses in women after a mammogram each year for ten years is 49.1%. Another study showed that when women were advised to biopsy the Sentinel Lymph Node (NLS), this reduced the risk of disease progression (breast cancer) [9]; moreover, other authors [10] cautioned against taking the results of breast cancer thermography as sufficient information for decision-making, but such images can be used by a powerful Computer Assist Device (CAD) for positive outcome.

A typical maximum transient thermal contrast during warming for a breast with a 10 mm tumor to a depth of 5 mm after being cooled for 1 min was presented in this article. At the time of reading of the thermal graph processes by the computer, the amplitude of the transient peak and its corresponding time, as well as the response time, were extracted from the maximum transient thermal contrasts for tumors of different diameters located at different depths. This analysis shows the peak temperature generated on the breast surface with the tumor. As shown in the diagram of (see Fig. 1), the area surrounding the tumor will produce more heat during the warm-up phase before falling back to a stable temperature.

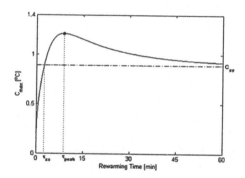

Fig. 1. A phase of breast warming [7].

Let's explored other important aspects:

A Genetic Factor for Breast Cancer: The article [11], the analysis and synthesis of 12 short- and medium-term breast cancer clinical research projects suggested 12 areas of research aimed at to improve detection rates and treatment of breast cancer disease. Our point of interest (Developing better tools for identifying genetically predisposed patients) was explored in the third part of this article, where they identified two essential genes (BRCA1 and BRCA2) in the genetic testing of patients with suggestive family history. It was found that confirmation of genetic predisposition could facilitate the implementation of risk reduction strategies. Besides, the use of new genetic testing tools, such as the high-risk hereditary breast cancer panel, should be accompanied by an appropriate interpretation of the results and their variants for better use in the clinical decision. Recent studies show that poly (ADP-ribose) polymerase inhibitors (PARPs) may be useful in treating tumors with BRCA1/2 mutations that develop breast cancer.

Stages of Breast Cancer: According to [12], Breast Cancer can be grouped into four stages describes as:

- **T1:** The breast tumor is 20 mm or less in its most extensive area, which is a little less than an inch. This category can be divided into four sub-categories depending on the size of the tumor:
  - T1a: a tumor has a size greater than 1 mm but less than or equal to 5 mm.
  - T1b: a tumor has a size greater than 5 mm but less than or equal to 10 mm.
  - T1c: a tumor has a size greater than 10 mm but less than or equal to 20 mm.
- T2: The tumor is more significant than 20 mm but not larger than 50 mm.
- T3: The tumor is more significant than 50 mm.
- T4: Cancer has grown into the chest wall. This category can be divided into four sub-categories depending on the size of the tumor:
  - T4a means that the tumor has developed in the chest wall.
  - T4b is when the tumor has developed in the skin.
  - T4c is a cancer that has developed in the chest wall and skin.
  - T4d is an inflammatory breast cancer.

3 Propose Model

For this work, the images were extracted from the Research Data Base (RDB) containing images of frontal thermograms, obtained using a FLIR SC-620 infrared camera with a resolution of 640 × 480 pixels [13]. The dataset includes images of individuals aged 29 to 85 years old. These images include breasts of different shapes and sizes, such as medium, wide and asymmetrical breasts (Table 1).

The statistic data of the subjects are presented in Table 2, with:

- Total number of subjects (N) = 67
- Total number of healthy/normal subjects (NH) = 43
- Total number of sick/abnormal subjects (NS) = 24

Table 1. Summary of subjects.

Age range	Total number of sick	Total number of healthy
[29–50]	11	07
[51–70]	12	31
[71–85]	01	05

The articles discussed earlier, let emphasize the importance of the image processing, which is currently not well achieve by Artificial intelligence (in comparison to Human being). This highlights the need for a better Computer Assist Device (CAD) that will help us to understand better the thermal images captured by our different thermal imaging cameras. In this context, a CAD will be a deep neural network with an KNN model as a classifier, as shown in Fig. 2 (assuming it is already formed), which will take the thermal images and classify them as non-cancerous (healthy) or cancerous with the possibility to see in which stage Breast cancer is. This model which is, in fact, an InceptionV3 with **k-Nearest Neighbors** (InceptionV3-KNN) has an extension to what we called " CancerStage".

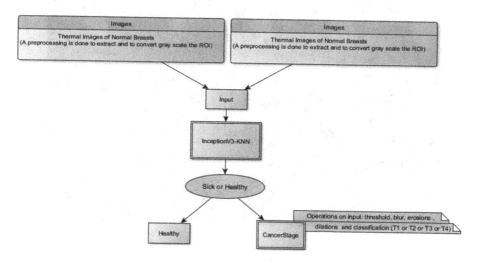

Fig. 2. Flow chart of the proposed model.

CancerStage: As shown (see Fig. 2), it is after KNN classifier and it is used when our InceptionV3-KNN, classifies a picture as Sick. This last is given to the next module "CancerStage "as input which will be thresholded to reveal light regions. Also, it will blur the resulting image, and a series of erosions and dilations will be done to remove any small blobs of noise as shown in Fig. 3.

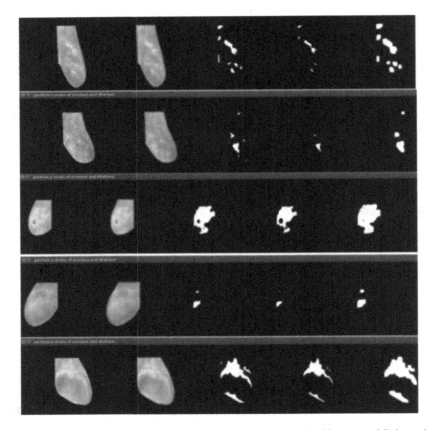

Fig. 3. The third images column wise show the result of the threshold (to reveal light regions), and the remain images show results of blur operation, erosions operation and dilations.

3.1   Pre-processing of Breast Thermal Images

We can subdivide this task as follows:

Pre-treatment of Thermal Images of the Breast: We chose to use a public dataset [14]. The thermal images of this data set were obtained following a dynamic protocol of taking a picture after cooling the breasts with air flow. During the process of restoring the thermal balance of the patient's body with the environment, the author of the dataset obtained 20 sequential images spaced 15 s apart. The pictures in their original input format (640 × 480) are huge for our DNN, so it is important to trim them to eliminate unwanted areas.

Obtaining the Region of Interest: From each grayscale image, the region of interest (ROI) was extracted. Each ROI image is converted into a matrix of features that will be processed, and the areas most likely to have cancer will be transferred to the next component entry.

Thus, pre-processing involves grayscale RGB conversion and image culture to remove unwanted regions such as the neck region, arms, and sub-membrane folding of the part.

3.2 Image Classification Framework

Considering the concept of transfer learning, we used a pre-trained Inception V3 model [15], which is powerful enough for feature extraction. Our DNN can be described as [10, 20, 10], which means there is a layer of 10 neurons, where each is connected to 20 neurons in the next layer, and similarly, each is connected to 10 neurons in the third layer. Also, we retrained the final classified layer so that it could determine cancer versus no cancer with considerable confidence (>0.65). If the last layer output had the confidence of <0.65 and 0.55, we submitted the matrix of features to our k-Nearest Neighbors (KNN) for output prediction [17]. During the training of our model, we set the learning rate at 0.0001, the epochs to 14, and the steps to 4000 (all these numbers were obtained through several experiments). Furthermore, our training was done using a sample of 64 breasts, including 32 that were healthy and 32 with some abnormality, where each breast had 20 sequential images. This resulted in 1062 images (after the pre-processing phase) used to train and test our network with respective repartitions of 0.8 and 0.2 as shown in Table 3. To validate the performed tests, we used 12 breasts that were entirely new to our model.

Table 2. Dataset repartition use for training

	Training	Testing	Total
Healthy	481	121	602
Sick	368	92	460
Total	849	213	**1062**

InceptionV3 Model: When training our InceptionV3 model (Figs. 2 and 3), we observed an increased accuracy and reduced entropy of our model after 3900 steps. Above this number of steps, the model will be over-adjusted, which will result in a decrease in accuracy (Fig. 4).

k-Nearest Neighbors (KNN): Our representation of extracted features via t-distributed stochastic neighbor embedding (tnse) indicates that the features are not fully classified. Also, two main groups can be distinguished: Healthy and Sick Breast. Our goal here is to add on top of our Deep Neural Network (DNN), another classifier to get a good accuracy. For this reason, we have performed through several experiments, a test of several classifiers as shown in Fig. 7. As a result, we found that KNN has the best performance (see Fig. 6D).

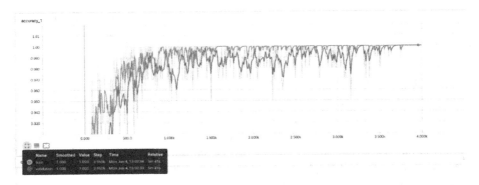

Fig. 4. Shows the increase in the accuracy, over the number of iterations, the training and the validation become stable after 3900 training steps.

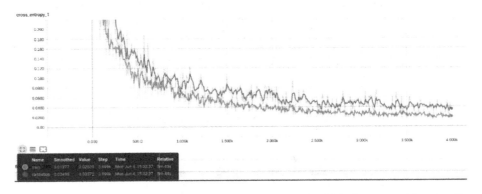

Fig. 5. The entropy decreases during the training of our model, which shows a positive sign (the ambiguity reduces in our model).

After the model is trained, a validation test is done by taking the breast images from the dataset of 12 new breasts (as mentioned earlier). Bearing in mind that for each breast, we have 20 sequential images; we submitted 480 images for the validation test to be organized according to Table 3. Figures 6 and 7 show the performance of our model.

During our experiments, we got a similar ROC curve using a Support Vector Machine (SVM) [16] and k-Nearest Neighbors (KNN). To clear any doubt and add some confidence in the prediction, we have implemented a probability calibration of all the classifiers that we could use on top of our DNN (see Fig. 7). Furthermore, we found that KNN in this case is a well calibrated classifiers as its curve can be assimilated to the diagonal (Perfectly Calibrated).

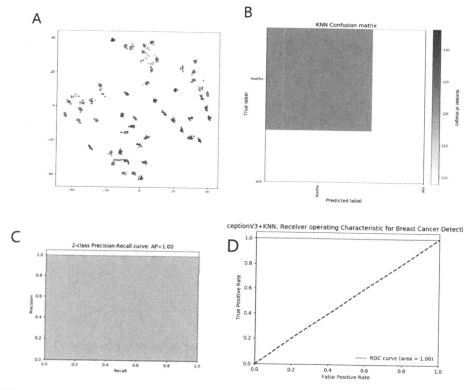

Fig. 6. Performance of our method: Confusion Matrix, Precision-Recall and 2 classes precision-recall of our InceptionV3-KNN. A shows a 2D representation of our feature map; B shows the confusion matrix of our InceptionV3-KNN model, we can see that our model can easily classify a breast (Sick or Healthy); C and D demonstrates the power our model.

Table 3. Dataset repartition use for validation test

	Testing
Healthy	300
Sick	180
Total	**480**

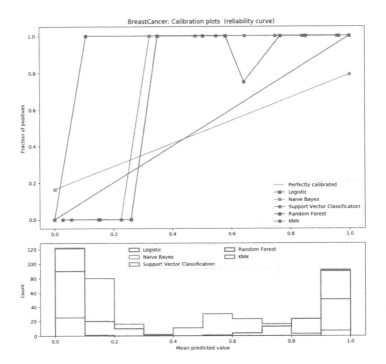

Fig. 7. Calibration plots of five Classifier applied on top of our InceptionV3

CancerStage: As mentioned before, the images classified as Sick by our InceptionV3-KKN will be further process by our module "CancerStage". It will identify after performing threshold, blur, and a series of erosions and dilations operations the areas containing a dense concentration of white pixel in our gray Scale image. Furthermore, the radius of the zones will be computed and compare according to T1–T4 properties as per Sect. 2. Figures 5, 6 and 7 show how efficient the module is (Figs. 8, 9 and 10).

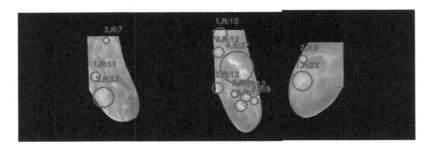

Fig. 8. Cancerous area identified and classified as Stage 2 (T2). The analysis reveals that the tumor is still located on one side of the breast.

Fig. 9. Cancerous area identified and classified as Stage 3 (T3). The analysis reveals that the tumor has spread to almost all the breast's tissues surrounding the nipple and it causes an abnormal heat.

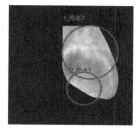

Fig. 10. Cancerous area identified and classified as Stage 4 (T4). The analysis reveals that the tumor has spread to all the breast's tissues causing an abnormal heat.

4 Conclusion

During the literature review, it appeared that work in the field of breast cancer detection from computer scientists' angle could be a valuable contribution to the field. With this in mind, the most common techniques used to detect breast cancer were presented, as well as their strengths and weaknesses. One method seemed to have a promising future, due to its non-immersive property. Infrared imaging coupled with a powerful Computer Assist Device (CAD) can lead to a very accurate tumor detector. In a future study, we will use a thermal sensitivity camera of 0.5 and will propose a model that can address the issue of breast cancer detection and classification using a 3D structure of the Breast.

Acknowledgement. The work and the contribution were supported by the SPEV project "Smart Solutions in Ubiquitous Computing Environments", 2019, University of Hradec Kralove, Faculty of Informatics and Management, Czech Republic.

References

1. National Breast Cancer Foundation: Breast Cancer Facts. www.nationalbreastcancer.org. https://www.nationalbreastcancer.org/breast-cancer-facts. Accessed 2016
2. Mambou, S., Maresova, P., Krejcar, O., Selamat, A., Kuca, K.: Breast cancer detection using infrared thermal imaging and a deep learning model. Sensors **18**, 2799 (2018)
3. Azam, H., Erika, D., Andrik, R., Kate, H., Zwiggelaar, R.: Deep learning in mammography and breast histology, an overview and future trends. Med. Image Anal. **47**, 45–67 (2018)
4. Andrik, R., Bryan, W., Philip, J., Hui, W., John, W.: Breast density classification using local quinary patterns with various neighbourhood topologies. J. Imaging **4**, 14 (2018)
5. Mambou, S., Maresova, P., Krejcar, O., Selamat, A., Kuca, K.: Breast cancer detection using modern visual IT techniques. In: Sieminski, A., Kozierkiewicz, A., Nunez, M., Ha, Q.T. (eds.) Modern Approaches for Intelligent Information and Database Systems. SCI, vol. 769, pp. 397–407. Springer, Cham (2018). https://doi.org/10.1007/978-3-319-76081-0_34
6. Amina, A., Susan, H., Anthony, J.: Potentialities of steady-state and transient thermography in breast tumour depth detection: a numerical study. Comput. Methods Programs Biomed. **123**, 68–80 (2016). ISSN 0169-2607
7. Boogerd, L.S.F., et al.: Laparoscopic detection and resection of occult liver tumors of multiple cancer types using real-time near-infrared fluorescence guidance. Surg. Endosc. **31**, 952–961 (2016)
8. Satish, G., Kandlikar, I.: Infrared imaging technology for breast cancer detection – current status. Int. J. Heat Mass Transf. **108**, 2303–2320 (2017)
9. Namikawa, T.: Recent advances in near-infrared fluorescence-guided imaging surgery using indocyanine green. Surg Today **45**(12), 1467–1474 (2015)
10. Kontos, M., Wilson, R., Fentiman, I.: Digital infrared thermal imaging (DITI) of breast lesions: sensitivity and specificity of detection of primary breast cancers. Clin. Radiol. **66**(6), 536–539 (2011)
11. Cardoso, F., et al.: Research needs in breast cancer. Ann. Oncol. (2016). https://doi.org/10.1093/annonc/mdw571
12. Breast Cancer: Stages. Cancer.Net. https://www.cancer.net/cancer-types/breast-cancer/stages
13. Unar-Munguía, M.: Economic and disease burden of breast cancer associated with suboptimal breastfeeding practices in Mexico. Cancer Causes Control **28**, 1381 (2017)
14. Lab, V.: A methodology for breast disease computer-aided diagnosis using dynamic thermography. Visual Lab. http://visual.ic.uff.br/en/proeng/thiagoelias/
15. Szegedy, C.: Going deeper with convolutions, pp. 1–9 (2015)
16. scikit-learn.org: sklearn.svm.LinearSVC. sklearn. http://scikit-learn.org/stable/modules/generated/sklearn.svm.LinearSVC.html
17. Mambou, S., Krejcar, O., Kuca, K., Selamat, A.: Novel cross-view human action model recognition based on the powerful view-invariant features technique. Future Internet **10**(9), 89 (2018)

Comparison of Numerical and Laboratory Experiment Examining Deformation of Red Blood Cell

Kristina Kovalcikova[✉], Ivan Cimrak, Katarina Bachrata, and Hynek Bachraty

Cell-in-Fluid Biomedical Modeling and Computations Group,
Department of Software Technologies,
Faculty of Management Science and Informatics,
University of Zilina, Zilina, Slovakia
kristina.kovalcikova@fri.uniza.sk

Abstract. In this work, we are dealing with a numerical model of red blood cell (RBC), which is compared to a real RBC. Our aim is to verify the accuracy of the model by comparing the behavior of simulated RBCs with behavior of the real cells in equivalent situations. For this comparison, we have chosen a microfluidic channel with narrow constrictions which was already explored in laboratory conditions with real RBCs. The relation between the cell deformation and its velocity is the element of comparison of simulated and real RBC. We conclude that the velocity-deformation dependence is similar in the two compared experiments.

Keywords: Red blood cell · Deformation · Numerical simulation

1 Introduction

We focus on modelling of cells in blood or similar fluids. Our long-time goal is to explore the behavior of blood as a complex unit, for example behavior of a cancerous blood, or to investigate the damage of red blood cells (RBCs) during their passage through artificial medical devices. For any use of the model, it is important to calibrate the behavior of RBCs in the blood, because it represents the majority of cells which are present in blood.

The model of the cell in fluid was developed and implemented in an open source environment Espresso [1, 2], with a lattice-Boltzmann method [3, 4]. The RBC is represented in the model by its surface – by multiple nodes which are placed on the surface of the RBC and connected by elastic bonds. These bonds are defined by five elastic moduli: stretching, bending, local area conservation, global area conservation, and global volume conservation. Each of these moduli has its own stiffness coefficient.

This work was supported by the Slovak Research and Development Agency (contract number APVV-15-0751) and by the Ministry of Education, Science, Research and Sport of the Slovak Republic (contract No. VEGA 1/0643/17).

© Springer Nature Switzerland AG 2019
I. Rojas et al. (Eds.): IWBBIO 2019, LNBI 11466, pp. 75–86, 2019.
https://doi.org/10.1007/978-3-030-17935-9_8

The calibration of these elastic coefficients for the RBC was based on a stretching experiment [5] and the values of these moduli were calibrated recently.

The aim of this work is to validate the calibration which was made with the stretching experiment. Once we have specified the values of elastic coefficients of the RBC, we have to verify whether the behavior of the RBC in our numerical simulations is comparable to the behavior of living RBC in different laboratory experiments. The laboratory experiment which we have chosen for this validation is described in more details in [6]. We have chosen this experiment because it deals with deformed cells. The comparison of behavior of deformed cells is a good way to verify the accuracy of calibrated elastic coefficients.

In this article, we firstly explain the design of the original laboratory experiment, in Sect. 2. After that, in Sect. 3, we talk about the way how the deformation-velocity dependency was determined from the laboratory experiment data. In the Sect. 4, the details of the numerical simulation are presented. After that, in Sect. 5, we present the results of the numerical experiment and their comparison with laboratory experiment results. Finally, in Sect. 6, we conclude our findings.

2   Design of the Executed Laboratory Experiment

The microfluidic device, which we have chosen for the validation of our numerical model, was explored in Osaka University. The cells in this device were deformed when they passed through the narrow constrictions in it (presented in Fig. 1 as Slit 1, Slit 2 and Slit 3). Furthermore, a video recording captured by a high-speed camera [7] is available from the experiment, which helps also to understand the behavior of the cells.

The height of the device was 3.5 μm. The dimensions of the biggest RBC in our numerical experiments were $7.82 \times 7.82 \times 2.57$ μm. This means that the vertical rotation of the cells in the experiment was limited, and that their deformation in narrow constrictions was well defined. In Figs. 1 and 2, the detailed geometry of the simulation box is presented, as it was used for the numerical experiment.

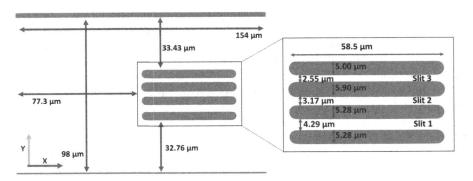

Fig. 1. Design and dimensions of the microfluidic device – view from above. Liquid in the channel is flowing from left to right, in direction of X-axis

Fig. 2. Design and dimensions of the microfluidic device – side view, the source of the flowing fluid is behind

The fluid used in the biological experiment was blood diluted with standard saline, at the blood - saline ratio of 1:50.

The blood used in the experiment came from five different male subjects. The distribution of cell-size in the blood was thus different for each sample. The diameter of RBCs ranged from 5.3 μm to 8.0 μm.

The flow of the fluid in the microfluidic device was controlled by a pressure control system, the constant flow was driven by a pressure difference of 0.7 kPa. The velocity of the cells in wide parts of the microfluidic device was approximately 0.6 μm/ms, as can be seen from illustrations in article [6].

3 Velocity vs Deformation Dependency

The cells of different diameters enter different constrictions in the experiment. Thus, each of the cell is deformed differently. This deformation have to be normalized if we want to compare different cells. The normalized cell deformation is defined the same way in [6]:

$$\varepsilon = \frac{D - w_i}{D} \tag{1}$$

Here D is the cell's diameter, and w_i is the width of the i-th constriction. In numerical simulation, all cells are perfectly circular in X- and Y- direction. The bigger is the value of ε, the bigger is the cell's deformation.

After that, the velocity of the cell moving through the constriction – the transit velocity – have to be normalized as well, with respect to the velocity of the fluid flow in the microfluidic device. The normalized transit velocity is defined as in [6]:

$$v_n = \frac{v_c}{v_f} \tag{2}$$

Here v_c is the velocity of the cell moving through the constriction, and v_f is the average velocity of the fluid in the reduced cross-section of the channel. The value v_f can be obtained easily from a numerical simulation, because we know exactly the value of the fluid velocity in any position of the simulation box. But we need to consider different approach. We need to be consistent with the approach used in interpretation of laboratory observation. In the laboratory condition, the velocity of the fluid in particular positions cannot be measured directly. Thus, to evaluate the velocity of the fluid v_f, we used the same approach as in [6]. The fluid velocity is considered to be same as the velocity of the undeformed cells flowing in it, in the widest part of the microfluidic device. With this assumption, and taking into account that the volumetric flow rate is the same for any cross-section of the channel, the velocity v_f can be obtained as follows:

$$v_f = v_{ff} \frac{w_{ff}}{w_{ff} - \sum w_w} \tag{3}$$

Here v_{ff} is the cell velocity in the widest part of the microfluidic device – before entering the constriction, w_{ff} is the total interior width of the microfluidic channel, and w_w is gradually the width of the four obstacles. Schematic interpretation of those dimensions and velocities is shown in Fig. 3.

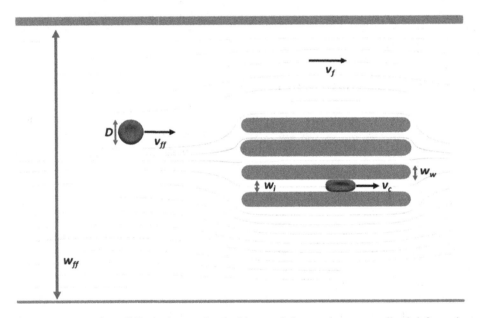

Fig. 3. Interpretation of dimensions and velocities needed to evaluate normalized deformation and normalized velocity of cells in experiment

The resulting deformation vs velocity dependency obtained in the laboratory experiment [6] was finally shown in graphs. For each cell, there is one data point in the velocity-deformation graph. The aim of the laboratory experiment was to show that the deformation-velocity dependence can be determined more precisely in experiment with multiple constrictions, in comparison with an experiment with only one constriction.

The results were used to determine the value of the correlation coefficient (R). For our purpose, we will not use these values, but we will concentrate to slopes of trend lines fitting the data in deformation-velocity graphs. The detailed comparison of laboratory and numerical results can be found in Sect. 5 (Results).

4 Description of Approach in Numerical Experiment

The numerical model of cells used in the simulation employs a triangular mesh with 642 nodes placed on the surface of the cell, connected with elastic bonds, see Fig. 4. This mesh was used for four cell-types with different dimensions. Detailed summary of those cells' dimensions can be seen in Table 1. Values of elastic coefficients, which define the elastic behavior of cells, were chosen to be the same for all of the four cell-types, and they are listed in Table 2. The interaction between the cells and the obstacles is defined by four coefficients. There is also an interaction between nodes within one cell, which is as well defined by another four coefficients. This kind of interaction prevents the overlapping of the cell's mesh with itself. It is important mainly in situations where a cell is entering a narrow space, and nodes from different parts of the mesh are getting very close. Values of these interaction coefficients can be found in Table 3. Almost all of them are the same for all types of cells, excepting the parameter "cut_off" for the internal cell interaction, which had to be smaller for the cell with diameter of 5 μm. Usually, there is also an interaction between different colliding cells, which have to be defined, but in this case the cells are not close enough to each other, so they do not collide with one another.

Fig. 4. Triangular mesh for a RBC model used in simulations

Table 1. Dimensions of the cells used in the simulations

Cell size	Diameter	Thickness
Size 1	7.82 μm	2.57 μm
Size 2	7 μm	2.30 μm
Size 3	6 μm	1.97 μm
Size 4	5 μm	1.64 μm

Table 2. Values of elastic coefficients of RBCs in simulations

Parameter	LB units	SI units
Stretching coefficient k_s	$5.56 \cdot 10^{-3}$ LN/Lm	$5.56 \cdot 10^{-6}$ N/m
Bending coefficient k_b	$7 \cdot 10^{-3}$ LNLm	$7 \cdot 10^{-18}$ Nm
Coefficient of local area conservation k_{al}	$2.3 \cdot 10^{-2}$ LN/Lm	$2.3 \cdot 10^{-5}$ N/m
Coefficient of global area conservation k_{ag}	0.7 LN/Lm	$7 \cdot 10^{-4}$ N/m
Coefficient of volume conservation k_v	0.9 LN/Lm^2	$9 \cdot 10^{2}$ N/m^2
Viscosity of cell membrane	0 Lm^2/Ls	0 m^2/s

Table 3. Values of interactions in simulations

Interaction type	a	n	Cutoff	Offset
Cell-wall interaction	$1 \cdot 10^{-4}$ [-]	1.2 [-]	$7.5 \cdot 10^{-2}$ Lm $7.5 \cdot 10^{-8}$ m	0
Internal cell interaction	$2 \cdot 10^{-3}$ [-]	1.5 [-]	0.5/0,35 Lm $5/3.5 \cdot 10^{-7}$ m	0

The fluid in the simulation was simulated with lattice spacing 0.25 μm. This distribution was chosen in order to obtain a good representation of the behavior of the cells and the fluid in the narrowest constriction of the microfluidic device. The sensibility study of the necessary discretization of the fluid is presented in [8]. We can see the distribution of the fluid nodes in the Fig. 5, with a detailed view of the smallest constriction.

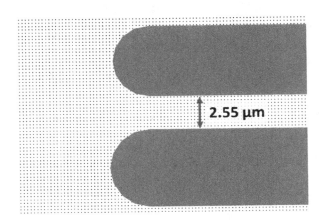

Fig. 5. Numerical model of the fluid, points representing the fluid are placed with lattice spacing 0.25 μm

Table 4. Numerical parameters of the fluid and other simulation parameters

Parameter	LB units	SI units
Density	1 Lkg/Lm^3	$1 \cdot 10^3$ kg/m^3
Kinematic viscosity	1 Lm^2/Ls	$1 \cdot 10^{-6}$ m^2/s
Friction coefficient	0.43 [-]	0.43 [-]
Lb-grid (distance between two adjacent points in fluid mesh)	0.25 Lm	$2.5 \cdot 10^{-7}$ m
Timestep	0.05 Ls	$5 \cdot 10^{-8}$ s
External fluid force	$4.5 \cdot 10^{-4}$ LN	$4.5 \cdot 10^{-13}$ N

The simulation box has rigid boundaries in Z and Y directions, and periodic boundary conditions in X direction. Numerical parameters of the fluid and other simulation parameters are listed in Table 4.

Simulations with such a fine mesh were quite time-consuming, that is why we decided not to take into account the hematocrit used in the laboratory conditions. Instead, we placed the cells into positions, where they were forced to pass through the constrictions. In this way we obtained the velocities of the cells in the constrictions. To obtain the velocity of the cell in the open part of the channel (v_{ff} in Fig. 3), we placed one cell to the upper part of the simulation box, above the narrow constrictions. This cell flows through the side part of the channel and moves closer to the center of the channel in the part without obstacles – and the value of v_{ff} was determined from this position of the cell, in the part of channel without obstacles.

The simulations with cells of diameter 7.82 μm, 7 μm and 6 μm had to be run twice: first time with cells passing through the middle constriction, the wide constriction (Slit 1 and Slit 2), and through the side part of the channel. The second time the simulation was run with one cell passing through the narrowest constriction (Slit 3), and one cell passing through the side part of the channel. The initial position for those simulations is presented in Fig. 6.

It was not possible to run all of the four cells in one simulation, because the cell designed to the narrowest constriction avoided it and entered to the side part of the channel. By comparing the two situations in the Fig. 6 we can notice, that the cell designed for the Slit 2 (3,19 μm) on the left and the cell designed for the Slit 3 (2,55 μm) would overlap, if they were in the same simulation. Only the simulation with cells with diameter of 5 μm was run with the four cells at the same time.

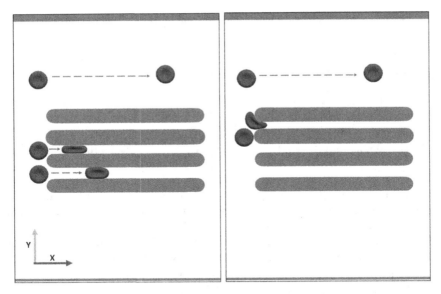

Fig. 6. Initial position of the cells for the numerical simulation, and their position after 300 steps of simulation. Example is shown for cell with diameter of 7 μm.

5   Results

In this chapter we will show how we treated data from numerical simulation, and we will show the comparison between the laboratory and numerical results.

We had to determine the values of average fluid velocity v_f in the reduced cross section, and the deformation of the cells moving through the constrictions ε. The both of them appeared to be ambiguous.

Firstly, let us talk about the average fluid velocity v_f in the reduced cross section (at the level of the constrictions). It was defined by velocity of an undeformed cell in wide part of the channel v_{ff}. However, this velocity was not equal for the four types of the simulated cells. The smallest cell with diameter of 5 μm was the fastest one. This phenomenon can be explained if we observe the cells from a side view, Fig. 7. The smallest cell is more influenced by the fast fluid in the middle of the channel width, while the biggest cell is influenced more by the fluid close to the upper and lower wall of the channel, which flows slower. The different velocities of this free cells, as well as the related average fluid velocities on the level of the reduced cross section, are listed in Table 5. Finally, for evaluation of results, we used the average value of the calculated fluid velocity.

After that, there is also ambiguity in determination of the deformation of the cells moving through the constrictions. The source of this ambiguity is presence of a gap between the cells and the obstacles. This gap is related with the cutoff parameter of cell-wall interaction. It is necessary for the numerical stability of the simulation. It is important mainly in situations, where a cell is colliding with an obstacle, and the gap between them is here to prevent their penetration. More detailed explanation of the cell-

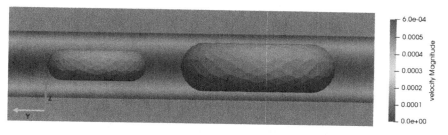

Fig. 7. Position of the undeformed cell in the fluid, 5 μm cell on the left, 7.82 μm cell on the right

Table 5. Velocities of undeformed cells and calculated average fluid velocity in the reduced cross-section, for cells with different size

Cells diameter	Velocity of the undeformed cell	Average velocity of the fluid in the reduced cross section
7.82 μm	$3.9 \cdot 10^{-4}$ m/s	$5.1 \cdot 10^{-4}$ m/s
7 μm	$4.3 \cdot 10^{-4}$ m/s	$5.5 \cdot 10^{-4}$ m/s
6 μm	$4.6 \cdot 10^{-4}$ m/s	$6.0 \cdot 10^{-4}$ m/s
5 μm	$4.9 \cdot 10^{-4}$ m/s	$6.4 \cdot 10^{-4}$ m/s
Average value		$5.7 \cdot 10^{-4}$ m/s

wall interaction mechanism is given in [4]. On the other hand, the width of the gap between the cell and the obstacle is present also when the cell has already entered into the constriction and it is flowing with steady deformation and steady velocity. The effective width of the constriction for such a cell is then smaller than the real size of the slit. We can consider it to be as follows:

$$w_{i\_effective} = w_{i\_real} - 2cut_{off} \qquad (4)$$

Where w_{i_real} is the real dimension of the constriction, and cut_{off} is size of the gap between the obstacle and the cell, defined by coefficient cutoff in the cell-wall interaction parameters. The difference between the w_{i_real} and $w_{i\_effective}$ is approximately 5% in this case. Interpretation of these dimensions can be found in Fig. 8. To evaluate the results of the simulation, we calculated the normalized deformation (Eq. 1) of the cell with both values of w_i, and then we took the average value of the two deformations.

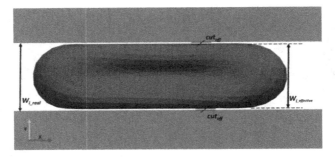

Fig. 8. Effective width of a constriction for cell in numerical simulation is smaller than the real width, because of the presence of the gap between the cell and the obstacle. The gap prevent the penetration of the cell into the obstacle

Finally, we obtained 11 measurements. For each constriction there were four cells with different dimensions which passed through. The only exception was the 7.82 μm cell, which did not pass through 2.55 μm constriction. It avoided this slit every time it passed around.

The obtained velocity-deformation dependence and its fit are presented in Fig. 9. The value of the velocity-deformation dependence, obtained from the equation of linear trend line fitted to the measured data, is −1.22. This value is compared with five correspondent values obtained from laboratory experiments, where experimental samples were seized from five different male subjects. The comparison with those velocity-deformation dependences is presented in Fig. 10 and Table 6.

We can notice that the absolute value of the velocity-deformation dependence is smaller in the numerical simulation than in the laboratory experiment. By omitting first and second datapoint from the graph in Fig. 9, we would obtain the fit with slope of −1.39.

We can see from Fig. 10 that the velocity of the cells in the numerical experiment is smaller than in the laboratory. The difference is visible mainly for cells passing through the widest constriction (4.29 μm). These cells have about 20–30% lower velocity than the cells in laboratory experiment. The other cells fit better with data from laboratory. The cells from the middle constriction (3.17 μm) have about 4% lower velocity in comparison with the real cells, and for the smallest constriction (2.55 μm), there is difference of about 2% between the velocity of real and numerical cells.

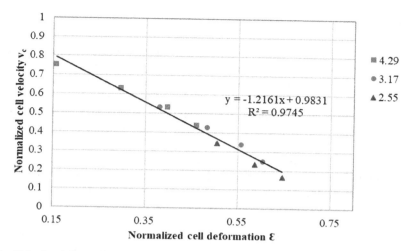

Fig. 9. Velocity-deformation dependence from numerical experiment. Squares are for cells passing through constriction with 4.29 μm, circles are for constriction 3.17 μm, and triangles are for constriction 2.55 μm

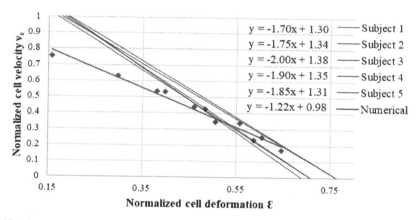

Fig. 10. Comparison of simulation data with velocity-deformation dependences from laboratory experiment with five samples

Table 6. Values of slope in the velocity-deformation dependence graphs from laboratory experiments, compared to the values obtained from numerical experiment. Values are rounded to the closest multiple of 0.05

Laboratory subject 1	Laboratory subject 2	Laboratory subject 3	Laboratory subject 4	Laboratory subject 5	Numerical simulation
−1.70	−1.75	−2.00	−1.90	−1.85	−1.20

6   Conclusion

In this work, we made a numerical simulation of red blood cells flow in a microfluidic device, which was formerly analysed in laboratory conditions. Our aim was to compare the deformation-velocity dependence of living and numerical red blood cells, in order to validate the previously calibrated model of the red blood cell.

We noticed that the laboratory experiment was easily reproducible, and the behavior of the cells was at the first sight similar to the behavior of the real cells in video [7].

We found out that the qualitative behavior is conserved – the cells in wider constrictions are moving faster than the cells in the narrower constrictions. For the quantitative comparison, the cells which velocity differs the most from the cells' velocity measured in laboratory conditions, have about 20–30% lower velocity. This concerns the cells passing through the widest of the three constrictions.

We can find in this result kind of analogy with the results from the laboratory experiment. Here, the cells passing through the widest constriction were the worst organized. The cells passing through two other constriction were organized better. It means the correlation coefficient of the linear trend line fitting the results from these two slits was higher than for the data from the widest slit.

This work indicates that the setting-up of the simulation parameters is convenable, at least for this specific experiment. One of the future options is thus the calibration of tumor cells and its validation. Then there is possibility of exploring the diagnostic microfluidic devices, which could help us to separate several kinds of cells, for example in function of their size and stiffness.

References

1. ESPResSo: Extensible Simulation Package for Research on Soft Matter. http://espressomd. org
2. ESPResSo sources: https://github.com/espressomd/espresso/tree/release-4.0.1
3. Cimrák, I., Gusenbauer, M., Jančigová, I.: An ESPResSo implementation of elastic objects immersed in a fluid. Comput. Phys. Commun. **185**(3), 900–907 (2014)
4. Cimrák, I., Jancigova, I.: Computational Blood Cell Mechanics: Road Towards Models and Biomedical Applications. Chapman & Hall/CRC Mathematical and Computational Biology, CRC Press, Boca Raton (2018). ISBN 9781351378666
5. Dao, M., Lim, C.T., Suresh, S.: Large deformation of living cells using laser traps. Acta Mater. **52**, 1837–1845 (2004)
6. Tsai, C.H.D., et al.: An on-chip RBC deformability checker significantly improves velocity-deformation correlation. Micromachines **7**, 176 (2016)
7. Tsai, C.H.D., et al.: Video from the experiment. www.mdpi.com/2072-666X/7/10/176/s1
8. Kovalčíková, K: Discretization of simulation elements – feasibility limits and accuracy of results. In: 2019 Proceedings of MIST Conference on Mathematics in Science and Technologies (2019, In press)

Levenberg-Marquardt Variants in Chrominance-Based Skin Tissue Detection

Ayca Kirimtat[1] , Ondrej Krejcar[1(✉)] , and Ali Selamat[1,2]

[1] Faculty of Informatics and Management,
Center for Basic and Applied Research, University of Hradec Kralove,
Rokitanskeho 62, 500 03 Hradec Kralove, Czech Republic
a.kirimtat@gmail.com, ondrej.krejcar@uhk.cz
[2] Malaysia Japan International Institute of Technology (MJIIT),
Universiti Teknologi Malaysia,
Jalan Sultan Yahya Petra, Kuala Lumpur, Malaysia
aselamat@utm.my

Abstract. Levenberg-Marquardt method is a very useful tool for solving nonlinear curve fitting problems; while it is also a very promising alternative of weight adjustment in feed forward neural networks. Forcing the Hessian matrix to stay positive definite, the parameter λ also turns the algorithm into the well-known variations: steepest-descent and Gauss-Newton. Given the computation time, the results achieved by these methods surely differ while minimizing the sum of squares of errors and with an acceptable accuracy rate in skin tissue recognition. Therefore in this paper, we propose the implementation of these variations in network training by a set of tissue samples borrowed from SFA human skin database. The RGB images taken from the set are converted into YCbCr color space and the networks are individually trained by these methods to create weight arrays minimizing the error squares between the pixel values and the function output. Consisting of hands on computer keyboards, the images are analyzed to find skin tissues for achieving high accuracy with low computation time and for comparison of the methods.

Keywords: Neural network · Levenberg-Marquardt · Gauss-Newton ·
Steepest descent · Skin tissue detection · YCbCr

1 Introduction

Optimizing the arrays reshaped from neural network weights, Levenberg-Marquardt algorithm (LM) [1] is a non-backpropagation approach for training; while firstly proposed for optimization of curve fitting. On the other hand, error minimization of curve fitting problems could easily be transformed into least squares of network weights; mostly in array-based inputs. As a batch process, the networks are trained not by backpropagation of single inputs; but of arrays instead.

Similar with back propagated neural networks, LM based networks are also iterative indeed; where the computation time widely differ by epoch number. While the epoch is increasing, the accuracy is expected to increase as well; however the feed

© Springer Nature Switzerland AG 2019
I. Rojas et al. (Eds.): IWBBIO 2019, LNBI 11466, pp. 87–98, 2019.
https://doi.org/10.1007/978-3-030-17935-9_9

forward outputs of LM based trained networks would create trusted interval in image processing. Therefore, computation time and accuracy are the major indicators of performance of LM and LM-variants, such as Gauss-Newton (GN) and steepest descent (SD). An extensive application area of these algorithms could be found in the literature.

In the study of Ibrahimy et al. [2], LM algorithm was used together with scaled conjugate gradient based back-propagation training algorithm for identifying hand motions. The optimal design of LM based neural network classifier achieved 88.4% success rate according to the results. Moreover, the reason for the use of two algorithms is faster training ability for pattern recognition problems. Kermani et al. [3] focused on the LM training neural network training to recognize the odor patterns with electronic nose. The authors conducted four different experiments consist of coffees, fragrances, hog farm air, and cola beverages. The experiments showed that LM method reached high accuracy ratios in terms of the odorants used for this research. Alpar [4] used LM method in the initial training network of keystroke recognition problem for minimizing the error in least squares fitting. Another study by Alpar [5] showed that LM algorithm as training algorithm was successfully operated for again minimizing the error in least squares fitting in pattern password authentication problem.

Schweiger et al. [6] used GN method for reconstructing images in diffuse optical tomography by implementing different strategies on the nonlinear inverse problem. GN-GMRES algorithm was applied to image reconstruction by making experimental measurements and two-dimensional circular test. In both cases, the authors simultaneously obtained the reconstruction of the spatial distribution of absorption and scattering parameters from frequency domain boundary data. Another study by Rubæk et al. [7] was conducted by implementation of GN method for breast cancer screening. The authors proposed a new algorithm and compared it to the GN method to reconstruct images. According to the results obtained from the experiments, the new algorithm reconstructed images in the same quality with fewer iterations.

Considering the study of Tanner and Wei [8], an alternative SD algorithm and scaled variant of alternative SD were introduced for the fixed-rank matrix completion problem. Through empirical evaluation of two algorithm, they are competitive with other matrix completion algorithms based on their computation time and ranking performance in image inpainting. Furthermore, the proposed algorithms are very effective in this kind of large-size problems. Another article by Fiori [9] described a Riemannian-steepest-descent method in order to compute the average out of a set of optical system transference matrices using a Lie-group averaging criterion function. The numerical experiments showed that the proposed algorithm converged the solutions in good agreement with the ones converged by the Harris' exponential-mean logarithm.

Given the literature survey on LM, GN and SD, skin tissue detection still lacks in the application area of these algorithms. Furthermore, the examples of the application area of these algorithms could be defined as hand detection, human machine interaction and biometrics [5]. Figure 1 basically explains the workflow of our experiment on skin tissue detection on hand step by step.

On the other hand, we operate YCbCr (Luminance, Chrominance) color space for image segmentation, since it is difficult to make skin tissue detection on RGB (Red, Green, Blue) image. Basically, the main difference between YCbCr and RGB format is

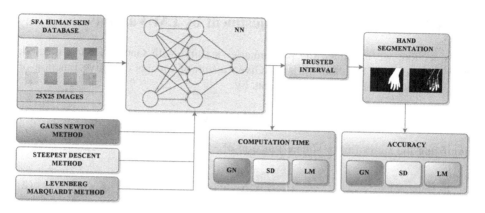

Fig. 1. Workflow of the skin tissue detection

the representation of colors. The colors in YCbCr format are bright and they have two color difference signals, yet the colors in RGB have red, green and blue signals. For instance, Kolkur et al. [10] developed a new human skin detection algorithm to easily recognize the skin pixels in given images. Their proposed algorithm found very promising results in the sense of accuracy and precision. Mandal and Baruah [11] performed a segmentation in YCbCr color space. Their segmented image is black and white colors, thus they used Cb component of YCbCr. The experimental results of their study provided good process for color image segmentation.

Briefly, in this paper we firstly utilize the skin matrices taken from SFA human skin database [12] as inputs of the identical neural network to adjust the weights by LM and variants. The Cb and Cr values of the images are extracted to form the arrays for training. In this stage we also collected the computation times given the epoch number and corresponding the method. The network consequently produces weight arrays to be saved and inserted into the network in succeeding feed forward operations to conduct trusted intervals of Cb and Cr. Finally, comparing within a pre-segmented image of a hand on a keyboard, the segmented images are separately created by the methods for finding the accuracy. Considering the workflow mentioned above and presented in Fig. 1, the paper starts with the mathematical foundations of preliminaries on the LM and the variants, which are given as GN and SD methods. Secondly, explaining the images in RGB format, yet it is insufficient for skin tissue detection, we also defined our images in YCbCr format. Thirdly, in training session, we focused on the structure of the network that we used for the calculation of computation time of LM and each variant. Finally, to achieve high accuracy detection in skin tissues, we conducted experiments using LM, and its variants namely GN and SD methods separately.

2 Preliminaries on Weight Adjustment by LM Variants

In this section, we introduce the preliminaries starting with Gauss-Newton method, which is the core of the other variants. Given a sum-of-squares objective function of $V(w)$, the GN method adjusts the weights by:

$$\Delta w = -\left[\nabla^2 V(w)\right]^{-1}\nabla V(w) \tag{1}$$

minimizing the sum of error square function of

$$V(w) = \sum_{i=1}^{N} e_i^2(w) \tag{2}$$

where $\nabla^2 V(w)$ is the Hessian matrix and $\nabla V(w)$ is the gradient. It could be converted to include Jacobian matrix by:

$$\nabla V(w) = J^T(w)e(w) \tag{3}$$

$$\nabla^2 V(w) = J^T(w)J(w) + S(w) \tag{4}$$

where

$$S(w) = \sum_{i=1}^{N} e_i(w)\nabla^2 e_i(w) \tag{5}$$

and J is the Jacobian matrix;

$$J(w) = \begin{bmatrix} \frac{\partial e_1(w)}{\partial w_1} & \frac{\partial e_1(w)}{\partial w_2} & \cdots & \frac{\partial e_1(w)}{\partial w_N} \\ \frac{\partial e_2(w)}{\partial w_1} & \frac{\partial e_2(w)}{\partial w_2} & & \frac{\partial e_2(w)}{\partial w_N} \\ \cdots & \cdots & \ddots & \cdots \\ \frac{\partial e_N(w)}{\partial w_1} & \frac{\partial e_N(w)}{\partial w_2} & \cdots & \frac{\partial e_N(w)}{\partial w_N} \end{bmatrix} \tag{6}$$

In this method, $S(w)$ is assumed as negligible; so the weight adjustment procedure could be rewritten as;

$$\Delta w = \left[J^T(w)J(w)\right]^{-1}J^T e(w) \tag{7}$$

The main assumption of GN method is that the objective function is quadratic in the parameters near the optimal solution. The steepest/gradient descent method is viable in minimization of the errors in neural network weights throughout the downhill direction, opposite to the gradient of the objective function and computed by:

$$\Delta w = \alpha J^T e(w) \tag{8}$$

The LM modification of GN method can be stated as;

$$\Delta w = \left[J^T(w)J(w) + \lambda I \right]^{-1} J^T e(w) \tag{9}$$

where I is the identity matrix and λ is an adaptive parameter making the Hessian positive-definite in each iteration. Depending on the selection of parameter λ, the LM algorithm is updated as GN or SD, namely: small vales of λ result in GN and large values in SD. Therefore, in this study, we only focus on the starting conditions given the parametric alterations to approximate the variants of LM in training session.

3 Preprocessing

The dataset consisting of skin samples is directly taken from SFA database where several sizes of samples are placed; yet we had chosen 5×5 samples for faster computation. Each pixel $p_{i,j,k}$ on the RGB image $P_{i,j,k}$ in the dataset could be mathematically denoted as:

$$p_{i,j,k} \in P_{i,j,k}(i = [1:w], j = [1:h], k = [1:3]) \tag{10}$$

where $w \in \mathbb{Z}^+$ is the width, $h \in \mathbb{Z}^+$ is the height of the image; $i \in \mathbb{Z}^+$ $j \in \mathbb{Z}^+$ represent row and column number and $k \in \mathbb{Z}^+$ is the layer of color channel of RGB. However, RGB color space is not adequate to detect skin tissues, therefore the images are converted into YCbCr color space by:

$$y_{i,j,1} \in Y_{i,j,1} = 16 + \left(65.481\, P_{i,j,1} + 128.553\, P_{i,j,2} + 24.966\, P_{i,j,3} \right) \tag{11}$$

$$y_{i,j,2} \in Y_{i,j,2} = 128 + \left(-37.797\, P_{i,j,1} - 74.203\, P_{i,j,2} + 112\, P_{i,j,3} \right) \tag{12}$$

$$y_{i,j,3} \in Y_{i,j,3} = 128 + \left(112\, P_{i,j,1} - 93.786\, P_{i,j,2} - 18.214\, P_{i,j,3} \right) \tag{13}$$

Given this conversion; the resulting Cb and Cr layers of training images could be represented as:

$$C_b = \begin{bmatrix} y_{1,1,2} & y_{1,2,2} & \cdots & y_{1,j,2} \\ y_{2,1,2} & y_{2,2,2} & \cdots & y_{2,j,2} \\ \cdots & \cdots & \ddots & \cdots \\ y_{i,1,2} & y_{i,2,2} & \cdots & y_{i,j,2} \end{bmatrix} \quad C_r = \begin{bmatrix} y_{1,1,3} & y_{1,2,3} & \cdots & y_{1,j,3} \\ y_{2,1,3} & y_{2,2,3} & \cdots & y_{2,j,3} \\ \cdots & \cdots & \ddots & \cdots \\ y_{i,1,3} & y_{i,2,3} & \cdots & y_{i,j,3} \end{bmatrix} \tag{14}$$

The matrices, however, should be reshaped into arrays to insert the neural network system as inputs. Therefore the inputs of the system could be rewritten as:

$$C_b = \left[y_{1,1,2}, y_{2,1,2}, \ldots y_{i,1,2}, \ldots, y_{1,j,2}, \ldots, y_{i,j,2} \right] \tag{15}$$

$$C_r = \left[y_{1,1,3}, y_{2,1,3}, \ldots y_{i,1,3}, \ldots, y_{1,j,3}, \ldots, y_{i,j,3} \right] \tag{16}$$

The arrays are individually utilized as the inputs of the neural network in the following training session.

4 Training Session

In this section, we focused on training the network with GN, SD and LM separately to achieve feed-forward results for constructing trusted intervals and to calculate the computation time of each algorithm. However, we didn't allocate any variants of algorithms; but changed the parameter of LM to reach GN and SD instead. Given the inputs as arrays of the skin tissues C_b and C_r, a 10-node neural network system is primed as presented in Fig. 2 below.

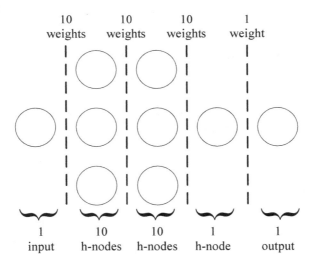

Fig. 2. Structure of the network

While the output of the training set is always set to 1, the inputs vary according to the tissues randomly selected from the training set. On the other hand, the weight arrays consisting of 31 weights are adjusted for each training image by a batch-training of the inputs with a selected epoch number. Therefore, depending on the number of selected images for the training set η, epoch number ε and the parameter λ that turns LM into GN-like or SD-like, the feed-forward values of the training set would vary as well as the operation times and accuracy. As an example, the feed-forward values of the inputs passed through the trained network with specified η, ε and λ are presented in Fig. 3 below.

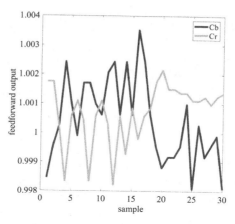

Fig. 3. Feed-forward values of the trained network: $\eta = 30$, $\varepsilon = 100$ and $\lambda = 0.1$

On the other hand, depending on the parameters η and ε, the operation times Δt drastically vary, which would be calculated only for training the network, from the very beginning until saving the weight matrices for each layer. The detailed experiments are presented in the following section with the mathematical foundations of accuracy which actually represents the image closeness between the reference image and resulting images.

5 Experimental Results of Skin Tissue Detection

Disregarding the Type I and II errors and differences among them, what we focus here is the accuracy of detection, which could be restated as the correctness of classification for each pixel on the images. Since the resulting images and the reference image are both binary, accuracy could be found by the binary classification of the pixels $r_{i,j}$ on the resulting images $R_{i,j}$ compared to the reference image $b_{i,j} \in B_{i,j}$. This algorithm also indicates the cosine of the planes of the images. Given the resulting image:

$$r_{i,j} \in R_{i,j}(i = [1{:}w], j = [1{:}h]) \tag{17}$$

the accuracy θ to be calculated by the image closeness algorithm could be stated as:

$$\theta = \left(\sum_{i=1}^{w} \sum_{j=1}^{h} b_{i,j} \cdot r_{i,j} \Big/ \sum_{i=1}^{w} \sum_{j=1}^{h} r_{i,j} \right) - \left(\sum_{i=1}^{w} \sum_{j=1}^{h} |b_{i,j} - r_{i,j}| / wh \right) \tag{18}$$

Therefore, via a trained network as presented in Fig. 3 with the specified parameter, any pixel on an image could be classified as "skin" or "not skin" in the resulting binary image. The protocol of the complete experiment initiates with turning the RGB images into YCbCr, picking the Cb and Cr layers out and determining the reference image, as presented in Fig. 4.

Fig. 4. Image processing protocol: from left to right: original RGB image $P_{i,j,k}$, YCbCr converted image $Y_{i,j,k}$, Cb layer $Y_{i,j,2}$, Cr layer $Y_{i,j,3}$ and reference binary image $B_{i,j,k}$. (Color figure online)

The last image in Fig. 4 is the reference binary image acquired by Adobe Photoshop, where the pixels of the hand on the keyboard are perfectly classified, therefore this image would be benchmarked with the resulting images to find the accuracy. Changing the parameters would train the network in different finite time and result in various weight matrices that will change the feed forward outputs of the pixels on the training set as well as in the images consisting of hands on the keyboard. The keyboard adds noise to the images triggering a hard challenge to detect skin tissues in Cb Cr layer due to similarity of the pixel values. However, the intersection of these layers could be found by training the network twice using the Cb and Cr values of the images in public SFA dataset images. Given this procedure, the results of the experiment using various parameters creating instances of LM are presented in Table 1.

We randomly assigned the values to variables for each experiment of each algorithm. The parametric results show that the image with the highest accuracy, which is 0.9716, was obtained by LM algorithm with $\eta = 500$, $\varepsilon = 50$ and $\lambda = 0.5$. On the contrary, the worst result belongs to GN variant with $\eta = 5$, $\varepsilon = 1000$ and $\lambda = \sim 0$. In addition, the experiment by SD variant with $\eta = 300$, $\varepsilon = 500$ and $\lambda = \sim \infty$ has the highest computation time, since the iteration number and sample size are higher than other experiments. Furthermore, the experiments by SD variant with $\eta = 12$, $\varepsilon = 100$ and $\lambda = \sim \infty$ and GN variant with $\eta = 5$, $\varepsilon = 100$ and $\lambda = \sim \infty$ have very similar image results and computation time. However, GN algorithm reached the almost same result with lower sample size than SD algorithm. Epoch number (ε) also affects the computation time of the algorithms; for instance in the experiment with SD, the computation time of the detection is 41.16 s, since the epoch number (ε) is 500.

According to Table 1, we used 500 samples at most to cope up with the redundantly large covariance. Indeed there are 3354 samples in the dataset including 1×1 and 5×5 matrix. Figure 5 shows the covariance diagrams of different sample sizes in different colors. From left to right, covariance is expanding redundantly, while sample sizes are increasing. Thus, we only showed experimental results up to 500 samples due to covariance expansion.

On the other hand, depending on the requirements, which could be minimization of false negatives or false positives, it is revealed that the best accuracy also brings a high false positive; while the others have more false negative. Therefore in finite time with less epoch using LM would be a preferable choice when generalized accuracy matters disregarding type I and II errors.

Table 1. Parametric results of the experiment

	η	ε	λ	$R_{i,j}$	θ	Δt (sec)
LM	30	100	0.1		0.8648	1.21
LM	20	50	0.9		0.8492	0.41
LM	**500**	**50**	**0.5**		**0.9716**	**11.99**
GN	100	50	~0		0.8723	2.26
GN	5	100	~0		0.7595	0.24
GN	5	1000	~0		0.4571	1.69
SD	300	500	~∞		0.9229	41.16
SD	200	200	~∞		0.9228	11.96
SD	12	100	~∞		0.7600	0.43

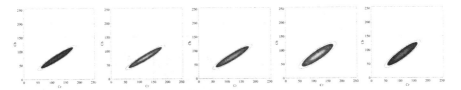

Fig. 5. Covariance diagrams of samples in dataset: from left to right: 10 samples in blue, 100 samples in turquoise, 500 samples in green, 1000 samples in yellow, 3000 samples in red. (Color figure online)

6   Conclusions and Discussions

In this study, we proposed Levenberg-Marquardt algorithm with its variations namely Gauss-Newton and steepest descent for skin tissue detection by network training. We used a set of skin tissue samples from SFA human skin database for the network training. To decrease the computation time of each variant, we used 5×5 images in RGB format, yet these original images are not sufficient for skin tissue detection. Thus, we converted these images into YCbCr image format and the matrices of the YCbCr images were also converted into arrays for training the neural network. In order to reach high accuracy in the binary images through set of experiments with each variant, Type I and Type II errors were used for the classification of each pixel on each binary image. By randomly assigning parameters that are given as epoch number, training set and lambda for LM and its variants GN and SD, the experiments were completed. The results showed that the image with the highest accuracy belongs to LM algorithm, yet it has a very high computation time when it is compared to other experiments. In addition, the worst case belongs to GN algorithm, and it has very low computation time with fewer samples.

As a conclusion, using alternative variants for feed forward neural network problems provides us to explore various accuracy ratios with higher and lower computation time in skin tissue detection. Moreover, one algorithm can reach almost the same solution as other algorithm can reach with different sample size. We would also increase the number of experiments, however the experiments in Table 1 are the ones that shows significant contrast and variations in terms of accuracy ratios and computation time. In addition, more than 500 sample sizes were not used for the experiments in Table 1, since the covariance are getting larger as seen in Fig. 5.

Since they still lack in the literature, more future studies are still needed in skin tissue detection and image segmentation problems using LM and its variants in network training. Moreover, these methods would be used in the field of medical imaging and biomedical diagnosis [13–16] to reveal regions sharing certain characteristics. In addition to these, since the application area of the LM algorithm is very extensive [17–20], the method that we used in this study could be implemented in these areas as well.

Acknowledgement. The work and the contribution were supported by the SPEV project "Smart Solutions in Ubiquitous Computing Environments", University of Hradec Kralove, Faculty of Informatics and Management, Czech Republic (under ID: UHK-FIM-SPEV-2019).

References

1. Levenberg, K.: A method for the solution of certain non-linear problems in least squares. Q. Appl. Math. **2**, 164–168 (1944)
2. Ibrahimy, M.I., Ahsan, M.R., Khalifa, O.O.: Design and optimization of Levenberg-Marquardt based neural network classifier for EMG signals to identify hand motions. Measur. Sci. Rev. **13**(3), 142–151 (2013)
3. Kermani, B.G., Schiffman, S.S., Nagle, H.T.: Performance of the Levenberg–Marquardt neural network training method in electronic nose applications. Sens. Actuators **110**, 13–22 (2005)
4. Alpar, O.: Keystroke recognition in user authentication using ANN based RGB histogram technique. Eng. Appl. Artif. Intell. **32**, 213–217 (2014)
5. Alpar, O.: Intelligent biometric pattern password authentication systems for touchscreens. Expert Syst. Appl. **42**(17–18), 6286–6294 (2015)
6. Schweiger, M., Arridge, S.R., Nissila, I.: Gauss-Newton method for image reconstruction in diffuse optical tomography. Phys. Med. Biol. **50**, 2365–2386 (2005)
7. Rubæk, T., Meaney, P.M., Meincke, P., Paulsen, K.D.: Nonlinear microwave imaging for breast-cancer screening using Gauss–Newton's method and the CGLS inversion algorithm. IEEE Trans. Antennas Propag. **55**(8), 2320–2331 (2007)
8. Tanner, J., Wei, K.: Low rank matrix completion by alternating steepest descent methods. Appl. Comput. Harmon. Anal. **40**, 417–429 (2016)
9. Fiori, S.: A Riemannian steepest descent approach over the inhomogeneous symplectic group: application to the averaging of linear optical systems. Appl. Math. Comput. **283**, 251–264 (2016)
10. Kolkur, S., Kalbande, D., Shimpi, P., Bapat, C., Jatakia, J.: Human skin detection using RGB, HSV and YCbCr color models. Adv. Intell. Syst. Res. **137**, 324–332 (2017)
11. Mandal, A.K., Baruah, D.K.: Image segmentation using local thresholding and YCbCr color space. Int. J. Eng. Res. Appl. **3**(6), 511–514 (2013)
12. Casati, J.P.B., Moraes, D.R., Rodrigues, E.L.L.: SFA: a human skin image database based on FERET and AR facial images. In: 2013 IX Workshop de Visão Computacional, Rio de Janeiro, Anais do VIII Workshop de Visão Computacional (2013)
13. Alpar, O., Krejcar, O.: Detection of irregular thermoregulation in hand thermography by fuzzy C-means. In: Rojas, I., Ortuño, F. (eds.) IWBBIO 2018. LNCS, vol. 10814, pp. 255–265. Springer, Cham (2018). https://doi.org/10.1007/978-3-319-78759-6_24
14. Alpar, O., Krejcar, O.: Superficial dorsal hand vein estimation. In: Rojas, I., Ortuño, F. (eds.) IWBBIO 2017. LNCS, vol. 10208, pp. 408–418. Springer, Cham (2017). https://doi.org/10.1007/978-3-319-56148-6_36
15. Alpar, O., Krejcar, O.: A new feature extraction in dorsal hand recognition by chromatic imaging. In: Nguyen, N.T., Tojo, S., Nguyen, L.M., Trawiński, B. (eds.) ACIIDS 2017. LNCS (LNAI), vol. 10192, pp. 266–275. Springer, Cham (2017). https://doi.org/10.1007/978-3-319-54430-4_26
16. Kirimtat, A., Krejcar, O.: Parametric variations of anisotropic diffusion and gaussian high-pass filter for NIR image preprocessing in vein identification. In: Rojas, I., Ortuño, F. (eds.) IWBBIO 2018. LNCS, vol. 10814, pp. 212–220. Springer, Cham (2018). https://doi.org/10.1007/978-3-319-78759-6_20
17. Sarabakha, A., Imanberdiyev, N., Kayacan, E., Khanesar, M.A., Hagras, H.: Novel Levenberg–Marquardt based learning algorithm for unmanned aerial vehicles. Inf. Sci. **417**, 361–380 (2017)

18. Amini, K., Rostami, F.: Three-steps modified Levenberg–Marquardt method with a new line search for systems of nonlinear equations. J. Comput. Appl. Math. **300**, 30–42 (2016)
19. Amini, K., Rostami, F.: A modified two steps Levenberg–Marquardt method for nonlinear equations. J. Comput. Appl. Math. **288**, 341–350 (2015)
20. Iqbal, J., Iqbal, A., Arif, M.: Levenberg–Marquardt method for solving systems of absolute value equations. J. Comput. Appl. Math. **282**, 134–138 (2015)

A Mini-review of Biomedical Infrared Thermography (B-IRT)

Ayca Kirimtat[1] (ID), Ondrej Krejcar[1](✉) (ID), and Ali Selamat[1,2] (ID)

[1] Faculty of Informatics and Management,
Center for Basic and Applied Research, University of Hradec Kralove,
Rokitanskeho 62, 500 03 Hradec Kralove, Czech Republic
a.kirimtat@gmail.com, ondrej.krejcar@uhk.cz
[2] Malaysia Japan International Institute of Technology (MJIIT),
Universiti Teknologi Malaysia,
Jalan Sultan Yahya Petra, Kuala Lumpur, Malaysia
aselamat@utm.my

Abstract. Infrared thermography (IRT) is a non-destructive imaging technique that is used for revealing temperature differences on the surfaces of the human body or objects. Once it is used for biomedical purpose, it measures the emitted radiation on the surfaces of the human body. We, in this research, present Biomedical Infrared Thermography (B-IRT) applications with various measurement methods, analysis types, analysis schemes and study types from the existing literature in a detailed literature matrix. A mini-review of 30 studies from the literature are summarized through focusing on substantial features and backgrounds. Finally, recent advances and future opportunities are also presented to highlight high potential use of IRT in biomedical applications.

Keywords: Review · Infrared thermography · Biomedicine · Infrared cameras

1 Introduction

Infrared thermography (IRT) is a non-destructive imaging technique and it is used for visualizing the emitted radiation on the surfaces of the objects and human body. Furthermore, IRT method measures the emitted radiation on the surfaces in the infrared spectrum between 0.9 and 14 μm as seen in detail in Fig. 1 [1, 2]. Biomedical Infrared Thermography (B-IRT) reveals the inhomogeneous skin and superficial tissue temperature through capturing the emitted radiation on the human body surface [3]. If temperature distribution on healthy human body is measured, most probably a high degree of thermal symmetry will be seen, yet if there is a thermal asymmetry in the same human body, it is a warning sign that there is something wrong with that body [4].

In the literature, several studies with various methodologies about B-IRT for monitoring diseases and treatments exist. IRT is quite powerful method for dealing with symptoms and disorders on human body, and it has broad application areas in biomedicine such as breast mass diagnosis, dry eye treatment, thermal map monitoring, grading of ankle injuries and diabetic foot diagnosis etc.

© Springer Nature Switzerland AG 2019
I. Rojas et al. (Eds.): IWBBIO 2019, LNBI 11466, pp. 99–110, 2019.
https://doi.org/10.1007/978-3-030-17935-9_10

For instance, Polidori et al. [12] used medical thermal imaging in the diagnosis and osteopathic management of back pain. A 50-year old woman was used as a subject, who had acute back pain syndrome. The thermographic results showed that medical thermal imaging is a high potential diagnosis tool for initial and treatment phases of the mentioned disease.

Another study by Albarran et al. [27] was conducted with noninvasive methodology for biomedical thermal imaging that provided beneficial information on facial expressions of participants. The facial expressions of the subjects were captured in terms of joy, sadness, anger, disgust and fear through obtaining regions of interests. According to the test results of 625 thermograms, this test with biomedical thermal imaging was 89.9% successful.

Oliveira et al. [34] evaluated the suitability of IRT as a high potential diagnostic tool of the distinct lesions grades, and thus, assessment of various thermographic values of the ankle region was carried out. The most significant result of this study was that thermographic analysis of ankle sprain injuries has high potential for clinical assessment, especially for acute incidences. Moreover, IRT provided faster diagnosis of diseases without the need for extensive equipment in the operation phase.

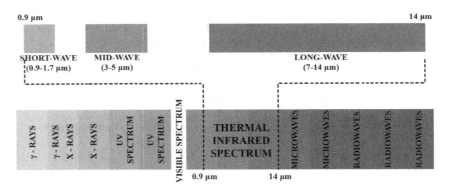

Fig. 1. Representation of the thermal infrared spectrum among other spectrums [2]

Therefore, in this research, total 30 studies with different analysis types, analysis schemes, study types and measurement methods are presented to feature the efficiency of IRT for the diagnosis of biomedical problems on human body. We also summarize the recent developments by recommending new perspectives for future studies and researches on IRT for the purpose of biomedical diagnosis. Furthermore, we categorize the outcomes of the previous studies in detail through a literature matrix. Summarizing this research, we, in Sect. 2, introduces the preliminaries of IRT and IR cameras for the evaluation of biomedical diseases. In Sect. 3, existing IRT methodologies for biomedical diagnosis are explained, whereas Sect. 4 indicates the applications and previous studies in the field of IRT for biomedicine in a literature matrix. Finally, the results of this study are given in Sect. 5, and Sect. 6 gives the conclusions and discussion parts.

2 Preliminaries of IRT and IR Cameras

In this section, we introduce the background of IRT and IR cameras for the first attempt to develop first thermal image and continuing thermal images. Around 1800s, the infrared light was discovered by William Herschel, however the infrared scanning devices were being used after 1940s. The first attempt to discover the first thermal image was made by Herschel in 1840 and the differential evaporation of a thin film was utilized by being exposed to a heat pattern [2, 5, 6]. In 1940s, the first infrared device was used for the purpose of night vision in military. After this date, the infrared devices used in the commercial markets nowadays were first produced in 1983 and opened up to a large application areas such as medical analysis, facial analysis, building diagnostics, fire and gas detection, and industrial applications [7]. As an example of biomedical application, Kirimtat and Krejcar [2], in their article, compared the results of FLIR and SEEK model infrared cameras for the purpose of revealing the temperature differences on an injured toe of the subject. Regarding the thermal images, various formats on the digital thermograms were also obtained through comparing the injured toe with the left toes of the subject.

According to Astarita et al. [8], on account of its scientific background, IRT will provide quantitative and qualitative measurements to new generation through technological developments in various research areas. Due to the technological developments, we necessitate new cameras rather than standard camera models in RGB or grayscale format. Furthermore, it is not easy to recognize objects in dark environments with standard camera models, thus infrared cameras which have passive sensors came to existence as a new solution to recognize objects between 3 and 14 μm infrared spectrums [7].

Infrared cameras has ability to capture the emitted radiation on the surfaces of human and objects and turned this radiation into thermal images, and they also reveal temperature differences on the surfaces [9]. Moreover, the infrared cameras in the current commercial market have focal plane array detectors and they are the third generation thermal camera models. These cameras exist in the commercial market with cooled and uncooled detectors [10].

3 Existing B-IRT Methodologies

The thermal camera records of the human body surfaces are analyzed using two different measurement methods namely qualitative and quantitative IRT. In quantitative IRT, which is more complicated than qualitative IRT, a temperature calibration should be performed by detecting objects or surfaces of known fixed temperatures. Quantitative results generally include temperatures, line plots, software tools and reports. On the other hand, the IR cameras that has only qualitative false color images of the surfaces without any analysis make qualitative measurements. Weigert et al. [11] conducted a study on body skin temperature changes after resistance exercise. While capturing thermal images of the participants, data analysis and time point calculation were made through quantitative measurement method. Temperature changes of each time point were also calculated by resulting delta values. On the other hand, a qualitative B-IRT was used in the study of Polidori et al. [12]. The main aim of this study was to only diagnose the basic temperature differences on the 50-year old woman body to easily recognize osteopathic management accuracy in back pain.

Furthermore, two different analysis schemes are used in B-IRT: active B-IRT and passive B-IRT. The main difference of these schemes is being exposed to an external stimulus or not being exposed to. For example, the study by Sarigoz et al. [15] was made in order to diagnose breast mass by B-IRT. For the measurements, all the thermal images of the participants were captured in a controlled room, which has a constant temperature degree and humidity without any external stimulus, thus this kind of study is called passive B-IRT. However, in the study of Figueiredo, a new technique, which was capable of estimating the coordinates of the geometric center of a tumor, was developed. The thermal images of the body were captured by infrared camera by being exposed to natural convection during the experiment, thus this study was done under non-steady state conditions due to natural convection.

4   Applications of B-IRT and Previous Studies

In this section, we present a detailed literature matrix for summarizing the most significant aspects of the previous studies from the literature on B-IRT. The tabulation that we prepare in this section contains comprehensive related works based on the evaluation criteria that we mentioned in the previous sections. They are measurement method, analysis scheme, and analysis type. Thermal images of each article is also inserted to the literature table in order to give some hints about the related work to the readers. A total number of 30 studies between 2014 and 2018 are categorized briefly in terms of the evaluation criteria that we mentioned. Table 1 is prepared and correlatively created by getting motivation from the review study of Kirimtat and Krejcar [2].

Therefore and according to Table 1, information graphics from the previous studies on B-IRT are composed and a detailed discussion is conducted referring the relevant studies in the following section.

5   Results and Discussion

Regarding the importance of IRT for biomedical diagnosis, the relevant studies for this mini-review were collected between 2014 and 2018 from the existing literature. The publication timeline of the previous studies is shown in Fig. 2 and according to this figure it is apparent that the publication number of B-IRT studies is increasing in recent years. Especially in 2018, there are various and high number of studies conducted through B-IRT.

The quantitative and qualitative B-IRT vary based on the investigation level of the mentioned diseases in the previous studies. Figure 3 shows the distribution of measurement methods used in the previous studies and according to this figure, only one of these studies was conducted through qualitative B-IRT [12], the other 29 were carried out using quantitative B-IRT. For instance, in [12], thermal imaging of the patient was made through qualitative measurements by only capturing the temperature differences on the body surfaces without taking any numerical measurements or giving any numerical reports (Figs. 6 and 7).

Table 1. Literature matrix on IRT for biomedical diagnosis

Thermal image	Year	Measurement method	Analysis scheme	Analysis type	Testing location	Study type	Infrared camera type	Other diagnosis methods
[11]	2018	Quantitative	Active	Body fat distribution	Laboratory	Experimental and numerical	FLIR A35	X-ray and body fat percentage
[12]	2018	Qualitative	Passive	Back pain relieving	Home	Experimental	FLIR SC620	N/A
[13]	2018	Quantitative	Active	Burn depth assessment	Laboratory	Experimental and numerical	Model TH, Therm-App	N/A
[14]	2018	Quantitative	Active	Diabetic foot diagnosis	Controlled room	Experimental and numerical	Thermographic system VarioCAM_	N/A
[15]	2018	Quantitative	Passive	Breast mass diagnosis	Controlled room	Experimental	FLIR ThermaCam E45	Breast ultrasound, mammography
[16]	2018	Quantitative	Passive	Diabetic foot diagnosis	Controlled room	Experimental and numerical	Thermographic system VarioCAM	N/A
[17]	2018	Quantitative	Passive	Diabetic foot diagnosis	Controlled room	Experimental	FLIR E60bx	N/A
[18]	2018	Quantitative	Active	Breast cancer detection	N/A	Experimental and numerical	FLIR SC620	N/A
[19]	2018	Quantitative	Active	Breast cancer detection	Laboratory	Experimental and numerical	FLIR T420	N/A
[20]	2018	Quantitative	Passive	Diabetic foot diagnosis	Controlled room	Experimental and numerical	FLIR E60	N/A
[21]	2018	Quantitative	Passive	Diabetic foot diagnosis	Controlled room	Experimental	FLIR SC7200	Ankle Brachial Pressure Index (ABPI) and Spectral Doppler Waveform
[22]	2018	Quantitative	Active	Breast cancer detection	Controlled room	Experimental and numerical	FLIR ThermaCAM T-400	Mammography
[23]	2017	Quantitative	Passive	Body fat distribution	Controlled room	Experimental and numerical	Fluke Ti400	Dual-Energy X-Ray Absorptiometry (DXA)

(continued)

Table 1. (*continued*)

Thermal image	Year	Measurement method	Analysis scheme	Analysis type	Testing location	Study type	Infrared camera type	Other diagnosis methods
[24]	2017	Quantitative	Passive	Body fat distribution	Controlled room	Experimental and numerical	Fluke Ti400	Anthropometrics and DXA
[25]	2017	Quantitative	Passive	Thermal profile	Controlled room	Experimental and numerical	FLIR T335	N/A
[26]	2017	Quantitative	Passive	Meibomian gland dysfunction evaluation	Controlled room	Experimental and numerical	IT-85	N/A
[27]	2017	Quantitative	Active	Human emotion detection	Controlled room	Experimental and numerical	FLIR A310	N/A
[28]	2017	Quantitative	Passive	Diabetic foot diagnosis	Controlled room	Experimental and numerical	FLIR E60	N/A
[29]	2017	Quantitative	Passive	Deep vein thrombosis diagnosis	Controlled room	Experimental and numerical	Testo 875i	N/A
[30]	2017	Quantitative	N/A	Diabetic foot diagnosis	N/A	Experimental and numerical	FLIR A320	N/A
[31]	2016	Quantitative	N/A	Pediatric musculoskeletal injuries	N/A	Experimental and numerical	FLIR E60	X-ray imaging
[32]	2016	Quantitative	Passive	Skin temperature	Controlled room	Experimental and numerical	AVIO TVS-700	Ultrasound Doppler
[33]	2016	Quantitative	Passive	Skin temperature	Laboratory	Experimental and numerical	Testo 875i	N/A
[34]	2016	Quantitative	Passive	Sprained ankle injuries	Controlled room	Experimental and numerical	FLIR E60 SC	Ultrasound imaging

(*continued*)

Table 1. (*continued*)

Thermal image	Year	Measurement method	Analysis scheme	Analysis type	Testing location	Study type	Infrared camera type	Other diagnosis methods
[35]	2015	Quantitative	Passive	Tuberculin skin temperature	Laboratory	Experimental and numerical	TiR32 FLUKE	N/A
[36]	2015	Quantitative	N/A	Dry eye diagnosis	N/A	Experimental and numerical	VarioTHERM	N/A
[37]	2015	Quantitative	N/A	Breast cancer detection	N/A	Experimental and numerical	N/A	N/A
[38]	2014	Quantitative	Active	Skin temperature	Laboratory	Experimental and numerical	ATIR-25	Dual energy X-ray absorptiometry (DXA)
[39]	2014	Quantitative	Passive	Dry eye diagnosis	Controlled room	Experimental and numerical	VarioTHERM	N/A
[40]	2014	Quantitative	Passive	Skin temperature	Laboratory	Experimental and numerical	AVIO TVS-700	N/A

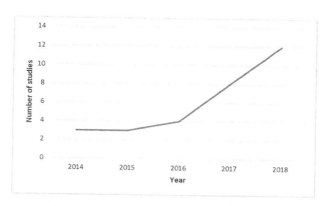

Fig. 2. The publication timeline of the previous studies.

The previous studies were also analyzed based on analysis type in other words disease type. For instance, according to Fig. 4 the most studied disease type is diabetic foot diagnosis. Secondly, breast cancer detection and skin temperature measurement

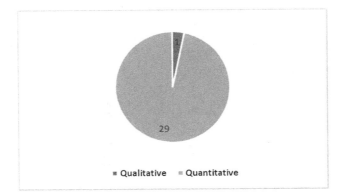

Fig. 3. Frequency of measurement methods used in the previous studies.

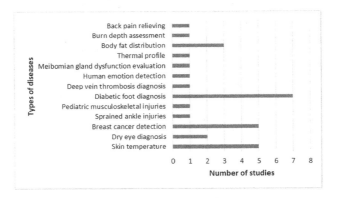

Fig. 4. Distribution of the previous studies based on disease type.

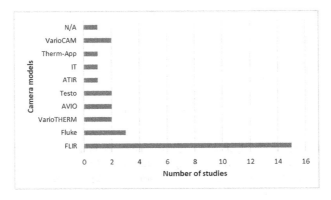

Fig. 5. Distribution of the previous studies based on infrared camera type.

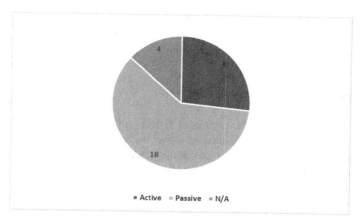

Fig. 6. Frequency of analysis schemes used in the previous studies.

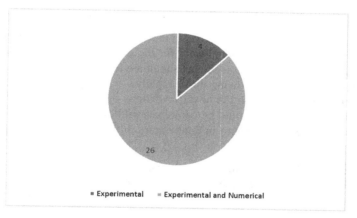

Fig. 7. Frequency of study type used in the previous studies.

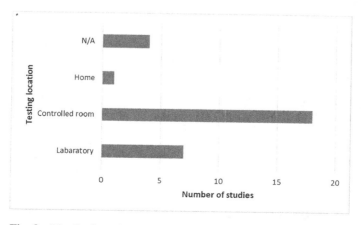

Fig. 8. Distribution of the previous studies based on testing location.

are also studied a lot in the previous studies. Body fat distribution was the third most studied disease type among other previous studies. Between Figs. 5 and 8, the other analysis results with info graphics can be seen.

6 Conclusions and Future Trends

In this study, we proposed a mini-review of B-IRT applications and previous studies within the field of research by collecting those studies based on their measurement methods, analysis schemes, analysis types and study types. Since the early diagnosis is very important in biomedical treatment, utilization of IRT tools would be primary alternative for the detection of abnormal surface detection of human body. Apart from this, importance of B-IRT mentioned and studied in the previous works is highlighted in detail for this research. In order to lead the scientific community on this subject, potential research areas and new perspectives are defined as below:

- More accurate data collection should be made within the field of B-IRT in order to avoid wrong disease diagnosis in the early stage of the treatments.
- Since there is a lack of video processing in the previous researches, capturing videos instead of images would be more useful for the diseases that proceed very fast.
- Progresses and advances in biomedicine should be followed up constantly since the medicine sector has been growing up so rapidly.
- B-IRT should keep step with the novel inventions on diseases in biomedicine.
- Various filtering methods should be combined with thermal imaging techniques in order to recognize abnormalities in temperature differences on human body.

Acknowledgement. The work and the contribution were supported by the SPEV project "Smart Solutions in Ubiquitous Computing Environments", University of Hradec Kralove, Faculty of Informatics and Management, Czech Republic (under ID: UHK-FIM-SPEV-2019).

References

1. Mollmann, K.P., Vollmer, M.: Infrared Thermal Imaging. WILEY-VCH, New York (2010)
2. Kirimtat, A., Krejcar, O.: A review of infrared thermography for the investigation of building envelopes: advances and prospects. Energy Build. **176**, 390–406 (2018)
3. Ring, E.F., Ammer, K.: Infrared thermal imaging in medicine. Physiol. Measur. **33**(3), 33–46 (2012)
4. Jiang, L.J., Ng, E.Y., Yeo, A.C., et al.: A perspective on medical infrared imaging. J. Med. Eng. Technol. **29**(6), 257–267 (2005)
5. Kylili, A., Fokaides, P.A., Christou, P., Kalogirou, S.A.: Infrared thermography (IRT) applications for building diagnostics: a review. Appl. Energy **134**, 531–549 (2014)
6. FLIR Systems: ThermaCAM B2: Operator's Manual. Flir Systems, Sweden (2005)
7. Gade, R., Moeslund, T.B.: Thermal cameras and applications: a survey. Mach. Vis. Appl. **25**, 245–262 (2014)
8. Astarita, T., Cardone, G., Carlomagno, G.M., Meola, C.: A survey on infrared thermography for convective heat transfer measurements. Opt. Laser Technol. **32**, 593–610 (2000)

9. Buzug, T.M., Schumann, S., Pfaffmann, L., Reinhold, U., Ruhlmann, J.: Functional infrared imaging for skin-cancer screening. In: Annual International Conference IEEE Engineering Medical and Biology – Proceedings, pp. 2766–2769 (2006)

10. Cardone, D., Merla, A.: New frontiers for applications of thermal infrared imaging devices: computational psychopshysiology in the neurosciences. Sensors **17**, 1042 (2017)

11. Weigert, M., Nitzsche, N., Kunert, F., Lösch, C., Schulz, H.: The influence of body composition on exercise-associated skin temperature changes after resistance training. J. Therm. Biol. **75**, 112–119 (2018)

12. Polidori, G., Kinne, M., Mereu, T., Beaumont, F., Kinne, M.: Medical infrared thermography in back pain osteopathic management. Complement. Ther. Med. **39**, 19–23 (2018)

13. Simmons, J.D., et al.: Early assessment of burn depth with far infrared time-lapse thermography. J. Am. Coll. Surg. **226**(4), 687–693 (2018)

14. Adam, M., et al.: Automated characterization of diabetic foot using nonlinear features extracted from thermograms. Infrared Phys. Technol. **89**, 325–337 (2018)

15. Sarigoz, T., Ertan, T., Topuz, O., Sevim, Y., Cihan, Y.: Role of digital infrared thermal imaging in the diagnosis of breast mass: a pilot study diagnosis of breast mass by thermography. Infrared Phys. Technol. **91**, 214–219 (2018)

16. Adam, M., et al.: Automated detection of diabetic foot with and without neuropathy using double density-dual tree-complex wavelet transform on foot thermograms. Infrared Phys. Technol. **92**, 270–279 (2018)

17. Picado, A.A., Martinez, E.E., Nova, A.M., Rodriguez, R.S., Martin, B.G.: Thermal map of the diabetic foot using infrared thermography. Infrared Phys. Technol. **93**, 59–62 (2018)

18. Cortes, M.A.D., et al.: A multi-level thresholding method for breast thermograms analysis using Dragonfly algorithm. Infrared Phys. Technol. **93**, 346–361 (2018)

19. Figueiredo, A.A.A., Fernandes, H.C., Guimaraes, G.: Experimental approach for breast cancer center estimation using infrared thermography. Infrared Phys. Technol. **95**, 100–112 (2018)

20. Silva, N.C.M., Castro, H.A., Carvalho, L.C., Chaves, E.C.L., Ruela, L.O., Iunes, D.H.: Reliability of infrared thermography images in the analysis of the plantar surface temperature in Diabetes Mellitus. J. Chiropr. Med. **17**(1), 30–35 (2018)

21. Gatt, A., et al.: The identification of higher forefoot temperatures associated with peripheral arterial disease in type 2 Diabetes Mellitus as detected by thermography. Prim. Care Diab. **12**, 3112–3318 (2018)

22. Kirubha, S.P.A., Anburajan, M., Venkatamaran, B., Menaka, M.: A case study on asymmetrical texture features comparison of breast thermogram and mammogram in normal and breast cancer subject. Biocatal. Agr. Biotechnol. **15**, 390–401 (2018)

23. Neves, E.B., Salamunes, A.C.C., de Oliveira, R.M., Stadnik, A.M.W.: Effect of body fat and gender on body temperature distribution. J. Therm. Biol **70**, 1–8 (2017)

24. Salamunes, A.C.C., Stadnik, A.M.W., Neves, E.B.: The effect of body fat percentage and body fat distribution on skin surface temperature with infrared thermography. J. Therm. Biol **66**, 1–9 (2017)

25. Del Estal, A., Brito, C.J., Galindo, V.E., de Durana, A.L.D., Franchini, E., Quintana, M.S.: Thermal asymmetries in striking combat sports athletes measured by infrared thermography. Sci. Sports **32**, e61–e67 (2017)

26. Su, T.Y., Ho, W.T., Chiang, S.C., Lu, C.Y., Chiang, H.K., Chang, S.W.: Infrared thermography in the evaluation of meibomian gland dysfunction. J. Formos. Med. Assoc. **116**, 554–559 (2017)

27. Albarran, I.A.C., Rangel, J.P.B., Rios, R.A.O., Hernandez, L.A.M.: Human emotions detection based on a smart-thermal system of thermographic images. Infrared Phys. Technol. **81**, 250–261 (2017)

28. Contreras, D.H., Barreto, H.P., Magdaleno, J.R., Bernal, J.A.G., Robles, L.A.: A quantitative index for classification of plantar thermal changes in the diabetic foot. Infrared Phys. Technol. **81**, 242–249 (2017)

29. Kacmaz, S., Ercelebi, E., Zengin, S., Cindoruk, S.: The use of infrared thermal imaging in the diagnosis of deep vein thrombosis. Infrared Phys. Technol. **86**, 120–129 (2017)

30. Etehadtavakol, M., Ng, E.Y.K., Kaabouch, N.: Automatic segmentation of thermal images of diabetic-at-risk feet using the snakes algorithm. Infrared Phys. Technol. **86**, 66–76 (2017)

31. Blasco, J.M., Sanchez, E.S., Martin, J.D., Sanchis, E., Palmer, R.S., Cibrian, R.: A Matlab based interface for infrared thermographic diagnosis of pediatric musculoskeletal injuries. Infrared Phys. Technol. **76**, 500–503 (2016)

32. Formenti, D., et al.: Dynamics of thermographic skin temperature response during squat exercise at two different speeds. J. Therm. Biol **59**, 58–63 (2016)

33. Balci, G.A., Basaran, T., Colakoglu, M.: Analysing visual pattern of skin temperature during submaximal and maximal exercises. Infrared Phys. Technol. **74**, 57–62 (2016)

34. Oliveira, J., Vardasca, R., Pimenta, M., Gabriel, J., Torres, J.: Use of infrared thermography for the diagnosis and grading of sprained ankle injuries. Infrared Phys. Technol. **76**, 530–541 (2016)

35. Fiz, J.A., et al.: Tuberculine reaction measured by infrared thermography. Comput. Methods Programs Biomed. **122**, 199–206 (2015)

36. Acharya, U.R., et al.: Automated diagnosis of dry eye using infrared thermography images. Infrared Phys. Technol. **71**, 263–271 (2015)

37. Mahmoudzadeh, E., Montazeri, M.A., Zekri, M., Sadri, S.: Extended hidden Markov model for optimized segmentation of breast thermography images. Infrared Phys. Technol. **72**, 19–28 (2015)

38. Marins, J.C.B., et al.: Time required to stabilize thermographic images at rest. Infrared Phys. Technol. **65**, 30–35 (2014)

39. Acharya, U.R., et al.: Diagnosis of response and non-response to dry eye treatment using infrared thermography images. Infrared Phys. Technol. **67**, 497–503 (2014)

40. Ludwig, N., Formenti, D., Gargano, M., Alberti, G.: Skin temperature evaluation by infrared thermography: Comparison of image analysis methods. Infrared Phys. Technol. **62**, 1–6 (2014)

Mathematical Modeling and Docking of Medicinal Plants and Synthetic Drugs to Determine Their Effects on Abnormal Expression of Cholinesterase and Acetyl Cholinesterase Proteins in Alzheimer

Shaukat Iqbal Malik[1(✉)], Anum Munir[1,2], Ghulam Mujtaba Shah[3], and Azhar Mehmood[2]

[1] Departments of Bioinformatics and Biosciences,
Capital University of Science and Technology, Islamabad, Pakistan
drshaukat@cust.edu.pk, anummunir786@yahoo.com
[2] Department of Bioinformatics, Government Post Graduate College Mandian,
Abbottabad, Pakistan
amabbassi71@gmail.com
[3] Department of Botany, Hazara University, Mansehra, Pakistan
gmujtabashah72@yahoo.com

Abstract. Alzheimer's is a neurodegenerative disease, typically begins slowly and after the passage of time gets worse. Around 70% of the cause of the disease is accepted to be hereditary with numerous genes generally involved. Objective: There are no proper medications for the disease. At present, the most acknowledged Alzheimer's treatment is cholinesterase inhibitor that inactivates the acetyl cholinesterase chemical to expand acetylcholine level in the brain. Medicinal plants are also used for Alzheimer's. Here, an In-silico attempt is made to compare the Medicinal plants with synthetic drugs to determine the efficacy of medicinal plants in the treatment of Alzheimer's, data is collected for both medicinal plants and commercial drugs. Interactions of compounds with proteins are determined and side effects are calculated and compared to find out the best among them. The mathematical modeling of all the drugs is done to determine their effects on proteins expression. It is observed that none of the synthetic drugs interacts with acetyl cholinesterase, while the chemical constituents of medicinal plants represent greater interactions with both cholinesterase and acetyl cholinesterase, as compared to the synthetic drugs. The mathematical modeling of compounds also confirmed the inhibitory effects of medicinal plants compounds on proteins, whereas the synthetic drugs showed an in increase in the expression level of proteins. On the basis of the results, it is suggested that these chemical constituents are better used as the remedy against Alzheimer's.

© Springer Nature Switzerland AG 2019
I. Rojas et al. (Eds.): IWBBIO 2019, LNBI 11466, pp. 111–127, 2019.
https://doi.org/10.1007/978-3-030-17935-9_11

Keywords: Alzheimer's · CHLE · Interactions · Target proteins · Mutations · Herbal drugs

1   Introduction

Alzheimer's disease (AD); or basically Alzheimer's, an unending neuro-degenerative infection that normally starts gradually and after the progression of time deteriorates. It is the significant purpose behind 60% to 70% of dementia. The most, generally perceived early symptom is an inconvenience in recalling the ongoing occasions; also known as short-term memory loss [1]. As the infection drives, symptoms occur as issues in talking, loss of motivation, confusion in remembering, mood changes, loss of self-care, and issues of behavior [2,3] Gradually the abilities of working are lost, in the end provoking to death [4]. Although the speed of illness progression can differ in patients, the typical life expectancy of the patients is three to nine years [5,6]. Around 70% of the reasons for the illness are hereditary. Different causes are a series of wounds in the head, hypertension or sadness. The disease is associated with plaques and tangles in the mind [7]. A conceivable investigation relies upon the recorded history of the ailment and mental testing with therapeutic imaging and blood tests to decide other possible causes [8].

There are no appropriate treatments to lessen the chances of the sickness [9–11] Treatment of conduct issues or psychosis as a result of dementia with antipsychotics is ordinary, but, not generally endorsed due to their routinely being less beneficial and increased chances of death [12,13]. Around 0.1% of the cases are familial kinds of autosomal, beginning before the age of 65 (Norton et al. 2014) called early beginning familial Alzheimer's. Most of autosomal familial AD can be credited due to changes in one of three genes, encoding amyloid precursor protein (APP) and presenilins 1 and 214 [14]. Most changes in the APP and presenilins genes result in an increased production of a protein A42, the foremost section of weak plaques [15]. A portion of the mutations just changes the extent among A42 and alternate structures, particularly A40 without enhancing A42 levels [16,17]. The best known risk variable of heredity is the legacy of the $\varepsilon 4$ allele of the apolipoprotein E (APOE) [18] Between 40 and 80% of people with AD have no less than one APOEe4 allele [19]. A few broad examinations have found 19 different regions in the gene that appear to cause the ailment. These genes include: ABCA7, BIN1, CASS4, CLU, CD2AP, CELF1, CR1, EPHA1, FERMT2, HLA-DRB5, INPP5D, MEF2C, MS4A, NME8, PICALM, PTK2B, SORL1, SlC24A4, and ZCWPW1 [20].

Changes in the TREM2 gene have been connected with a 3 to 5 time higher chances of developing up Alzheimer's disease [21,22]. Apolipoprotein E (apo-E) 6 is a generally small protein (34kDa, 299 amino acids) that has been gradually uncovering its biologic privileged insights in cardiovascular and neurological ailments. It additionally builds major hazard for Alzheimer and lessens the age of onset and is connected with the expanded seriousness or progression of various other neuro-degenerative issues [23] At present, the most acknowledged AD treatment technique is cholinesterase inhibitors that can inactivate the acetyl cholinesterase (AChE) chemical with a specific end goal to expand acetylcholine levels in the brain [24] Acetyl cholinesterase inhibitors include rivastigmine, tacrine, donepezil, and galantamine though methyl-D-aspartate receptor etc. Usually, there is no cure for AD, but to relief disease symptoms [25].

Numerous scientific investigations have been done on the therapeutic herbs. Herbs have tumor avoidance and anti-inflammatory agents that may be used as a piece of the AD treatment. Alzheimer's patients have an insufficiency of acetylcholine [26]. Anti-inflammatory herbs may diminish aggravation of the cerebral tissue in Alzheimer's. Some Ayurvedic herbs like Guduchi, Yashtimadhuk, Padma (Nelumbo nucifera), Vacha, Convolvulus pluricaulis, Shankhpushpi, Pancha-Tikta-Ghruta Gugguli, Amalaki, Musta Arjun, Amalaki, Ashwagandha, Galo Satva, Kutaj, and others are sublime herbs for diminishing the cerebrum cell degeneration expedited by Alzheimer's. They upgrade the ability of the brain to work, give strength when used reliably [27]. In this Research work, an In-silico endeavor is made to compare the Medicinal plants and industrially accessible medications to decide the adequacy of therapeutic plants in the treatment of AD and the modeling and simulations of compounds are performed to determine their effects on proteins.

2   Materials and Methods

2.1   Ethno Medicinal Data Collection

The ethno medical data about the Medicinal herbs: Acorus Calamus, Cruciferous, Centella Asiatica, Corydalis Ternata, Curcuma Longa, Emblica Officinalis, Evolvulus alsinoides, Gingko Biloba, Glycyrrhiza Glabra, and Huperzia Serrata, were obtained through local inhabitants of Hazara Division, Pakistan by conducting surveys and semi structured interviews, Approximately the informers were between the age of 30–70 years, with a comprehensive traditional understanding of the advantageous wild plants. The medicinal plants are shown in Fig. 1.

a)   b)   c)   d)   e)

f)   g)   h)   i)   j)

Fig. 1. The medicinal plants that are used to cure Alzheimer's (a) Acorus Calamus, (b) Cruciferous, (c) Centella Asiatica, (d) Corydalis Ternata, (e) Curcuma Longa, (f) Emblica Officinalis, (g) Evolvulus Alsinoides, (h) Gingko Biloba, (i) Glycyrrhiza Glabra, and (j) Huperzia Serrata

2.2 Collection of Effective Chemical Constituents of the Medicinal Plants

During the interview process, the questions were asked to obtain information about the effective parts of the medicinal plants; the effective chemical constituents of these plants were identified and downloaded, from the PubChem database. All the chemical constituents belong to the flower, seed and stem parts of the plants.

2.3 Identification of Marketed Drugs Used to Cure AD

The information about the commercially available drugs: Donepezil, Galantamine, Memantine, Rivastigmine, and Tacrine used to cure AD were obtained through Alzheimer's Association. The chemical structures of those marketed drugs were downloaded from the PubChem Database.

2.4   Identification of Drug Targets

The targeted genes and proteins of the commercially available drugs were identified through Balestra web server, it is the online server used to predict the binding interactions between the drug and its target [28]. The mutated 3D protein structures of the target genes were downloaded from the PDB database; it is a repository containing X ray crystallographic and magnetic resonance 3D structures of proteins [29].

2.5   Docking of All Chemical Compounds

The chemical structures of commercially available drugs and the effective constituents of medicinal plants were docked with target proteins in Auto dock. The auto dock is a complete collection of docking tools, used to calculate how well the ligand compound is bind with the receptor molecule in a docked complex.

2.6   Analysis of Binding Interactions

After the docking, the binding interactions of medicinal plants and commercially available drugs were checked with those downloaded proteins through protein ligand interaction profile (PLIP) server [30]. And the Ligplot tool. The ratio of the interactions of both the commercially available drug data and chemical constituents of medicinal plants data were compared with each other, to identify the best molecule that can cure Alzheimer's.

2.7   Identification of Side Effects Produced by the Marketed Drugs and Medicinal Plants on the Human Body

The side effects produced by the selected marketed drugs and the chemical constituents of plants on the human body were identified through the drugs.com. Drugs.com is actually, an online pharmaceutical compendium which delivers the information about the drugs to the customers and healthcare specialists.

2.8   Modeling of All Compounds with Effective Proteins to Determine Their Effects

Drug-receptor complex model was built in Simbiology toolbox of Matlab, to identify the inhibitory effects of drugs on proteins expression, to simulate the time course profiles of both constituents of medicinal plants and synthetic drugs, protein levels, it is necessary to provide certain parameters to the model. The parameters of the proteins in the form of molecular weight, gravy index, theoretical pI, and composition were obtained through Protparam tool available at https://web.expasy.org/protparam/. The molecular weights of drugs and their maximum tolerated doses in the form of LD50 values were obtained through Protox Server.

2.9   Parameters Estimation and Validation of Model

All the parameters of the model were estimated by fitting the developed model by using experimental time-course data, and ODE solver 23 of Matlab, several differential equations were produced for the model. Several dosing variants were developed for all the drugs doses and simulations were performed.

3   Result and Discussions

Therapeutic plants have been renowned and exploited from the beginning of universe. Plants make numerous chemical constituents that are useful for biological purposes, such as, prevention against pests, parasites, fungi, and plant eating vertebrates. In this research, the effective chemical constituents identified in the selected medicinal plants are shown in Table 1.

Table 1. The effective chemical constituents of medicinal plants identified through literature and their activities

Plants	Chemical constituents	Activity
Acorus Calamus	Asarone	Capable of improving memory power and intellect
Cruciferous	Sulforaphane	Decreases brain edema
Centella Asiatica	Acetic Acid	Can be used as brain tonic, cognition, and anti-anxiety
Corydalis Ternata	Protropine	Anti-cholinesterase and anti-amnesic
Curcuma Longa	Curcumin	Protects against synaptic dysfunctions
Emblica Officinalis	Phyllemblin	Anti cholinesterase activity
Gingko Biloba	Bilobalide	Produces neural atrophy
Glycyrrhiza Glabra	Glycyrrhizin	Improves learning and memory
Huperzia Serrata	Huperzine	Anti cholinesterase activity
Evolvulus alsinoides	Betaine	Memory enhancing agent

The Table 1 demonstrated that each chemical constituent has the ability to improve the mental health or to stop the abnormal expressions of cholinesterase that results in the Alzheimer's Basically, the chemical constituents in the plants mediate their effects for the human body through systems like those formally comprehended for the synthetic drugs; in this way, herbal medications don't vary gigantically from conventional drugs with respect to how they work. This empowers therapeutic herbs to have helpful pharmacology [31,32]. At present, only five medicines are available to treat the psychological issues of AD.

Among them, four are acetyl cholinesterase inhibitors: donepezil, galantamine, rivastigmine, tacrine, and the other: Memantine is a NMDA receptor opponent. The benefit from their use is lesser and they don't ease the patients [33,34]. The target is anything within the body of the living organism to which any external entity binds. The target proteins of these drugs identified through Balestra web are shown in Table 2.

Table 2. The Targeted proteins of the commercially available drugs: tacrine, rivastigmine, galantamine and donepezil and Memantine, used to cure Alzheimer's

Protein name	Symbol	Structure
5-hydroxytryptamine receptor 2A	5H2A	
Acetyl cholinesterase	ACES	
Cholinesterase	CHLE	
D(2) dopamine receptor	DRD2	
Glutamate receptor	NMDZ	
Acetylcholine receptor	ACH	

It was observed that ACES and CHLE proteins were common targets of all the drugs, however, the drug Galantamine demonstrated ACH also as its target, NMDZ and DRD2 were predicted as the targets of Memantine, 5H2A is the target of Dopenzil. Determination of drug target interaction system is a basic stride in the medication disclosure pipeline [35]. It is both tedious and exorbitant to decide compound-protein connections or potential medication target associations by examinations alone. As a supplement, several computational methodologies have been created for dissecting and foreseeing drug-protein interactions. The most normally utilized are docking simulations [36–39], writing content mining [40] and consolidating compound structure, genomic succession, and 3D structure data [41]. In this project interactions of target proteins with medicinal plants and drugs were determined with the help of docking simulations. All the commercially available drugs were docked with their respective target proteins, while the chemical constituents of plants were docked with the acetyl cholinesterase and cholinesterase proteins. The greater the interactions the safer is the compound, as a drug candidate. In molecular docking, docking is a procedure which ascertains the favored position of one particle to a second when bound to one another with the end goal to make a stable complex [42,43]. The interactions among the drugs and proteins are shown in Fig. 2.

In the analysis of docking results, it was observed that none of the drug forms any interacting complex with acetyl cholinesterase ACES protein, however, the ACES protein was the known target for each drug. Among all the docked complexes the Galantamine demonstrated more interactions with CHLE protein than

a)   b)   c)

d)   e)   f)

g)   h)   i)

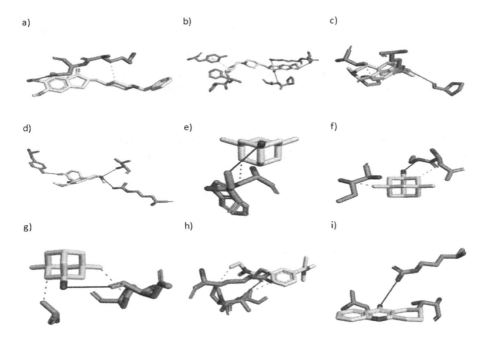

Fig. 2. The interactions analysis, among the drugs and their target proteins through the PLIP server. (a) Dopenzil interactions with 5H2A protein, (b) Dopenzil interactions with CHLE Protein, (c) Galantamine interactions with ACH protein, (d) Galantamine interactions with CHLE protein, (e) Memantine interactions with CHLE protein, (f) Memantine interactions with DRD2 protein, (g) Memantine interactions with NMDZ protein, (h) Rivastigmine interactions with CHLE protein, (i) Tacrine interactions with CHLE protein

the other drugs. The World Health Organization (WHO) analyzes that 80% of the inhabitants in some Asian and African countries uses herbal medications for the number of health issues. The herbal therapeutics used in this study can be created from a seed of a particular plant or collected from nature at no expense. A large number of herbal pharmaceutics have a long history of usage by doctors and clinicians, including opium, cerebral pain drug, digitalis, and quinine. More than 7,000 medicines in the pharmacopeia are gotten from plants [44]. In this research work, an In-silico approach is used to determine the interactions of medicinal plants, used to treat AD and the comparison of Interactions among the docked complexes of medicinal plants and commercially available drugs. The interactions of chemical constituents with ACES and CHLE proteins are shown in Figs. 3 and 4.

From the Figs. 3 and 4, it is confirmed that the medicinal plants demonstrated more interactions with ACES and CHLE protein, which confirmed their efficacy to treat the Alzheimer's, instead of commercially available drugs.

a)   b)   c)

d)   e)   f)

g)   h)   i)

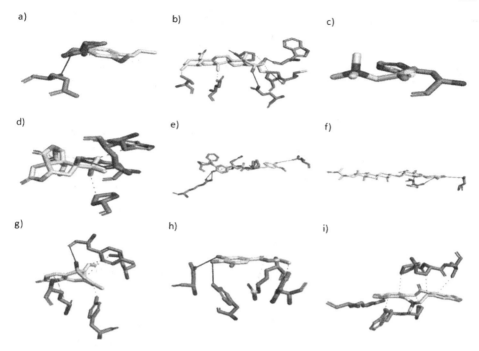

Fig. 3. Interactions of the chemical constituents of medicinal plants with ACES protein. (a) Alpha asarone interactions with ACES protein, (b) Aciatic acid interactions with ACES protein, (c) Betaine interactions with ACES protein, (d) Bilobalide interactions with ACES protein, (e) Curcumin interactions with ACES protein, (f) Glycirrhizic acid interactions with ACES protein, (g) Hupperzine interactions with ACES protein, (h) Phyllemblin interactions with ACES protein, (i) Protopine interactions with ACES protein

These plants have no side effects as compared to the marketed drugs. The comparison between the interacted amino acid residues of the marketed drugs and chemical constituents are shown in Table 3.

The Table 3 demonstrates that the chemical constituents represent more interactions with the pocket amino acids of mutated protein targets, as compared to the interactions of marketed drugs with the amino acids of target proteins. The side effects of all the marketed drugs and medicinal plants were identified, the chemical constituents of medicinal plants demonstrated very few side effects only in the form of mild head ache, cold sore throat, whereas the marketed drugs demonstrate alto of side effects. The side effects produced by the marketed drugs are shown in Table 4.

All the drugs have a lot of side effects as compared to the chemical constituents of medicinal plants that might lead to a serious condition, if not handled at the time; therefore, herbal medicines are better than the synthetic drugs. On the basis of these results, it is suggested that these chemical constituents are

Table 3. The comparison of interactions among the docked complexes of marketed drugs along with the interactions of chemical constituents of medicinal plants

Drug	Protein	AA	Distance	Plant	Protein	AA	Distance
Dopenzil	5H2A	LYS	3.75	Alpha- Asarone	ACES	VAL	3.15
		ASP	3.51		CHLE	TYR	3.49
	CHLE	PRO	3.36				
	PHE		3.80				
Galantamine	ACH	VAL	3.27	Aciatic Acid	ACES	PRO	3.95
		PRO	2.69			THR	2.47
	CHLE	ARG	3.77			GLU	3.78
		VAL	3.74			PRO	3.88
		TYR	2.82		CHLE	GLN	3.65
						PHE	3.00
						TYR	2.54
						TRP	2.18
						TYR	3.76
Memantine	CHLE	TRP	3.88	Bilobalide	ACES	PRO	3.09
	DRD2	ASP	3.96			TRP	2.71
		ILE	3.71		CHLE	GLN	3.22
	NMDZ	PRO	3.98			THR	3.76
		THR	3.00			PHE	3.42
Rivastigmine	CHLE	VAL	3.36	Circumin	ACES	PRO	2.75
						VAL	3.89
					CHLE	GLU	3.12
						ARG	3.05
Tacrine	CHLE	VAL	2.87	Glychrrhizate	ACES	MET	2.88
						GLY	3.29
						ASP	3.36
					CHLE	GLN	3.69
						TRP	3.92
						PHE	2.91
				Hupperzine	ACES	TYR	3.15
						LEU	3.10
						TYR	2.69
					CHLE	MET	3.69
						ASN	2.80
						HIS	3.35
						VAL	3.18
				Phyllemblin	ACES	TRP	3.18
					CHLE	ALA	3.79
						PHE	3.69
						HIS	3.82
				Protopine	ACES	PRO	3.34
						TRP	3.26
						PRO	3.91
						GLN	3.69
						VAL	3.81
					CHLE	MET	3.68

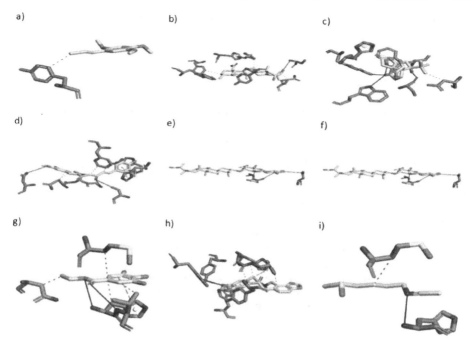

Fig. 4. Interactions of the chemical constituents of medicinal plants with ACES protein. (a) Alpha asarone interactions with ACES protein, (b) Aciatic acid interactions with ACES protein, (c) Betaine interactions with ACES protein, (d) Bilobalide interactions with ACES protein, (e) Curcumin interactions with ACES protein, (f) Glycirrhizic acid interactions with ACES protein, (g) Hupperzine interactions with ACES protein, (h) Phyllemblin interactions with ACES protein, (i) Protopine interactions with ACES protein

better used as the remedy against Alzheimer's. None of the drugs can cure the disease; these drugs only result in the ease in the symptoms of the disease for a shorter period of time. To further confirm the efficacy of medicinal plants the drug receptor complex model was developed and simulations were performed to determine the inhibitory effects of drugs on proteins. The simulation results of commercially available drugs with their respective proteins are given in Fig. 5.

The simulations time was set to 30 days cycle as default and fixed dosage criteria of drugs obtained through DrugBank was applied for the simulations. From Fig. 5 it is clearly observed that all the drugs causes over expression of their targeted proteins instead of inhibiting their mutated expression, demonstrated their inefficiency as better drugs. The simulation results of chemical plants with their respected proteins are shown in Fig. 6a and b.

Table 4. The side effects produced by marketed drugs used to cure Alzheimer's on the human body

Drugs	Side effects
Donepezil	vomiting, diarrhea, loss of appetite, muscle pain, fatigue, asleep problems
Galantamine	Tiredness, dizziness, nuisance, distorted vision, runny nose, hopelessness, sleep problems, vomiting, stomach pain, loss of craving, weight loss, unfriendly taste in mouth
Memantine	vomiting, diarrhea, constipation, loss of appetite, faintness, drowsy feeling, weight loss, inflammation in hands or feet, fast heart-beat, informal staining or blood loss, uncommon faintness, joint pain, nervousness, anger
Rivastigmine	upset stomach, diarrhea, weight loss, paleness, vertigo, inflammation in hands or feet, joint pain, cough, runny or stuffy nose, enhanced sweating, asleep issues
Tacrine	slight nausea, diarrhea, troubled stomach, weight loss, urinating more than usual, tension, unhappy mood, skin rash, sweating, temperature or chills, runny nose, cough, faintness, lethargy, tired feeling, joint or muscle pain

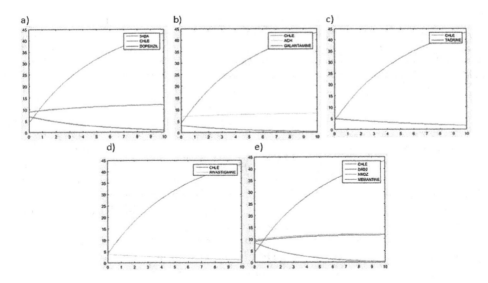

Fig. 5. The simulations of marketed drugs with proteins, (a) Dopenzil effect on 5H2A and CHLE, (b) effect of Galantamine on ACH and CHLE, (c) Tasrine effect on CHLE, (d) Rivastigmine on CHLE, and (e) effects of memantine on CHLE, DRD2, and NMDZ protein, X-axis shows concentration level while y-axis shows the time

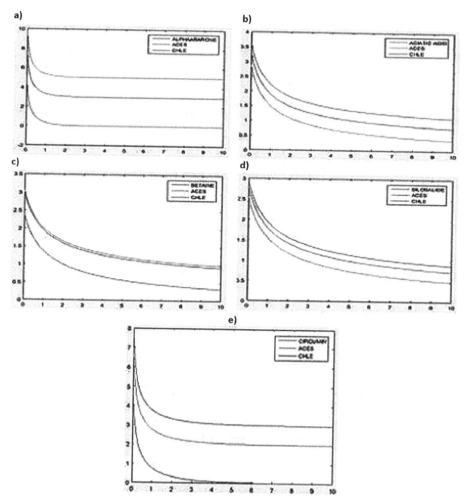

Fig. 6. The simulation results of chemical constituents of medicinal plants with CHLE and ACES protein, (a) alpha asarone effect on ACES and CHLE protein, (b) effect of Aciatic acid on both ACES and CHLE, (c) Betaine on ACES and CHLE, (d) Bilobalide on CHLE and ACES, (e) effect of Circumin on CHLE and ACES protein, X-axis shows concentration level while y-axis shows the time

For some of the plants constituents the dosage criteria was retrieved through drug bank and for other the normal dosage was found from protox server. All the chemical constituents inhibited the abnormal expression of ACES and CHLE proteins (Fig. 6a, b), the simulations results proved that these constituents have inhibitory effects on the ACES and CHLE protein, by using these compounds as

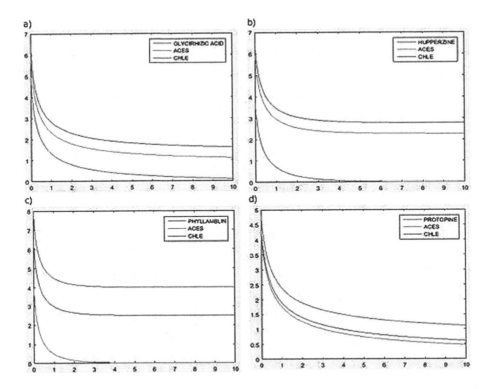

Fig. 7. The simulation results of chemical constituents of medicinal plants with CHLE and ACES protein, (a) Glycirrhizic acid effect on ACES and CHLE protein, (b) effect of Hupperzine on both ACES and CHLE, (c) Phyllemblin on ACES and CHLE, (d) Protopine on CHLE and ACES, X-axis shows concentration level while y-axis shows the time

a treatment option for Alzheimer, better results can be obtained. Four differential equations were produced for the model given below

$$d(drug)/dt = 1/[drugReceptorComplex] * (-DRUGACTIVITY * Drug * ACES * CHLE) \quad (1)$$

$$d(ACES)/dt = 1/[drugReceptorComplex] * (-DRUGACTIVITY * Drug * ACES * CHLE) \quad (2)$$

$$d(COMPLEX)/dt = 1/[drugReceptorComplex] * (DRUGACTIVITY * Drug * ACES * CHLE) \quad (3)$$

$$d(CHLE)/dt = 1/[drugReceptorComplex] * (-DRUGACTIVITY * Drug * ACES * CHLE) \quad (4)$$

4   Conclusion

In this Research work, an In-silico attempt was made to compare the Medicinal plants with commercially available drugs to determine the efficacy of medicinal plants in the treatment of AD. During the results, it was observed that none of the marketed drug forms any interacting complex with acetyl cholinesterase ACES protein, however, the ACES protein was the known target for each drug. The medicinal plants demonstrated more interactions with ACES and CHLE protein, which confirmed their efficacy to treat the Alzheimer's, instead of commercially available drugs. The modeling of all the compounds with their respected proteins also confirmed the efficacy of medicinal plants as better remedy against Alzheimer. Therefore, it is suggested that these chemical constituents are better used as the remedy against Alzheimer's. None of the drugs can cure the disease; these drugs only result in the ease in the symptoms of the disease for a shorter period of time. The herbal medicines do not have any side effects; they can be easily available to common people and are better than marketed drugs.

References

1. Burns, A., Iliffe, S.: Alzheimer's disease. BMJ **338**, b158 (2009). https://doi.org/10.1136/bmj.b158
2. Dementia Fact sheet No. 362: World Health Organization (2015)
3. About Alzheimer's Disease: Symptoms. National Institute on Aging (2011)
4. Querfurth, H.W., LaFerla, F.M.: Alzheimer's disease. N. Engl. J. Med. **362**(4), 329–344 (2010). https://doi.org/10.1056/NEJMra0909142
5. Todd, S., Barr, S., Roberts, M., Passmore, A.P.: Survival in dementia and predictors of mortality: a review. Int. J. Geriatr. Psychiatry **28**(11), 1109–24 (2013). https://doi.org/10.1002/gps.3946
6. Ballard, C., Gauthier, S., Corbett, A., et al.: Alzheimer's disease. Lancet **377**(9770), 1019–1031 (2011). https://doi.org/10.1016/S0140-6736(10)61349-9. PMID 21371747
7. Dementia diagnosis and assessment. National Institute for Health and Care Excellence (NICE) (2014)
8. More research needed on ways to prevent Alzheimer's, panel finds: National Institute on Aging (2006)
9. Thompson, C.A., Spilsbury, K., Hall, J., Birks, Y., Barnes, C., Adamson, J.: Systematic review of information and support interventions for caregivers of people with dementia. BMC Geriatr. **7**, 18 (2007). https://doi.org/10.1186/1471-2318-7-18
10. Forbes, D., Forbes, S.C., Blake, C.M., Thiessen, E.J., Forbes, S.: Exercise programs for people with dementia. Cochrane Database of Syst. Rev. **4** (2015)
11. National Institute for Health and Clinical Excellence. Low-dose antipsychotics in people with dementia. National Institute for Health and Care Excellence (NICE) (2014)
12. Information for Healthcare Professionals: Conventional Antipsychotics. US Food and Drug Administration (2008)
13. Norton, S., Matthews, F.E., Barnes, D.E., Yaffe, K., Brayne, C.: Potential for primary prevention of Alzheimer's disease: an analysis of population-based data. Lancet Neurol. (2014). http://www.thelancet.com/journals/laneur/article/PIIS1474-4422(14)70136-X/fulltext. Accessed 2 Feb 2017
14. Waring, S.C., Rosenberg, R.N.: Genome-wide association studies in Alzheimer disease. Arch. Neurol. **65**(3), 329–34 (2008). https://doi.org/10.1001/archneur.65.3.329

15. Selkoe, D.J.: Translating cell biology into therapeutic advances in Alzheimer's disease. Nature **399**(6738 Suppl), A23–A31 (1999). https://doi.org/10.1038/19866. PMID 10392577
16. Borchelt, D.R., Thinakaran, G., Eckman, C.B., Lee, M.K., Davenport, F., et al.: Familial Alzheimer's disease-linked presenilin 1 variants elevate A1-42/1-40 ratio in vitro and in vivo. Neuron **17**(5), 1005–13 (1996). https://doi.org/10.1016/S0896-6273(00)80230-5
17. Shioi, J., et al.: FAD mutants unable to increase neurotoxic Abeta 42 suggest that mutation effects on neurodegeneration may be independent of effects on Abeta. (2007). PubMed - NCBI. https://www.ncbi.nlm.nih.gov/pubmed/17254019
18. Strittmatter, W.J., et al.: Apolipoprotein E: high-avidity binding to beta-amyloid and increased frequency of type 4 allele in late-onset familial Alzheimer disease. Proceedings of the National Academy of Sciences **90**(5), 1977–1981 (1993). https://doi.org/10.1073/pnas.90.5.1977
19. Mahley, R.W., Weisgraber, K.H., Huang, Y.: Apolipoprotein E4: a causative factor and therapeutic target in neuropathology, including Alzheimer's disease. Proc. Nat. Acad. Sci. U.S.A. **103**(15), 5644–5651 (2006). https://doi.org/10.1073/pnas.0600549103
20. Lambert, J.C.: Meta-analysis of 74,046 individuals identifies 11 new susceptibility loci for Alzheimer's disease. Nat. Genet. **45**(12), 1452–1458 (2013). https://doi.org/10.1038/ng.2802. PMC 3896259
21. Jonsson, T., et al.: Variant of TREM2 associated with the risk of Alzheimer's disease. N. Engl. J. Med. **368**(2), 107–116 (2013). https://doi.org/10.1056/NEJMoa1211103
22. Guerreiro, R., et al.: TREM2 variants in Alzheimer's disease. N. Engl. J. Med. **368**(2), 117–127 (2013). https://doi.org/10.1056/NEJMoa1211851
23. Mahley, R.W.: Apolipoprotein E: remarkable protein sheds light on cardiovascular and neurological diseases. Clin. Chem. **63**(1), 14–20 (2017). https://doi.org/10.1373/clinchem.2016.255695
24. Jivad, N., Rabiei, Z.: A review study on medicinal plants used in the treatment of learning and memory impairments. Asian Pac. J. Trop. Biomed. **4**(10), 780–789 (2014). https://doi.org/10.12980/APJTB.4.2014APJTB-2014-0412
25. Orhan, I., Aslan, M.: Appraisal of scopolamine-induced antiamnesic effect in mice and in vitro antiacetylcholinesterase and antioxidant activities of some traditionally used Lamiaceae plants. J. Ethnopharmacol. **122**(2), 327–332 (2009)
26. Kim, H.G., Oh, M.S.: Herbal medicines for the prevention and treatment of Alzheimer's disease. Curr. Pharm. Des. **18**(1), 57–7 (2012)
27. Singhal, A., Bangar, O., Naithani, V.: Medicinal plants with a potential to treat Alzheimer and associated symptoms. Int. J. Nutr. Pharmacol. Neurol. Dis. **2**(2), 84 (2012). https://doi.org/10.4103/2231-0738.95927
28. Cobanoglu, M.C., Oltvai, Z.N., Taylor, D.L., Bahar, I.: BalestraWeb: efficient online evaluation of drug-target interactions. Bioinformatics **31**(1), 131–133 (2015). https://doi.org/10.1093/bioinformatics/btu599
29. Munir, A., Azam, S., Khan, Z., Mehmood, A., Mehmood, A., et al.: Structure-based pharmacophore modeling, virtual screening and molecular docking for the treatment of ESR1 mutations in breast cancer. Drug Des. **5**, 137 (2016). https://doi.org/10.4172/2169-0138.1000137
30. PLIP: Fully automated protein-ligand interaction profiler — Nucleic Acids Research — Oxford Academic (n.d.). Accessed 26 Jan 2017
31. Tapsell, L.C., Hemphill, I., Cobiac, L., et al.: Health benefits of herbs and spices: the past, the present, the future. Med. J. Aust. **185**(4 Suppl), S4–24 (2006)

32. Lai, P.K., Roy, J.: Antimicrobial and chemopreventive properties of herbs and spices. Curr. Med. Chem. **11**(11), 1451–1460 (2004). https://doi.org/10.2174/0929867043365107. PMID 15180577

33. Whitehouse, P.J.: The end of Alzheimer's disease - from biochemical pharmacology to ecopsychosociology: a personal perspective. Biochem. Pharmacol. **88**(4), 677–681 (2014). https://doi.org/10.1016/j.bcp.2013.11.017

34. Birks, J., Harvey, R.J.: Donepezil for dementia due to Alzheimer's disease. Cochrane Database Syst. Rev. **1**, CD001190 (2006). https://doi.org/10.1002/14651858.CD001190.pub2

35. Birks, J.S., Evans, J.G.: Rivastigmine for Alzheimer's disease. Cochrane Database Syst. Rev. **4**, CD001191 (2015). https://doi.org/10.1002/14651858.CD001191.pub3. ISSN 1469-493X

36. He, Z., et al.: Predicting drug-target interaction networks based on functional groups and biological features. PLoS ONE **5**(3), e9603 (2010). https://doi.org/10.1371/journal.pone.0009603

37. Cheng, A.C., Coleman, R.G., Smyth, K.T., Cao, Q., Soulard, P., et al.: Structure-based maximal affinity model predicts small-molecule druggability. Nat. Biotechnol. **25**, 71–75 (2007)

38. Rarey, M., Kramer, B., Lengauer, T., Klebe, G.: A fast flexible docking method using an incremental construction algorithm. J. Mol. Biol. **261**, 470–489 (1996)

39. Choc, K.C.: Structural bioinformatics and its impact to biomedical science. Curr. Med. Chem. **11**, 2105–2134 (2004)

40. Chou, K.C., Wei, D.Q., Zhong, W.Z.: Binding mechanism of coronavirus main proteinase with ligands and its implication to drug design against SARS. Biochem, Biophys. Res. Commun. **308**, 148–151 (2003). (Erratum: ibid 2003, vol. 310, 675)

41. Zhu, S., Okuno, Y., Tsujimoto, G., Mamitsuka, H.: A probabilistic model for mining implicit 'chemical compound-gene' relations from literature. Bioinformatics **21**(Suppl 2), ii245–ii251 (2005)

42. Yamanishi, Y., Araki, M., Gutteridge, A., Honda, W., Kanehisa, M.: Prediction of drug-target interaction networks from the integration of chemical and genomic spaces. Bioinformatics **24**, i232–i240 (2008)

43. Lengauer, T., Rarey, M.: Computational methods for biomolecular docking. Curr. Opin. Struct. Biol. **6**(3), 402–406 (1996). https://doi.org/10.1016/S0959-440X(96)80061-3

44. Interactive European Network for Industrial Crops and their Applications. Summary Report for the European Union. QLK5-CT-2000-00111. Archived from the original on 22-12-2013 (2005)

Definition of Organic Processes via Digital Monitoring Systems

Svetlana Martynova[1(✉)] and Denis Bugaev[2]

[1] Institute of Philosophy of Human,
The Herzen Pedagogical State University of Russia, Moika Emb. 48,
191186 Saint-Petersburg, Russia
svetlanus.martinova@yandex.ru
[2] Institute of Archaeology, The Warsaw University,
Krakowskie Przedmieście str., 26/28, 00-927 Warsaw, Poland
denisbugaev@yahoo.com
https://www.herzen.spb.ru/, https://www.uw.edu.pl/

Abstract. Digital monitoring systems in medical practice is one of the instruments for determination of organic processes. Various kinds of organic processes such as childbirth, growing, development, eating, ageing, dying can be witnessed by digital monitoring systems. It aims to establish knowledge about working bodies and minimize spreading of illness or negative processes. Systems are witnessing about human actions, about functioning of invisible organs and very little changes in human organisms. Witnessing has links with measuring and judgemental codes in computer programs. Sensors convert mode of body to numeric parameters, measure statement of organism and compare it with necessary levels. As a result, digital monitoring systems mark changes and transgression of norms in individual organic processes.

In contemporary reaching of certain numeric parameters becomes in medicine a strategy of organic processes' improvement. Devices' witnessing connected with measuring and judgemental codes is an instrument for it. Doctors or patients look at results of bodies functioning and make their own decisions based on them. Specificity of witnessing of organic processes by digital monitoring systems is in its digital nature. Digital code is a basement for witnessing and measuring. It is created to save normal body's functioning. So digital code presents new reality (order), which helps understand the statement of organism. In this paper, we clear the question how digital code correlates with principles of body's functioning. What are the positive and negative aspects of digital determining? How digital monitoring systems can advance functioning of organism and what they need to pay back?

Keywords: Digital code · E-health · Fetal monitoring ·
Adherence monitoring · Organism's functioning · Body's creativity

The reported study was funded by RFBR according to the research project № 19-011-00899.

I. Rojas et al. (Eds.): IWBBIO 2019, LNBI 11466, pp. 128–135, 2019.
https://doi.org/10.1007/978-3-030-17935-9_12

1   Actuality

Contemporary sociocultural and technical transformations engender the necessity of the reflection addressed to organic processes. Now digital and net worlds spread of a virtual image of organic. Digital reality makes influence on organic processes – advises what a man/woman should do. Computer technologies observe human bodies, offer their version of the best flow of organic processes, compare computer model and human organisms and so do what a man/woman cannot do. We have new reality, when electronic machines are instruments of determination of organic processes. Biomedia is a reflection of these processes. According to philosopher and biologist Thaker's point of view it "emphasis on the ways in which an intersection between genetic and computer "codes" can facilitate a qualitatively different notion of the biological body—one that is technically enhanced, and yet still fully "biological" [1].

Establishing systems of interpretation and action is inartificial for media including one of the DMS. For purposes of this paper, it is necessary to distinguish conflicts of interests between positions and settings. There are two positions: individual and superindividual. In addition, there are settings: routine medical practice, cosmic medicine, sports one, disaster one, battlefield one, and so on. Individual position is fluent and depends on personal will, for instance, to win a competition, to participate in war, or to avoid such activities according to his/her medical status. Superindividual position depends on group's frames in which it functions, such as interests of runaway group or national ones, and on national priorities, such as to be a strong country in sport, in military force, or in the Eurovision.

Situations in settings also differ. Only very rich in instruments and resources everyday medicine theoretically will try to keep life and health of everyone in the same manner. In all other cases there are some accents to do so on more valuable for the setting groups, such as public persons, officers, children, women and so on. To discuss our theme we are forced to keep this in our minds to accept existence of anchor points in the DMS encoding.

Digital monitoring systems (DMS) in medical practice are one of the instruments for determination of organic processes. Individually a man/woman uses devices to comfort his/her life. Devices use data about working bodies to minimize spreading of illness or negative processes (fatness, low musculoskeletal status). Reaching of this goal means establishing of human power under organic processes. A man/woman thinks such devices to be instruments for improvement of organic processes, because they can help to decide what to do in medical practice. We want to know how digital code correlates with principles of body's functioning. What are the positive and negative aspects of digital determining? How the DMS can advance functioning of organism?

2   Methods

We can define several methods for researching ability of determination by the DMS. For understanding of different projections of bodies in medical practice, we appeal to the A. Mol's conception of body multiplicities. This conception provides the idea about limitation of bodies by focusing of investigating interest. Either it explains irreducibility different focuses of bodies to each other. We address to the theory of digital media for explanation of digital code in monitoring systems. For critical researching ability to witness and determine of bodies by digital monitoring systems we appeal to the ability of organic processes to correlate with each other.

3 Material

Devices have witnessing function. For purposes of this article witnessing means receiving information from concrete place due to essence in there, transferring this data to other places and describing, what has happened. Witnessing for a long time was only a man/woman's prerogative. There are many projections of human witnessing in judgment, in religion etc. Since invention of photograph in the XIX century machines systematically, become up–to–date improved witnesses. They can complete their functions continuously and emotionless in all settings both dangerous (war, catastrophes) and even unreachable for a human (micro world, Space). Their emotionless and unbiased but totally controlled ability of judgment is their benefits.

Ability of witnessing is applied by autonomous program in the processes of machine evolution. At the beginning of their existence, systems were analogue and only human interpreted incoming data. Up–to–date digital devices are opposite of it. They are intelligible and have ability to define a statement of organism according to special autonomous program. According to this program, monitoring systems measure statement of organism comparing it with numeric data and show the result. There are many different statements of organism for monitoring. For instance, smartphone-based biomedical tools include imaging flow cytometers, immunochromatographic diagnostic test readers, bacteria/pathogen sensors, blood analyzers for complete blood count, and allergen detectors [2], contact lens sensors for the diagnosis of glaucoma by measuring intraocular pressure [3], detector of vascular reactivity [4] and so on. Digital data collected by one device now can easily circulate among other ones to make a precise decision, for instance, circulation of visual and other data among dental services.

Witnessing of organism statement by the DMS have positive affect for medical management. John J. Mastrototaro says, that wearables, sensors, and mobile apps aim to "transition routine care from more costly environments (hospital, specialists) to less burdensome environments (local pharmacy, home), allowing the specialists to focus on patients with genuine need, while using alternative approaches for the management of patients with noncritical issues" [2]. So borders between out– and inpatients is now changing to establish, may be, remote ones, who will have their DMS-based diagnosticians at home.

This aim correlates with strategies of modern medicine to reach certain judgment parameters by more precise and widespread monitoring and adherence of patients. Witnessing and measuring of ones determines status of human organism – health, illness, death. Doctors or patients can look at results of organisms' functioning and make their own decision based on them.

Necessary parameters are reachable by different ways – organism can sustain itself or doctors can improve it by medical intervention. We are planning that monitoring devices will take a part in direct correction of organic processes. They will be joined with other devices to put medicaments into organism, to control patients' adherence. Up today this function is not wide spread. However, despite of it the DMS determine potential of organic processes and lead to more medical interventions.

For interpreting how organic processes are determined by the DMS, we should appeal to digital code as special autonomous program. Digital code is basement for witnessing it creates new reality. The code establishes the order to protect optimal organism's functioning. This means that digital code reflects the statement of organism and connection of bodies. According to this digital code establishes numeric system for the statement of organism and fixes changes in this system referring to changes in organism. If one of the organs is bad, system can fix it via special program linking organs and their places in judgment system.

Digital code either has not only its own program, but also creative ability. It creates the order and defines processes according to the code. Human statement in such a situation is measured by device's logic, but not by logic of organism. Physiological and psychological bondages not on organism's self-regulation, but on the DMS, have dependence-producing potential, which is vitally important for further ranging of medical adherence.

Organism has its own logic and can find new routes for realization its functions, create new abilities. In situation of infecting and further recuperation, he/she becomes immune. It means that he/she got protective functions by himself and creates new order. One organ can help another one when it loses functionality and become destroyed (losing of eye ability). It is compensatory function. In case of losing some functionality other functions become stronger (replacement of visual ability by tactile sensation, if some parts of brain are damaged other function alike).

Absence of organ and its functions leads for searching by organism alternative routes for its future existence. For instance, new veins grow on the arm in case of transplantation previous ones to heart. Strongly advanced atherosclerosis not always leads to hobble and renal dysfunction (illness may flow slowly and metabolism adopts to it) [5]. Organisms can create alternative routes and this ability is out of digital code and artificial improvement. These types of organism working are not random. For understanding what organic processes are, we appeal to philosopher, physician and biologist I. Kant's theory. Kant suggests our knowledge about organic processes including two aspects: we can say about them due to appealing to mechanical laws and due to teleology ability.

Teleology allows saying about organic processes in terms of means and goals. It fixes the difference between matter and nature. Teleology ability explains that the organism has creative ability. It means that it has reproductive and replacing abilities. Biologist and philosopher Kant wrote: "Hence one wheel in the watch does not produce the other, and, still less, does one watch produce other watches, by utilizing, or organizing, foreign material; hence it does not of itself replace parts of which it has been deprived, nor, if these are absent in the original construction, does it make good the deficiency by the addition of new parts; nor does it, so to speak, repair it own defects. But these are all things which we are justified in expecting from organized nature. – An organized being is, therefore, not a mere machine" [6].

Knowing about organic processes as means and goals and organisms' creative abilities is the basements for medicine. Kant suggested that doctors must believe that nature or human body has rationality and all of the organs are connected with each other. Therefore, if we cannot be sure that organic processes are goals and means, we should not to make a decision in a medicine [7].

What does it mean for contemporary? We should note that there were only romantic beginnings of machine reality and no digital and net worlds in Kant's times. It is in apposite with the methods of this article. We may suggest retrospection to Kant's conception being as a clue for innovations in medical sphere. It is necessary to understand what happen in medical practice when creative ability of organism as attributive characteristics of all organic processes is not in accordance with digital monitoring systems.

Up-to-date researcher A. Mol speaks about multiple body and appeals to medical practice. She points on the difference between representation of body in clinic and its representation in pathologists' laboratory [8]. According to her article, projections of body make new illness body. We agree with this conception. It is impossible to find only one adequate representation of statement of real organs. The DMS cannot witness and determine statement of the organism in all spheres.

The DMS also cannot regret transformation of organic processes and establishing of it in accordance to a new order. For instance, these systems can determine destroyed organ, but it cannot represent teleological transferring a function of destroyed organ to other organ(s). So statement of organism is witnessed as abnormal, because one organ is no able to fulfill its function, but really other organ(s) fulfill(s) this function or not.

The DMS can witness and determine infected organism, but they cannot show getting a protective functions (in this case witnessing of pathogens, high temperature and bad blood count cannot represent of organic processes – is it normal or abnormal, organism have illness with temperature and bad blood count and it needs help or it is fighting successfully with viruses). The DMS can determine statement of organism without its creative ability.

Result of witnessing of bad significances is medical intervention. Understanding of body's creativity must be instrument of correction of medical conclusions, based on device's witnessing. In addition, understanding of body's creativity should be used in adherence monitoring. Researchers explain the goal of this monitoring next way. Wai Yin Lam and Paula Fresco remark that "Non–adherence leads to poor health outcomes and increased healthcare costs. Improving medication adherence is, therefore, crucial and revealed on many studies, suggesting interventions can improve medication adherence" [9]. However, we should note, that adherence monitoring has no purpose to show how organs are integrally functioning, but aims only to improve some their systems. For this purpose, the DMS mark when a man/woman has taken inside his/her medicaments, effects of drugs and other activities. Before it the doctor observed the patient, defined the doze for him and the patient agreed to fulfill these recommendations in accordance with media helping. It is not right, if doctor has only plan to reach expected results, but he does not expect on organism's creative ability. Understanding of this body's ability should fill this gap.

Our critical researching of the DMS does not means that we contradict with it. We would like to pay attention that it is very important to explain what has happened and note to another order made by an organism at present or in future. Doctors should not hurry up with medical interventions, if it is possible. Understanding of body's creativity can be provided not only by doctors, medicine staff, but also by devices.

The first step on this way is witnessing of self-regulative ability of organism in different conditions. It means that digital monitoring systems measure not only due to digital code, but also due to principles of organisms' functioning. There are some examples for it in modern medical practice. One of the good examples is self-powered systems of integrated sensors and technologies (ASSIST). Veena Misra supposes, "Beyond activity monitoring, ASSIST wearable platforms can provide real-time measurements of ozone and volatile organic compounds in the environment and critical corresponding health signals, such as wheezing, heart rate, electrocardiogram (ECG), and pulse oximetry. This functionality can address the needs of asthma management by providing users immediate assessment of respiration burden and environmental triggers, leading to rapid treatment and enabling effective correlation between health and the environment" [2]. Determination of organic processes is flexible, when the DMS represent influence of environment on organism.

This strategy is also applicable by other DMS. They are able to witness, measure and compare statements of organism during the day and night, influence of different drugs on organic processes. I agree that actual question is how to personalize the remote patient monitoring (RPM). Researches suppose that RPM should "include patient-reported health related quality of life (HRQOL), symptom severity, satisfaction with care, resource utilization, hospitalizations, readmissions, and survival. There is little data investigating the impact of RPM on these outcome measures. It may strengthen the interventions if they are developed directly in partnership with end-users – i.e. patients themselves" [10]. All of these strategies of witnessing help to examine body's creativity. It helps to make right chose in process of human healing.

Furthermore, there are too much situations, when we cannot hope to body's creativity. In critical conditions' monitoring, he/she has not enough time to begin new order without intervention. There are risks that organic processes will have finished before they can start self-healing and self-improvement algorithms. Notwithstanding, in this situation we can trust only actions aimed at monitoring and fixating the statement of organism continuously. Devices are special instrument for it. They at once (automatic definition) can help to decide what doctors or patients would do in medical practice. It is good because in this case we do not rely on body's creativity and can correct the statement of organism immediately. Using of the DMS is one of the instruments to produce some analogue algorithm of such an ability by medical intervention. In addition, it detects what organ is out of order.

We mean, that there are positive aspects of using the DMS in medical practice. For understanding achievements of witnessing by them, we appeal to fetal monitoring. In 1966 it was established and today there are many critical articles about this instrument in medical practice. Problematic for analogue systems is interpretation of results, digital ones are out of this problem. Digital electronic fetal monitoring can define statement of woman and child automatically.

We appeal to patent US 9,693,690 B2, named Digital Electronic Fetal Heart Rate and Uterine Contraction Monitoring System. System can provide not only significances, but also signals of dangerous situation, what need to do and it can point on wrong doctor's actions and so on. Stewart Bruce Ater, the author of this patent, writes the following: "The system can provide automatic alarms and discrete methods to relay information to doctors and nurses and medical providers for the mother whenever average and median rest intervals and other parameters exceed user adjustable preset safety limits. The system can provide automatic alarms enabling medical providers to make treatment decisions and interrupt injurious conditions before brain injuries occur, such as discontinuing oxytocin infusion or deciding to perform a Cesarean section. The system can provide automatic pausing of an oxytocin infusion pump to a mother until medical providers respond to detected elevated risks" [11].

The DMS are new method for preventing pathology of the organ or organic process. These systems are the mechanism to optimize processes of childbirth, which is almost free from doctor's position. We hope that they will reflect dangerous statement immediately and so it can help to improve childbirth.

The DMS are used beyond the limits of body's creativity tended to avoid sequelae. Can the DMS define approaches of organic processes' improvement? What happens if we agree that the DMS are right, and we immediately try to improve the significances? Doctors have not enough time to wait for the organism's actions to improve him/herself in critical situations. They make a decision due to machine data and understand that mechanical intervention to the processes is predominantly right chose in these cases (using of oxytocin or Cesarean section). Doctors decide to use mechanical instruments for reaching of necessary results. These results mean, that waiting of organisms' creativity have finished. Using of mechanical instruments correlate with correction of organism to save life.

4 Perspectives

In contemporary human organisms often are not able to regulate themselves. Contemporary medicine provides strong opportunities both for artificial preservation of human life, such as lung ventilation, assisted circulation etc. Such animated being without response from his/her reason evokes among society euthanasia disputes in local legislations, because death just without this staff becomes fast, painless and 'humanistic'. Criminals and other problematic social groups also are under request for these disputes. It is fair to assume that medically supervised anabiosis can become a form of detention (like as in the action film "Demolition Man", 1993).

How we can replace body's creativity not negatively, but by the best way? Scientists would like to create devices, which can witness and improve body's statement inside of organism via drug delivery, which is applicable all settings excluding professional sport one. It means that devices improve organic processes according to methods of improvement intervention with digital code programming it.

How it can correlate with body's creativity? Will organism (or its digital version) be able to fulfill self-regulation in future? In case, if no, decision-making centers will receive a powerful instrument to control and to discipline population. It will mean new types of medical benefits (health, youth, and immortality) for loyalty for social ideals and detentions (just 'switching off' some medical options).

Also much easier will become 'switching' between settings: routine medical practice, cosmic medicine, sports one, disaster one, battlefield one, and so on. Such 'switches' in the DMS and drug delivery systems can be a reasonable instrument for surviving through critical situations, because they are much faster and obligatory for organism than some complicated mental processes, reasons and individual decisions.

5 Results

Organism has creative abilities, so it is important to combine devices and doctor's position. Doctor observes the organism for definition if medical intervention is necessary. The DMS represent influence of environment on organism, witnesses and compare different statements of organism during the day and night, influence of different drugs on organic processes. All of these strategies of witnessing help to examine body's creativity. It is actual besides critical conditions of man/woman. Combination of monitoring system and devices for improvement of organism also should include knowing about body's creativity. Optimal position is media improvement of one's functioning before returning self-regulative function to it.

References

1. Thaker, E.: Biomedia (Electronic Mediations), p. 6. University of Minnesota Press, Minneapolis, London (2004)
2. Munos, B., et al.: Mobile health: the power of wearables, sensors and apps to transform clinical trials. Ann. NY Acad. Sci. **1375**(1), 3–18 (2016). https://nyaspubs.onlinelibrary.wiley.com/doi/full/10.1111/nyas.13117
3. Farandos, N.M., et al.: Contact lens sensors in ocular diagnostics. Adv. Healthcare Mater. **4**(6), 792 (2015)
4. Naghavi, M., et al.: New indices of endothelial function measured by digital thermal monitoring of vascular reactivity: data from 6084 patients registry. Int. J. Vasc. Med. (2016). https://www.hindawi.com/journals/ijvm/2016/1348028
5. Mol, A.: Mnoghestvennoe telo: Ontologiya v medicinskoy praktike. Pisarev, A., Gavrilenko, S. (trans.), pp. 81–83. Hile-Press, Perm (2017)
6. Kant, I.: Critique of Judgement. Meredith, J.C. (trans.), p. 202. Oxford University Press, NY (2007)
7. Kant, I.: Spor fakul'tetov. Arzakanyan, I.D., Levina, M.I. (trans.) Kalinnikov, L.A. (ed.), pp. 64–66. Kaliningrad State University, Kaliningrad (2002)
8. Mol, A.: Mnoghestvennoe telo: Ontologiya v medicinskoy praktike. Pisarev, A., Gav-rilenko, S. (trans.), pp. 69–70. Hile-Press, Perm (2017)
9. Lam, W.Y., Fresco, P.: Medication adherence measures: an overview. BioMed. Res. Int. (2015). https://www.hindawi.com/journals/bmri/2015/217047
10. Noah, B., et al.: Impact of remote patient monitoring on clinical outcomes: an updated meta-analysis of randomized controlled trials. NPJ Digit. Med. - Nat. (2018)
11. Ater, S.B.: Digital electronic fetal heart rate and uterine contraction monitoring system. Patent. https://patents.google.com/patent/US9693690

Detection of Pools of Bacteria with Public Health Importance in Wastewater Effluent from a Municipality in South Africa Using Next Generation Sequencing and Metagenomics Analysis

Anthony Ayodeji Adegoke[1,2], Emmanuel Adetiba[3,4(✉)],
Daniel T. Babalola[5], Matthew B. Akanle[3,5], Surendra Thakur[6],
Anthony I. Okoh[7], and Olayinka Ayobami Aiyegoro[8]

[1] Department of Microbiology, Faculty of Science, University of Uyo,
P.M.B 1018, Uyo, Akwa Ibom State, Nigeria
[2] Department of Biochemistry and Microbiology, University of Fort Hare,
Alice 5700, South Africa
[3] Department of Electrical and Information Engineering, Covenant University,
Canaanland, P.M.B 1023, Ota, Nigeria
emmanuel.adetiba@covenantuniversity.edu.ng
[4] HRA, Institute for Systems Science, Durban University of Technology,
P.O. Box 1334, Durban, South Africa
[5] Centre for Systems and Information Services (CSIS), Covenant University,
Canaanland, P.M.B 1023, Ota, Nigeria
[6] KZN e-Skills CoLab, Durban University of Technology, Durban, South Africa
[7] Applied and Environmental Microbiology Research Group,
Department of Biochemistry and Microbiology, University of Fort Hare,
Alice 5700, South Africa
[8] GI Microbiology and Biotechnology Unit,
Agricultural Research Council - Animal Production,
Private Bag X02, Irene 0062, South Africa

Abstract. Wastewater effluents are always accompanied with possibilities for human health risks as diverse pathogenic microorganisms are harboured in them, especially if untreated or poorly treated. They allow the release of pathogens into the environment and these may find its way into food cycle. This paper reports the findings of our research work that focused on the characterization of microorganisms from a municipal final wastewater effluent that receives bulk of its spent water from a research farm. High throughput sequencing using Illumina MiSeq apparatus and metagenomics analysis showed a high abundance of microbial genes, which was dominated by Bacteria (99.88%), but also contained Archaea (0.07%) and Viruses (0.05%). Most prominent in the bacterial group is the Proteobacteria (86.6%), which is a major phylum containing wide variety of pathogens, such as Escherichia, Salmonella, Vibrio, Helicobacter, etc. Further analysis showed that the Genus Thauera occurred in largest amounts across all 6 data sets, while Thiomonas and Bacteroides propionicifaciens also made significant appearances. The presence of some of the detected bacteria like

© Springer Nature Switzerland AG 2019
I. Rojas et al. (Eds.): IWBBIO 2019, LNBI 11466, pp. 136–146, 2019.
https://doi.org/10.1007/978-3-030-17935-9_13

Corynebacterium crenatum showed degradation and/or fermentation in the effluent, which was evidenced by fouling during sampling. Notable pathogens classified with critical criteria by World Health Organization (WHO) for research and development including *Acinetobacter* sp., *Escherichia coli,* and *Pseudomonas* sp. in the effluent were being released to the environment. Our results suggest a potential influence of wastewater effluent on microbial community structure of the receiving water bodies, the environment as well as possible effects on the individuals exposed to the effluents. The evidences from the results in this study suggest an imminent public health problem that may become sporadic if the discharged effluent is not properly treated. This situation is also a potential contributor of antimicrobial resistance genes to the natural environments.

Keywords: Metagenomics · Illumina · MiSeq · Genus Thauera · *Acinetobacter* sp. · Human health risks

1   Introduction

Microorganisms are ubiquitous and they exist in diverse communities. These microbes play very important roles in proper functioning of the ecosystem and they are central in the flow of biogeochemical cycles. On the other hand, some of them are implicated in several diseases that plague the human population. Microbial pathogens that cause diseases in humans include; viruses, bacteria, fungi, protozoa and helminthes [1, 2]. Wastewater provides a safe haven for such pathogens to multiply because it contains various materials that are rich to support microbial growth [3]. Illustrative examples include the death of several humans exposed to Vibrio Cholerae bearing wastewater in London in mid-19th century [3]. Epidemiological evidence in Norway also linked the outbreaks of legionnaires disease to contaminated wastewater [4]. Although, wastewater and animal farm have been reported to contain useful bacterial diversities, they have also been reported as the source of various pathogens of immense public health importance [5–7].

Adegoke and Okoh [6] reported the detection of Methicillin Resistant Staphylococcus Species (MRSS) bearing mecA genes from some animal farms in Nkonkobe municipality of South Africa. Stevik et al. [5] as well as Cai and Zhang [8] also reported the presence of Escherichia coli, Mycobacterium tuberculosis, Pseudomonas aeruginosa, Salmonella enterica, Shigella flexneri, Staphylococcus aureus and Vibrio cholerae, etc. in wastewater. Furthermore, several viruses have also been detected by some researchers as reported in the literature. Thus, the detection of these pathogens may provide a source of surveillance that gives early warnings ahead of disease outbreaks, which may be consequent on the release of wastewater to the environment [9, 10].

It is apparent that the general assessment of microbial population that are inherent in municipal wastewater is highly imperative [8]. In codicil, the methods that are deployed for such assessment should allow broad detection of both cultured and uncultured microbes. However, most of the existing methods target one or few pathogens, consequently providing less than the required information on the diversities of microorganisms in wastewater samples [11]. Methods such as colony count [12, 13], PCR [13, 14], qPCR [15] and Microarray [16] are frequently deployed albeit their limitations in assessing uncultured microbes and providing comprehensive information on the microbial population.

The emergence of metagenomics analysis has addressed the limitations of the stated traditional approaches for the assessment of both cultured and uncultured microbial populations [17, 18]. It is becoming easier to carry out direct genetic analysis of genomes in an environmental sample with a view to determining the potential health hazards they can constitute [19]. The beauty in assessing hidden diversity of microscopic life as well as the huge potentials in understanding the entire biosphere make metagenomics analysis a widely sort after tool [20–22]. Besides the determination of diversities, it is possible to determine the role of each of the microorganisms in their niche environments [23]. As both taxonomic and functional diversities of microbes are assessed, the inference drawn forthwith could be relevant for diagnosis and therapy administration especially when human pathogens are involved [18, 24–27]. This kind of dependable procedure for monitoring pathogens is imperative for setting guidelines on restricted and unrestricted treated wastewater reuse [28]. This is important because treated wastewater to be released to the environment (for purposes of irrigation and drinking water production) is expected to be pathogen-free. This paper presents a report on the metagenomics analysis of wastewater effluent from a municipal site in South Africa. In Sect. 2, we outline the materials and methods, Sect. 3 presents the result and discussion while the conclusion is drawn in Sect. 4.

2   Materials and Methods

2.1   Study Site and Sample Collection

The sample site is close to the Agricultural Research Council- Animal Production (ARC-AP), South Africa. The site is situated about 25 km south of Pretoria (25°52′S 28°13′E/25.867°S 28.217°E/− 25.867; 28.217 in Gauteng) adjacent to the village of Irene in the suburb of Centurion. The area has a typical Highveld climate (altitude 1,523 m), with hot days and cool nights in summer and moderate winter days with cold nights. Winter maximum daytime temperature is around 20 °C dropping to a crisp average minimum of 5 °C, and warm to hot summers (October to April) tempered by late-afternoon showers often accompanied by extreme thunder and lightning. Hailstorms are not uncommon, but a serious hailstorm has not happened for many years. Soil structure in the area is mostly coarse, with pores of 0.7 mm to 2 mm in size. Soil in ARC-AP is loamy with mostly silt soil, which is richer in nutrients and is able to retain water for long periods.

Expended water was collected as a composite sample from the final wastewater effluent. The effluent contains wastewaters from the residential surrounding suburbs and a research farm. Water sample (1000 mL) was collected in glass bottle cleaned with dilute Nitric acid (HNO3) and detergent followed by rinsing with sterile distilled water. The sample was transported on ice to the laboratory and processed.

2.2   Collection of Samples and Extraction of Environmental DNA

The sample was collected according to the protocol described in Gonzalez-Martinez et al. [29] with modifications. 76 subsets of wastewater effluent samples were collected to form a composite sample. The sample was kept at 4 °C until they reached the laboratory. Then mixed sample of about 5000 mL was centrifuged in batches at 3500 rpm during 10 min at room temperature until a mixture of 1000 mL was achieved. Further centrifugation under the same condition was done for separation of biomass and water. The pelleted biomass was kept at −20 °C for subsequent DNA extraction procedure. The DNA extraction was done using the FastDNA SPIN Kit for water (MP Biomedicals, Solon, OH, USA) and the FastPrep apparatus following the instructions given by the manufacturer. The integrity of the DNA was determined using Thermo NanoDrop 1000 Spectrophotometer. The DNA pools were then kept at −20 °C and sent to ARC- Biotechnology Platform Laboratory for sequencing.

2.3   Illumina MiSeq Sequencing

The DNA pools were subjected to sequencing procedure using the Illumina MiSeq apparatus and the Illumina MiSeq Reagent v3. This protocol was done three times for each DNA pool to independently identify Bacteria, Archaea, Viruses and Fungi OTUs. The primer pairs 28FF-519R (5'-GAGTTTGATCNTGGCTCAG-3' and 5'-GTNTT-ACNGCGGCKGCTG-3'), 519F-1041R (5'-CAGCMGCCGCGGTAA-3' and 5'-GGC-CATGCACCWCCTCTC-3') and ITS1F-ITS2 (5'-CTTGGTCATTTAGAGGAAG-TAA-3' and 5'-GCTGCGTTCTTCAT CGATGC-3') were chosen for the amplification of the hypervariable regions V1-V3 of 16S rRNA gene of Bacteria, the hypervariable regions V4-V6 of 16S rRNA gene of Archaea, and ITS region of Fungi, respectively. The raw reads of this study has been deposited in the Sequence Read Archive (SRA) under the accession number SRP159184.

2.4   Data Analysis

MetaPhlAn2 was utilised to analyse and identify the microbial composition of the microbial communities as depicted in the sequenced wastewater effluent data [30]. The reads were analysed and combined to form a merged abundance table. This table was edited and viewed using LibreOffice Calc. A heatmap showing the abundance profiles of the most abundant 50 microbes was generated using Hclust2. A cladogram showing taxonomic relatedness was captured using GraphlAn [31]. This was done by rendering trees and annotating them with microbial names and relative abundances. Several charts, showing specific comparisons based on various clade were generated using Krona [32].

3   Results and Discussions

3.1   Microbial Composition of Wastewater Sample

The analysis of the microbial samples showed that the environment was dominated by Bacteria (99.88%). However, the sample also showed that Archaea (0.07%) and Viruses (0.05%) were present in very small proportion. Figures 1 shows graphical illustrations of the wastewater sample. This illustration was generated using krona with the microbes' distributions shown in Table 1 for clarity.

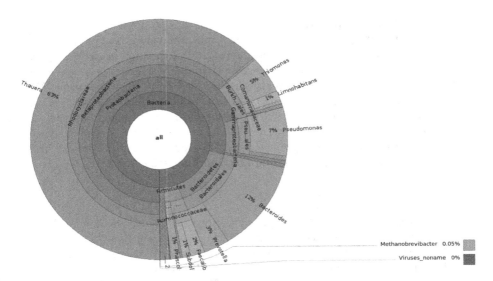

Fig. 1. Microbial composition of the wastewater sample

Table 1. Microbial abundance by Kingdom based on MetaPhlAn2 Analysis

ID	Set 1(%)	Set 2(%)	Set 3(%)	Set 4(%)	Set 5(%)	Set 6(%)
k_Bacteria	99.84544	99.88687	99.89423	99.94428	99.46088	100
k_Archaea	0.15456	0.11313	0	0.05572	0.15536	0
k_Viruses	0	0	0.10577	0	0.38376	0

Upon further analysis, it was shown that about 5 phyla were present within the Bacterial population namely: Actinobacteria (0.5%), Bacteroidetes (9%), Firmicutes (3.3%), Proteobacteria (86.6%) and Spirochaetes (0.6%). Proteobacteria is a major phylum containing wide variety of pathogens, such as Escherichia, Salmonella, Vibrio, Helicobacter and etc. [33]. Their large detection in this analysis is a pointer to the fact that large numbers of pathogens in the effluents are being released from the farm site to the environment.

A total of 40 different genera were identified with the genera Thauera making up 74% of the entire population (Table 2). This comes as no surprise as Thauera is a denitrifying bacteria playing a crucial role in the wastewater ecosystem [34]. Thauera plays an important role in the removal of nitrogen nitrate and other aromatic compounds [35]. For this reason, Thauera is usually detected in most wastewater treatment samples [36, 37]. The composition of the 10 most abundant species in the effluent is shown in Table 2.

Another notable genera worth noting among the bacterial population is the genera Thiomona making about 5% of the bacterial population (Fig. 2). Studies by Ryoki et al. [38] have shown that thiomona is useful for hydrogen sulfide removal through oxidation [38, 39].

Table 2. Microbial composition of the 10 most abundant species

ID	Set 1(%)	Set 2(%)	Set 3(%)	Set 4(%)	Set 5(%)	Set 6(%)	Set 7(%)	Set 8(%)	Set 9(%)
Thauera unclassified	74.78797	73.7054	74.45052	72.75747	73.01384	75.59471	73.71421	73.76752	71.86082
Bacteroides propionicifaciens	6.24441	5.90474	6.0248	6.35372	6.09813	7.02906	8.09416	6.17204	6.34544
Thiomonas unclassified	3.78366	5.52679	5.62429	6.79345	5.4099	3.39223	10.7756	5.32803	5.73392
Pseudomonas caeni	3.53141	3.66291	3.75899	3.76593	3.62164	3.84445	2.92225	3.6477	3.78147
Limnohabitans unclassified	1.43182	1.2971	1.39118	1.27389	1.29692	1.23933	0.32113	1.35369	1.56769
Prevotella copri	1.42776	1.22748	1.5482	1.27939	1.37822	1.34788	1.26741	1.14799	1.29424
Subdoligranulum unclassified	1.32637	0.83955	1.02524	1.28996	0.86505	1.44284	0	0.7471	1.19192
Faecalibacterium prausnitzii	0.98493	0.96129	1.01895	0.95755	1.03276	0.91465	0	0.83229	1.03483
Acinetobacter unclassified	0.88567	0.93844	0.78184	0.34917	1.18873	0.61385	0	0.67223	0.68754
Phascolarctobacterium succinatutens	0.81715	0.77118	0.58456	0.59888	0.61381	0.56615	0	0.66996	0.70073

The viruses present in the effluent are composed majorly of the genus Siphoviridae and Gammaretrovirus. The Kingdom Archaea was found to consist of only the genera Methanobrevibacter. To aid further analysis, a heat map was generated using Hclust2 (Fig. 3). This made it possible to view the various organisms and their relative abundances in one glance. With the map showing the degree of abundance, it became easy to trace the most populous species present. The map was shortened to the 50 most abundant species to allow for visibility. The results are the same as those earlier presented - the Genus Thauera occurred in largest amounts across all the data sets. Thiomonas and Bacteroides_propionicifaciens also made significant appearances.

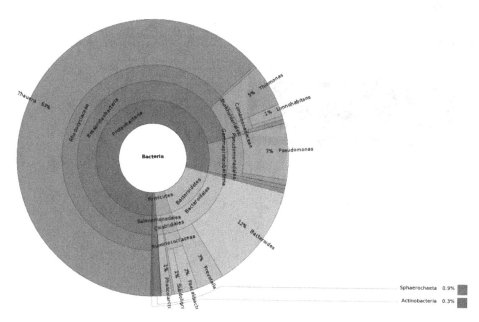

Fig. 2. Microbial composition of Bacteria present using Krona

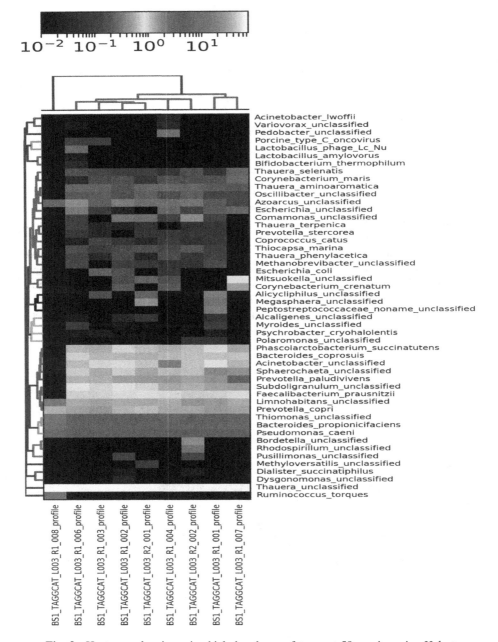

Fig. 3. Heat map showing microbial abundance of topmost 50 species using Hclust

The detection of Acinetobacter sp., Escherichia coli, Pseudomonas sp, etc. are significant as strains of these bacteria have been listed with critical criteria for research and development by WHO [40]. Besides that, Corynebacterium crenatum detected in these effluent samples shows there would be active fermentation and fouling of the environment where these effluent is introduced, since Corynebacterium crenatum is notorious for outstanding fermentation [41].

WHO [28] clearly highlights the health risks to humans that are associated with the release of wastewater effluent containing pathogens for reuse on farmers, farm workers, residents in the neighbouring households where the wastewater is reused, as well as the sellers and consumers of crops irrigated with the wastewater. Both WHO [28] and United States Environmental Protection Agency (USEPA) [42] disallow the presence of pathogens in wastewater that are meant for reuse. These pathogens could be taken up by crops, get internalized and then passed to consumers of such crops [13, 43].

4 Conclusion

Using metagenomic analysis, this study was able to provide insight into the microbial community present in the wastewater effluent. We were also able to use various bioinformatics tools to provide graphical illustrations that aided our interpretation. The results revealed that a large percentage of the microorganisms present were bacteria and we were able to view their diversity. A huge part of this bacterial presence was directly involved in the wastewater ecology and had major roles in the breakdown of compounds present. The presence of potential opportunistic pathogens showed that the effluents did not meet both WHO and USEPA guidelines for final effluents, which may be reused in the environment, especially when the sampling location is surrounded by highly cultivated farmlands. This is because the presence of these pathogens may constitute high health risks to exposed individuals.

Acknowledgements. The authors would like to acknowledge the Agricultural Research Council- Animal Production, Irene, South Africa for the provision of the research grants to carry out this study.

References

1. Chen, L., et al.: VFDB: a reference database for bacterial virulence factors. Nucleic Acids Res. 33(Suppl_1), D325–D328 (2005)
2. Okoh, A.I., Odjadjare, E.E., Igbinosa, E.O., Osode, A.N.: Wastewater treatment plants as a source of microbial pathogens in receiving watersheds. Afr. J. Biotech. 6(25), 2932–2944 (2007)
3. Bates, A.J.: Water as consumed and its impact on the consumer—do we understand the variables? Food Chem. Toxicol. 38, S29–S36 (2000)
4. Olsen, J.S., et al.: Alternative routes for dissemination of Legionella pneumophila causing three outbreaks in Norway. Environ. Sci. Technol. 44(22), 8712–8717 (2010)

5. Stevik, T.K., Aa, K., Ausland, G., Hanssen, J.F.: Retention and removal of pathogenic bacteria in wastewater percolating through porous media: a review. Water Res. **38**(6), 1355–1367 (2004)
6. Adegoke, A.A., Okoh, A.I.: Species diversity and antibiotic resistance properties of Staphylococcus of farm animal origin in Nkonkobe Municipality, South Africa. Folia Microbiol. **59**(2), 133–140 (2014)
7. Stenström, T.A., Okoh, A.I., Adegoke, A.A.: Antibiogram of environmental isolates of Acinetobacter calcoaceticus from Nkonkobe Municipality, South Africa. Fresenius Environ. Bull. **25**, 3059–3065 (2016)
8. Cai, L., Zhang, T.: Detecting human bacterial pathogens in wastewater treatment plants by a high-throughput shotgun sequencing technique. Environ. Sci. Technol. **47**(10), 5433–5441 (2013)
9. Trout, D., Mueller, C., Venczel, L., Krake, A.: Evaluation of occupational transmission of hepatitis A virus among wastewater workers. J. Occup. Environ. Med. **42**(1), 83 (2000)
10. Hellmér, M., et al.: Detection of pathogenic viruses in sewage gave early warning on hepatitis A and norovirus outbreaks. Appl. Environ. Microbiol. **80**(21), 6771–6781 (2014)
11. Gilbride, K.A., Lee, D.Y., Beaudette, L.A.: Molecular techniques in wastewater: understanding microbial communities, detecting pathogens, and real-time process control. J. Microbiol. Methods **66**(1), 1–20 (2006)
12. Wen, Q., Tutuka, C., Keegan, A., Jin, B.: Fate of pathogenic microorganisms and indicators in secondary activated sludge wastewater treatment plants. J. Environ. Manag. **90**(3), 1442–1447 (2009)
13. Adegoke, A.A., Faleye, A.C., Stenström, T.A.: Residual antibiotics, antibiotic resistant superbugs and antibiotic resistance genes in surface water catchments: public health impact. Phys. Chem. Earth **105**, 177–183 (2018)
14. Toze, S.: PCR and the detection of microbial pathogens in water and wastewater. Water Res. **33**(17), 3545–3556 (1999)
15. Lee, D.Y., Shannon, K., Beaudette, L.A.: Detection of bacterial pathogens in municipal wastewater using an oligonucleotide microarray and real-time quantitative PCR. J. Microbiol. Methods **65**(3), 453–467 (2006)
16. Lee, D.Y., Lauder, H., Cruwys, H., Falletta, P., Beaudette, L.A.: Development and application of an oligonucleotide microarray and real-time quantitative PCR for detection of wastewater bacterial pathogens. Sci. Total Environ. **398**(1-3), 203–211 (2008)
17. Bowler, C., Karl, D.M., Colwell, R.R.: Microbial oceanography in a sea of opportunity. Nature **459**(7244), 180 (2009)
18. George, I., Stenuit, B., Agathos, S., Marco, D.: Application of metagenomics to bioremediation. Metagenomics: Theory Methods Appl. **1**, 119–140 (2010)
19. Thomas, T., Gilbert, J., Meyer, F.: Metagenomics-a guide from sampling to data analysis. Microb. Inform. Exp. **2**(1), 3 (2012)
20. Mendes, L.W., Kuramae, E.E., Navarrete, A.A., Van Veen, J.A., Tsai, S.M.: Taxonomical and functional microbial community selection in soybean rhizosphere. ISME J. **8**(8), 1577 (2014)
21. Andreote, F.D., et al.: The microbiome of Brazilian mangrove sediments as revealed by metagenomics. PLoS One **7**(6), e38600 (2012)
22. Marco, D. (ed.): Metagenomics: Current Innovations and Future Trends. Horizon Scientific Press, Poole (2011)
23. Martins, L.F., et al.: Metagenomic analysis of a tropical composting operation at the São Paulo Zoo Park reveals diversity of biomass degradation functions and organisms. PLoS One **8**(4), e61928 (2013)

24. Adetiba, E., Olugbara, O.O., Taiwo, T.B.: Identification of pathogenic viruses using genomic cepstral coefficients with radial basis function neural network. In: Pillay, N., Engelbrecht, A.P., Abraham, A., du Plessis, M.C., Snášel, V., Muda, A.K. (eds.) Advances in Nature and Biologically Inspired Computing. AISC, vol. 419, pp. 281–291. Springer, Cham (2016). https://doi.org/10.1007/978-3-319-27400-3_25

25. Adetiba, E., Olugbara, O.O.: Classification of eukaryotic organisms through cepstral analysis of mitochondrial DNA. In: Mansouri, A., Nouboud, F., Chalifour, A., Mammass, D., Meunier, J., ElMoataz, A. (eds.) ICISP 2016. LNCS, vol. 9680, pp. 243–252. Springer, Cham (2016). https://doi.org/10.1007/978-3-319-33618-3_25

26. Adetiba, E., Badejo, J.A., Thakur, S., Matthews, V.O., Adebiyi, M.O., Adebiyi, E.F.: Experimental investigation of frequency chaos game representation for in silico and accurate classification of viral pathogens from genomic sequences. In: Rojas, I., Ortuño, F. (eds.) IWBBIO 2017. LNCS, vol. 10208, pp. 155–164. Springer, Cham (2017). https://doi.org/10.1007/978-3-319-56148-6_13

27. Adetiba, E., et al.: Alignment-free Z-curve genomic cepstral coefficients and machine learning for classification of viruses. In: Rojas, I., Ortuño, F. (eds.) IWBBIO 2018. LNCS, vol. 10813, pp. 290–301. Springer, Cham (2018). https://doi.org/10.1007/978-3-319-78723-7_25

28. World Health Organization-WHO: Guidelines for the Safe Use of Wastewater, Excreta and Greywater, vol. I-IV. World Health Organization, Geneva (2006)

29. Gonzalez-Martinez, A., et al.: Comparison of bacterial communities of conventional and A-stage activated sludge systems. Sci. Rep. **6**, 18786 (2016)

30. Truong, D.T., et al.: MetaPhlAn2 for enhanced metagenomic taxonomic profiling. Nat. Methods **12**(10), 902 (2015)

31. Asnicar, F., Weingart, G., Tickle, T.L., Huttenhower, C., Segata, N.: Compact graphical representation of phylogenetic data and metadata with GraPhlAn. PeerJ **3**, e1029 (2015)

32. Ondov, B.D., Bergman, N.H., Phillippy, A.M.: Interactive metagenomic visualization in a Web browser. BMC Bioinform. **12**(1), 385 (2011)

33. Sharmin, F., Wakelin, S., Huygens, F., Hargreaves, M.: Firmicutes dominate the bacterial taxa within sugar-cane processing plants. Sci. Rep. **3**, 3107 (2013)

34. Cantafio, A.W., Hagen, K.D., Lewis, G.E., Bledsoe, T.L., Nunan, K.M., Macy, J.M.: Pilot-scale selenium bioremediation of San Joaquin drainage water with Thauera selenatis. Appl. Environ. Microbiol. **62**(9), 3298–3303 (1996)

35. Liu, B., et al.: Thauera and Azoarcus as functionally important genera in a denitrifying quinoline-removal bioreactor as revealed by microbial community structure comparison. FEMS Microbiol. Ecol. **55**(2), 274–286 (2006)

36. Shanghai Jiaotong University: A specificity molecule method on function florae in industrial wastewater treatment. China 200610116628(3) (2007)

37. Jiang, X., Mingchao, M.A., Jun, L.I., Anhuai, L.U., Zhong, Z.: Bacterial diversity of active sludge in wastewater treatment plant. Earth Sci. Front. **15**(6), 163–168 (2008)

38. Ryoki, A., Hirooka, K., Nakai, Y.: Middle-thermophilic sulfur-oxidizing bacteria Thiomonas sp. RAN5 strain for hydrogen sulfide removal. J. Air Waste Manag. Assoc. **62**(1), 38–43 (2012)

39. Arsène-Ploetze, F., et al.: Structure, function, and evolution of the Thiomonas spp. genome. PLoS Genet. **6**(2), e1000859 (2010)

40. WHO: WHO priority pathogens list for R&D of new antibiotics, May 2017. http://www.who.int/medicines/publications/WHO-PPL-Short_Summary_25Feb-ET_NM_WHO.pdf

41. Kang, A., Lee, T.S.: Converting sugars to biofuels: ethanol and beyond. Bioengineering **2**(4), 184–203 (2015)

42. USEPAL: Guidelines for Water Reuse. U.S. Environmental Protection Agency Office of Wastewater Management Office of Water, Washington, D.C. (2012). https://www3.epa.gov/region1/npdes/merrimackstation/pdfs/ar/AR-1530.pdf

43. Adegoke, A.A., Madu, C.E., Aiyegoro, O.A., Stenström, T.A.: Antibiogram and beta lactamase genes among cefotaxime resistant E. coli from wastewater treatment plant. Scientific Reports (2018, in press)

Computational Genomics/Proteomics

Web-Based Application for Accurately Classifying Cancer Type from Microarray Gene Expression Data Using a Support Vector Machine (SVM) Learning Algorithm

Shrikant Pawar[1,2]([⊠])

[1] Department of Computer Science, Georgia State University,
34 Peachtree Street, Atlanta, GA 30303, USA
spawar2@gsu.edu
[2] Department of Biology, Georgia State University,
34 Peachtree Street, Atlanta, GA 30303, USA

Abstract. Intelligent optimization algorithms have been widely used to deal complex nonlinear problems. In this paper, we have developed an online tool for accurate cancer classification using a SVM (Support Vector Machine) algorithm, which can accurately predict a lung cancer type with an accuracy of approximately 95%. Based on the user specifications, we chose to write this suite in Python, HTML and based on a MySQL relational database. A Linux server supporting CGI interface hosts the application and database. The hardware requirements of suite on the server side are moderate. Bounds and ranges have also been considered and needs to be used according to the user instructions. The developed web application is easy to use, the data can be quickly entered and retrieved. It has an easy accessibility through any web browser connected to the firewall-protected network. We have provided adequate server and database security measures. Important notable advantages of this system are that it runs entirely in the web browser with no client software need, industry standard server supporting major operating systems (Windows, Linux and OSX), ability to upload external files. The developed application will help researchers to utilize machine learning tools for classifying cancer and its related genes. Availability: The application is hosted on our personal linux server and can be accessed at: http://131.96.32.330/login-system/index.php.

Keywords: Cancer · Microarray · Support Vector Machine (SVM)

1 Introduction

Many machine learning algorithms such as random forest, k-nearest neighbor, neural network, and SVM (Support Vector Machine) have been applied to the classification study of bioinformatics. SVM has distinctive advantages in handling data with high dimensionality and a small sample size. Expression levels of genes can be analyzed using Microarray experiments. In this technique RNA is isolated from different tissues which is labeled and hybridized to the arrays [1]. The expression levels of treatment sample can be compared to control sample to differences in gene expression levels

© Springer Nature Switzerland AG 2019
I. Rojas et al. (Eds.): IWBBIO 2019, LNBI 11466, pp. 149–154, 2019.
https://doi.org/10.1007/978-3-030-17935-9_14

amongst two treatments. Microarrays can be of two type: single channel where only one dye, say red or green dye is used and two channel microarray where tow dyes at same time are used, one for control (CY5) and one for treatment (CY3). Different array spots give different fluorescence intensity values for each gene which can be analyzed with different Bioconductor package in R to perform normalization and statistical analysis.

Normalization is needed prior to making biological comparisons. It is required as the RNA used for hybridization can be of different quantities or there can different labeling conditions for different probes or the expression level scanning may be biased. Normalization essentially adjusts the expression levels of probes to reference probes [1]. Figure 1 is the one channel microarray lung cancer data file after normalization. After normalization a significant change in expression levels with correction of standard deviation and variance within and between samples states the importance of normalization. Four lung cancer samples were read in R and box plots were made on files to measure standard deviation and variance on raw data and RMA normalized data, a significant variance stabilization is seen after normalization as shown in Fig. 1.

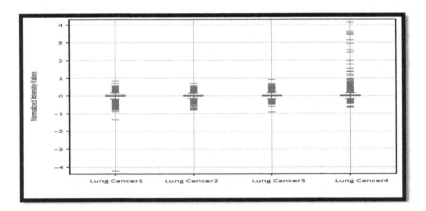

Fig. 1. Four lung cancer samples were read in R and box plots were made on files after RMA normalization to measure standard deviation and variance.

A SVM is a machine learning technique which categorizes new samples based on supervised learning of training data. A hyperplane is defined which forms a minimum distance in training sample which is used for classification [2]. Support Vector Machine makes classifications based on non-probabilistic binary linear classifier [3]. It has been shown that SVM's can significantly increase accuracy compared to traditional query refinement schemes [3]. Although, several groups have shown importance of machine learning tools in classifying big data, there still remains a constant need to develop a user friendly tool do perform its application. This paper summarizes a developed functional user tool, which can input a one channel microarray data and predict accurately the cancer type with machine learning technique (Fig. 2).

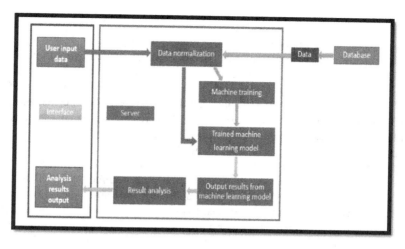

Fig. 2. Depicts the work flow for this project. User inputs a microarray data, it is downloaded in database, normalized, analyzed by SVM, and the results of which thrown on HTML page.

2   Materials and Methods

a. Data Collection

103 one channel microarray lung cancer samples (10,000 genes per channel) (GPL590) were downloaded from Gene Expression Omnibus (GEO) database. Similar number of normal lung tissue microarray data accompanied cancer data. The downloaded raw data needs normalization for variance stabilization.

b. Normalization

Normalization was tried using two techniques mas5.0 and RMA normalization. RMA normalization was performed for the samples [4]. It initially reads the affymetrix data using AFFY package in R language, we then perform Mas5.0 normalization and expression levels are extracted using exprs() function. Since it's a no logarithmic transformed data we transform that to logarithm to base 2 and export that to excel file for further analysis.

c. Machine Learning Analysis

From sklearn.svm module was imported with classification parameters as C = 1.0, cache_size = 200, class_weight = None, coef0 = 0.0, decision_function_shape = None, degree = 3, gamma = 'auto', kernel = 'rbf', max_iter = −1, probability = True, random_state = None, shrinking = True, tol = 0.001, verbose = False to perform SVM on normalized data [5]. Classification is done using support vector machines (SVMs). These are set of supervised learning methods useful with high dimensional spaces where dimensions are greater than number of samples [6]. It is also memory efficient and uses subset of training points in the decision function [6]. Initially a training set of 103 cancer samples was trained and results were stored on server space on local machine. This data will be used to perform predictions on query data.

d. Setting up a Server Space and developing a Graphical User Interface (GUI)

GUI was developed by importing module Tkinter. 200×100 dimension frame was made for uploading files from the user. Server was developed with Python using BaseHTTPServer module, other classes like re, os, Classifier, Constants, normalize were declared in main class MyHandler. Three main methods are declared, def do_GET(self) responding to a GET request, def do_POST(self) calls normalization and classifier function and def predict_data(self, test_data) responds to a GET request. Def get_html_response_after_prediction(self, result, fileName) method writes the results on HTML page after predictions have been made. BaseHTTPServer module creates and listens at the HTTP socket, *Re* module accepts regular expression, *Os* module allows to interface with operating system. Class constants defines HOST_NAME = '10.241.213.126', PORT_NUMBER = 9000 and paths for trained data. Based on the user specifications, we chose to write this suite in Python CGI and based on a MySQL database. It enables the set-up of single or a multi-users access controls. A Linux server hosts the application and database. It has been developed on a Mac OS X operating system using MySQL as the relational database management system and Python as the scripting language. The hardware requirements of suite on the server side are moderate. The server we utilized is GSU Orion with CentOS 6.7 64-bit, 6x IBM System x3850 x5, Intel Xeon Processor E7-4850, 4 CPUs (10 cores per CPU), 2.0 GHz processors, 512 GB RAM and 2 TB of scratch storage for jobs. The database has a column text-field table which is updated interactively with the user. Further, the uploading of information is standardized as certain parameters like date has to be inputted in specific format only which helps in retrieval process. Checking of data type (float, integer, text, Boolean) and dates is important before putting into database. Errors with dates or invalid parameters or wrong data type cause a halt of workflow. Bounds and ranges have also been considered and needs to be used according to the user instructions.

e. Security

We have implemented strong security measures for authentication of SQL server. Kerberos protocol uses a number of encrypted messages to authenticate SQL server and the passwords are not passed across the network. Authentication is more reliable and managing it can be reduced by leveraging active directory groups for role- based access to SQL server. The sysadmin (sa) account is vulnerable when it exits unchanged so we have disabled the sa account on the SQL server instance. We chose to give options of complex passwords for sa and all other SQL-server-specific logins on SQL server and checked in the 'Enforce password expiration' and 'Enforce password policy' options for sa. We haven't allowed to explicit grant control server permission because logins with this permission get full administrative privileges. For permissions to users, an built-in fixed server roles and database roles or creating own custom server roles and database roles can be achieved. Guest user exists in every user and system database, which is a potential security risk in a lock down environment because it allows database access to logins who don't have associated users in the database. We have restricted this access and also accesses to user and system stored procedures. Furthermore, we have used common specific TCP ports (excluding 1433 and 1434) instead of dynamic ports. SQL server browser service is only running on SQL servers, and secure SQL server error logs and registry keys using NTFS permissions are utilized as they can provide great deal of information about the SQL server instance and installation.

3 Results and Conclusions

An HTML page is made to provide options to the user for uploading data. Once the user uploads a .CEL file it calls functions in server, does machine learning and uploads results in same HTML page. A sample lung cancer .CEL file (GSM159355) was uploaded by user as a query sample and our prediction analysis gave a prediction of 94.88% to be a lung cancer type, a robust multichip analysis was performed and the results are shown in Table 1. This application is a primary skeleton to dig complex gene patterns via SVM [7–9]. Such open source skeletons with large server space, interface and algorithm implementations are currently unavailable and this is an attempt to provide it to the user in a friendly and cost effective interface, which will further be expanded to apply machine learning algorithms to understand complex gene patterns [10].

Table 1. A robust multichip analysis was performed on lung cancer chips and the prediction analysis (accuracy) results recorded as follows.

Cancer type	GSM series ID	Accuracy (%)
Lung	GSM159355	94.88
Lung	GSM1322678	93.08
Lung	GSM1322679	92.11
Lung	GSM1322680	91.05
Lung	GSM1322681	92.00
Lung	GSM1322682	93.10
Lung	GSM1322683	94.99
Lung	GSM1322684	90.11
Lung	GSM1322685	92.95

4 Future Scope

The expansion of training set is needed to improve prediction accuracies. Also increasing cancer types (breast, brain, liver etc.) can make this program usable for different groups [11]. Enhancement of the current pipeline is needed in terms of incorporation of other machine learning algorithms [12]. The current pipeline seems applicable in future to predict unknown genes involved in different cancer types using machine learning tools.

Acknowledgements. Support from the Georgia State University Information Technology Department (GSU IT) for server space is gratefully acknowledged.

Author's Contributions. SP conceived, designed the study and critically revised the manuscript. SP developed, tested the software and also did the setup of SQL database and server space.

Funding. No external funding for developing this suite has been utilized.

Competing Interests. The authors declare that they have no competing interests.

References

1. Quackenbush, J.: Microarray data normalization and transformation. Nat. Genet. 496–501 (2002). https://doi.org/10.1038/ng1032
2. OpenCV. Introduction to Support Vector Machine (2014). http://docs.opencv.org/2.4/doc/tutorials/ml/introduction_to_svm/introduction_to_svm.html
3. Ben-Hur, A., Horn, D., Siegelmann, H., Vapnik, V.: Support vector clustering. J. Mach. Learn. Res. 2, 125–137 (2001)
4. Gautier, L., Cope, L., Bolstad, B.M., Irizarry, R.A.: Affy - an R package for the analysis of Affymetrix GeneChip data at the probe level. Bioinformatics (2003)
5. Smola, A.J., Schölkopf, B.: A Tutorial on support vector regression. Stat. Comput. 14(3), 199–222 (2004)
6. Scikit-learn developers. Support Vector Machines (2014). http://scikit-learn.org/stable/modules/svm.html
7. Pawar, S., Ashraf, M., Mujawar, S., Mishra, R., Lahiri, C.: In silico identification of the indispensable quorum sensing proteins of multidrug resistant Proteus mirabilis. Front. Cell. Infect. Micro-Biol. 8, 269 (2018)
8. Ashraf, M., et al.: A side-effect free method for identifying cancer drug targets. Sci. Rep. (2018)
9. Lahiri, C., Pawar, S., Sabarinathan, R., Ashraf, M.I., Chand, Y., Chakravortty, D.: Interactome analyses of Salmonella pathogenicity islands reveal SicA indispensable for virulence. J. Theor. Biol. 363, 188–197 (2014)
10. Lahiri, C., Shrikant, P., Sabarinathan, R., Ashraf, M., Chakravortty, D.: Identifying indispensable proteins of the type III secretion systems of Salmonella enterica serovar Typhimurium strain LT2. BMC Bioinform. 13(Suppl. 12), SA10 (2012)
11. Pawar, S., Ashraf, M., Mehata, K., Lahiri, C.: Computational identification of indispensable virulence proteins of Salmonella typhi CT18. Curr. Top. Salmonella Salmonellosis (2017)
12. Pawar, S., Davis, C., Rinehart, C.: Statistical analysis of microarray gene expression data from a mouse model of toxoplasmosis. BMC Bioinform. 12(Suppl. 7), SA19 (2011)

Associating Protein Domains with Biological Functions: A Tripartite Network Approach

Elena Rojano[1], James Richard Perkins[2(✉)], Ian Sillitoe[3], Christine Orengo[3], Juan Antonio García Ranea[1,2], and Pedro Seoane[2]

[1] Department of Molecular Biology and Biochemistry, University of Malaga, Bulevar Louis Pasteur, 31, 29010 Malaga, Spain
{elenarojano,ranea}@uma.com
[2] CIBER de Enfermedades Raras, ISCIII, Av. Monforte de Lemos, 3-5. Pabellon 11. Planta 0, 28029 Madrid, Spain
{jimrperkins,seoanezonjic}@uma.com
[3] Department of Structural and Molecular Biology, University College London, Gower Street, London WC1E 6BT, UK
{i.sillitoe,c.orengo}@ucl.uc.uk

Abstract. Protein domains are key determinants of protein function. However, a large number of domains have no recorded functional annotation. These domains of unknown function (DUFs) are a recognised problem and efforts have been made to remedy this situation, including the use of data such as structural and sequence similarity and annotation data such as that of Gene Ontology (GO) and The Enzyme Commission.

Here, we present a new approach based on tripartite network analysis to assign functional terms to DUFs. We combine functional annotation at the protein level, taken from GO, KEGG, Reactome and UniPathway, with structural domain annotation, taken from the CATH-Gene3D resource. We validate our method using 10-fold cross-validation and find it performs well when assigning annotation from the UniPathway, Reactome and GO resources, but less well for KEGG. We also explored using a finer functional subclassification of CATH superfamilies (FunFams) but these families were found to be too specific in this context.

Keywords: Network analysis · FunFam · CATH · Protein structure · Structural domains · Functional annotation

1 Introduction

Protein domains are conserved regions of proteins with specific functions that tend to fold into independent three-dimensional units. They can bind to other

J. R. Perkins is a postdoctoral researcher within the CIBERER research network for rare diseases.

I. Rojas et al. (Eds.): IWBBIO 2019, LNBI 11466, pp. 155–164, 2019.
https://doi.org/10.1007/978-3-030-17935-9_15

molecules, play a role in metabolism or regulation, help maintain the structure of cells, or act as a binding sites for various types of ligands. Taken together, the different domains confer the functionality of a protein. A single protein can contain multiple domains which can be similar or distinct, moreover, the same domain can occur in multiple proteins that have very different functions [1]. However, there are many domains for which the functions and/or the biological processes in which they are involved are unknown, referred to as domains of unknown function (DUFs) [2].

Many efforts have been made to separate proteins into their constituent domains and classify them based on sequence and structure [3]. Resources for this purpose first predict their location in proteins and place them into superfamilies, in accordance with conserved evolutionary features. Then, as these superfamilies may contain domains with different functions, they can be further clustered into functionally coherent subgroups (families), for example by using proteins that contain these domains and that have functional annotations. This procedure is performed under the assumption that proteins with similar structural domains should have similar functional annotations.

One of the most well-known protein domain classification systems, the CATH resource, is based on structural information of proteins [4]. It uses a hierarchical system to classify the structural domains of proteins in the Protein Data Bank (PDB) [5] into increasingly specific categories, using a combination of computational tools and expert knowledge. In brief, the deposited protein is split in its structural domains, which are then classified using a hierarchical system [6,7].

CATH includes four different levels of classification from more general to more specific: class, architecture, topology and homology [4]. The most general class level classifies domains according to secondary structure, for example alpha-helices or beta-sheets. The next class level in the hierarchy is architecture, which is based on 3D packing of the secondary structures. Next, how these structures are further organised gives rise to the topology layer. Finally, if domains are considered to be evolutionary related, i.e. homologous, they are classified at the most-specific level, known as homologous superfamily [4]. However, these superfamilies can contain thousand of domains, making it difficult to assign a specific function. To solve this, the CATH team have developed multiple algorithms that aim to cluster superfamilies into functional subfamilies, also known as the FunFams [8]. Briefly, for each superfamily this functional classifier establishes relationships between the domains that are part of it, according to sequence similarity to domains with assigned function [6].

Here we propose a novel network-based analysis method to assign annotations to DUFs. We use a tripartite network to calculate the association degree between protein domains and functional annotations, based on the number of proteins they share and the specificity of the connections. This procedure has been performed for eight different networks, built using annotations from GO [9], KEGG [10], Reactome [11] and UniPathway [12] for CATH homologous superfamilies and for FunFams, using the NetAnalyzer software [13]. We compared the results for the different networks 10-fold cross-validation procedure.

2   Methods

2.1   Domains Data and Protein Annotations Sources

The aim of this work was to annotate protein domains with biological functions through the use of tripartite networks linking domains, proteins and functions. The construction of these networks involved combining protein-domain annotation data and protein-function annotation data.

We obtained all available human proteins from UniProt (Version 278, Swiss-Prot Release 2018_11), including Swiss-Prot and TrEMBL and discarding all the fragments belonging to incomplete proteins. Of these filtered proteins, we obtained their annotations from GO, KEGG, Reactome and UniPathway.

We then obtained domain annotation for these proteins from the CATH database where available. Domain annotation was obtained at the structural level, based on membership of homologous superfamilies, and add the functional level based on membership of the FunFams, which are sub-classifications of homologous superfamilies based on function [14]. Given that the same gene can give rise to multiple protein isoforms, many of which will contain similar domains, we chose to use the union of all annotated domains available for each gene/protein when building the network.

2.2   Domain-Protein-Function Tripartite Network Analysis and Validation

We constructed eight tripartite networks with three types of nodes: proteins, domains and functional annotations by combining the protein-function and protein-domain pairs. Four of these networks relate homologous superfamilies to annotations in GO, KEGG, Reactome and UniPathway; the other four relate FunFams with the same annotations. The tripartite network model representation is shown in Fig. 1.

The eight networks were analysed using the network analysis tool developed in our research group, NetAnalyzer [13], using to calculate the association between nodes the hypergeometric index. This index has been used previously to perform annotate structural protein domains in GO [15]. The hypergeometric index calculates the log-transformed probability of finding an equal or greater interaction overlap than the one observed between a given domain and functional annotation [13,16]:

$$H_{AB} = -log \sum_{i=|N(A)\cap N(B)|}^{min(|N(A),N(B)|)} \frac{\binom{|N(A)|}{i} \cdot \binom{n_y - |N(A)|}{|N(B)|-i}}{\binom{n_y}{|N(B)|}}$$

Where $N(A)$ is the number of domain nodes, $N(B)$ is the number of functionally annotation nodes, n_y corresponds to the total number of proteins and i is the number of shared protein nodes between $N(A)$ and $N(B)$.

To validate the associations of our network analysis method, we used a 10-fold cross-validation for each of the eight networks and calculated their PR values.

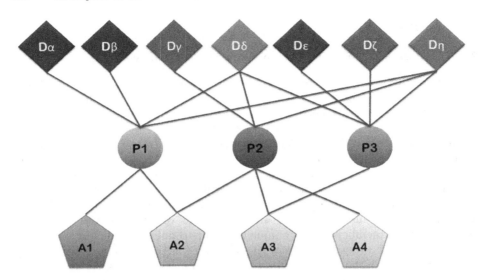

Fig. 1. Tripartite network model that represents the domain nodes (diamonds, D$\alpha - \eta$), which are either FunFams or homologous superfamilies, the protein nodes (circles, P1-3) and the annotation nodes (pentagons, A1-4). The idea is to associate domains with function based on the number of proteins that link them, relative to how many other domains and functions these proteins connect.

In this way, we could determine which annotation (GO, KEGG, Reactome or UniPathway) - domain (FunFams or superfamilies) relations were more consistent. For this, we removed a tenth of the domain-protein-function tuples from the network and assumed these to be the ground truth, i.e. we assumed the domain-function connections in these to be certain. We then calculated association values between domains and functions for the remaining 90% of the tuples and used the ground truth domain-function connections to validate these associations and produce precision-recall curves. This validation procedure has been described previously in [13].

3   Results

3.1   Annotation Coverage

We obtained a total of 96,620 proteins from UniProt, of which 75,222 (77.86%) are annotated in one or more of the functional annotation databases. Of these, we could find domain annotation for 38,616 proteins, of which 16,076 had an annotation in GO, 13,196 in KEGG, 8,174 in Reactome, and 1,053 in UniPathway. These values are shown in Fig. 2 and the overlap between annotated proteins is shown in Fig. 3.

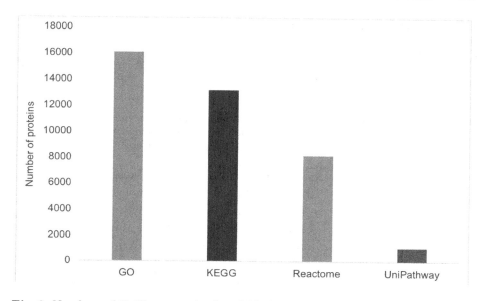

Fig. 2. Numbers of UniProt proteins for which domain information is available, annotated in GO (41.63%), KEGG (34.17%), Reactome (21.16%) and UniPathway (2.72%).

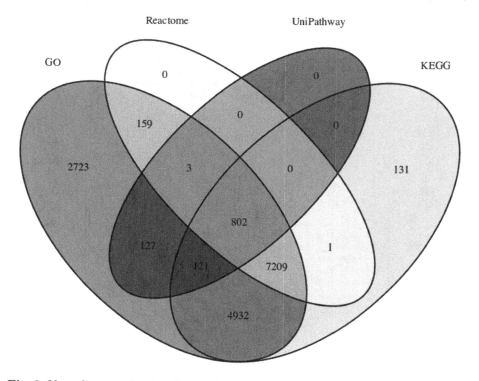

Fig. 3. Venn diagram showing the overlap in terms of annotation between GO, KEGG, Reactome and/or UniPathay.

3.2   Tripartite Network Analysis

The protein-domain and protein-function annotations were combined to build
8 distinct networks following the procedure represented in Fig. 1. Full details of
these networks are described in Table 1.

Table 1. Distinct proteins and annotations for each annotation source, the number of
distinct domains in each of the networks, FunFams and Superfamilies, and the average
number of domains per protein. Dom/prot: domains per protein.

Type	Proteins	Annotations	FunFams		Superfamilies	
			Count	Dom/prot	Count	Dom/prot
GO	16,076	16,438	11,199	1.62	1,220	1.34
KEGG	13,196	12,475	10,928	1.86	1,210	1.44
Reactome	8,174	1,715	7,915	1.92	1,110	1.51
UniPathway	1,053	395	1,134	1.76	333	1.58

NetAnalyzer [13] was then used to calculate associations between the domains
and functions in these networks by calculating the hypergeometric index. We
assessed how well we were able to predict function for the different domain

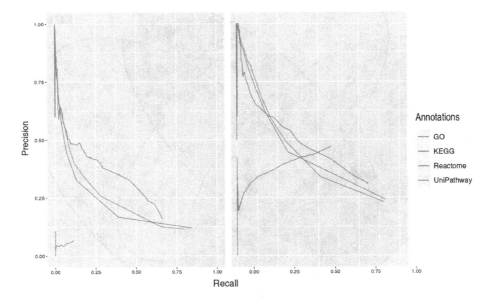

Fig. 4. Precision and recall curves that compare the different types of annotations (GO,
KEGG, Reactome and UniPathway) associated with FunFams (left) and superfamilies
(right).

Table 2. Summary of the five highest domain-function associations calculated between superfamily-Reactome and FunFam-Reactome functional annotations. Assoc. val.: association value. P-val.: P-value.

Annotation (Reactome ID)	Domain (Superfamily ID)	Assoc. val.	P-val.
Immunoregulatory interactions between a Lymphoid and a non-Lymphoid cell (198933)	Immunoglobulins (2.60.40.10)	163.84	1.44e−164
Rho GTPase cycle (194840)	GTPase -GAP domain (1.10.555.10)	118.89	1.28e−119
Nuclear Receptor transcription pathway (383280)	Retinoid X Receptor (1.10.565.10)	118.44	3.73e−119
NRAGE signals death through JNK (193648)	DBL homology domain (DH-domain) (1.20.900.10)	112.33	4.67e−113
Nuclear Receptor transcription pathway (383280)	Erythroid Transcription Factor GATA-1, subunit A (3.30.50.10)	103.13	7.41e−104
Annotation (Reactome ID)	Domain (FunFam ID)	Assoc. val.	P-val.
Generic Transcription Pathway (212436)	Early growth response (23446)	219.02	9.54e−220
Keratinization(6805567)	Keratin, type II cytoskeletal I (18190)	104.83	1.47e−105
Formation of the cornified envelope (6809371)	Keratin, type II cytoskeletal I (18190)	93.11	7.76e−94
Fc epsilon receptor (FCERI) signaling (2454202)	Light chain variable region (136068)	79.44	3.63e−80
CD22 mediated BCR regulation (5690714)	Light chain variable region (136068)	79.09	8.12e−80

types and different annotation sources using ten-fold cross-validation, by removing a tenth of the data, considering the domain-function links in this tenth as the ground truth, and calculating precision recall curve based on the association scores of these left out links compared to the remaining association values. The curves are shown in Fig. 4. We see that in all cases, GO, Reactome and UniPathway show high precision at low recall, which then drops quickly in the case of FunFams, and less quickly for homologous superfamilies. Interestingly, performance is broadly similar for all three. KEGG shows low precision at low recall, somewhat surprisingly this increases gradually as recall increases.

The top associations between superfamilies/FunFams and Reactome functional annotations are shown in Table 2. Despite giving marginally the best performance, we have not focused on UniPathway due to the low number of associations in comparison with superfamily-GO (341,045) and superfamily-Reactome (20,092), and because the resource is no longer supported. Results are sorted in accordance with the most significant P-value, calculated as follows: 10^{-HyI}. Due to the extension of this article, we excluded from this summary the rest of domain-function associations calculated with GO and KEGG.

4   Discussion and Conclusions

In this work we have used a novel approach based on tripartite network analysis to infer the function of domains of unknown function (DUFs), based on the function of the proteins in which they are found, in terms of GO, KEGG, Reactome and UniPathway annotation. DUFs remain a key problem, as most efforts to assign function focussing on whole proteins, rather than their constituent domains [2].

We have focussed on homologous superfamily domains taken from CATH, and FunFams, which further subdivide homologous superfamilies into functionally coherent subfamilies.

We see high precision for GO, UniPathway and Reactome, especially at low recall, showing that high hypergeometric index based values between domains and functions are likely to give reliable annotation, although further experimental validation of this method on domains of unknown function will be necessary to fully validate these results. The method appears to perform slightly better for UniPathway than for other annotation sources, however as shown in Table 1, this method is able to provide annotation for a much smaller number of proteins.

Notably, performance was particularly poor for KEGG domains, remarkably these results showed an increase in precision alongside an increase in recall. This suggests that as the domain-function association value threshold was decreased, that is we consider a greater number of domains as correctly annotated, we actually find a higher proportion of true positives. This is somewhat counterintuitive - it appears that our domain-function prediction improves as we reduce the threshold, suggesting a greater association score is actually less likely to find true associations. This may be due to the type of data stored in KEGG - these are generally metabolic pathways, which are likely to require many different types of domains in order to continually modify substrate and product. This is in contrast to Reactome, which includes a much wider range of pathways, including for example signalling pathways which may be likely to include more similar domains.

The other clear pattern that can be seen in the precision-recall graphs in Fig. 4 is that the precision for the FunFams drops off much more quickly than for the homologous superfamilies. In this study, we chose to use the FunFams in their most specific sense. Therefore, it is to be expected that it would be more difficult as the level of annotation is very specific. Moreover, as the FunFams are essentially subsets of homologous superfamilies, there are a much larger number of them. Taken together, this makes the prediction problem harder: you are trying to predict more specific function for a much larger number of domains. In the future, we will explore using them in a broader sense as in the work of [14].

Further analysis of the network results shown that the strongest association values between a Reactome term (R-HSA-212436) and a FunFam (3.30.160.60-ff-23446) is 219.685. This Reactome term corresponds to "Generic Transcription Pathway". By its part, the FunFam 3.30.160.60-ff-23446 appears in 474 different CATH proteins in the superfamily "Classic Zinc Finger". This high association value for both terms have a biological sense: most of genes that code for zinc finger-binding domains have been enriched in the GO term transcription cofactor activity, amongst other functions [17].

GO has both inferred *in silico* and experimentally-supported annotations of molecular functions, cellular components and biological processes [9]. Future work would further dissect the GO annotations, and assess performance separately for molecular function, biological process, and cellular component annotations. It is tempting to speculate that molecular function might be show an improved performance, as similar domains tend to perform similar functions. Indeed, previous work annotating domains with function has tended to focus on Molecular Function GO categories [15].

Acknowledgements. The authors would like to thank the team of Prof. Christine Orengo for their technical support during the realization of this work. The authors thank the Supercomputing and Bioinnovation Center (SCBI) of the University of Malaga for their provision of computational resources and technical support (http://www.scbi.uma.es/site).

Fundings. This work was supported by The Spanish Ministry of Economy and Competitiveness with European Regional Development Fund [SAF2016-78041-C2-1-R], the Andalusian Government with European Regional Development Fund [CTS-486] and the University of Malaga (Ayudas del I Plan Propio). The CIBERER is an initiative from the Institute of Health Carlos III. James Richard Perkins holds a research grant from the Andalusian Government (Fundacion Progreso y Salud)[PI-0075-2017]. Elena Rojano is a researcher from the Plan de Formacion de Personal Investigador (FPI) supported by the Andalusian Government.

References

1. Ponting, C.P., Russell, R.R.: The natural history of protein domains. Ann. Rev. Biophys. Biomol. Struct. **31**(1), 45–71 (2002)
2. Bateman, A., Coggill, P., Finn, R.D.: DUFs: families in search of function. Acta Crystallogr. Section F: Struct. Biol. Crystallization Commun. **66**(10), 1148–1152 (2010)
3. Dawson, N., Sillitoe, I., Marsden, R.L., Orengo, C.A.: The classification of protein domains. In: Methods in Molecular Biology, pp. 137–164 (2017)
4. Sillitoe, I., et al.: CATH: comprehensive structural and functional annotations for genome sequences. Nucleic Acids Res. **43**(Database issue), D376–D381 (2015)
5. Rose, P.W., et al.: The RCSB protein data bank: integrative view of protein, gene and 3D structural information. Nucleic Acids Res. **45**(D1), D271–D281 (2017)
6. Dawson, N.L., et al.: CATH: an expanded resource to predict protein function through structure and sequence. Nucleic Acids Res. **45**(D1), D289–D295 (2017)
7. Lewis, T.E., et al.: Gene3D: extensive prediction of globular domains in proteins. Nucleic Acids Res. **46**(D1), D435–D439 (2018)
8. Rentzsch, R., Orengo, C.A.: Protein function prediction using domain families. BMC Bioinform. **14**(Suppl. 3), 1–14 (2013)
9. Carbon, S., et al.: Expansion of the gene ontology knowledgebase and resources: the gene ontology consortium. Nucleic Acids Res. **45**(Database issue), D331–D338 (2017)
10. Ogata, H., Goto, S., Sato, K., Fujibuchi, W., Bono, H., Kanehisa, M.: KEGG: Kyoto encyclopedia of genes and genomes. **27**(1), 29–34 (1999)

11. Fabregat, A., et al.: The Reactome pathway knowledgebase. Nucleic Acids Res. **44**(Database issue), D481–D487 (2018)
12. Morgat, A., et al.: UniPathway: a resource for the exploration and annotation of metabolic pathways. Nucleic Acids Res. **40**(Database issue), D761–D769 (2012)
13. Rojano, E., Seoane, P., Bueno-Amoros, A., Perkins, J.R., Garcia-Ranea, J.A.: Revealing the relationship between human genome regions and pathological phenotypes through network analysis. In: Rojas, I., Ortuño, F. (eds.) IWBBIO 2017. LNCS, vol. 10208, pp. 197–207. Springer, Cham (2017). https://doi.org/10.1007/978-3-319-56148-6_17
14. Das, S., Lee, D., Sillitoe, I., Dawson, N.L., Lees, J.G., Orengo, C.A.: Functional classification of CATH superfamilies: a domain-based approach for protein function annotation. Bioinformatics **31**(21), 3460–3467 (2015)
15. Lopez, D., Pazos, F.: Gene ontology functional annotations at the structural domain level. Proteins: Struct. Funct. Bioinform. **76**(3), 598–607 (2009)
16. Bass, J.I., Diallo, A., Nelson, J., Soto, J.M., Myers, C.L., Walhout, A.J.: Using networks to measure similarity between genes: association index selection. Nat. Methods **10**(12), 1169–1176 (2013)
17. Cassandri, M., et al.: Zinc-finger proteins in health and disease. Cell Death Discov. **3**, 17071 (2017)

Interaction of ZIKV NS5 and STAT2 Explored by Molecular Modeling, Docking, and Simulations Studies

Gerardo Armijos-Capa[1,2], Paúl Pozo-Guerrón[1,2], F. Javier Torres[1,2], and Miguel M. Méndez[1,2(✉)]

[1] Instituto de Simulación Computacional (ISC-USFQ),
Universidad San Francisco de Quito,
Diego de Robles y Vía Interoceánica, Quito 17-1200-841, Ecuador
mmendez@usfq.edu.ec
[2] Departamento de Ingeniería Química,
Grupo de Química Computacional y Teórica (QCT-USFQ),
Universidad San Francisco de Quito,
Diego de Robles y Vía Interoceánica, Quito 17-1200-841, Ecuador

Abstract. ZIKV NS5 has been associated with inhibition of type I IFN during the host antiviral response. The protein-protein interaction may promote the proteasomal degradation of STAT2, although the entire mechanism is still unknown. In this study, a three-dimensional model of the full STAT2 protein (C-score = −0.62) was validated. Likewise, the top scored docked complex NS5-STAT2 is presented among several other models; the top model shows a total stabilizing energy for the complex of $-77.942 \, \mathrm{kcal \, mol^{-1}}$ and a Gibbs binding free energy of $-4.30 \, \mathrm{kcal \, mol^{-1}}$. The analysis of the complex has revealed that the interaction is limited to three domains known as N-terminal from STAT2 and Mtase-Thumb from NS5; both located in the ordered regions of these proteins. Key residues involved in the interaction interface that showed the highest frequency among the models are stabilized by electrostatic interactions, hydrophobic interactions, salt bridges, and ionic interactions. Therefore, our findings support the experimental preliminaries observations reported in the literature and present additional structural characterization that will help in the drug design efforts against ZIKV NS5.

Keywords: Zika virus · NS5 · STAT2 · Molecular dynamics · Docking

1 Introduction

Since 2015, the outbreak of Zika virus (ZIKV) in Central and South America has become a worldwide health concern. ZIKV infection is associated with congenital diseases such as microcephaly and rare but severe complications in adults such as

Supported by Universidad San Francisco de Quito.

I. Rojas et al. (Eds.): IWBBIO 2019, LNBI 11466, pp. 165–176, 2019.
https://doi.org/10.1007/978-3-030-17935-9_16

the Guillain-Barré syndrome [1,2]. Zika virus is a mosquito-borne flavivirus that has a genome of single-strand positive RNA of 10 kb that encodes ten proteins namely the envelope (E), membrane precursor (PrM), and capsid (C) which contribute to the viral particles. It also encodes for seven nonstructural (NS) proteins (NS1, NS2A, NS2B, NS3, NS4A, NS4B, and NS5) which contribute to viral replication [2]. Among NS proteins, NS5 is the largest Zika protein with a weight of ~103 kDa and consists in two principal domains. The first one is known as the methyltransferase (Mtase) which is related to the decrease of host innate immune response and promotes the translation of the polyprotein through the addition of 5' RNA cap structure. The second one is the RNA-dependent RNA polymerase (RdRp) required for the initiation and elongation of RNA synthesis [3]. The viral response at human cells is set up by intracellular pattern recognition receptors (PRRs) proteins. They recognize viral pathogen marks such as viral RNA, DNA or protein in the type I interferon (IFN) signaling. This recognition induces activation of innate immune signaling, leading to the up-regulation of IFN-stimulated genes (ISGs). Flaviviruses such as West Nile virus (WNY), Dengue virus (DENV) or Yellow Fever virus (YFV) use various strategies of IFN antagonism [4]. However, flaviviruses share replication strategies based on the formation of a polyprotein which can inhibit transcriptional activation of IFNs and ISGs during virus infection [5–7]. In this vein, NS5 is considered as a potent and specific antagonist of type I IFN signaling [4]. ZIKV NS5 has been associated with inhibition of type I IFN during the host antiviral response, because it may promote the proteasomal degradation of STAT2. This role has been substantiated by different experimental studies which have determined that strains of ZIKV antagonize type I IFN where NS5 reduces the STAT2 level and prevents the translocation of STAT2 from the cytoplasm to nuclei in immunoprecipitation assays in 293 T cells [4]. Consequently, there is inhibition of the IFN induction of the ISG [4]. In others test, STAT2 levels have been compared in cells expressing each of the two functional domains of NS5. The comparison has shown that with only the Mtase domain expressed, the levels of STAT2 degradation were higher. While with only RdRp domain expressed, the levels of STAT2 degradation were negligible [7]. On the other hand, the Mtase activity in full-length NS5 has not shown to be necessary for the STAT2 degradation, suggesting that other regions of the protein NS5 may contribute to degradation mechanism [7]. The interaction with other proteins in the pathway of type I IFN were reviewed through other essays. It has shown that reduced STAT2 levels associated to ZIKV infection are independent of the presence of proteins such as ubiquitin ligase UBR4 or ubiquitin-specific protease USP 18; hence the interaction between STAT2-NS5 is directly related with degradation of STAT2, although the complete mechanisms is still unknown [4–7]. The combination of different computational tools such as molecular modeling, molecular docking, and molecular dynamic (MD) simulation permit to study a binary complex contributing to provide a clear protein-protein binding mechanism that could lead to significant advances in the understanding of NS5-STAT2 complex. Therefore, the aim of the present study is to elucidate the potential role and mechanism of the NS5-STAT2 interaction involved in pathway of type I IFN signaling during ZIKV infection from a computational approach.

2 Methods

Molecular modeling and quality check. The NS5 three-dimensional model (PDB code: 5TMH) was obtained from the protein data bank. However, only two fragments (PDB codes: 5OEN and 2KA4) have been reported for STAT2. Three-dimension models were generated through algorithms of protein threading and homology alignment programs known as I-TASSER and Phyre2. In order to evaluate the NS5 and STAT2 models quality the programs Verify-3D, ERRAT, and PROCHECK were used. STAT2 experimental fragments reported in the database were employed to perform a structural alignment with the STAT2's model through TM-align. **MD simulation and clustering.** GROMACS 5.1.2 [8] was used to perform the MD simulation of NS5 and STAT2 as well as NS5-STAT2 complex. MD simulations were conducted by using AMBER 03 force field. The systems were established using the following features; cubic boxes filled with SPC216 water molecules, TIP3P water, Na^+ and Cl^- counterions were added to neutralize the system with a concentration of 0.1 M in order to simulate physiological conditions of the cells and periodic boundary conditions. PME was used for non-bonded interactions such as electrostatic interaction and van der Waals with a cut-off of 12 Å and a 2 fs time step during the simulation. The energy minimization was obtained through the steepest-descent algorithm, and the maximum force of the system was set to 100 kJ $(mol\,nm)^{-1}$ on all atoms. NVT and NPT ensembles were equilibrated using Berendsen thermostat and Nose-Hoover thermostat for 500 ps. Parrinello-Rahman barostat was employed to maintain the pressure isotropically with a value of 1.0 bars and compressibility of $4.5 \times 10^{-5} bar^{-1}$. The systems were submitted to 50 ns of production which was initialized using output data retrieved from previously run equilibration simulation at 310 K and 1 atm. Besides, all bonds length containing hydrogen were constrained using the Linear Constraint Solver (LINCS) algorithm. Gromacs utilities was used to analyze MD trajectory, and the charts were plotted using Grace. In order to explore the structural conformations generated from MD trajectories, a clustering was made. The clustering is based on Root Mean Square Distance (RMSD) of Cα atoms with a cut-off of 2.2 Å for each trajectory through the GROMOS clustering algorithm which is executed in the *gmx cluster* tool of GROMACS 5.1.2. A representative structure of each protein's trajectory was extracted in order to be used in an ensemble docking. **Docking and Gibbs binding free energy.** Docking complex was performed using ClusPro 2.0 and Pydock. Each binary complex selected was quantified by the binding strength of the protein-protein interface, Gibbs binding free energy (ΔG_{Bind}) and interface area through the programs PPCheck, FoldX and PISA, respectively. **Protein-protein interaction and electrostatic calculation.** Protein Interaction Calculator (PIC) web server and COCOMAPS were used to analyzed the protein-protein interaction between binary complexes. The lowest energy structure of

NS5, STAT2, and NS5-STAT2 complexes were used to perform the electrostatic calculations through the Poisson-Boltzmann (PB) equation implemented in the APBS Program as plug-in added in the program PyMOL 2.2.2.

3   Results and Discussion

3.1   Molecular Modeling and Validation of STAT2

The three-dimensional model of STAT2 was generated using the amino acid sequence by two different approaches namely protein threading of I-TASSER and homology of Phyre2. Through, I-TASSER five models were obtained. The best model has been chosen according to the highest C-score whose value of -0.62 suggests a correct topology. Likewise, the best model of Phyre2 has been obtained according to the identity percentage and confidence values which were 41% and 100%, respectively. This model was built with a template of the non-phosphorylated STAT1 (PDB code: C1YVIB). Both the best model of I-TASSER and Phyre2 were compared by structural quality to select the model to be used in the next analyses. The results have shown that STAT2's model of I-TASSER obtains a quality factor of 70.74% better than Phyre2's model with a value of 56.65% according to Verify-3D. With ERRAT the trend was similar, because I-TASSER's model presents better value than Phyre2's model, for this latter model an unacceptable value of 21.69% unlike to the value of 77.80% achieved by I-TASSER's model. Ramachandran plots were obtained by PROCHECK where Phyre2's model has reached more appropriated values for favored and allowed regions than I-TASSER's model since gathers values of 75% and 17.2%, respectively. These latter values are higher than the I-TASSER's model with values of 72.6% and 21.6%. However, Phyre2's model has gathered a higher percentage in the disallowed regions indicating that the stereochemical quality is inferior to I-TASSER'S model. Based on these outcomes, we selected the I-TASSER's model because their residues are positioned in the favorable three-dimensional structure environment in agreement with the experimental functional domains as shown in Fig. 1. Modeling by homology is usually more accuracy wherein stereo-chemical restrains and segments matching are then considered. However, it is a limited method because a template must have a high identity with target sequence. Generally, the accuracy of homology models under to 30% is due to alignment error leading to incapability to generate structures that fit with the target sequences [9,10]. Otherwise, modeling by threading is capable to build a model without a template. It represents an improvement when the target sequence is unknown or has segments (surface loops) that are misaligned with template sequences. Therefore, this latter is considered more successful in contrast to an approach by homology [9,10]. Additionally, STAT2's model has been compared with experimental fragments of STAT2 in order to observe the structural similarity. Each fragment (2KA4 and 5OEN) was aligned with STAT2's model where the fragment 5OEN has been matched by 168 residues with a TM-score of 0.86 in the Coiled-coil domain while the fragment 2KA4 has been paired by 39 residues with a TM-score of 0.34 in the Transactivation domain.

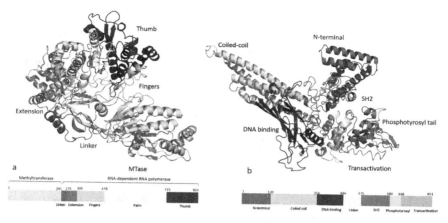

Fig. 1. Three-dimensional structures of (a) NS5 and (b) STAT2 model. The functional domains of each model are shown in different color. NS5: Mtase = Methyltransferase domain (1–264 aa), Linker domain (265–275 aa), Extension (275–304 aa), Fingers domain (305–477 aa), Palm domain (478–714 aa) and Thumb domain (715–903 aa) STAT2: N-terminal domain (1–138 aa) involved in dimerization/-tetramerization, Coiled-coil domain (139–315 aa) involved in interaction with other proteins, DNA binding domain (316–485 aa), Linker domain (486–574 aa), SH2 domain (575–679 aa), Phosphotyrosyl tail segment (680–697 aa), and Transactivation domain (698–851 aa). (Color figure online)

This latter one has hardly coincided with the STAT's model, because it is in a disorder region with a coiled-coil as secondary structure. Nevertheless, the STAT2's model obtained is suitable to be used as a reference protein model of STAT2.

3.2   Analysis of Docking Complex NS5-STAT2

In order to understand the biological functions as well as interactions of these proteins, it is essential to explore the conformational changes in short time lapses at the atomic level. MD simulation is a tool that describes atomic motion, structural properties and thermodynamic behavior of the systems at equilibrium [11,12]. Besides, MD simulation is employed to improve the virtual screening in the docking processes since it can be considered a step of refining such models. This refining provides flexibility and structural rearrangements to proteins allowing to obtain in overall a more realistic structure of each model since the methods of protein-protein docking are usually supported by the shape complementarity [13–16]. Therefore, NS5's x-ray structure and STAT2's model were subjected to 50 ns MD simulation protocol defined in the method section to minimize and equilibrate at near physiological conditions. From the trajectories of NS5 and STAT2, it was performed a clustering analysis to explore the conformation of proteins generated by MD simulation. According to the RMSD cut-off detailed in the methods section, the dominant clusters for NS5 and STAT2 constitute

~75% of total protein structures. Representative structures were extracted from these cluster as targets to build an ensemble docking [13]. Each structure has the largest stabilizing energy which is considered close to a native structure [17]. In order to explore the interaction between these proteins, docking analyses have been performed through ClusPro and Pydock. At this stage, two troubles arise to predict the correct solution, the first one is the scoring functions of ClusPro and Pydock and second one is the binding site which is unknown [14]. Although some algorithms are able to score a list of possible structures, however they are not reliable to discriminate false positives, that is, complexes with a high score but with a low rank [14]. Therefore, in order to classify docked complexes we used the software PPCheck with a process of discrimination by functional domains, and ΔG_{Bind}. PPCheck quantifies the strength of protein-protein interaction using pseudo-energies. The outcomes have shown that the total stabilizing energy for Pydock's models is positive ($\Delta E > 0$) while that for ClusPro's models are negative ($\Delta E < 0$). The total stabilizing energy values that have negative energy tendency are related with the increase of the number of interface residues [18]. Hence, negative energy suggests that there is a contact interface between proteins. In contrast, the interaction is deficient when the contact areas are almost null in systems with positive energy. Among of them, the ClusPro's models 4, 7 and 9 have obtained the largest total stabilizing energy with a normalized energy per residue closed to $-6 kJ\,mol^{-1}$. Therefore, they have fallen in a correct docking pose [18]. Cluspro's models have also shown to interact via known domains in both proteins. A test of total stabilizing energy and normalized energy per residue among the N-terminal domain from STAT2 and Mtase and Thumb domains from NS5 have been performed. The models 1, 4, and 7 have demonstrated that the interaction between these domains has gathered the largest total stabilizing energy. However, the interaction in the model 9 is only between the N-terminal and Mtase domains and presents a negligible total stabilizing energy (see Table 1). Likewise, the models 1, 4 and 7 have reached the largest normalized energy per residue close to $-6 kJ\,mol^{-1}$ showing that these models involve a correct docked position [18]. In contrast to the other models that do not show interaction among these domains. In order to understand the difference between the model of Cluspro and Pydock was used the surface area. It provides features in the protein-protein interaction which displays a high degree of structural complementarity and chemical complementarities [19]. As a result, the interface areas of Cluspro's models ($\overline{x} = 1973.59\,Å^2$) are larger than any Pydock's model ($\overline{x} = 1068.03\,Å^2$). Similar results to our ClusPro's models have been found in a published set of 75 crystal structures that present an interface area average of $2000\,Å^2$ in each member. This average is considered a specific protein-protein interaction with high complementarity [19,20]. Likewise, It has been observed that protein-protein interactions have large contact surfaces (1500–$3000\,Å^2$). In contrast, the contact area between small molecules and proteins targets has been estimated between 300 to $1000\,Å^2$ [21]. The ΔG_{Bind} has been calculated using empirical force field of FoldX for the atomic coordinates of Cluspro's models [22]. In this initial analysis, each model has achieved a positive value of interaction energy $\Delta G_{Bind} > 0$. It means that the docked complexes may be considered as unstable structures since it denotes energy-unfavorable coupling between both proteins. This unstable state between proteins NS5-STAT2 can also be considered as a non-covalent interaction with negative cooperativity since the affinities between ligand-receptor are decreased. Then, the docked complex is less well bonded and their atoms exhibit major internal motions [23]. A negative cooperativity also suggests that the docked complexes will need a great amount of energy (highly endothermic) to couple because the binding between NS5 and STAT2 is not a process spontaneous. Therefore, $\Delta G_{Bind} > 0$ is related with a favorable change entropy [23]. Additionally, the non-covalent interaction with a negative cooperativity in the interface between NS5-STAT2 may be influenced by coupling sites geometry. A coupling of non-covalent interaction with positive cooperative causing a structural loosening in the docked complex because the contact distance between ligand-receptor is greater [23]. The models 1, 7, 4, and 9 show the largest binding energy (ΔG_{Bind}) in comparison with the remaining models. However, we decided to take into consideration only models 1, 4, and 7 to be subjected to MD simulation because they showed interaction in the Mtase and Thumb domains with the N-terminal domains in the NS5-STAT2 complex.

Table 1. Results of total stabilizing energy and normalized energy per residue for ten ClusPro's models based on the functional domains of NS5 and STAT2. The functional domains involved in the interaction of complex were the N-terminal from STAT2 and Mtase and Thumb from NS5.

Models	Total stabilizing energy (kcal mol^{-1})		Normalized energy per residue (kJ mol^{-1})	
	N-Ter/Mtase	N-Ter/Thumb	N-Ter/Mtase	N-Ter/Thumb
Model 1	-23.2	-27.192	-2.7	-1.9
Model 2	-	-	-	-
Model 3	-	-	-	-
Model 4	-19.116	-48.282	-3.33	-3.61
Model 5	-	-	-	-
Model 6	-9.037	-	-1.18	-
Model 7	-59.388	-39.959	-3.5	-3.89
Model 8	-3.774	-	-0.88	-
Model 9	-1.114	-	-4.66	-
Model 10	-	-	-	-

3.3 Contact Map and Electrostatic Analysis

In order to confirm the interaction among Mtase and Thumb with N-terminal domains, a contact map of intramolecular interactions of the NS5-STAT2 complex is illustrated in Fig. 2. The distance range between two proteins has been marked with red, yellow, green and blue for 7 Å, 10 Å, 13 Å, and 16 Å, respectively. Largest regions of two partners that are in contact have been located in

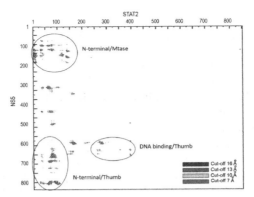

Fig. 2. Contact map of NS5 and STAT2 shows the intermolecular contacts at reducing distances which are red = 7 Å, yellow = 10 Å, green = 13 Å and blue = 16 Å. The circles identify the domains related such as N-terminal/Mtase, N-terminal/Thumb and Thumb/DNA binding. (Color figure online)

the map. The regions associated corresponds to first residues of Mtase domain (1–200) and the last residues of Thumb domain (650–750) in the NS5 while that for region of STAT2 is in the residues of N-terminal domain (1–150) and a smaller interaction region in the residues from 300 to 400 that correspond to DNA binding domain. The contact map confirms our previous results since the protein-protein interaction is mainly focused on in the domains described. An electrostatic potential analysis has been performed because electrostatic interactions are favored for the protein-protein interaction as well as the stabilization in a complex. The electrostatic role depends on the type of hetero- or homo-complexes, that is, a complex is formed by different or identical proteins which have net charge (positive or negative). In our case, the complex NS5-STAT2 is a heterocomplex which carries an opposite net charge that leads to the attraction between proteins. However, the arrangement of heterocomplex will be limited by residues distribution that will change global net charge of each protein at short distances in the interface [24]. Figure 3 shows the NS5-STAT2 complex. In the region of the N-terminal domain of STAT2, their buried residues are highly polar. In contrast, the residues located in the cavity of NS5 have a hydrophobic character with few polar groups around. Studies suggest that polar residues in the interface favorably contribute in two ways; the first, through a specific association between proteins, and second improving the stabilization of complexes since the interacting forces do not need to be strong for the formation of complexes. Moreover, regions in the protein with polar groups are usually sites known as hot spots which are crucial for a better affinity between proteins [24].

3.4   MD Simulation of NS5-STAT2 Complexes

The models 1, 4, and 7 have been minimized to optimize the complex geometry and check the Gibbs binding free energy (ΔG_{Bind}) again. ΔG_{Bind} has been

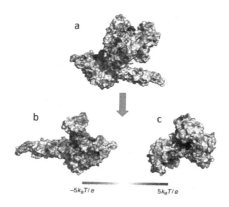

Fig. 3. A view of electrostatic surface potential for (a) NS5-STAT2 complex, (b) STAT2 and (c) NS5. Red is negative charge, blue is positive charge, and white is neutral. (Color figure online)

calculated for representative conformations (snapshots) of each trajectory in the models 1, 4, and 7, in order to verify the behavior of the complex NS5-STAT2. According to the clustering of MD trajectories, each model has obtained 6001 structures which have been clustered in 33, 22, and 21 groups for models 1, 4, and 7, respectively [25]. Hence, the results have shown that ΔG_{Bind} of model 1 and 7 ($\Delta G_{Bind} < 0$) are $-4.30\,\text{kcal}\,\text{mol}^{-1}$ and $-1.67\,\text{kcal}\,\text{mol}^{-1}$, respectively while for model 4 ($\Delta G_{Bind} > 0$) is $0.27\,\text{kcal}\,\text{mol}^{-1}$. The coupling between the proteins is energetically favorable in models 1 and 7 in opposition to model 4. The same range of values of Gibbs binding free energy $\Delta G_{Bind} < 0$ are also obtained in experimental measurements of binding affinity for protein-protein complexes tested on a benchmark of 144 structures [26]. Likewise, a study with an empirical approach to calculate the Gibbs binding free energy based on three variables of the interface in complexes has estimated values almost close to model 1 [27]. These latter results have demonstrated that applying MD simulation to the system is possible to improve the values of Gibbs binding free energy. On the other hand, the initial atomic coordinates of models have shown to reach a $\Delta G_{Bind} > 0$ meaning that docked complexes has non-covalent interactions with negative cooperativity. However, after a MD simulation, the models 1 and 7 have gathered a $\Delta G_{Bind} < 0$ which benefit the non-covalent interactions with positive cooperativity [23]. This change in the cooperativity is acceptable in the same system. The motion of atoms of the protein by the MD simulation produces a new stable state that has a net effect in the thermodynamic parameters [12, 23]. Hence, non-covalent interaction with positive cooperativity is related to an exothermic binding which allows an increment in the bonding ligand-receptor. The improving in the binding is associated with a favorable enthalpy and adverse in entropy because with a strong coupling the internal motions of the complex is reduced [23].

3.5   Interaction NS5-STAT2

The protein-protein interaction is mediated by domain-domain interactions. In our analyses, it has been identified that the interaction is given by the N-terminal domain from STAT2 and Mtase and Thumb domains from NS5 [28]. The interactions in the interface of NS5-STAT2 are stabilized by electrostatic interactions, hydrophobic interactions, salt bridges, and ionic interactions. In the case of NS5, 33 residues are in the contact area with an elevated frequency in all models. Moreover, 16 of them have presented one o more types of interactions. The NS5 residues involved are Arg-163, Arg-175, Arg-37, Arg-57, Arg-681, Arg-84, Arg-856, Glu-149, Leu-847, Lys-105, Lys-331, Pro-108, Pro-857, Trp-848, Val-335, and Val-336. These residues are located in the Mtase and Thumb domains. Regarding STAT2, 55 residues have a high frequency where 19 of them are involved with at least one type of interaction. The residues are Arg-796, Arg-88, Asp-77, Asp-794, Asp-850, Asp-93, Glu-40, Glu-715, Glu-722, Glu-79, Glu-801, Glu-804, Glu-814, His-85, Leu-684, Leu-691, Leu-727, Leu-81, and Lys-89. They have been located in the N-terminal domain. Studies of hot-spots in the binding sites of protein-protein interfaces have located a particular enrichment of Trp, Try and Arg, as well as a high presence of polar residues [29]. In contrast, hydrophobic residues as Val and Leu are associated to interfaces largely hydrophobic and nonpolar surface areas [29]. In our case, Arg and Glu are the hydrophilic residues with the largest presence in the interface of both NS5 and STAT2. Besides, hydrophilic residues as Lys, Pro, and Asp are also displayed in the interface of both proteins. Other studies have concluded that residues as Trp, Met, Try, Phe, Cys, and Ile are frequently in the binding interfaces. Residues as Tyr, Trp, His, and Cys have been traced in high-affinity interfaces in comparison with low-affinity interfaces [30]. His and Trp are present in the interface of NS5-STAT2 complex. Hence the interface shows a high-affinity. However, residue as Lys is located in low-affinity interface. In our outcomes, this last residue is found in the interface of NS5-STAT2 complex which will counteract a possible high-affinity in the NS5-STAT2 complex [30]. As was mentioned before, NS5-STAT2 is a heterocomplex which has differences properties associated with amino acid composition, contact sites, and interface area [31]. Hence, studies of homo- and heterocomplexes have revealed that the amino acid interface composition is different between them. The residues in heterocomplex have a greater prevalence and propensity to be in the interface are Leu, Val, Ile, Arg, Tyr, Trp, Met, and Phe [31]. These types of residues are also prevalent in our findings except for Ile, Met, and Phe.

4   Conclusions

The computational approach has permitted to analyze the different structural and dynamics features of NS5, STAT2 and the interaction of NS5-STAT2 complex. The STAT2's structure from I-TASSER was the selected model according to quality analyses. Moreover, it displays a high correlation with experimental fragments of STAT2. Three docked complexes provided by ClusPro showed interaction on three domains (N-terminal/Mtase and N-terminal/Thumb) enriched with polar residues. MD simulations on the three docked complexes benefits the interaction between proteins because the behavior of proteins is affected by internal atomic motions. Thus, they has shown a $\Delta G_{Bind} < 0$ where the best docked complex have a ΔG_{Bind} of $-4.30\,\text{kcal}\,\text{mol}^{-1}$. On the other hand, the NS5-STAT2 docked complex has revealed that the key interacting residues are stabilized by electrostatic interaction, hydrophobic interaction, salt bridges, and ionic interaction. Therefore, this study sheds light in the interaction of NS5-STAT2 as support of the experimental studies and the development of drugs against ZIKV NS5.

Acknowledgments. Our thankful to Universidad San Francisco de Quito for the use of the High Performance Computing System-USFQ.

References

1. Cox, B.D., Stanton, R.A., Schinazi, R.F.: Predicting Zika virus structural biology: challenges and opportunities for intervention. Antiviral Chem. Chemother. **24**(3–4), 118–126 (2015)
2. Wang, B., et al.: The structure of Zika virus NS5 reveals a conserved domain conformation. Nat. Commun. **8**, 14763 (2017)
3. Zhao, B., et al.: Structure and function of the Zika virus full-length NS5 protein. Nat. Commun. **8**, 1–9 (2017)
4. Grant, A., et al.: Zika virus targets human STAT2 to inhibit type I interferon signaling. Cell Host Microbe **19**(6), 882–890 (2016)
5. Arimoto, K., et al.: STAT2 is an essential adaptor in USP18-mediated suppression of type I interferon signaling. Nat. Struct. Mol. Biol. **24**(3), 279–289 (2017)
6. Bowen, J.R., et al.: Zika virus antagonizes type I interferon responses during infection of human dendritic cells. PLoS Pathog. **13**(2), e1006164 (2017)
7. Kumar, A., et al.: Zika virus inhibits type-I interferon production and downstream signaling. EMBO Rep. **17**(12), 487–524 (2016)
8. Abraham, M., Hess, B., van der Spoel, D., Lindahl, E.: GROMACS User Manual version 5.0.7 (2015). www.Gromacs.org
9. Fiser, A.: Template-based protein structure modeling. Methods Mol. Biol. **673**, 1–20 (2010)
10. Schwede, T., Sali, A., Eswar, N.: Protein structure modeling. In: Schwede, T., Peitsch, M.C. (eds.) Computational Structural Biology: Methods and Applications, chap. 1, pp. 1–33. World Scientific Publishing Co., Pte. Ltd., Danvers (2008)
11. Vlachakis, D., Bencurova, E., Papangelopoulos, N., Kossida, S.: Current state-of-the-art molecular dynamics methods and applications, 1 edn, vol. 94. Elsevier Inc. (2014)
12. Karplus, M., Petsko, G.A.: Molecular dynamics simulations in biology. Nature **347**(6294), 631–639 (1990)
13. Hospital, A., Goñi, J.R., Orozco, M., Gelpi, J.: Molecular dynamics simulations: advances and applications. Adv. Appl. Bioinf. Chem. **8**, 37–47 (2015)
14. Halperin, I., Ma, B., Wolfson, H., Nussinov, R.: Principles of docking: an overview of search algorithms and a guide to scoring functions. Proteins Struct. Funct. Genet. **47**(4), 409–443 (2002)

15. Brooijmans, N., Kuntz, I.D.: Molecular recognition and docking algorithms. Ann. Rev. Biophys. Biomol. Struct. **32**(1), 335–373 (2003)

16. Smith, G.R., Sternberg, M.J.: Prediction of protein-protein interactions by docking methods. Curr. Opin. Struct. Biol. **12**(1), 28–35 (2002)

17. Sikosek, T., Chan, H.S.: Biophysics of protein evolution and evolutionary protein biophysics. J. R. Soc. Interface **11**(100), 20140419–20140419 (2014)

18. Sukhwal, A., Sowdhamini, R.: Oligomerisation status and evolutionary conservation of interfaces of protein structural domain superfamilies. Mol. BioSyst. **9**(7), 1652–1661 (2013)

19. Elcock, A.H., Sept, D., Mccammon, J.A.: Computer simulation of protein protein interactions. J. Phys. Chem. B **105**, 1504–1518 (2001)

20. Keskin, O., Tuncbag, N., Gursoy, A.: Characterization and prediction of protein interfaces to infer protein-protein interaction networks. Curr. Pharm. Biotechnol. **9**(2), 67–76 (2008)

21. Smith, M.C., Gestwicki, J.E.: Features of protein-protein interactions that translate into potent inhibitors: topology, surface area and affinity. Expert Rev. Mol. Med. **14**, 1–24 (2012)

22. Schymkowitz, J., Borg, J., Stricher, F., Nys, R., Rousseau, F., Serrano, L.: The FoldX web server: an online force field. Nucleic Acids Res. **33**(Suppl. 2), 382–388 (2005)

23. Williams, D.H., Stephens, E., O'Brien, D.P., Zhou, M.: Understanding noncovalent interactions: ligand binding energy and catalytic efficiency from ligand-induced reductions in motion within receptors and enzymes. Angew. Chem. Int. Ed. **43**(48), 6596–6616 (2004)

24. Zhang, Z., Witham, S., Alexov, E.: On the role of electrostatics on protein-protein interactions. Phys. Biol. **8**(3), 035001 (2011)

25. Snyder, P.W., Lockett, M.R., Moustakas, D.T., Whitesides, G.M.: Is it the shape of the cavity, or the shape of the water in the cavity? Eur. Phys. J.: Spec. Top. **223**(5), 853–891 (2014)

26. Vreven, T., Hwang, H., Pierce, B.G., Weng, Z.: Prediction of protein-protein binding free energies. Protein Sci. **21**(3), 396–404 (2012)

27. Ma, X.H., Wang, C.X., Li, C.H., Chen, W.Z.: A fast empirical approach to binding free energy calculations based on protein interface information. Protein Eng. **15**(8), 677–681 (2002)

28. Brito, A.F., Pinney, J.W.: Protein-protein interactions in virus-host systems. Front. Microbiol. **8**(Aug), 1–11 (2017)

29. Ma, B., Elkayam, T., Wolfson, H., Nussinov, R.: Protein-protein interactions: structurally conserved residues distinguish between binding sites and exposed protein surfaces. Proc. Nat. Acad. Sci. **100**(10), 5772–5777 (2003)

30. Erijman, A., Rosenthal, E., Shifman, J.M.: How structure defines affinity in protein-protein interactions. PLoS ONE **9**(10), e110085 (2014)

31. Talavera, D., Robertson, D.L., Lovell, S.C.: Characterization of protein-protein interaction interfaces from a single species. PLoS ONE **6**(6), e21053 (2011)

Entropy-Based Detection of Genetic Markers for Bacteria Genotyping

Marketa Nykrynova[1][(✉)], Denisa Maderankova[1], Vojtech Barton[1],
Matej Bezdicek[2], Martina Lengerova[2], and Helena Skutkova[1]

[1] Department of Biomedical Engineering, Brno University of Technology,
Technicka 12, 616 00 Brno, Czech Republic
{nykrynova,maderankova,barton,skutkova}@vutbr.cz
[2] Department of Internal Medicine - Hematology and Oncology, University Hospital
Brno, Brno, Czech Republic
{Bezdicek.Matej,Lengerova.Martina}@fnbrno.cz

Abstract. Genotyping is necessary for the discrimination of bacteria
strains. However, methods such as multilocus sequence typing (MLST)
or minim typing (mini-MLST) use a combination of several genes. In this
paper, we present an augmented method for typing *Klebsiella pneumoniae* using highly variable fragments of its genome. These fragments were
identified based on the entropy of the individual positions. Our method
employs both coding and non-coding parts of the genome. These findings
may lead to decrease in the number of variable parts used in genotyping
and to make laboratory methods faster and cheaper.

Keywords: Genotyping · Entropy · *Klebsiella pneumoniae* ·
Minim typing

1 Introduction

Bacteria typing is a process used to identification of relationships between microbial isolates. Knowing the relations between different bacteria strains of the same
species is necessary to understand the outbreak emergencies [1]. Typing of bacteria also helps us determine the source of the specific bacterial clones and their
transmission routes [2].

In the past, phenotypic and genotypic methods were used for bacterial typing.
Nowadays, mainly the genotypic methods are used because they can distinguish
strains more accurately, especially closely related ones [3].

Choosing appropriate typing methods for analysis is based on our goals as
all methods have some advantages and disadvantages [4,5]. Weaknesses of currently used typing methods include low reproducibility, huge time demands and
technical complexity [6]. Also, molecular typing methods are not capable of distinguishing closely related bacterial population, e.g. in one hospital. The only
typing method which can precisely determinate bacteria strains is whole genome

© Springer Nature Switzerland AG 2019
I. Rojas et al. (Eds.): IWBBIO 2019, LNBI 11466, pp. 177–188, 2019.
https://doi.org/10.1007/978-3-030-17935-9_17

sequencing (WGS) [1,7]. However, it cannot be used for common clinical practice due to its cost.

Thus, a combination of molecular typing methods with bioinformatics methods is very promising research area possibly addressing some of the aforementioned challenges.

Nowadays, pulsed-field electrophoresis (PFGE) is considered as a gold standard among typing methods despite some of its disadvantages, such as time demands (analysis takes dozens of hours), technical complexity and low discriminatory power for similar strains [8]. Genomic DNA is digested with restriction endonucleases, and fragments of DNA are separated on an agarose gel by electrophoresis [9]. The obtained gel with banding patterns is dyed with fluorescent colour and images are captured and analysed [6]. PFGE has high discriminatory power and reproducibility.

Although WGS is capable of determination of bacteria strains the cost remains a problem. In order to avoid significant cost multilocus sequence typing (MLST) is used. It is based on the sequencing the fragments (450–500 bp) of up to seven housekeeping genes which are essential for primary cellular functions. For each gene, to unique sequences (alleles) are assigned different allele numbers. The combination of these seven numbers is known as sequence type (ST). Thus, each isolate is clearly characterized [10]. Sequences of alleles and ST profile are saved in accessible database. The main advantage is that data are unambiguous due to the standardised nomenclature and data can be compared worldwide [11].

Because the sequencing of seven housekeeping genes is still expensive for routine use, the new method called minim typing (mini-MLST) which is derived from MLST was created. It consists of two main steps. In the first one, from six to eight fragments of seven housekeeping genes are amplified by PCR. Fragments of housekeeping genes are used because they are present in almost all organisms of given species. The second part consists of a high resolution melting analysis (HRM) [12]. As a result, we obtain in each gene from two to six different melting curves and when 50% of DNA is denatured the melting temperature is determined [13]. Different melting curves are called Melt Alleles, that are named according to the number of GC bases in the amplified region. Combination of melt alleles from each gene defines so-called melt type. The main advantage is a low cost which is about 10–20% of MLST [14] and low time demands.

In our project we worked with genomes of *Klebsiella pneumoniae* because the growing antibiotic resistance of this bacteria is problem not only in the Czech Republic but also worldwide. Evaluation of our method was performed on *Klebsiella pneumoniae* but it can be adapted for any bacteria.

Goal of our project is to use bioinformatic methods to detect new genetic markers. They are variable parts of genome sequences that can be used for genotyping of bacterial strains in a lab.

2   Materials and Methods

Our project aims to discover highly variable sequences for genotyping and consists of several steps which are described in the following paragraphs. First, data from sequencer must be assembled. Then, entropy for each position in genomes is calculated. In the next step, variable fragments of genomes are established based on entropy calculation. Evaluation of located variable fragments is done by phylogenetic analysis and melting temperature calculation. Block diagram of the proposed process is shown in Fig. 1.

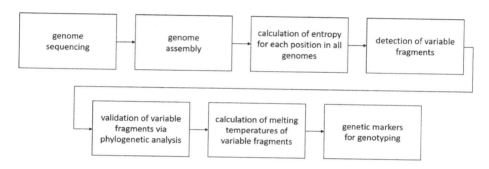

Fig. 1. Block diagram of the proposed method.

2.1   Genome Assembly

Once we obtained sequenced genomes, assembly of the genomes must be done. Genomes of *Klebsiella pneumoniae* were assembled using reference genome *Klebsiella pneumoniae subsp. pneumoniae NTUH-K2044* which was obtained from NCBI database. Specifically, the reads from sequencer were assigned to the reference genome. As an assembly algorithm, we employed Burrows-Wheeler Aligner MEM because it is recommended for fast and accurate assembly of Illumina reads [15]. Then, Samtools [16] was used for removing unmapped reads, reads with poor quality and to remove PCR and optical duplicates. In the last step, consensus sequences were created.

2.2   Entropy Calculation of Bacterial Genomes

Entropy H_i [17] of each position in genome was calculated as

$$H_i = (-\sum_a f_{a,i} \cdot \log_2 f_{a,i}) \cdot e_n, \tag{1}$$

where e_n is correction factor, $f_{a,i}$ is frequency of occurrence of one of four nucleotides ($a = A, C, G, T$) at one position and can be calculated as

$$f_{a,i} = \frac{n_a}{n}, \tag{2}$$

where n_a is number of occurrences of given nucleotide and n is number of characters at single positions.

If N appears at certain position that means a sequencer was not able to determinate nucleotide base with sufficient confidence. In such case 1/4 is added to each frequency of occurrence for all nucleotides.

Figure 2 exhibits an example calculation of entropy of eight sample sequences.

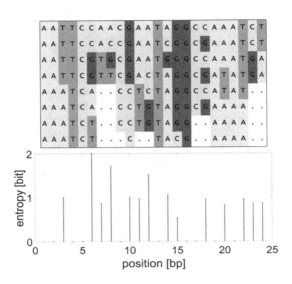

Fig. 2. On the top: Example of multiple sequence alignment of eight sequences. On the bottom: Example of entropy for eight sequences for each position.

2.3   Correction Function

Correction function e_n for a small number of sequences or when there were gaps in sequences was calculated as

$$e_n = \frac{M - 1}{2 \cdot \ln 2 \cdot n}, \tag{3}$$

where M is number of nucleotides which can be found in DNA and n is number of characters at one position (in our case from 0 to 24). Then, feature scaling normalization was used. As a result, we obtained a correction function which is depicted in the Fig. 3.

Thus, for a low number of characters at given position we obtain a small multiplication factor, i.e. entropy value of a given position is decreased. As we can see in the Fig. 2 at position 10 and 11 already a slight change can be observed in entropy graph.

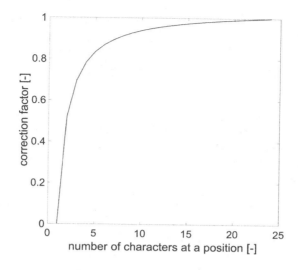

Fig. 3. Correction function.

2.4 Evaluation by Phylogenetic Analysis

To determine if selected variable parts of the genomes are capable of distinguishing between bacterial melt types, the phylogenetic analysis takes place [18]. For each pair of sequences, we use newly identified variable parts and calculate the proportional distance which is then used to create a distance matrix.

In the next step, evolution distances d based on commonly used Kimura model [19] are calculated as

$$d = -\frac{1}{2}\ln(1 - 2P - Q) - \frac{1}{4}\ln(1 - 2Q), \tag{4}$$

where P is number of transitions (change of purine nucleotide to another purine or pyrimidin nucleotide to another pyrimidine) and Q is number of transversion (change of purine nucleotide to pyrimidin nucleotide or vice-versa) and it is also calculated for each pair of sequences.

Using the calculated evolutionary distances the phylogenetic tree is constructed.

2.5 Calculation of Melting Temperatures

For calculation of melting temperatures the Nearest Neighbors [20] method was used. As was mentioned in introduction, melting temperatures and melt type are derived during minim typing procedure. In our method, discrimination based on temperatures is necessary because selected variable parts will be used in minim typing that is based on DNA melting. Melting temperatures T_m are calculated as

$$T_m = \frac{\Delta H}{A + \Delta S + R\ln(\frac{C}{4})} - 273.15 + 16.6\log[Na^+], \tag{5}$$

where ΔH is enthalpy change and ΔS is entropy change and they are defined for each combination of nucleotide pair [21], A is constant of $-0.0108 \; kcal \cdot K^{-1} \cdot mol^{-1}$, R is gas constant and C is oligonucleotide concentration (for calculation value $0.5 \, \mu M$ was used according to our laboratory protocol).

2.6   Dataset

Klebsiella pneumoniae is a member of the *Enterobacteriaceae* family and it belongs to Gram-negative bacteria [22]. The average size of the genome is 5.5 Mbp, and it encodes approximately 5500 genes, where 2000 genes are common among most strains [23].

In this paper, we worked with 24 genomes of *Klebsiella pneumoniae* which were obtained by the Department of Clinical Microbiology (University Hospital Brno) of 4 different melt types which were determined by minim typing. Names and melt types of all genomes are shown in Table 1.

Table 1. Genomes of *Klebsiella pneumoniae* and their melt types.

Genome	Melt type	Genome	Melt type	Genome	Melt type	Genome	Melt type
S01	98	S07	29	S13	23	S19	61
S02	29	S08	98	S14	23	S20	61
S03	98	S09	29	S15	23	S21	61
S04	98	S10	98	S16	23	S22	61
S05	29	S11	29	S17	23	S23	61
S06	98	S12	29	S18	23	S24	61

In genomes of *Klebsiella pneumoniae* we can find five housekeeping genes which are used for typing via minim typing. These genes, their fragment lengths and proteins which they encode are described in Table 2. We worked with housekeeping genes as they remain stable during evolution. However, variability in this genes can help us distinguish bacteria strains even over long periods of time [3].

3   Results

3.1   Genes Used for Minim Typing

In the first part of the project, our goal was to identify fragments of 5 housekeeping genes which are used for bacteria typing via minim typing and to calculate their entropy.

Genes fragments were located in 24 genomes. Then, entropy for each fragment was calculated and it is depicted in Fig. 4. As we can see, the fragments are only slightly variable across genomes. Only fragment of *phoE* contains three nucleotides changes and fragment *rpoB* contains two changes. In three other

Table 2. Genes and their fragments used for minim typing, their lengths and products.

Gene	Gene length [bp]	Fragment length [bp]	Protein
infB	896	19	Translation initiation factor IF-2
mdh	312	19	Malate dehydrogenase
phoE	351	41	Outer membrane pore protein E
rpoB	1342	89	DNA-directed RNA polymerase subunit beta
tonB	243	64	Protein tonB
		84	

cases (*infB*, *tonB*) only one nucleotide change was detected. Moreover, for fragments of gene *mdh* no variability was detected. Thus, the genes used for mini-MLST are variable only slightly if at all.

3.2 Determination of Variable Parts in Genomes

In the second part of the project, our goal was to find variable parts of genomes which can be used for genotyping instead of currently used fragments in minim typing.

Entropy for all positions in genomes was calculated. The obtained graph is depicted on Fig. 5. Because we want to find variable parts of genomes which contain more than one variable position, entropy in a floating window was calculated and it is also depicted on Fig. 5. The size of the window was twenty nucleotides with overlap of one nucleotide.

From the graph of entropy, the variable parts were established. For this purpose, the thresholding was used. As a value of threshold 0.45 was set empirically. After application of the thresholding 6 peaks were found. Variable sequences which correspond to peaks are examined.

3.3 Evaluation of Newly Identified Variable Fragments

For each variable region which was identified the phylogenetic analysis is used. As a result, we obtained phylogenetic trees.

Three out of six fragments, established from entropy calculation, can distinguish between all four melt types. One variable area is found only in ten genomes and defines two melt types which are actually present. Last two variable regions can determinate two clusters instead of four hence, they cannot be used. Example of one variable part of genomes and its phylogenetic tree is depicted on Fig. 6.

In the last part, for four variable regions, the melting temperatures were calculated. Melting temperatures of each variable part and its melt types are in Table 3. Calculation of temperatures is used because the variable parts of sequences will be used in minim typing which is based on melting temperatures. Thus, if two melt types have the same temperature of melting, the mini-MLST

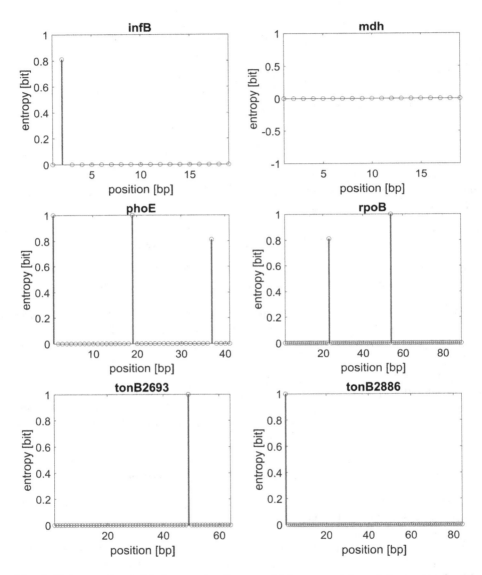

Fig. 4. Entropy of variable fragments of genes which are used for minim typing for 24 genomes of *Klebsiella pneumoniae.*

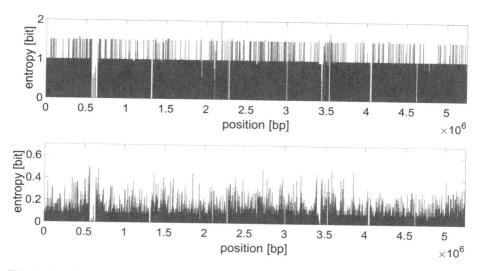

Fig. 5. On the top: Entropy of 24 genomes of *Klebsiella pneumoniae* calculated for each position. On the bottom: Entropy of 24 genomes of *Klebsiella pneumoniae* calculated in floating window.

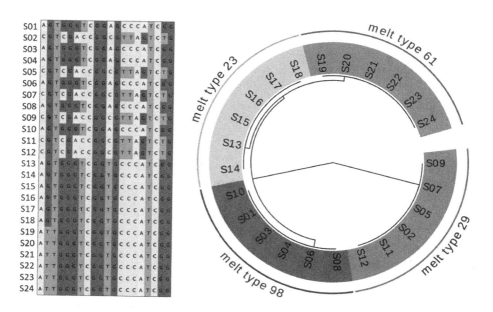

Fig. 6. On the left: Alignment of sequences of one variable part. On the right: Phylogenetic tree of one variable part created by Evolview [25].

could not distinguish between them. Example of possible output from minim typing is on Fig. 7. Melting curves are based on data obtained by uMELT [24].

Table 3. Melting temperatures of chosen variable parts for each melt type.

Number of variable region of sequences	Position in genomes	Melting temperatures [°C]			
		Melt type 98	Melt type 29	Melt type 23	Melt type 61
2	565541..565560	76.9	67.3	71.4	77.0
3	674053..674072	74.7	70.1	75.7	75.2
4	2722769..2722788	56.6	-	53.2	-
6	3856957..3856976	73.9	73.0	68.8	77.5

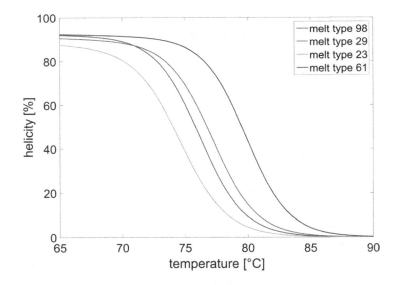

Fig. 7. Melting curves of variable part no. 6.

4   Conclusion

In this paper, we present a new approach for the entropy-based determination of highly variable parts of genomes which can be used for genotyping of *Klebsiella pneumoniae*. Thus, we are capable of distinguishing between melt types using only one variable part from genome instead of using six fragments of five housekeeping genes.

The results of the proposed methodology were compared to data obtained from minim typing and evaluated using phylogenetic analysis. Based on entropy calculation we established four variable fragments which are capable of distinguishing between four melt types of *Klebsiella pneumoniae*. Also, melting temperatures for each variable sequence were calculated to verify whether variable parts can be used in a lab.

The proposed signal-based approach seems promising due to its speed and whole genome range. We are capable of using non-coding parts which may have even higher variability than coding parts. However, the proposed method is highly sensitive to the quality of sequence data and its assembly. Also, during entropy calculation we are not able to distinguish deletions in sequences from gaps in the assembly. This is due to small size of the dataset that was used. Unfortunately, the problem can be fixed only by using bigger number of genomes.

The newly identified variable fragments may be used for typing bacteria by mini-MLST, or they can extend the current set of gene fragments used in mini-MLST to deliver more precise result.

Acknowledgments. This work was supported by grant project of the Czech Science Foundation [GACR 17-01821S].

References

1. Ranjbar, R., Karami, A., Farshad, S., Giammanco, G.M., Mammina, C.: Typing methods used in the molecular epidemiology of microbial pathogens: a how-to guide. New Microbiol. **37**, 1–15 (2014)
2. Ruppitsch, W.: Molecular typing of bacteria for epidemiological surveillance and outbreak investigation. Die Bodenkultur: J. Land Manag. Food Environ. **67**, 199–224 (2016)
3. Castro-Escarpulli, G., et al.: Identification and typing methods for the study of bacterial infections : a brief review and mycobacterial as case of study. Arch. Clin. Microbiol. **7**, 1–10 (2015)
4. van Belkum, A., et al.: Guidelines for the validation and application of typing methods for use in bacterial epidemiology. Clin. Microbiol. Infect. **13**, 1–46 (2007)
5. Olive, D.M., Bean, P.: MINIREVIEW principles and applications of methods for DNA-based typing of microbial organisms. J. Clin. Microbiol. **37**, 1661–1669 (1999)
6. Foley, S.L., Lynne, A.M., Nayak, R.: Molecular typing methodologies for microbial source tracking and epidemiological investigations of gram-negative bacterial foodborne pathogens. Infect. Genet. Evol. **9**, 430–440 (2009)
7. Kao, R.R., Haydon, D.T., Lycett, S.J., Murcia, P.R.: Supersize me: how whole-genome sequencing and big data are transforming epidemiology. Trends Microbiol. **22**, 282–291 (2014)
8. Sabat, A.J., Budimir, A., Nashev, D., Sá-Leão, R., van Dijl, J.M., Laurent, F.: Overview of molecular typing methods for outbreak detection and epidemiological surveillance. Eur. Commun. Dis. Bull. **18**, 20380 (2013)
9. Wang, X., King Jordan, I., Mayer, L.W.: A phylogenetic perspective on molecular epidemiology. Mol. Med. Microbiol. **1–3**, 517–536 (2014)
10. Sullivan, C.B., Diggle, M.A., Clarke, S.C.: Multilocus sequence typing: data analysis in clinical microbiology and public health. Mol. Biotechnol. **29**, 245–254 (2005)

11. Li, W., Raoult, D., Fournier, P.E.: Bacterial strain typing in the genomic era. FEMS Microbiol. Rev. **33**, 892–916 (2009)
12. Tong, S.Y.C., Giffard, P.M.: Microbiological applications of high-resolution melting analysis. J. Clin. Microbiol. **50**, 3418–3421 (2012)
13. Andersson, P., Tong, S.Y.C., Bell, J.M., Turnidge, J.D., Giffard, P.M.: Minim typing - a rapid and low cost MLST based typing tool for *Klebsiella pneumoniae*. PLoS ONE **7**, 1–7 (2012)
14. Brhelova, E., et al.: Validation of Minim typing for fast and accurate discrimination of extended-spectrum, beta-lactamase-producing *Klebsiella pneumoniae* isolates in tertiary care hospital. Diagn. Microbiol. Infect. Dis. **86**, 44–49 (2016)
15. Li, H.: Aligning sequence reads, clone sequences and assembly contigs with BWA-MEM. In: Antimicrobial Resistance & Infection Control (2013)
16. Li, H., et al.: The sequence alignment/map format and SAMtools. Bioinformatics **25**, 2078–2079 (2009)
17. Schneider, T.D., Stephens, R.M.: Sequence logos: a new way to display consensus sequences. Nucleic Acids Res. **18**, 6097–6100 (1990)
18. Nykrynova, M., Maderankova, D., Bezdicek, M., Lengerova, M., Skutkova, H.: Bioinformatic tools for genotyping of Klebsiella pneumoniae Isolates. In: Pietka, E., Badura, P., Kawa, J., Wieclawek, W. (eds.) ITIB 2018. AISC, vol. 762, pp. 419–428. Springer, Cham (2019). https://doi.org/10.1007/978-3-319-91211-0_37
19. Nei, M., Kumar, S.: Molecular Evolution and Phylogenetics. Oxford University Press, New York (2000)
20. Sigma-Aldrich: Oligonucleotide Melting Temperature. Merck, Darmstadt (2018)
21. Breslauer, K.J., Frank, R., Blöcker, H., Marky, L.A.: Predicting DNA duplex stability from the base sequence. In: Proceedings of the National Academy of Sciences of the United States of America, vol. 83, pp. 3746–3750 (1986)
22. Martin, R.M., et al.: Molecular epidemiology of colonizing and infecting isolates of *Klebsiella pneumoniae*. MSphere **1** (2016)
23. Wyres, K.L., Holt, K.E.: Klebsiella pneumoniae population genomics and antimicrobial-resistant clones. Trends Microbiol. **24**, 944–956 (2016)
24. Dwight, Z., Palais, R., Wittwer, C.T.: uMELT: prediction of high-resolution melting curves and dynamic melting profiles of PCR products in a rich web application. Bioinformatics **27**, 1019–1020 (2011)
25. Zhang, H., Gao, S., Lercher, M.J., Hu, S., Chen, W.-H.: EvolView, an online tool for visualizing, annotating and managing phylogenetic trees. Nucleic Acids Res. **40**, W569–W572 (2012)

Clustering of *Klebsiella* Strains Based on Variability in Sequencing Data

Vojtech Barton[1]([✉]), Marketa Nykrynova[1], Matej Bezdicek[2], Martina Lengerova[2], and Helena Skutkova[1]

[1] Department of Biomedical Engineering, Brno University of Technology, Technicka 12, 616 00 Brno, Czech Republic
{barton,nykrynova,skutkova}@vutbr.cz
[2] Department of Internal Medicine - Hematology and Oncology, University Hospital Brno, Brno, Czech Republic
{Bezdicek.Matej,Lengerova.Martina}@fnbrno.cz

Abstract. Genotyping is a method necessary to distinguish between strains of bacteria. Using whole sequences for analysis is a computational demanding and time-consuming approach. We establish a workflow to convert sequences to a numerical signal representing the variability of sequences. After segmentation and using only parts of the signals, they have still enough information to form a topologically according to the clustering structure.

Keywords: Genotyping · *Klebsiella pneumoniae* · Variability signal · Numerical processing

1 Introduction

Genotyping is a method used to distinguish between bacteria strains and to identify relationships between them. Genotyping of bacteria also can reveal more information about the origin of infection and its spreading [1].

Current methods are robust enough to distinguish between closely related strains. Most of the information used for typing is carried by variability in genomes. To identify the variability it is not necessarily needed to know the exact sequence of the analyzed organism. Typing methods concentrate on differences between DNA sequences of the organisms [2].

To obtain sequences of genomes we can use Whole genome sequencing (WGS) techniques, which produce a huge amount of data. Most of the outputs are text-like files [3]. To process these data it is suitable to convert them into some numerical representations, though it is more convenient to work with numbers. We can use already established signal analysis methods [4].

The WGS is not commonly used in clinical practice due to its cost. The computational methods can find characteristical segments. From knowing its sequence it could be enough to reconstruct the whole phylogenetic

I. Rojas et al. (Eds.): IWBBIO 2019, LNBI 11466, pp. 189–199, 2019.
https://doi.org/10.1007/978-3-030-17935-9_18

information about the whole organism. To concentrate further analysis only on these segments can decrease the amount of data handled, improve time-efficiency a decrease in the cost of analysis.

2   Materials and Methods

The project aims to distinguish between the types of *Klebsiella pneumoniae* strains. We want to identify parts of genomes, which are variable enough between the groups to differentiate the strains into different melt types.

We obtain several genomes of *Klebsiella pneumoniae* collected from Department of Internal Medicine - Hematology and Oncology of University Hospital Brno. All of these genomes were sequenced on Illumina Mi-Seq machine. All samples were prepared with standard laboratory procedures and were sequenced to obtain around 100 reads of depth coverage. All of the used genomes are summarized in Table 1. All together we obtain 48 samples of strains of 8 different melt types.

2.1   Mapping and Filtering Reads

All sequenced genomes were mapped to the same reference sequence. We decided to use the complete genome sequence of *Klebsiella pneumoniae subsp. pneumoniae* strain NTUH-K2044 obtained from NCBI database. The reference genome can be found by the reference NC_012731.1 [6].

Mapping of paired-end reads datafiles to reference sequence was done by the BWA aligner with *Mem* algorithm [7], which is designed to map short reads to the reference. There was no need to adjust the default settings.

After mapping the reads, there was a need to tidy up the reads alignment. As first, there was a need to throw away all unmapped reads. Then we discarded all reads identified as duplicates. These reads result from PCR process during the sequencing and they are a bearer of redundant information, which can result in increasing of sequencing error rate.

We also remove all reads with mapping quality bellow twenty, to avoid using poorly mapped reads.

We also keep only the pairs of reads, which both mates were successfully mapped. All of the filterings was performed with Samtools package [8].

All data files were after that inspect in Qualimap tool [9]. We discover that even after filtering there were a coverage depth of our samples around 100 basecalls per position. An example report is shown in Fig. 1. We consider it high enough to proceed to further analysis.

After all these steps, we convert our file to *vcf* format, which is a position oriented data file. We excluded ale basecallings with Phred quality [10] above twenty, so only high-quality bases are included. Phred quality twenty means that there is a probability of only 1% that the basecall is wrong. That probability can be calculated from Eq. (1), where P is the probability of incorrect basecall and Q is assigned Phred quality score.

$$P = 10^{\frac{-Q}{10}} \tag{1}$$

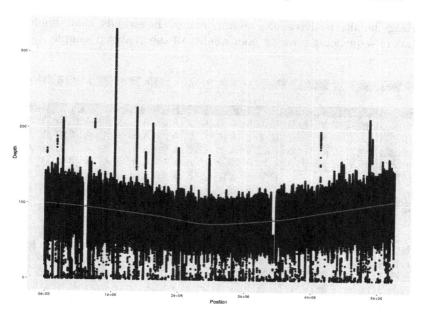

Fig. 1. Total per base coverage depth of S01_M98_S321_KP30 genome. By a blue line is shown the trend of depth value. (Color figure online)

2.2 Identification of Variable Positions

From this point, we set up another set of filters to export only variable positions from each genome. For our purposes by variable we mean the positions which are certainly different from the reference sequence. There are two types of variability [11] in our data files. One is caused by the difference between the genome and the reference sequence. The other type is intrastrain variability, which means that there are two or more alternating alels which got solid support in sequencing data. This type is shown in Fig. 2 in the fifth column from the right. There is a moreless equal ratio of cytosine and guanine basecalls.

The intrastrain variability is rather hard to identify and distinguish from sequencing errors. Especially when it can occur only in a small ratio of cells and then there is a rather small propagation of it to the sequencing data.

To avoid working with false positive variabilities we establish a set of rules to filter our data. We worked only with single nucleotide variants, so we discarded all variabilities identified as insertions or deletions.

To avoid using intrastrain variable regions, we excluded all variabilities which have the support of less than half a basecalls on its specific positions.

In pursuit of enhancing the quality of our dataset, we looked at all variable spots and count the basecalls origins from forward reads and reverse reads. The ratio of these reads should be equal to one. We decided to set the threshold for this parameter to be between 30/70. If there are significantly more basecalls supporting the alternative alleles on only one of the strands, that should imply some sort of sequencing error or poor quality of the analysed sample.

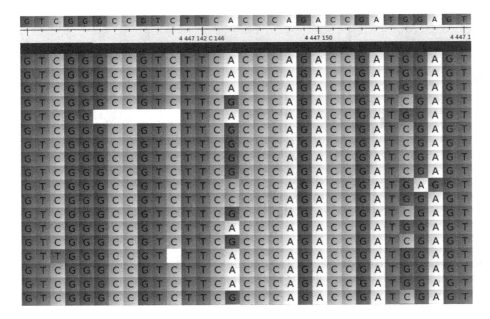

Fig. 2. Showing reads alignment to the reference sequence. On the top, there is a consensual sequence from reads assembly.

After applying all the filters mentioned above, we obtain a set of variable position for each sample. The summary is shown in Table 1.

2.3   Comparison of Variable Positions

Since all samples were mapped to the same sequence, there is no need to align the data.

All our genomes were converted into the signals of the same length. The length of the signals is defined by the length of the reference sequence used to map the reads. The samples of the signals are defined as the value of the ratio of occurrence of an alternative allele on their positions. All positions that are not marked as variable are set to zero.

Then we could perform some signal analysis, to determine the similarity of variability occurrence in our samples. We decided to measure the similarity of our samples by correlation of signals. We made this decision because it is simple and fast computation, and robust enough to take into account the character of the signals, which should be unique for every samples.

Table 1. *Klebsiella pneumoniae* strains used for analysis.

Name	Melt type	Num. of var. positions
S01_M98_S321_KP30	98	1503
S02_M29_S433_EB76	29	1219
S03_M98_S321_VYT9399	98	1263
S04_M98_S321_EB355	98	1591
S05_M29_S433_JIP9915	29	1377
S06_M98_S321_EB137	98	1552
S07_M29_S433_KP58	29	1355
S08_M98_S321_JCH10888	98	1448
S09_M29_S433_EB346	29	1130
S10_M98_S321_EB362	98	1373
S11_M29_S433_EB360	29	1334
S12_M29_S433_JIP10551	29	1248
S13_M23_S29_EB193	23	1486
S14_M23_S1271_HBC10480	23	1444
S15_M23_S1271_K176	23	1371
S16_M23_S1271_KP231	23	1638
S17_M23_S1271_KP255	23	1214
S18_M23_S1271_KP268	23	1492
S19_M61_S405_EB78	61	1369
S20_M61_S405_EB338	61	1301
S21_M61_S405_EB345	61	1155
S22_M61_S405_KP177	61	1476
S23_M61_S405_KP215	61	2643
S24_M61_S405_KP225	61	1886
S25_M61_S405_KP68	61	2346
S26_M61_S405_HBC2725	61	1495
S27_M14_S323_EB206	14	996
S28_M14_S323_EB50	14	1441
S29_M98_S321_EB362	98	1462
S30_M98_S321_EB368	98	1104
S31_M14_S323_EB146	14	1447
S32_M14_S323_EB156	14	2013
S33_M16_S23_EB359	16	132
S34_M16_S23_EB298	16	323
S35_M61_S405_KP502	61	7493
S36_M61_S405_KP416	61	10830
S37_M29_S433_KP387	29	1119
S38_M29_S433_KP378	29	1560
S39_M29_S433_KP312	29	1543
S40_M29_S433_KP264	29	2688
S41_M23_S1271_KP268	23	1350
S42_M29_S433_KP344	29	1322
S43_M23_S1271_EB364	23	1332
S44_M61_S405_EB378	61	3468
S45_M44_S458_EB358	44	2147
S46_M44_S458_EB178	44	1524
S47_M95_S1646_EB203	95	1398
S48_M29_S433_EB54	29	1521

That coefficient was appointed to every pair of our sample signals. From the correlation, we could determine the distance metrics as is shown by Eq. (2).

$$Distance = \frac{(1 - correlation)}{2} \tag{2}$$

Using this method, we rescale the distance matrices to a scale from 0 to 1.

Build upon the defined distances between signals, we can compute a phylogenetic tree to show the similarity of our samples. We decided to use UPGMA (Unweighted Pair Group Method with Arithmetic Mean) agglomerative hierarchical clustering method provided with our distance matrix based on correlations.

Establishing the dendrogram from the full-length signals gives us a robust reference structure.

A dendrogram is a similar to phylogenetic tree. Obtaining the tree from a signal is computationally less time-consuming way than construct the tree by some letter-based method, but robust enough to be considered valid.

After estimating the reference tree. We split our signals to pieces corresponding to CDS segments of the reference sequence.

With each segment we perform the same process to obtain a dendrogram as with full-length signal. After every tree construction process, the dendrogram is compared with the reference tree by the branch length score defined by Kuhner and Felsenstein [12] method.

For all of the sites and their constructed phylogenetic trees, we estimate the distance by Kuhner and Felsenstein. This method takes regard not even to topological order of tree branches, but also their lengths. We consider selecting the genes with the minimum distance length as the best segments to carry enough information to correctly cluster the samples.

3   Results

We analyzed all the provided samples, in total 48 in-house sequenced genomes. Firstly there were identification of all variable positions in all samples. This information was then converted to numerical signal and we compute the correlogram of all sample signals. Correlogram is shown in Fig. 3.

We can clearly see the close similarity between the samples of the same melt type. The correlation coefficients between more distant strains are significantly lower.

Based on this correlation matrix, we convert it into a distance matrix and perform a clustering method to visualize the similarity between the samples. We decided to use the UPGMA agglomerative hierarchical clustering method. The method constructs a rooted dendrogram as shown in Fig. 4.

We consider the output of the UPGMA clustering method to be our reference structure.

Without consideration of the length of the branches, we could see that the variability signals provided enough information to distinguish between the melt types of our samples.

After the signals segmentation according to CDS regions provided from the reference sequence, we perform the same signal analysis and sample clustering. The analysis reveals 36 CDS regions of the total 5037, which variability signals are enough to reconstruct the dendrogram with the same topological structure as the one from the full-length signal.

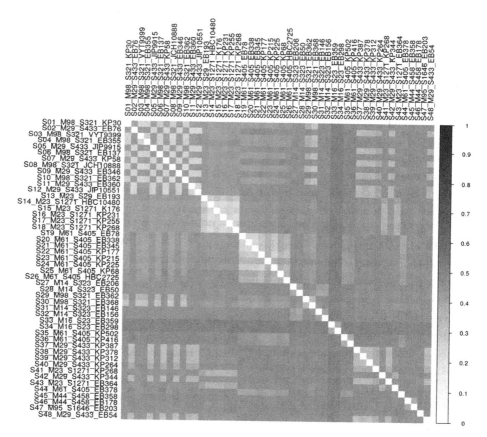

Fig. 3. Correlogram of occurrence of non-reference bases in analysed strains of *Klebsiella pneumoniae*

For all of these CDS we compute the Kuhned and Felsenstein distance [12]. All the CDS regions that have distance lower then 15% are shown in Table 2. These CDS regions we consider to be enough variable to distinguish the type of strains. And also their variability signal provided enough information to reconstruct the phylogenetic tree.

Table 2. Selected CDS regions with topology distance less then 15%.

Product	Protein_id
DUF1738 domain-containing protein	WP_014908068.1
transcriptional regulator	WP_012737276.1
TIGR03751 family conjugal transfer lipoprotein	WP_012737281.1
TIGR03750 family conjugal transfer protein	WP_012737285.1
TIGR03745 family integrating conjugative elementmembrane protein	WP_012737286.1
TIGR03758 family integrating conjugative elementprotein	WP_012737287.1
DUF3577 domain-containing protein	WP_012737296.1
carbon starvation protein	WP_002887454.1
dihydrolipoyl dehydrogenase	WP_002888731.1
MBL fold metallo-hydrolase	WP_004176966.1
5-oxoprolinase subunit PxpB	WP_004176857.1
hypothetical protein	
aliphatic sulfonates ABC transporter ATP-bindingprotein	WP_004150838.1
peptide ABC transporter substrate-binding protein	WP_041937769.1
glucans biosynthesis glucosyltransferase MdoH	WP_002898967.1
NADP-dependent isocitrate dehydrogenase	WP_004150800.1
tryptophan synthase subunit beta	WP_004148107.1
efflux transporter outer membrane subunit	WP_014907680.1
fructuronate reductase	WP_014907648.1
LysR family transcriptional regulator	WP_002904495.1
HlyD family secretion protein	WP_014907565.1
ABC transporter substrate-binding protein	WP_004184165.1
DUF2236 domain-containing protein	WP_004176040.1
MFS transporter	WP_002907773.1
DNA-binding transcriptional activator MhpR	WP_004175672.1
phosphoenolpyruvate synthase	WP_002909055.1
P-type DNA transfer ATPase VirB11	WP_015874997.1
CidB/LrgB family autolysis modulator	WP_002912867.1
xylose isomerase	WP_002914200.1
UbiX family flavin prenyltransferase	WP_002915099.1
hypothetical protein	WP_002915104.1
DEAD/DEAH family ATP-dependent RNA helicase	WP_004150943.1
aspartate 1-decarboxylase autocleavage activatorPanM	WP_002920810.1
NAD(P)/FAD-dependent oxidoreductase	WP_004150109.1
DUF3237 domain-containing protein	WP_002922963.1
4-oxalomesaconate tautomerase	WP_004186236.1

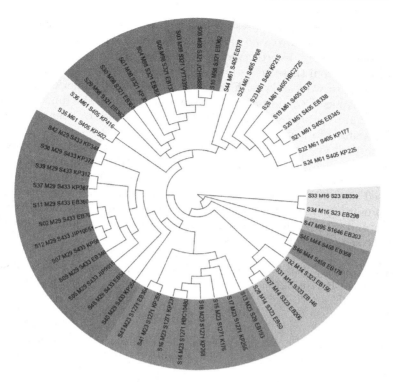

Fig. 4. Phylogenetic tree constructed by the UPGMA method from distance method based on correlation of occurrence of non-reference bases in analysed *Klebsiella pneumoniae* strains.

4   Conclusion

We provided a new approach to analyze and identify melt types of *Klebsiella pneumoniae* strains. Our approach is based on the identification of variable regions between the strains. The variability is estimated from sequencing data mapped to the same reference sequence.

This approach eliminates the need for assessment of a complete sequence for each individual genome. We are working with a variability signal, thus we eliminate the need to know the exact sequence. We work with a numerical signal of the probability of the alternation on the position in the genome, for further analysis we eliminate the need to preserve what type of substitution it is.

Working with the numerical representation of genomes is less computationally demanding and more time-saving then letter-based methods. In general letter-based methods are much more computationally demanding. With our numerical representation there are no need to ascertain the consensus sequences for all genomes. We simply mapped all to the chosen reference genome, and analyze the variability occurred.

Our signal representation preserves enough information to reconstruct the topologically correct dendrogram of analyzed samples. We showed that the information about topology is preserved even in some of the genes. Working with signals provided from these genes leads to the same results.

We propose to work only with segments of proposed numerical representations, which can save computational sources and decrease the dimension of the data files we are working with, yet still be robust enough to bear the majority of the clustering information.

Our method works with sequencing data, but not with the consensus sequences of the samples. It is concentrated only on the information about variability signal amongst the samples. We then working with much less data, then letter-based methods. Nevertheless we have shown that this numerical conversion can preserve enough information about the phylogeny to distinguish between melt types and with much less computional demands and without the knowledge of complete sequences of all samples.

Acknowledgments. This work was supported by grant project of the Czech Science Foundation [GACR 17-01821S].

References

1. Tagini, F., Greub, G.: Bacterial genome sequencing in clinical microbiology: a pathogen-oriented review. Eur. J. Clin. Microbiol. Infect. Dis. **36**(11), 2007–2020 (2017)
2. Zankari, E., et al.: Genotyping using whole-genome sequencing is a realistic alternative to surveillance based on phenotypic antimicrobial susceptibility testing. J. Antimicrob. Chemother. **68**(4), 771–777 (2012)
3. Levy, S.E., Myers, R.M.: Advancements in next-generation sequencing. Annu. Rev. Genomics Hum. Genet. **17**, 95–115 (2016)
4. Mendizabal-Ruiz, G., Román-Godínez, I., Torres-Ramos, S., Salido-Ruiz, R.A., Morales, J.A.: On DNA numerical representations for genomic similarity computation. PloS one **12**(3), e0173288 (2017)
5. Nykrynova, M., Maderankova, D., Bezdicek, M., Lengerova, M., Skutkova, H.: Bioinformatic tools for genotyping of Klebsiella pneumoniae isolates. In: Pietka, E., Badura, P., Kawa, J., Wieclawek, W. (eds.) ITIB 2018. AISC, vol. 762, pp. 419–428. Springer, Cham (2019). https://doi.org/10.1007/978-3-319-91211-0_37
6. Wu, K.-M., Li, L.-H., Yan, J.-J., Tsao, N., Liao, T.-L., Tsai, H.-C., et al.: Genome sequencing and comparative analysis of Klebsiella pneumoniae NTUH-K2044, a strain causing liver abscess and meningitis. J. Bacteriol. **191**(14), 4492–4501 (2009). https://doi.org/10.1128/JB.00315-09
7. Li, H., Durbin, R.: Fast and accurate short read alignment with Burrows-Wheeler transform. Bioinformatics **25**(14), 1754–1760 (2009). https://doi.org/10.1093/bioinformatics/btp324
8. Li, H., Handsaker, B., Wysoker, A., Fennell, T., Ruan, J., Homer, N., et al.: The sequence alignment/map format and SAMtools. Bioinformatics **25**(16), 2078–2079 (2009). https://doi.org/10.1093/bioinformatics/btp352
9. Okonechnikov, K., Conesa, A., García-Alcalde, F. Qualimap 2: advanced multi-sample quality control for high-throughput sequencing data. Bioinformatics https://doi.org/10.1093/bioinformatics/btv566

10. Ewing, B., Hillier, L.D., Wendl, M.C., Green, P.: Base-calling of automated sequencer traces using Phred 1 accuracy assessment. Genome Res. **8**(3), 175–185 (1998). https://doi.org/10.1101/gr.8.3.175

11. Pinto, M., Borges, V., Antelo, M., Pinheiro, M., Nunes, A., Azevedo, J., et al.: Genome-scale analysis of the non-cultivable Treponema pallidum reveals extensive within-patient genetic variation. Nature Microbiol. **2**(1), 112–119 (2017). https://doi.org/10.1038/nmicrobiol.2016.190

12. Kuhner, M.K., Felsenstein, J.: A simulation comparison of phylogeny algorithms under equal and unequal evolutionary rates. Mol. Biol. Evol. **11**(3), 459–468 (1994)

Addition of Pathway-Based Information to Improve Predictions in Transcriptomics

Daniel Urda[1]($\boxtimes$), Francisco J. Veredas[2], Ignacio Turias[1], and Leonardo Franco[2]

[1] Departamento de Ingeniería Informática, EPS de Algeciras,
Universidad de Cádiz, Cádiz, Spain
daniel.urda@uca.es
[2] Departamento de Lenguajes y Ciencias de la Computación, ETSI de Informática,
Universidad de Málaga, Málaga, Spain

Abstract. The diagnosis and prognosis of cancer are among the more critical challenges that modern medicine confronts. In this sense, personalized medicine aims to use data from heterogeneous sources to estimate the evolution of the disease for each specific patient in order to fit the more appropriate treatments. In recent years, DNA sequencing data have boosted cancer prediction and treatment by supplying genetic information that has been used to design genetic signatures or biomarkers that led to a better classification of the different subtypes of cancer as well as to a better estimation of the evolution of the disease and the response to diverse treatments. Several machine learning models have been proposed in the literature for cancer prediction. However, the efficacy of these models can be seriously affected by the existing imbalance between the high dimensionality of the gene expression feature sets and the number of samples available, what is known as the curse of dimensionality. Although linear predictive models could give worse performance rates when compared to more sophisticated non-linear models, they have the main advantage of being interpretable. However, the use of domain-specific information has been proved useful to boost the performance of multivariate linear predictors in high dimensional settings. In this work, we design a set of linear predictive models that incorporate domain-specific information from genetic pathways for effective feature selection. By combining these linear model with other classical machine learning models, we get state-of-art performance rates in the prediction of vital status on a public cancer dataset.

Keywords: Next-generation sequencing · Machine learning ·
Problem-specific information · Predictive modelling

1 Introduction

Personalized medicine, also known as *P4 medicine*—which is predictive, preventive, personalized and participatory–, is a recent field of the medicine that aims to establish personalized diagnoses and follow-ups for patients in order to fit the more appropriated and effective treatments [1]. In the case of cancer, personalized medicine aims at using heterogeneous data obtained from a single patient to predict the evolution

© Springer Nature Switzerland AG 2019
I. Rojas et al. (Eds.): IWBBIO 2019, LNBI 11466, pp. 200–208, 2019.
https://doi.org/10.1007/978-3-030-17935-9_19

of the disease, the patient's future vital status, the probability of suffering from new events of the disease (relapses), the response to different treatments, etc. This data can be of very different nature, such as medical images, histo-pathological samples or, more recently, genomics data from different sequencing techniques—DNA microarrays, RNASeq, miRNASeq, DNA methylation data, etc.—or even multi-omics data that combine genomics, transcriptomics, proteomics and epigenomics information.

Numerous machine learning studies have been proposed in the last years to tackle the problem of cancer diagnosis and prediction, using many different supervised and unsupervised approaches—such as logistic regression, Bayesian networks, support vector machines, neural networks, decision trees, etc.—on gene expression data [2, 3]. However, the efficacy of these machine learning approaches can be seriously affected by the typical existing imbalance between the large number (P) of input variables (i.e. gene expression features) and the small number (N) of samples. In the last years, many different techniques have been proposed to counteract the effects of what is known as the curse of dimensionality ($P >> N$) [4], most of them aimed at reducing the number of characteristics to get a more compact feature space, either by manual procedures (feature engineering methods) or by automatic feature selection techniques [5]. For cancer diagnosis and prediction, these feature selection algorithms allow not only to reduce the dimensionality of the input space but also to identify those genes with the highest correlation with the predicted output. Those genes constitute the so-called biomarkers or genetic signatures, which are of an extreme importance due to their clinical implications [6].

Of all the machine learning models published in the literature for cancer prediction, linear models have the major advantage of being more interpretable than other more complex ones, considered sometimes as "black-box" models. Although the performance of linear models (when compared to non-linear techniques) could be harmed by their intrinsic "linearity" restrictions, the use of domain-specific information has been proved useful to boost the performance of multivariate linear predictors in high dimensional settings [7,8]. In this sense, one could group features of the input space which may interact as a block. In particular, the analysis of gene expression data allows to use prior information available at well-known public resources like the KEGG database [9]. KEGG organizes genes into pathways according to molecular interaction networks [10,11], and this grouping structure is a valuable information that should be taken into account when developing predictive models in order to help us overcoming potentially negative effects of the curse of dimensionality.

In this work, we use the public TCGA-BRCA cancer dataset to design a set of linear predictive models that incorporate domain-specific information from genetic pathways. The resulting information-enriched linear models improve the overall performance given by a gold-standard Lasso model used as baseline linear predictor. Finally, by using the proposed linear models as efficient feature selection algorithms—which select not only genes (input features) but also gene groups (pathways)—for other classical machine learning models, we get state-of-art performance rates in the prediction of vital status on the TCGA-BRCA dataset.

The rest of the paper is organized as follows. Section 2 presents the predictive models tested and the methodology used in this paper. Next, the gene expression dataset used and the results obtained from the analysis are included in Sect. 3 and, finally, some conclusions are provided in Sect. 4.

2 Methodology

This study aims to use group variable information in order to improve predictions. In this sense, a well-known model has been used as baseline or reference model together with a set of machine learning models which somehow allows to incorporate this kind of information. Next, the different models as well as the methodology used to include the group variable information are described.

Baseline Model

A standard Lasso (Least Absolute Shrinkage and Selection Operator) model [12] was the reference model used to measure the improvement achieved in terms of predictive performance. Lasso adds a l_1-regularization penalty to the classification error in such a way that allows to perform variable selection during the estimation process. In this sense, Eq. 1 shows the objective which is minimized in the Lasso problem:

$$\hat{\boldsymbol{\beta}}_\lambda = \arg\min_{\boldsymbol{\beta}} \ ||\boldsymbol{y} - f(\boldsymbol{\beta}X)||_2^2 + \lambda ||\boldsymbol{\beta}||_1 \tag{1}$$

where X is the input matrix, $\boldsymbol{y}$ the outcome vector and $\boldsymbol{\beta}$ is the vector of coefficients which are estimated. In a classification setting like this one, f corresponds to the logistic function which shrinks any numeric input value to a number between 0 and 1. The first term in Eq. 1 corresponds to the classification error while the second term is the l_1-penalty. The amount of regularization applied is controlled by λ which is a hyper-parameter of the model typically learned through cross-validation.

Adding Pathway-Based Information to Lasso

The baseline model (Lasso) is limited in the sense that no prior information is used during the estimation process. However, it is possible to retrieve and use a priori information from a free-public[1] well-known resource such as the KEGG database [9]. KEGG organizes genes into pathways according to molecular interaction networks [10, 11] and allows us to built-in the grouping genes information into a linear model such as the group lasso [13]. This model is an extension of Lasso to do variable selection on predefined disjoint groups of variables as shown in Eq. 2:

$$\hat{\boldsymbol{\beta}}_\lambda = \arg\min_{\boldsymbol{\beta}} \ ||\boldsymbol{y} - f(\boldsymbol{\beta}X)||_2^2 + \lambda \sum_{g=1}^{G} ||\boldsymbol{\beta}_{I_g}||_2 \tag{2}$$

where I_g is the index set belonging to the g-th group of variables.

However, our problem requires a group lasso implementation which allows to define overlapping groups of variables since the same gene may appear in one or more pathways [14]. The R package *grpregOverlap* [15] which implements several types of penalties allow us to test the different settings listed below:

---

[1] https://www.genome.jp/kegg/.

- Group selection: selects important pathways, and not genes within the pathway, i.e., within a pathway, coefficients will either be all zero or all nonzero. The package implements three different penalties which are tested in this work: *grLasso*, *grMCP* and *grSCAD* (further details regarding differences on each penalty can be found in [15]).
- Bi-level selection: carries out variable selection at the pathway level as well as the level of individual genes, i.e. selecting important pathways as well as important genes of those pathways. The package implements two different penalties which are tested in this work: *gel* and *cMCP* (further details regarding differences on each penalty can be found in [15]).

Other Predictive Models

In addition to the previous models, a third setting was considered in this work which uses the group lasso with *cMCP* penalty to select relevant pathways and, within each one, relevant genes as a feature selection method, to then apply a set of well-known machine learning models with the retained genes. The following machine learning models were considered:

- Generalized linear models: it includes a classical linear model with different types of regularization. Three different options were considered: *Lasso* (l_1-penalty: $||\boldsymbol{\beta}||_1$), *Ridge* (l_2-penalty: $||\boldsymbol{\beta}||_2$), and *Elnet* (mix of l_1 and l_2 penalties: $\alpha||\boldsymbol{\beta}||_1 + (1 - \alpha)||\boldsymbol{\beta}||_2$, where $\alpha \in [0, 1]$ controls the amount applied of each penalty).
- *K*-Nearest Neighbours (*kNN*): it is a distance-based method which calculates predictions for a new sample based on the predictions of its *k*-nearest neighbours, where $k \in [1..N]$, according to a similarity measure (Euclidean distance, Mahalanobis distance, or any other) [16].
- Random forest (*RF*): it is a tree-based bagging ensemble for which multiple decision trees are fitted to different views of the observed data. Random forest predictions are made by averaging the individual predictions provided by the multiple decision trees [17].
- Support Vector Machines (*SVM*): a kernel-based method (linear and radial basis ones were considered in this paper) that maps the original input data into a new space where predictions can be made more accurately [18].

Therefore, the combination of group lasso with the *cMCP* penalty as filtering method followed by each of these predictive models lead to seven different settings: *cMCP+Lasso*, *cMCP+Ridge*, *cMCP+Elnet*, *cMCP+kNN*, *cMCP+RF*, *cMCP+SVMlin* and *cMCP+SVMrbf*. The R package *mlr* [19] provides an implementation of all the machine learning models used, permitting to test the different settings within the analysis.

3   Results

The Breast Cancer Adenocarcinoma (BRCA) dataset which is freely available at The Cancer Genome Atlas (TCGA) website[2] was used in this study. This dataset consists of

---

[2] https://cancergenome.nih.gov/.

1212 samples of patients, each of them described by more than twenty thousand genes. This dataset has been previously pre-processed to take into account batch effects and normalized through the RSEM procedure [20]. Additionally, an extra post-processed step was carried out to remove those genes with a constant expression level across the samples. The remaining genes were *log*-transformed, $log_2(gexpr + 1)$, to make their distribution look as close as possible to a normal distribution, thus resulting in a total of 20021 genes. Moreover, each of the samples in the dataset has a label that represents the event of interest or outcome to predict, which corresponds to the vital status of a given patient (0 = "*alive*", 1 = "*dead*") at a fixed time t. Out of the 1212 samples, the dataset contains 199 controls (or alive) and 1013 cases.

The analysis was carried out performing 10-fold cross-validation [21], thus partitioning the entire dataset in 10 folds of equal sizes in order to estimate the performance of each model. In this sense, models were fitted in 9 folds and tested in the unseen test fold left apart within an iterative procedure that rotates the train and test folds used. The Area Under the Curve (AUC) was used to measure the goodness of a given model fitted to data. With respect to the tune of models' hyper-parameters, the R package *mlrMBO* [22] was used to perform a Bayesian optimization within the train set. This package implements a Bayesian optimization of black-box functions which allows to find faster an optimal hyper-parameters setting in contrast to traditional hyper-parameters search strategies such as grid search (highly time consuming when more than 3 hyper-parameters are tuned) or random search (not efficient enough since similar or non-sense hyper-parameters settings might be tested).

Table 1 shows the average AUC performance, standard deviation and number of genes retained by the different models tested over the test sets of the cross-validation setting. In terms of AUC, models which consider the use of pathway-based information

Table 1. Average AUC test results, standard deviation and number of genes retained by the set of different models tested after performing 10-fold cross-validation.

Model	AUC	#Genes
Lasso	0.659 ± 0.065	259.5
grLasso	0.681 ± 0.059	6094.1
grMCP	0.692 ± 0.057	6492.1
grSCAD	0.671 ± 0.063	6240.4
gel	0.695 ± 0.065	2787.9
cMCP	0.696 ± 0.052	371.2
cMCP+Lasso	0.668 ± 0.047	
cMCP+Ridge	**0.702 ± 0.033**	
cMCP+Elnet	0.691 ± 0.032	
cMCP+kNN	0.653 ± 0.061	371.2
cMCP+RF	0.663 ± 0.059	
cMCP+SVMlin	**0.702 ± 0.034**	
cMCP+SVMrbf	0.698 ± 0.038	

Table 2. Pathways with more than 10 genes retained by the *cMCP* model fitted on the complete BRCA dataset.

Pathway ID	Description	#Genes	#Selected
hsa00980	Metabolism of xenobiotics by cytochrome P450	76	11
hsa04010	MAPK signaling pathway	295	14
hsa04024	cAMP signaling pathway	198	12
hsa04060	Cytokine-cytokine receptor interaction	294	27
hsa04062	Chemokine signaling pathway	189	11
hsa04080	Neuroactive ligand-receptor interaction	277	17
hsa04144	Endocytosis	244	13
hsa04151	PI3K-Akt signaling pathway	354	12
hsa04390	Hippo signaling pathway	154	11
hsa04740	Olfactory transduction	448	46
hsa05165	Human papillomavirus infection	339	19
hsa05166	Human T-cell leukemia virus 1 infection	255	13
hsa05168	Herpes simplex infection	185	14
hsa05200	Pathways in cancer	526	24
hsa05202	Transcriptional misregulation in cancer	186	11
hsa05203	Viral carcinogenesis	201	12
hsa05204	Chemical carcinogenesis	82	12

during the estimation process (*grLasso*, *grMCP*, *grSCAD*, *gel*, *cMCP*) are able to outperform the reference model *Lasso*. In the worst case scenario, the quantitative improvement of these kind of models is 0.012 (*grSCAD* versus *Lasso*) while in the best case scenario this improvement increases up to 0.037 points better (*cMCP* versus *Lasso*). Moreover, a slightly better performance is obtained when fitting a *Ridge* or *SVMlin* model to the genes retained by the *cMCP* filtering method (up to 0.702 of AUC). With respect to the standard deviations, it turned out that settings where *cMCP* is used as filtering method followed by the machine learning models tested presented approximately 50% less variability in the results (around 0.03, except Lasso, kNN and RF) than Lasso or group Lasso (around 0.06) with any of the penalties tried. For the group lasso settings, it is important to remark that 4 out of the 5 penalties led to substantially big genetic signatures (in the order of thousands) which may suggests not to use these settings if the main interest is to find a small and robust genetic signature with high prediction capabilities.

Regarding the pathways retained by the *cMCP* model, Table 2 presents those pathways for which more than 10 genes were kept (17 pathways in total). On one hand, it can be seen that a remarkable selection was carried out by the model discarding many of the genes involved in these pathways. On the other hand, many of these pathways have been previously linked to breast cancer. For instance, *hsa04740* and *hsa04080* were found to have mutational significance in [23], *hsa04060* and *hsa05200* were cited

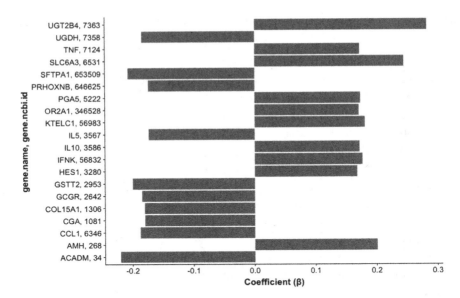

Fig. 1. Summary of the top-20 selected genes according to the $|\beta|$ coefficients of the *cMCP+Ridge* model fitted to the complete dataset.

in [24, 25] respectively as associated to breast cancer in the early stage, while *hsa04010* was found to be associated with aggressive breast cancer in [26].

Although biological interpretation of the results is beyond the scope of this paper, authors would like to highlight the possibility of going deeper in the analysis and, whenever an interpretable machine learning model was used, try to check the contribution of each gene to the given outcome. For instance, Fig. 1 shows the top-20 genes retained by the model *cMCP+Ridge* according the absolute value of the β coefficients. As *Ridge* was used in the end of the pipeline, one could open a discussion where genes with a positive coefficient would be negative for the patient survival while a negative coefficient will have right the opposite effect. Even more, the magnitude of the coefficients would be also indicative of the strength or weakness of these effects.

4   Conclusions

In this paper, a public TGCA-BRCA cancer dataset was used to design a set of linear predictive models that incorporate domain-specific information from genetic pathways. In particular, the group lasso model with 5 different regularization penalties were used (*grLasso, grMCP, grSCAD, gel, cMCP*), three of them allowing to perform pathway selection and the other two for applying bi-level selection (both pathways and genes within the pathways are selected), in order to test their performance compared to a standard *Lasso* model. Moreover, the *cMCP* setting was additionally used as filtering method and the retained genes were further used to fit several machine learning models (*Lasso, Ridge, Elnet, kNN, RF, SVMlin, SVMrbf*).

The results of the analysis showed that the *cMCP+Ridge* or *cMCP+SVMlin* models obtained the best performance in terms of AUC (0.044 points higher than the baseline *Lasso*). However, their performance is slightly better than simply using the *cMCP* model (just 0.006 points higher). Nevertheless, the settings where *cMCP* was used as filtering method followed by any of the machine learning models considered in this work presented approximately 50% less variability in the results than the other models (both the baseline *Lasso* or any of the group lasso settings). In addition, almost all of the group lasso settings ended up with a high number of genes retained (in the order of thousands), what may cause difficulties when trying to find the smallest and more robust genetic signatures with good predictive capabilities. Therefore, the filtering approach followed by classical machine learning models may be more suitable for this purposes.

In future work, authors will continue exploring the benefits of using pathway-based information to improve predictions based on gene expression data. To this end, ensemble approaches or deep learning models could potentially be explored as more powerful tools which may help to boost predictions even further.

Acknowledgments. This work is part of the coordinated research projects TIN2014-58516-C2-1-R, TIN2014-58516-C2-2-R and TIN2017-88728-C2 from MINECO-SPAIN which include FEDER funds.

References

1. Aronson, S.J., Rehm, H.L.: Building the foundation for genomics in precision medicine. Nature **526**(7573), 336–342 (2015)
2. Kourou, K., Exarchos, T.P., Exarchos, K.P., Karamouzis, M.V., Fotiadis, D.I.: Machine learning applications in cancer prognosis and prediction. Comput. Struct. Biotechnol. J. **13**, 8–17 (2015)
3. Bashiri, A., Ghazisaeedi, M., Safdari, R., Shahmoradi, L., Ehtesham, H.: Improving the prediction of survival in cancer patients by using machine learning techniques: experience of gene expression data: a narrative review. Iran. J. Public Health **46**(2), 165–172 (2017)
4. Johnstone, I.M., Titterington, D.M.: Statistical challenges of high-dimensional data. Philos. Trans. A Math. Phys. Eng. Sci. **367**(1906), 4237–4253 (2009)
5. Saeys, Y., Inza, I., Larrañaga, P.: A review of feature selection techniques in bioinformatics. Bioinformatics **23**(19), 2507–2517 (2007)
6. van't Veer, L.J., et al.: Gene expression profiling predicts clinical outcome of breast cancer. Nature **415**(6871), 530–536 (2002)
7. Simon, N., Friedman, J., Hastie, T., Tibshirani, R.: A sparse-group lasso. J. Comput. Graph. Stat. **22**(2), 231–245 (2013)
8. Urda, D., Jerez, J.M., Turias, I.J.: Data dimension and structure effects in predictive performance of deep neural networks. In: New Trends in Intelligent Software Methodologies, Tools and Techniques, pp. 361–372 (2018)
9. Kanehisa, M., Goto, S.: KEGG: kyoto encyclopedia of genes and genomes. Nucleic Acids Res. **28**(1), 27–30 (2000)
10. Kanehisa, M., Sato, Y., Kawashima, M., Furumichi, M., Tanabe, M.: KEGG as a reference resource for gene and protein annotation. Nucleic Acids Res. **44**(D1), D457–D462 (2016)
11. Kanehisa, M., Furumichi, M., Tanabe, M., Sato, Y., Morishima, K.: KEGG: new perspectives on genomes, pathways, diseases and drugs. Nucleic Acids Res. **45**(D1), D353–D361 (2017)

12. Tibshirani, R.: Regression shrinkage and selection via the lasso: a retrospective. J. R. Stat. Soc.: Ser. B (Stat. Methodol.) **58**(1), 267–288 (1996)
13. Meier, L., Van De Geer, S., Bühlmann, P.: The group lasso for logistic regression. J. R. Stat. Soc.: Ser. B (Stat. Methodol.) **70**(1), 53–71 (2008)
14. Jacob, L., Obozinski, G., Vert, J.P.: Group lasso with overlap and graph lasso. In: Proceedings of the 26th Annual International Conference on Machine Learning, pp. 433–440 (2009)
15. Zeng, Y., Breheny, P.: Overlapping group logistic regression with applications to genetic pathway selection. Cancer Inf. **15**(1), 179–187 (2016)
16. Cover, T., Hart, P.: Nearest neighbor pattern classification. IEEE Trans. Inf. Theor. **13**(1), 21–27 (2006)
17. Breiman, L.: Random forests. Mach. Learn. **45**(1), 5–32 (2001)
18. Rossi, F., Villa, N.: Support vector machine for functional data classification. Neurocomputing **69**(7), 730–742 (2006)
19. Bischl, B., et al.: mlr: machine learning in R. J. Mach. Learn. Res. **17**(170), 1–5 (2016)
20. Li, B., Dewey, C.N.: RSEM: accurate transcript quantification from RNA-Seq data with or without a reference genome. BMC Bioinf. **12**(1), 323 (2011)
21. Kohavi, R.: A study of cross-validation and bootstrap for accuracy estimation and model selection. In: Proceedings of the 14th International Joint Conference on Artificial Intelligence, IJCAI 1995, vol. 2, pp. 1137–1143 (1995)
22. Bischl, B., Richter, J., Bossek, J., Horn, D., Thomas, J., Lang, M.: mlrMBO: A Modular Framework for Model-Based Optimization of Expensive Black-Box Functions (2017)
23. Dees, N.D., et al.: MuSiC: identifying mutational significance in cancer genomes. Genome Res. **8**, 1589–98 (2012)
24. Shimomura, A., et al.: Novel combination of serum microrna for detecting breast cancer in the early stage. Cancer Sci. **107**(3), 326–334 (2016)
25. Zhao, H., Shen, J., Medico, L., Wang, D., Ambrosone, C.B., Liu, S.: A pilot study of circulating miRNAs as potential biomarkers of early stage breast cancer. PLoS ONE **5**(10), 1–12 (2010)
26. Chen, G.Q., Zhao, Z.W., Zhou, H.Y., Liu, Y.J., Yang, H.J.: Systematic analysis of microRNA involved in resistance of the MCF-7 human breast cancer cell to doxorubicin. Med. Oncol. **27**(2), 406–415 (2010)

Common Gene Regulatory Network for Anxiety Disorder Using Cytoscape: Detection and Analysis

Md. Rakibul Islam[1], Md. Liton Ahmed[1], Bikash Kumar Paul[1,2], Sayed Asaduzzaman[1,2], and Kawsar Ahmed[2(✉)]

[1] Faculty of Science and Information Technology, Software Engineering, Daffodil International University (DIU), Ashulia, Savar, Dhaka 1342, Bangladesh
[2] Department of Information and Communication Technology, Mawlana Bhashani Science and Technology University, Santosh, Tangail, Bangladesh
Kawsar.ict@mbstu.ac.bd

Abstract. Data mining, computational biology and statistics are unified to a vast research area Bioinformatics. In this arena of diverse research, protein - protein interaction (PPI) is most crucial for functional biological progress. In this research work an investigation has been done by considering the several modules of data mining process. This investigation helps for the detection and analyzes gene regulatory network and PPI network for anxiety disorders. From this investigation a novel pathway has been found. Numerous studies have been done which exhibits that a strong association among diabetes, kidney disease and stroke for causing most libelous anxiety disorders. So it can be said that this research will be opened a new horizon in the area of several aspects of life science as well as Bioinformatics.

Keywords: Bioinformatics · Genomics · PPI network · Regulatory network · Data analysis · Diabetes · Kidney disease · Stroke · Anxiety

Abbreviations

DB = Diabetes
KD = Kidney Disease
ST = Stroke
AN = Anxiety

1 Introduction

Bioinformatics is a hybrid science that provides biological information storage by a combine of biological data and techniques. Bioinformatics plays such monumental accost to gain very important information about diseases which are helpful for drug design. Anxiety disorder is the fastest growing disorder in the world. At present anxiety

© Springer Nature Switzerland AG 2019
I. Rojas et al. (Eds.): IWBBIO 2019, LNBI 11466, pp. 209–218, 2019.
https://doi.org/10.1007/978-3-030-17935-9_20

disorder is a threat for human being. Today the total calculated amount of people surviving with anxiety disorders in the world is around 264 million in ref [1]. Anxiety is one of the most leading cause of inability worldwide. About 75% of people with anxiety remain without any kind of treated in developing countries with almost 1 million people every year taking their own lives by anxiety. In addition, 1 in 13 globally suffers from anxiety in ref [2]. Diabetes is also one of the most growing disease in the world. Diabetes is a principal cause of many health related problems as like as blindness, kidney failure, heart attacks, stroke and lower limb amputation. In 1980, 108 million peoples in the world are affected by diabetes, but in 2014 this amount is huge in number 422 million people around the world are now suffering with diabetes in ref [3, 4].

Besides, Kidney disease is also globally growing deathly disease. Each year many people died without any treatment because they do not have access to affordable treatment. It is estimated that worldwide 10% of the population is attacked by chronic kidney disease. In the world over 2 million peoples at present receiving treatment with dialysis or a kidney transplant to remain alive in ref [5]. Stroke is a proviso where the blood supply to the brain is crumble up. Stroke has already propagate pestilence ratio. Today in the world 15 million people suffer a stroke each year and 5.8 million people die from it. Stroke is accountable for more deaths yearly than AIDS, tuberculosis and malaria both combined in ref [6].

This research investigates with Anxiety, Diabetes, Kidney Disease and Stroke whose are directly or indirectly associated with each other. For the first time we tried to build PPI and Regulatory Interaction networks between common genes. This paper is designed in 5 individual section.

2 Background

A very few previous research work has been adopted on Bioinformatics that involves analyzing liable genes, creating PPI network and build a common pathway for selected 4 diseases. This presented research is a descendent of all prior research which is aimed to work with 4 diseases, they are Anxiety, Diabetes, Kidney Disease and Stroke. In this research work, the liable genes for these diseases are found out and then create a PPI network and Regulatory Network by using various Bioinformatics tools.

Several Bioinformatics tools were used to build Interaction network & PPI Network among liable genes are discussed in the article [7]. Anxiety is one of the fastest growing disorder in this world that may cause of harms in the future. It has recently been shown that anxiety disorder has a strong relation with diabetes and closely related genetically. That means there is a significant genetic overlap between the two diseases in ref [8]. Anxiety is the root of many other mental diseases were shown in an article in ref [9].

Diabetes is a disease that befalls when your blood glucose, also known as blood sugar, is too exalted. Around 422 million of people are affected by this disease. Diabetes can raise may other major diseases such like as Stroke and many more in ref [10]. Anxiety disorders are common within patients with diabetes and seem to declaim in a tangible ratio of incident in ref [11]. In 2014, the total number of 52,189 people exhibited end-stage renal disease with diabetes with diabetes as the fundamental reason in ref [12].

Worldwide 1 in every 10 is influenced by chronic kidney disease, and above millions passed away each year because they do not have access to affordable treatment (World Kidney Day: Chronic Kidney Disease. 2015). For kidney disease, diabetes is a major risk factor (for kidney disease). Anxiety correlated significantly with kidney disease in ref [13].

Stroke is a major cause of impotence. Stroke is one of the top leading causes of death in developing countries. Stroke is a disease that affects the arteries prominent to and within the brain. It is the No. 5 reason of death and a leading cause of incapability in the United States (American Stroke Association). Diabetes patients have higher risk to face stroked in ref [14].

3 Proposed Methodology

In this research work a small number of modules have been performed. These modules are aiming for processing, filtering, mining, common gene finding etc. to ensure the expected goal. Each and every steps of this work has been illustrated concisely below. All steps of this methodology are inside through Subsects. 3.1 to 3.6.

3.1 Gene Collection

The NCBI (National Center for Biotechnology Information) is a vital resource to freely accessible and downloadable online gene database and a huge collection for Bioinformatics services and tools. Based on the behavior of several data, they stored data in varied databases. For example, Different kinds of genes are searchable from Gene database storage. For some aspects PubMed, OMIM and Gene Bank databases are also used for gene collection. In this research project genes associated with Anxiety, Diabetes, Kidney Disease & Stroke are collected from the NCBI gene database.

3.2 Preprocessing and Gene Filtering

In the prior step all the associated genes with Anxiety, Diabetes, Kidney Disease and Stroke are collected. But in this immediate step only preprocess and collect only those genes that are for Homo sapiens. That is, all the collected genes are filtered and only responsible genes for human diseases are kept for analysis. The genes are downloaded in the progressive order of their weight. For each of the four diseases, genes are downloaded and stored individually in a text file.

3.3 Gene Linkage

Gene linkage step ascertains the interrelated genes among the selected diseases. The interrelated genes between KD and DB; KD and AN; AN and ST; DB and AN; DB and ST; ST and KD; ST, KD and AN; ST, DB and AN; ST, DB and KD; ST, DB, AN and KD are identified and collected.

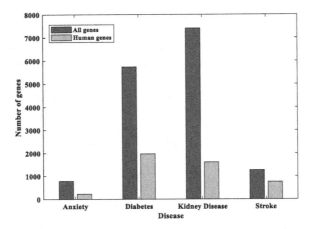

Fig. 1. Collected genes for specific diseases where X axis denotes anxiety, diabetes, kidney disease and stroke, Y axis denoting the total number of the genes for corresponding diseases; Blue are for all species and yellow area for *Homo sapiens*. (Color figure online)

3.4   Gene Mining

Data mining deftness is used mainly for making exact data for work. Gene mining is one of the most essential part of this research because single omission can backtrack an important gene that can cause for wrong result. Only linkage gene files genes are mined in this step, from this mining the top 100 genes are taken and saved them in a text file.

3.5   PPI Network

PPI network or Protein- Protein Interaction network is used to represent the directly and indirectly connected gene and protein interaction among the interrelated genes of the selected diseases. Cytoscape, a very known and popular, reliable application for Bioinformatics is used for this type of project purpose. In this step, from the interrelated common genes, PPI networks and common pathways are created by using Cytoscape application.

3.6   Regulatory Network

Gene Regulatory Network (GRN) is used to express the connection between genes each other. Cytoscape is a reliable application to find out regulatory network. In this research project we use Cytoscape to represent the regulatory network of common genes.

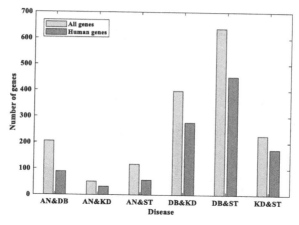

Fig. 2. Cross linkage between selected diseases

4 Results and Discussion

The main goal of this research is to design PPI and Regulatory network for selected 4 diseases. To fulfill the desired goal a few successive methods are accomplished as described in proposed methodology part. In this step, the final outcome of each step will be demonstrated and discussed.

4.1 Gene Collection and Filtering

The collected liable genes without any preprocessing and filtering are estimated as 789 for anxiety, 5737 for diabetes, 7409 for kidney disease and 1263 for stroke and after preprocessing as well as filtering the analogous genes for *Homo sapiens* are 228 for anxiety, 1972 for diabetes, 1608 for kidney disease and 753 for stroke. Before and after filtering the amount of genes for individual disease is shown by using bar plot as shown in Fig. 1.

4.2 Gene Linkage and Filtering

The cross linkage between the two diseases and among diseases are performed in this step. Firstly without filtering results are counted for the linkage between the two diseases as 398 for KD and DB, 51 for KD and AN, 117 for AN and ST, 205 for DB and AN, 638 for DB and ST and 227 for ST and KD and after filtering the corresponding genes for Homo sapiens are 277 for KD and DB, 32 for KD and AN, 57 for AN and ST, 89 for DB and AN, 452 for DB and ST and 175 for KD and ST. Linkage between the 2 diseases before filtering and after filtering genes amount is shown in Fig. 2. By bar plot. Resembling procedure are followed for cross linkage in among

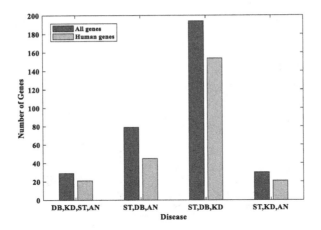

Fig. 3. Cross linkage among 4 diseases

diseases, results for without filtering are counted as 30 for AN, KD, ST; 79 for AN, DB, ST; 194 for DB, ST, KD and 29 for AN, DB, KD, ST and after filtering the corresponding genes for human are 21 for AN, KD, ST; 45 for AN, DB, ST; 154 for DB, ST, KD and 21 for AN, DB, KD, ST. Figure 3. Illustrate about before and after filtering amount of genes by bar plot.

4.3   Gene Mining

Gene mining is one of the most significant phase of this research work. After performing cross linkage among all 4 selected diseases 21 common genes are found. Corresponding genes are sorted in increasing order of their weight. After this phase, gene mining is applied for collecting top 50 weighted genes and 11 genes are found as a result. These 11 common genes are TNF, IL6, TGFB1, ADIPOQ, CRP, BDNF, PON1, SOD1, ICAM1, IL8, and AGT.

4.4   PPI network and Regulatory Interaction

PPI network is delegation of the physical contacts between proteins. It displays the protein interaction and common pathway among the interrelated genes. Cytoscape is an open source application project for Bioinformatics, which is used for building the PPI network and Biomolecular interaction networks with high throughput expression data and various states of molecular via a unified conceptual framework. PPI and regulatory interaction networks are used to illustrate the directly and indirectly connected gene and protein interactions (Figs. 4, 5, 6 and 7).

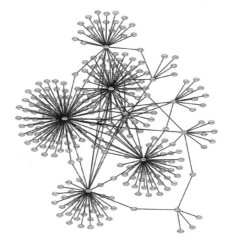

Fig. 4. PPI network for 11 common genes

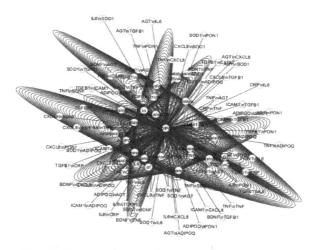

Fig. 5. Gene regulatory network for 11 common genes

4.5   Co-expression Network and Physical Interaction Pathway

Co-expression network usually is an undirected graph, this network displayed about significant correlation between a pair of genes. Two genes are connected if they were found to interact in a protein-protein interaction study. GeneMANIA is an open source online tool for Bioinformatics work. It is used to find out genetic interactions, co-expression and physical interaction pathways for an input dataset. From the 7 responsible common genes GeneMania creates co-expression network and physical interaction pathways which shown in Figs. 8 and 9 respectively.

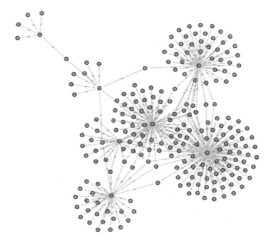

Fig. 6. PPI network 7 common genes.

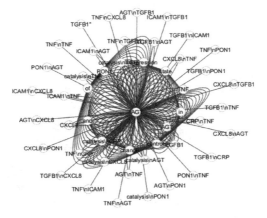

Fig. 7. Gene regulatory network among 7 common genes.

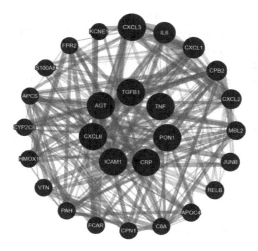

Fig. 8. Gene co-expression network among 7 common genes using GeneMania.

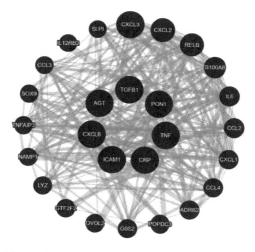

Fig. 9. Physical interaction pathways among 7 common genes using GeneMania.

5 Conclusion

The marvelous contribution on Bioinformatics has been found for modern genomics revolution. With the advancement of the computational biology and Bioinformatics some indomitable task makes easier. In the presented work, analyze on diseases has been performed to identify a common pathway PPI. PPI network, regulatory network, co-expressions leads to design a drug for certain diseases. An analysis is made on anxiety disorder which closely related with diabetes, kidney disease and stroke also. Analysis on a disease means analyzing the diseases genes, this study analyzes all the 4 diseases genes. Based on the analysis PPI network and Regulatory Interaction networks both are originate by showing the interrelated genes among the 4 diseases by using biological authentic tool Cytoscape version 3.6.1. Moreover the gene co expression for 7 common genes and pathway is constructed using GeneMANIA. It is obligate to know about liable genes for a certain disease for design a drug. PPI and Regulatory Networks are shows the cross linkage genes, common genes, interacted genes. Physical Interaction Pathways of liable genes are also assure the scheme to a common drug design. In future work of the research is to design a common drug for the selected 4 diseases. It can be considered this research will be provide a significant information for drug design.

References

1. WHO: Depression and Other Common Mental Disorders Global Health Estimates (2017)
2. Barlow, D.H.: Anxiety and Its Disorders: The Nature and Treatment of Anxiety and Panic. Guilford Publications, New York City (2013)
3. Anxiety and Depression Association of America (2017)
4. Mathers, C.D., Loncar, D.: Projections of global mortality and burden of disease from 2002 to 2030. PLoS Med. 3(11), e442 (2006)
5. National Kidney Foundation (2015)
6. World Stroke Organization (2015)
7. Klingström, T., Plewczynski, D.: Protein–protein interaction and pathway databases, a graphical review. Briefings Bioinf. 12(6), 702–713 (2010)
8. Habib, N., et al.: Design regulatory interaction network for anxiety disorders using R: a bioinformatics approach. Beni-Suef Univ. J. Basic Appl. Sci. 7(3), 326–335 (2018)
9. Spitzer, R.L., et al.: A brief measure for assessing generalized anxiety disorder: the GAD-7. Arch. Intern. Med. 166(10), 1092–1097 (2006)
10. Browner, W.S., Lui, L.-Y., Cummings, S.R.: Associations of serum osteoprotegerin levels with diabetes, stroke, bone density, fractures, and mortality in elderly women. J. Clin. Endocrinol. Metab. 86(2), 631–637 (2001)
11. Lustman, P.J., Clouse, R.E.: Depression in diabetic patients: the relationship between mood and glycemic control. J. Diabetes Complications 19(2), 113–122 (2005)
12. National Diabetes Statistics Report (2017)
13. Lee, Y.-J., et al.: Association of depression and anxiety with reduced quality of life in patients with predialysis chronic kidney disease. Int. J. Clin. Pract. 67(4), 363–368 (2013)
14. Ridker, P.M.: Inflammatory biomarkers and risks of myocardial infarction, stroke, diabetes, and total mortality: implications for longevity. Nutr. Rev. 65, S253–S259 (2007)

Insight About Nonlinear Dimensionality Reduction Methods Applied to Protein Molecular Dynamics

Vinicius Carius de Souza$^{(\boxtimes)}$ ⓘ, Leonardo Goliatt ⓘ,
and Priscila V. Z. Capriles ⓘ

Federal University of Juiz de Fora, Juiz de Fora, MG, Brazil
vinicius.carius@ice.ufjf.br,
{leonardo.goliatt,priscila.capriles}@ufjf.edu.br

Abstract. The advance in molecular dynamics (MD) techniques has made this method common in studies involving the discovery of physicochemical and conformational properties of proteins. However, the analysis may be difficult since MD generates a lot of conformations with high dimensionality. Among the methods used to explore this problem, machine learning has been used to find a lower dimensional manifold called "intrinsic dimensionality space" which is embedded in a high dimensional space and represents the essential motions of proteins. To identify this manifold, Euclidean distance between intra-molecular C_α atoms for each conformation was used. The approaches used were combining data dimensionality reduction (AutoEncoder, Isomap, t-SNE, MDS, Spectral and PCA methods) and Ward algorithm to group similar conformations and find the representative structures. Findings pointed out that Spectral and Isomap methods were able to generate low-dimensionality spaces providing good insights about the classes separation of conformations. As they are nonlinear methods, the low-dimensionality generated represents better the protein motions than PCA embedding, so they could be considered alternatives to full MD analyses.

Keywords: Molecular dynamics · Manifold · Clustering algorithm

1 Introduction

Since its inception, molecular dynamics (MD) techniques have suffered important modifications leading to the simulation of complex and relevant systems with hundreds of different atoms [15]. In a biological context, MD simulation has proved to be suitable to study transition states and predictions of physicochemical and geometric properties of proteins, the key to characterize molecules functions [22]. Previous studies had shown that MD simulations can be applied to generate an ensemble of conformations for docking studies to protein structures with inaccessible or poorly defined binding sites [12].

© Springer Nature Switzerland AG 2019
I. Rojas et al. (Eds.): IWBBIO 2019, LNBI 11466, pp. 219–230, 2019.
https://doi.org/10.1007/978-3-030-17935-9_21

Despite its usefulness, MD analysis may be difficult as many conformations are generated and classifying them demands a lot of time and knowledge about protein behavior [20]. So, artificial intelligence have been auspicious to detect and classify conformations from a set of trajectories, at no risk of loss of information on the protein dynamics [8,20,23]. Such techniques include unsupervised methods like clustering algorithm that attempts to partition data set into groups with similar features without prior knowledge. Every feature used as input for a clustering algorithm is considered a coordinate in a space with n dimensionality.

When applied to protein conformations, different physicochemical and conformational properties can be used as input for clustering. However, the high dimensionality of conformational space, noise, and other factors lead to the well-known curse of dimensionality in statistical pattern recognition, preventing homogeneous clusters from forming. Therefore, the use of dimensionality reduction (DR) methods with subsequent clustering of trajectories from MD simulations have been able to reduce noise and generate more homogeneous clusters [34].

Principal component analysis (PCA) has been applied to analyze data from MD since 1991 [16]. It is very important to obtain a set of orthogonal vectors which are considered the "essential subspace" and are able to capture the largest amplitude of protein motions from a set of trajectories [6]. Although PCA is probably the best known linear technique able to acquire information about complex dynamics, by intrinsically incorporating a dimensional hyperplane, the low-dimensional embedding obtained may be distorted [5,11,31] when PCA is applied to the nonlinear space of conformational protein changes. The reason why nonlinear machine learning techniques have been proposed to pinpoint the underlying manifold structure so as to analyze the space explored by MD simulations and solve the inherent problem of PCA [24].

This study aims to contribute to new approaches of MD analysis, placing the AutoEncoder, t-SNE and Spectral embedding in the context of the of DR methods applied to MD simulation analyses. Combining these methods with the clustering Ward algorithm to identify representative structures, a comparative analysis of six different DR methods (including linear and nonlinear) was performed.

2   Materials and Methods

2.1   Data Set for Clustering the MD Trajectory

Euclidean distances were calculated between intra-molecular C_α for every 501 conformations from MD simulation (at 310 K and 510 K temperatures) of calmodulin (PDB 1CLL), previously published in [27], to obtain a Euclidean distance matrix (EDM) $E_{i_{n \times n}}$, where i is the matrix index and n is total the number of residues. The upper triangular part of each EDM was converted to 1D vector. Each 1D vector was added to a feature matrix $M_{m \times p}$, where m is the number of conformations and p is the number of distances that describe the

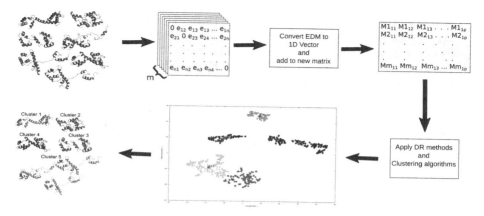

Fig. 1. Protein representation in computational experiments.

conformational fluctuations. The feature matrix was used as input for clustering and dimensionality reduction methods carried out in this paper (Fig. 1).

2.2 Dimensionality Reduction (DR)

In order to reduce the noise of data and find a new space of coordinates that represent the protein fluctuations, the following six methods were used:

(a) **Multidimensional scaling (MDS)** is a technique applied to nonlinear DR, which builds a projection in a lower dimensional space of n points in Euclidian space. In this new space obtained by MDS, elements are represented by points whose respective distances best approximate the initial distance [10,33]. In this work, was used an MDS variation called *Metric Multi-dimensional Scaling* (mMDS). This method minimizes the cost function called "Stress" which is a measure for the deviation from monotonicity between the distances d_{ij} and the observed dissimilarities [29],

$$S = \sqrt{\frac{\sum_{ij} \left(d_{ij} - d_{ij}^* \right)^2}{\sum_{ij} d_{ij}^2}} \tag{1}$$

where d_{ij} and d_{ij}^* are the predicted and target distances, respectively.

(b) **Isometric feature mapping (Isomap)** is considered an extension of MDS idea, incorporating the geodesic distances induced by a neighborhood graph embedded in the classical scaling [14,30]. This method performs three steps. Step 1, the algorithm determines the neighbors on the manifold M of each point in the input space X, based on the distances $d_X(ij)$. Using these distances, the method constructs an edge-weighted neighborhood graph G, where the weight of each edge is equal to the Euclidean distance $d_X(ij)$. Step 2, Isomap estimates the geodesic distances $d_M(ij)$ between all pairs of points on the manifold M by computing their shortest path in the graph G. Step 3, lower-dimensional embedding is computed applying MDS method to the matrix of graph distances $D_G = \{d_G(ij)\}$ [30].

(c) **t-distributed Stochastic Neighbor Embedding (t-SNE)** is a variation of Stochastic Neighbor Embedding (SNE), which converts the high-dimensional Euclidean distances between points from the data set into Gaussian joint probabilities that represent similarities [18]. This method performs two main steps: Step 1, t-SNE starts by converting the high-dimensional Euclidean distances between points in the initial space into conditional probabilities p_{ij} (Eq. 2) that represent their similarities.

$$p_{ij} = \frac{p_{i|j} + p_{j|i}}{2N} \tag{2}$$

where p_{ij} is proportional to the similarity of objects x_i and x_j as follows:

$$p_{j|i} = \frac{\exp\left(-\|x_i - x_j\|^2/2\sigma_i^2\right)}{\sum_{k \neq i} \exp\left(-\|x_i - x_k\|^2/2\sigma_i^2\right)} \tag{3}$$

Step 2, a similar probability distribution q_{ij} (Eq. 4) is defined over the points in the low-dimensional map using a heavy-tailed Student-t distribution and minimizes the gradient of the Kullback-Leibler divergence between P and the Student-t based joint probability distribution Q [18].

$$q_{ij} = \frac{\left(1 + \|y_i - y_j\|^2\right)^{-1}}{\sum_{k \neq l} \left(1 + \|y_k - y_l\|^2\right)^{-1}} \tag{4}$$

(d) **Spectral Embedding (SE)** uses the spectral decomposition of the Laplacian graph generated by the similarity matrix of the data [2], being considered a discrete approximation of the low dimensional manifold in the high dimensional space [2]. SE algorithm calculates an affinity matrix $A \in \mathbb{R}^{nxn}$ defined by $A_{ij} = \exp\left(-\|s_i - s_j\|^2/2\sigma^2\right)$ if $i \neq j$, and $A_{ii} = 0$. After this, the method defines a diagonal matrix D whose (i,i)-element is the sum of i-th rows of A, and constructs the laplacian matrix L, where $L = D^{-1/2} A D^{-1/2}$. Applying a spectral decomposition, the k largest eigenvectors of L are chosen and form the matrix $X = [x_1, x_2, \ldots, x_k] \in \mathbb{R}^{nxk}$, by stacking the eigenvectors in columns. Finally, the matrix Y is obtained from X by renormalizing each row of X to have unit length [19].

(e) **Principal Component Analysis (PCA)** is a linear method that uses the singular value decomposition of the data to project them to a linear subspace attempting to maintain most of the variability of the data [1,26].

(f) **AutoEncoder (AE)** is a type of artificial neural network that is able to learn representations from data sets in an unsupervised manner [32]. Architecturally, an autoencoder consists of two parts, the encoder and the decoder. The encoder maps an input $x_i \in \mathbb{R}^{d_x}$ to a hidden layer $y_i \in \mathbb{R}^{d_N}$ with reduced dimensionality, using a function g, as described bellow.

$$y_i = g(W x_i) \tag{5}$$

where g can be any function for a linear projection (*e.g.* identity function) or for a nonlinear mapping (*e.g.* sigmoid function). The parameter W is a $d_y x d_x$ weighted matrix. The decoder allows reconstructing $x_i' \in \mathbb{R}^{d_x}$ from the hidden layer with low-dimension y_i,

$$x_i' = f(W'y_i) \tag{6}$$

where W' is another $d_y x d_x$ weighted matrix defined as W^T. The function f is similar to g and can be a function for linear or nonlinear projection [32]. The Fig. 2 shows a structure of autoencoder built in this study. During training, we used 150 epochs and a batch size value equals 128. In addition, we also used the mean squared error (mse) as loss function and the optmizer Adaptive Moment Estimation (Adam).

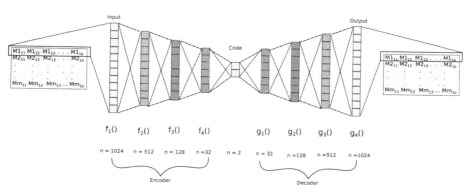

Fig. 2. Autoencoder representation. An autoencoder with 4 layers for encoder and decoder. For the encoder part, two functions was used: Exponential Linear Unit (ELU) from f_1 to f_3, and a linear function for f_4. For the decoder part, from g_1 to g_3 was used ELU and for g_4 was used a sigmoid function. In this figure, n represents the number of neurons in each layer.

2.3   Clustering Approach

Ward algorithm was used to cluster the protein conformations. Previous studies showed that this algorithm has generated good results when applied to data from MD simulations [9,27]. Ward is an agglomerative hierarchical clustering method that minimizes the total within-cluster variance for each pair of cluster centers or medoids found after merging. At the first step, all clusters contain a single point and then iterative steps are performed until all points are merged in the homogeneous group or that a condition of grouping is satisfied.

Ward algorithm is a K dependent method, which requires to provide the number of clusters *a priori*. In this work, we performed the elbow method [3] to find the best number of clusters (K number) for the data set after running each dimensionality reduction method. To evaluate unsupervised classification performed by Ward, we calculated different validation measures: Calinski-Harabasz index (CH) [4], Davies-Bouldin index (DBI) [7], Fowlkes Mallows Index (FMI) [13], and Silhouette score [25]. The Fig. 3 shows a flowchart developed here.

Another analysis performed was the Root-Mean-Square-Deviation (RMSD) between the lower energy structure obtained by weighted Histogram Analysis Method (WHAM) and the medoids found using Ward clustering algorithm.

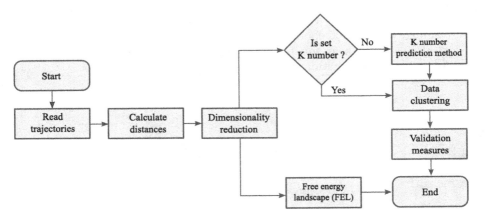

Fig. 3. Flowchart of the methodology. All steps were implemented using python v2.7. The different DR methods and Ward clustering algorithm were implemented using scikit-learn package v0.18 [21].

2.4   Statistical Evaluation

All tests were run thirty times to calculate the statistical differences between them. An Anderson-Darling test was performed to verify the normality of the answer variables. As non-parametric analysis, the Kruskal-Wallis followed by Dunn's *post-hoc* test were run to verify the statistical differences between the clustering scores calculated, using a significance set at the 5%. All computational experiments were carried out on intel® core^{TM} i7 860 2.8 GHz processor, 8 Gbytes (Gb) of RAM, HD of 860 Gb, operating system Fedora release 23.

2.5   Free Energy Landscape

In order to have a free energy landscape (FEL) for each conformational set from MD simulations, we applied the Weighted Histogram Analysis Method (WHAM) [17]. This method is based on the fact that given a set of discrete states of a molecule, is possible to obtain a histogram with discrete bins that provide a relative probability of a state occurs along the trajectory [17]. So, the higher density of states in a histogram region provides the insight that the probability of this set representing an energy basin is greater. The WHAM idea is derived from statistical mechanics and the function of free energy, considering a state ξ, is defined by

$$F\left(\xi\right) = -k_B T \ln Z\left(\xi\right) \qquad (7)$$

so that $Z\left(\xi\right)$ is a partition function that is proportional to the density of the states in bins and is given by

$$Z\left(\xi\right) = \int e^{\beta U(\xi)} d\Omega \qquad (8)$$

where $\beta = 1/k_B T$, k_B is the Boltzmann constant, T is the temperature in Kelvin, and $U(\xi)$ is the potential energy. Considering a reduced space (intrinsic space), found by DR methods applied to the internal coordinates of the protein conformations, the FEL can be generated by inverting the probability distribution ($\hat{P}$) of points (states) of a multidimensional histogram obtained from the n principal components ($\{\boldsymbol{\Psi}\}_{i=2}^{k+1}$) [11], as follows:

$$F = -k_B T \ln \hat{P}\left(\{\boldsymbol{\Psi}\}_{i=2}^{k+1}\right) + C \qquad (9)$$

3   Results and Discussion

When clustering methods are applied to bio-molecule simulation data, it is expected that generated partitions include similar conformations which represent transition and meta-states of the system. Approaches which combine dimensionality reduction (DR) and clustering methods are able to generate homogeneous clusters when applied to molecular dynamics (MD) simulations, which assists the analysis of protein conformational fluctuations [34]. In this paper, we performed a comparative analysis of different DR approaches combined with the Ward clustering algorithm, applied to MD data.

Figure 4 shows the clustering results for the Ward algorithm using different DR methods applied to data from 1CLL simulations at 310 K and 510 K. In 310 K simulation, all methods found 3 or 4 clusters, whereas in 510 K simulations it was found a range from 4 to 6 clusters. It was slightly different from previous manual analyses in which was observed 2 and 8 clusters for 310 K and 510 K simulations, respectively [27]. Probably, it is due to the overlapping between some classes in a lower dimensionality space, maybe because the data sets were reduced to only two dimensions and some information was lost.

To evaluate the class prediction quality of each approach, the internal clustering validation metrics were analyzed, which are based on the intrinsic information of the data: Calinski-Harabasz index (CH), Davies-Bouldin index (DBI) and Silhouette score. In 310 K simulations, it was observed that the Isomap method had the best values of quality, followed by SE (Fig. 5). However, in 510 K simulations, the best CH and DBI values were obtained by Spectral and Isomap, but for the silhouette result, Spectral and PCA presented best values (Fig. 5).

Internal validation metrics provide an important insight into the quality of different clustering approaches. However, unsupervised methods are more liable

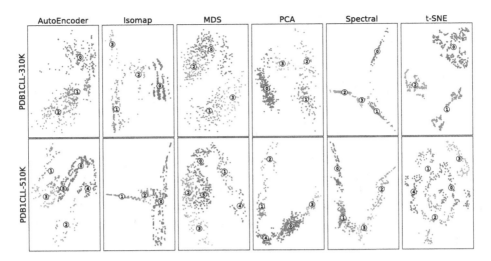

Fig. 4. Visualization of reduced space (2D) obtained from different DR method applied to 501 calmodulin (PDB 1CLL) conformations from MD simulations. Clusters were obtained by Ward algorithm.

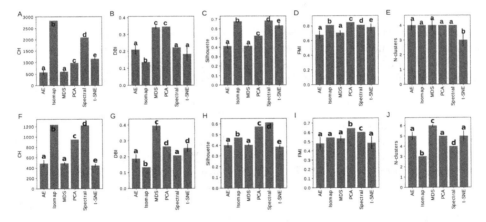

Fig. 5. Comparative analysis of the clustering evaluation metrics between DR approaches. Plots from A to E show metrics for 310 K simulations and those from F to J are results for 510 K simulations. Bars with the same letter were statistically similar, according to Dunn's test with a significance level of 5%.

to misclassification errors than supervised methods, especially for protein data sets. So, external measures based on previous knowledge about data have been used to evaluate clustering algorithms. Therefore, to verify the performance of each method when compared to the manual analyses, the Fowlkes Mallows Index (FMI) was calculated [13] using manual analyses as the reference (data not shown). In 310 K simulations, Isomap and PCA showed high FMI values, whereas AE and MDS presented low values and were not statistically different (Fig. 5D). In 510 K simulations, PCA had the best value of FMI followed by Spectral (Fig. 5I).

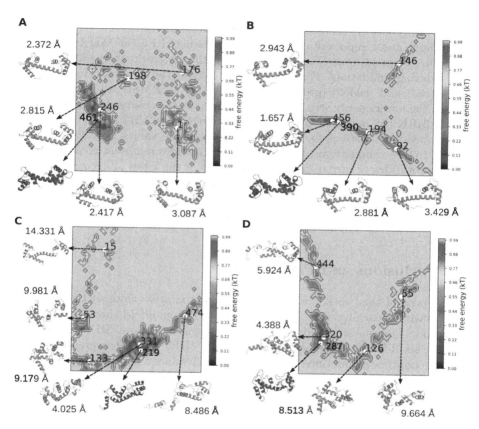

Fig. 6. Free Energy Landscape (FEL) for calmodulin (PDB 1CLL) and the representative structures predicted by the Ward algorithm for PCA and Spectral methods. Figures A and B show simulations at 310 K, whereas figures C and D show the FEL obtained for simulations at 510 K. Here the FEL was calculated using the Weighted Histogram Analysis Method (WHAM). In grey are represented the cluster medoids and in red are structures of lower energy value. (Color figure online)

According to our results, Spectral obtained good values in all evaluation measures. So, Free Energy Landscape (FEL) generated by PCA (classic method of DR) and Spectral (Fig. 6) were compared. In general, PCA gives more basins than Spectral, which could be explained by the fact that barriers and basins are influenced by the coordinate(s) in lower dimension space generated by different machine learning techniques. In addition, considering that the energy basins are regions of a greater density of states, we could say that the spectral method was able to group more conformations for the same cluster than PCA, which generates fewer regions of minima in the FEL. A similar result has been reported by a previous study in which PCA found more energy minima than Isomap, even though the separation is less clear [28]. Our results also revealed that the FEL generated by PCA has more energy barriers than Spectral, which is highlighted by comparing both methods applied to 310 K simulations.

Although most representative structures predicted by the Ward algorithm are not in energy minima (Fig. 6), they reflect the general patterns assumed by protein during simulations. For example, Fig. 6 shows the structures in minimum energy (red) and medoid (grey) states. According to RMSD values calculated between the lower energy structure and medoids, it was observed that in simulations at 310 K the minimum and maximum values were 2.372 Å and 3.087 Å for PCA embedding, whereas for Spectral embedding the values were 1.657 Å and 3.429 Å. In these simulations, the highest RMSD values represent conformations in which rotations occurred in helices and loops of the lobes or even rotation in the central helix. For simulations at 510 K, the minimum and maximum RMSD values were 4.025 Å and 14.331 Å for PCA embedding and 4.388 Å and 9.664 Å for Spectral embedding. As expected in these simulations, the higher values of RMSD represent misfolded structures due to the high temperature.

4   Conclusions and Perspectives

The purpose of this work was to perform a comparative analysis between different machine learning approaches to find out the manifold that characterizes the protein motions and representative conformations from MD trajectories. For this, six different dimensionality reduction (DR) methods were applied to internal coordinates obtained from Euclidean distances between C_α atoms of structures, and the "intrinsic dimensionality space" found was used as input for agglomerative Ward algorithm to group similar conformations and identify those considered representatives within each cluster.

The results show that when considering the best values of external and internal validation metrics, Spectral and Isomap arise good alternatives to explore the conformational space of proteins from MD simulation, although these methods have failed to predict the K groups expected. Considering the K number predicted, AE, PCA and MDS methods presented the best performance. Another significant finding from this study is that AutoEncoder was able to identify lower dimension in a way that similar conformations were close, showing as a promising alternative to current DM analyzes.

Acknowledgment. The authors thank the Graduate Program in Computational Modeling from Federal University of Juiz de Fora and the Brazilian agencies FAPEMIG (grant 01606/15), CNPq (grant 429639/2016-3) and CAPES for the financial support.

References

1. Abdi, H., Williams, L.J.: Principal component analysis. Wiley Interdisc. Rev. Comput. Stat. **2**(4), 433–459 (2010)
2. Belkin, M., Niyogi, P.: Laplacian eigenmaps for dimensionality reduction and data representation. Neural Comput. **15**(6), 1373–1396 (2003)
3. Bholowalia, P., Kumar, A.: EBK-means: a clustering technique based on elbow method and k-means in WSN. Int. J. Comput. Appl. **105**(9) (2014)
4. Caliński, T., Harabasz, J.: A dendrite method for cluster analysis. Commun. Stat.-Theory Methods **3**(1), 1–27 (1974)
5. Das, P., Moll, M., Stamati, H., Kavraki, L.E., Clementi, C.: Low-dimensional, free-energy landscapes of protein-folding reactions by nonlinear dimensionality reduction. Proc. Nat. Acad. Sci. **103**(26), 9885–9890 (2006)
6. David, C.C., Jacobs, D.J.: Principal component analysis: a method for determining the essential dynamics of proteins. In: Livesay, D. (ed.) Protein Dynamics, pp. 193–226. Springer, Heidelberg (2014). https://doi.org/10.1007/978-1-62703-658-0_11
7. Davies, D.L., Bouldin, D.W.: A cluster separation measure. IEEE Trans. Pattern Anal. Mach. Intell. **2**, 224–227 (1979)
8. De Paris, R., Frantz, F.A., Norberto de Souza, O., Ruiz, D.D.: wFReDoW: a cloud-based web environment to handle molecular docking simulations of a fully flexible receptor model. BioMed Res. Int. **2013** (2013)
9. De Paris, R., Quevedo, C.V., Ruiz, D.D., de Souza, O.N.: An effective approach for clustering inha molecular dynamics trajectory using substrate-binding cavity features. PLoS ONE **10**(7), e0133172 (2015)
10. Dokmanic, I., Parhizkar, R., Ranieri, J., Vetterli, M.: Euclidean distance matrices: essential theory, algorithms, and applications. IEEE Sig. Process. Mag. **32**(6), 12–30 (2015)
11. Ferguson, A.L., Panagiotopoulos, A.Z., Kevrekidis, I.G., Debenedetti, P.G.: Nonlinear dimensionality reduction in molecular simulation: the diffusion map approach. Chem. Phys. Lett. **509**(1–3), 1–11 (2011)
12. Ferreira, L.G., dos Santos, R.N., Oliva, G., Andricopulo, A.D.: Molecular docking and structure-based drug design strategies. Molecules **20**(7), 13384–13421 (2015)
13. Fowlkes, E.B., Mallows, C.L.: A method for comparing two hierarchical clusterings. J. Am. Stat. Assoc. **78**(383), 553–569 (1983)
14. Ghodsi, A.: Dimensionality reduction a short tutorial. Department of Statistics and Actuarial Science, University of Waterloo, Ontario, Canada, vol. 37, p. 38 (2006)
15. Hospital, A., Goñi, J.R., Orozco, M., Gelpí, J.L.: Molecular dynamics simulations: advances and applications. Adv. Appl. Bioinf. Chem. AABC **8**, 37 (2015)
16. Ichiye, T., Karplus, M.: Collective motions in proteins: a covariance analysis of atomic fluctuations in molecular dynamics and normal mode simulations. Proteins Struct. Function Bioinf. **11**(3), 205–217 (1991)
17. Kumar, S., Rosenberg, J.M., Bouzida, D., Swendsen, R.H., Kollman, P.A.: The weighted histogram analysis method for free-energy calculations on biomolecules. I. the method. J. Comput. Chem. **13**(8), 1011–1021 (1992)
18. Maaten, L.V.D., Hinton, G.: Visualizing data using t-SNE. J. Mach. Learn. Res. **9**(Nov), 2579–2605 (2008)
19. Ng, A.Y., Jordan, M.I., Weiss, Y.: On spectral clustering: analysis and an algorithm. In: Advances in Neural Information Processing Systems, pp. 849–856 (2002)
20. Paris, R.D., Quevedo, C.V., Ruiz, D.D., Souza, O.N.D., Barros, R.C.: Clustering molecular dynamics trajectories for optimizing docking experiments. Comput. Intell. Neurosci. **2015**, 32 (2015)

21. Pedregosa, F., et al.: Scikit-learn: machine learning in python. J. Mach. Learn. Res. **12**, 2825–2830 (2011)

22. Phillips, J.L., Colvin, M.E., Newsam, S.: Validating clustering of molecular dynamics simulations using polymer models. BMC Bioinf. **12**(1), 445 (2011)

23. Quevedo, C.V., De Paris, R., Ruiz, D.D., De Souza, O.N.: A strategic solution to optimize molecular docking simulations using fully-flexible receptor models. Expert Syst. Appl. **41**(16), 7608–7620 (2014)

24. Rohrdanz, M.A., Zheng, W., Clementi, C.: Discovering mountain passes via torchlight: methods for the definition of reaction coordinates and pathways in complex macromolecular reactions. Ann. Rev. Phys. Chem. **64**, 295–316 (2013)

25. Rousseeuw, P.J.: Silhouettes: a graphical aid to the interpretation and validation of cluster analysis. J. Comput. Appl. Math. **20**, 53–65 (1987)

26. Shlens, J.: A tutorial on principal component analysis. arXiv preprint arXiv:1404.1100 (2014)

27. de Souza, V.C., Goliatt, L., Goliatt, P.V.C.: Clustering algorithms applied on analysis of protein molecular dynamics. In: 2017 IEEE Latin American Conference on Computational Intelligence (LA-CCI), pp. 1–6. IEEE (2017)

28. Stamati, H., Clementi, C., Kavraki, L.E.: Application of nonlinear dimensionality reduction to characterize the conformational landscape of small peptides. Proteins Struct. Function Bioinf. **78**(2), 223–235 (2010)

29. Steyvers, M.: Multidimensional scaling. In: Encyclopedia of Cognitive Science, pp. 1–7 (2002)

30. Tenenbaum, J.B., De Silva, V., Langford, J.C.: A global geometric framework for nonlinear dimensionality reduction. Science **290**(5500), 2319–2323 (2000)

31. Teodoro, M.L., Phillips Jr., G.N., Kavraki, L.E.: A dimensionality reduction approach to modeling protein flexibility. In: Proceedings of the Sixth Annual International Conference on Computational Biology, pp. 299–308. ACM (2002)

32. Wang, W., Huang, Y., Wang, Y., Wang, L.: Generalized autoencoder: a neural network framework for dimensionality reduction. In: Proceedings of the IEEE Conference on Computer Vision and Pattern Recognition Workshops, pp. 490–497 (2014)

33. Weinberger, K.Q., Saul, L.K.: Unsupervised learning of image manifolds by semidefinite programming. Int. J. Comput. Vis. **70**(1), 77–90 (2006)

34. Wolf, A., Kirschner, K.N.: Principal component and clustering analysis on molecular dynamics data of the ribosomal L11·23S subdomain. J. Mol. Model. **19**(2), 539–549 (2013)

Computational Systems for Modelling Biological Processes

HIV Drug Resistance Prediction with Categorical Kernel Functions

Elies Ramon[1(✉)], Miguel Pérez-Enciso[1,2], and Lluís Belanche-Muñoz[3]

[1] Centre for Research in Agricultural Genomics (CRAG),
CSIC-IRTA-UAB-UB, Campus UAB, 08193 Bellaterra, Spain
elies.ramon@cragenomica.es
[2] Institució Catalana de Recerca i Estudis Avançats (ICREA),
Passeig de Lluís Companys 23, 08010 Barcelona, Spain
[3] Computer Science Department, Technical University of Catalonia,
Carrer de Jordi Girona 1-3, 08034 Barcelona, Spain

Abstract. Antiretroviral drugs are a very effective therapy against HIV infection. However, the high mutation rate of HIV permits the emergence of variants that can be resistant to the drug treatment. In this paper, we propose the use of categorical kernel functions to predict the resistance to 18 drugs from virus sequence data. These kernel functions are able to take into account HIV data particularities, as are the allele mixtures, and to integrate additional knowledge about the major resistance associated protein positions described in the literature.

Keywords: HIV · Drug resistance prediction · Categorical kernel · SVMs

1 Introduction

HIV is a retrovirus that infects human immune cells, causing a progressive weakening of the immune system. When untreated, the affected person develops acquired immunodeficiency syndrome (AIDS), which leads to a rise of opportunistic infections and, finally, death. HIV has infected more than 35 million people worldwide and is considered a global pandemic [1]. Despite the efforts, to date there is no definitive cure that eradicates the virus from the organism. However, the lifespan and quality of life of many people that live with HIV have expanded greatly thanks to antiretroviral therapy. Antiretroviral drugs lower the virus level in blood by targeting different stages of the virus life cycle. The most important classes of antiretroviral drugs are protease inhibitors (PIs), which target the protease, and nucleoside and non-nucleoside reverse transcriptase inhibitors (NRTIs and NNRTIs, respectively) which target the reverse transcriptase.

Some of the main reasons why HIV is so difficult to fight are its short life cycle (1–2 days), high replication rate (10^8–10^9 new virions each day), and high mutation rate (10^{-4}–10^{-5} mutations per nucleotide site per replication cycle) caused because reverse transcriptase lacks proofreading activity. This permits the

© Springer Nature Switzerland AG 2019
I. Rojas et al. (Eds.): IWBBIO 2019, LNBI 11466, pp. 233–244, 2019.
https://doi.org/10.1007/978-3-030-17935-9_22

fast emergence of new HIV variants, some of which may be resistant to the drug treatment [2]. These variants can be transmitted, and some studies show that ≈10% of patients who had never been on antiretroviral therapy carry at least one resistant HIV [3]. Therefore, it is advisable to do a resistance test before the treatment to find the best drug choice [2,4], especially in developing countries, as recommended by the WHO and the International AIDS Society-USA Panel [3]. A resistance test can be performed in vitro, obtaining HIV samples from the patient and using them to infect host cells cultured in presence of increasing levels of drug concentration. Then, the virus susceptibility is obtained empirically as the IC50 [4]. Another strategy is to infer the HIV variant resistance from its sequence. Genome sequencing is cheaper, faster and more widely available than performing an in vitro drug susceptibility test, allowing the prediction of the drug resistance from the virus sequence using machine learning methods [5].

HIV data has some traits that pose a whole set of challenges to the resistance prediction. The first approaches to the problem were rule-based: study the mutational profile of the HIV variant to look for known major drug-associated resistance mutations (lists of these mutations can be found in reviews like [2]). Some examples are the Stanford HIVdb, Rega or ANRS softwares [5]. However, the aforementioned high mutation rate of the virus hinders this approach, as it favors the emergence of large numbers of new resistance mutations and complex mutational patterns. Another challenge is the presence of mixtures of alleles (normally two, rarely three or four) in most clinical samples, in at least one position of the viral sequence. In the case of HIV, it is well known that this event indicates that there are two or more virus variants in the patient's blood [4]. Mixtures introduce ambiguity in the genotype-phenotype correlation and a problem of technical nature: the vast majority of machine learning methods are not able to deal with this multiallelic codes. To our knowledge, it has not yet been presented an algorithm prepared to handle directly allele mixtures without some sort of previous pre-processing of the data (e.g., removing the affected individuals or keeping only one allele of the mixture).

In this paper, we propose the use of specific kernel functions adapted to the HIV intricacies and able to take into account the categorical nature of the sequence data. Kernels are mathematical functions with some interesting properties. They can be coupled with numerous machine learning algorithms (the so-called kernel methods) and provide a framework to deal with data of virtually any type (vectors, strings, graphs, etc.). They can also encode complementary knowledge about a problem, as long as some mathematical conditions are satisfied [7]. In fact, most of the efforts in the kernel field are devoted to define new functions that capture as much relevant information as possible from data [8]. The output of a kernel function is called a kernel matrix, which is a squared, symmetric positive semidefinite matrix, containing the kernel function evaluations between all the dataset individuals. This pairwise comparison matrix represents the dataset and is the input of the kernel method. Our aim using specific kernel functions that acknowledge the aforementioned HIV data particularities is twofold: on the one hand, extract more information of the dataset to make better predictions; on the other hand, reduce pre-processing, thus preserving the data integrity and lowering the risk of inserting spurious patterns.

2   Materials and Methods

2.1   Datasets and Data Pre-processing

The Genotype-Phenotype Stanford HIV Drug Resistance Database [9] is a public dataset with sequences from HIV isolates and its relative susceptibility to several antiretroviral drugs. The drugs are eight protease inhibitors (ATV, DRV, FPV, IDV, LPV, NFV, SQV and TPV) and two classes of reverse transcriptase inhibitors: six NRTIs (3TC, ABC, AZT, D4T, DDI and TDF) and four NNRTIs (EFV, ETR, NVP and RPV). The dataset offers the relative virus resistance (compared to that of the wild type virus) to each drug, and the sequence of the protein (translated from gene sequence) targeted by this drug: protease or reverse transcriptase, respectively.

We downloaded the High Quality Filtered Datasets from Stanford webpage (version date: 2014-9-28). The data is split in three databases (PI, NRTI and NNRTI), which contain between 1500–1800 HIV isolates. Protein sequence length is 99 amino acids in the case of PI database and 240 in the case of NRTI and NNRTI databases. We took each polymorphic protein position as a (categorical) predictor variable, while drug resistance value (continuous) was the target variable. Since the distributions of resistances are highly skewed, we used the log-transformed values. We deleted all sequences with mutations that changed protein length (protein truncations, insertions and deletions). These events affected less than 5% of HIV sequences. Also, we removed all sequences containing one or more missing values. Missing values are present in protein sequence as well as in the target variables, because not all HIV isolates have been tested for all drugs. The final number of data instances for each drug is offered in Table 1.

Table 1. Final number of HIV isolates per drug.

Drug	Data size	Drug	Data size	Drug	Data size
ATV	971	SQV	1452	DDI	1229
DRV	599	TPV	690	TDF	990
FPV	1414	3TC	1209	EFV	1397
IDV	1460	ABC	1230	ETR	472
LPV	1247	AZT	1221	NVP	1405
NFV	1501	D4T	1228	RPV	174

2.2   Methods

We compared the performance of a nonlinear, non-kernel method (Random Forests or RF) to a kernel method: SVMs (Support Vector Machines). SVMs can be either linear or nonlinear, depending on the kernel used. The Linear kernel is the simplest of all kernel functions, given by the inner product of two vectors in input space, $\boldsymbol{x}$ and $\boldsymbol{y}$:

$$k_{\text{Lin}}(\boldsymbol{x}, \boldsymbol{y}) = \boldsymbol{x}^T \boldsymbol{y}. \tag{1}$$

To use it with categorical variables, vectors need to be recoded as a binary expansion representation (*1-out-of-m* code) [8]. We used this kernel as the linear method of reference and compared it with the following nonlinear kernel functions:

RBF Kernel. One of the most frequently used kernels. It is defined as:

$$k_{\text{RBF}}(\boldsymbol{x}, \boldsymbol{y}) = e^{-\gamma ||\boldsymbol{x} - \boldsymbol{y}||^2}, \tag{2}$$

where $||\boldsymbol{x} - \boldsymbol{y}||$ is the Euclidean distance between the two vectors, and $\gamma > 0$ is a hyperparameter that has to be optimized. As in the case of the linear kernel, the original data had to be recoded. RBF is a widely accepted default method [7,8], so we used it as a benchmark to compare with the categorical kernels.

Univariate Overlap Kernel. The most basic categorical kernel. This kernel assigns a similarity of 1 if the two individuals are identical and 0 otherwise [8].

$$k_{\text{Ov}}(x_i, y_i) = \begin{cases} 1 & \text{if } x_i = y_i \\ 0 & \text{if } x_i \neq y_i \end{cases}. \tag{3}$$

In our case, x_i and y_i represent the alleles of a given protein position i in two HIV isolates.

Univariate Jaccard Kernel. The Jaccard index measures the similarity between two finite sets and is a valid kernel function [10]. We used it to handle allele mixtures. Letting again i denote a given protein position in two HIV isolates –so that x_i and y_i are non-empty sets of amino acids– then:

$$k_{\text{Jac}}(x_i, y_i) = \frac{|x_i \cap y_i|}{|x_i \cup y_i|}. \tag{4}$$

When $|x_i| = |y_i| = 1$ (that is: neither of the individuals have an allele mixture at that position), the Jaccard kernel reduces to Overlap. Unlike Overlap, this kernel function can deal simultaneously with allele mixtures and categorical data.

Multivariate Kernels. As we are dealing with whole protein sequences, we can aggregate all the univariate Overlap and Jaccard evaluations as follows:

$$k_{\text{Multi}}(\boldsymbol{x}, \boldsymbol{y}) = \sum_{i=1}^{d} w_i\, k(x_i, y_i). \tag{5}$$

Where d is the number of categorical variables (i.e. protein positions), $k(x_i, y_i)$ each univariate kernel evaluation at the i-th position between two HIV isolates, and w_i the weight of that position. Weights are nonnegative and sum to one. This results in a valid kernel function, since the product of a positive scalar and a kernel is a kernel, and the sum of kernels is also a kernel.

We considered two approaches concerning the weights. The simplest is to assign an equal weight $1/d$ to all variables, thus computing the mean. However, it is well known that not all protein positions contribute equally to the virus resistance [2]. Therefore, we explored the potential advantage of including this information into the kernels, using a measure of variable importance as weight. To this effect, we used the RF mean decrease in node impurity to weigh the kernels and compared the results with using simply the mean.

RBF-Like Categorical Multivariate Kernels. To ensure that the only difference between categorical kernels and RBF was the categorical part, we introduced an exponential and a hyperparameter, in a way analogous to (2):

$$k_{\text{cat}}(\boldsymbol{x}, \boldsymbol{y}) = e^{-\gamma} e^{\gamma \sum_{i=1}^{d} w_i\, k(x_i, y_i)}. \tag{6}$$

This is also a valid kernel function, since the exponential of a kernel gives another kernel. $e^{-\gamma}$ normalizes the kernel matrix, keeping the evaluations between 0 and 1. The final versions of Multivariate Overlap and Jaccard kernels are obtained by replacing the $k(x_i, y_i)$ term by (3) or (4), respectively. Also, we modified the standard Linear and RBF definitions to include the position weighting:

$$k_{\text{lin}}(\boldsymbol{x}, \boldsymbol{y}) = \sum_{i=1}^{d} w_i\, x_i\, y_i \tag{7}$$

$$k_{\text{RBF}}(\boldsymbol{x}, \boldsymbol{y}) = e^{-\gamma \sum_{i=1}^{d} w_i (x_i^2 - y_i^2)} \tag{8}$$

Thus, as all the kernels have a weighted counterpart, we can assure a fair comparison between the categorical and the non-categorical kernels. In our analyses, we compared Overlap and Jaccard computed with (6) with unweighted Linear (1) and RBF (2), and their weighted versions (7) and (8).

2.3 Experimental Setup and Model Tuning

All the analyses were implemented in R statistical computing language [6]. To assess the performance of the methods used, each database was split at random into two partitions: training set (60% of the database) and test set (40%). Hyperparameter optimization was done by a 10 × 10 cross-validation on the training set. Once the best hyperparameter was found, the final model was built using the whole training set. To assess the model performance, the NMSE (Normalized Mean Square Error) between the actual and the predicted drug resistances of the test set was computed.

We repeated the whole process 40 times, each time with different 60/40 randomly split training/test partitions, to obtain an error distribution. Kernel complementary information (position weights in this case) was calculated from the training set only. Note that only the Jaccard kernel can directly handle allele mixtures; for the rest of kernels and the RF, we generated 40 versions of the database randomly sampling one allele at a time. Then, the 40 versions were used to compute all the models except Jaccard, which could deal directly with the database without further preprocessing. This way we can ensure an honest comparison between Jaccard and the rest of kernels and methods.

3   Results

As expected, HIV protein sequences showed great variability. As many as 93% of the protease positions were polymorphic, and among these, the number of different observed alleles varied between 2–16. In the case of reverse transcriptase, 85% of the positions were polymorphic and the allele range varied between 2–14. Also, we noticed that 57% of the sequences have at least one allele mixture.

We offer the error distribution boxplots for three representative drugs from each one of the databases: FPV (PI database, Fig. 1), DDI (NRTI database, Fig. 2) and NVP (NNRTI database, Fig. 3). Also, we summarize the full results for all 18 drugs in Tables 2 and 3, which show the mean and standard error of the NMSE distributions corresponding to the nine methods tested.

3.1   Performance Overview

The test error value varied greatly across drugs and methods. The best prediction was achieved for 3TC, with an average NMSE between 0.05–0.11 depending on the method. The drug with worst prediction error was RPV, with an average NMSE ranging 0.48–0.66. We noticed that methods applied to drugs with lower number of data instances (especially RPV, ETR and TPV; but also TDF and to some extent DRV) had considerably worse performance. In the PI database, errors across all drugs were fairly similar and around 0.13–0.17 (with the sole exception of TPV). In return, predictive performances of reverse transcriptase inhibitors were far more erratic across methods. Overall, the best method was the SVM with the Jaccard kernel (either in its weighted or in its unweighted version), which achieved the best performance in 17 out of 18 drugs.

3.2   Comparison Between Methods (Unweighted Case)

Linear vs Nonlinear Kernels. Nonlinear kernels performed better than the linear kernel in almost all drugs, with the only exception of ETR. In D4T, RBF behaved slightly worse than the linear kernel, but the other nonlinear kernels remained better.

Non-categorical vs Categorical Kernels. Categorical kernels outperformed RBF in all cases but two (EFV and ETR). In the case of RPV, RBF was better

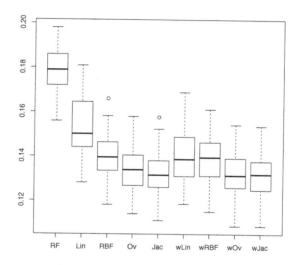

Fig. 1. Test error distribution for drug FPV.

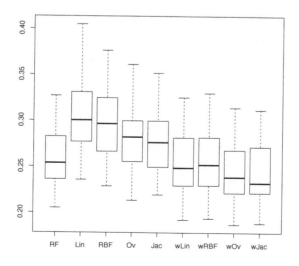

Fig. 2. Test error distribution for drug DDI.

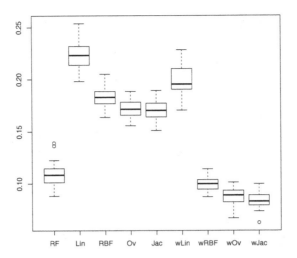

Fig. 3. Test error distribution for drug NVP.

Table 2. Mean test NMSE for all drugs and methods. In bold, best error for each drug.

Drug	RF	Lin	RBF	Ov	Jac	wLin	wRBF	wOv	wJac
ATV	0.188	0.141	0.132	0.128	**0.125**	0.138	0.139	0.133	0.133
DRV	0.196	0.195	0.164	0.161	**0.156**	0.160	0.158	0.159	0.158
FPV	0.179	0.154	0.140	0.135	**0.132**	0.139	0.140	0.133	**0.132**
IDV	0.165	0.147	0.130	0.127	**0.123**	0.129	0.132	0.124	0.124
LPV	0.132	0.101	0.097	0.091	**0.089**	0.094	0.092	0.090	**0.089**
NFV	0.163	0.155	0.132	0.126	0.122	0.128	0.124	0.120	**0.119**
SQV	0.189	0.163	0.151	0.143	0.139	0.145	0.141	0.137	**0.133**
TPV	0.470	0.474	0.350	0.346	**0.339**	0.406	0.383	0.372	0.361
3TC	0.056	0.112	0.079	0.065	0.064	0.079	0.068	0.051	**0.050**
ABC	0.163	0.191	0.150	0.151	0.148	0.172	0.144	0.139	**0.137**
AZT	0.222	0.227	0.196	0.194	0.190	0.222	0.193	0.189	**0.187**
D4T	0.262	0.261	0.264	0.253	0.249	0.241	**0.234**	0.242	0.240
DDI	0.260	0.307	0.298	0.280	0.277	0.255	0.255	0.245	**0.244**
TDF	0.385	0.376	0.365	0.352	0.348	0.355	0.341	0.335	**0.332**
EFV	0.147	0.140	0.133	0.134	0.134	0.122	0.107	0.104	**0.101**
ETR	0.406	0.374	0.386	0.400	0.391	0.375	0.349	0.328	**0.314**
NVP	0.109	0.223	0.183	0.171	0.170	0.200	0.100	0.088	**0.084**
RPV	0.600	0.658	0.581	0.607	0.578	0.529	0.537	0.501	**0.478**

Table 3. Standard error of test NMSE for all drugs and methods.

Drug	RF	Lin	RBF	Ov	Jac	wLin	wRBF	wOv	wJac
ATV	0.015	0.014	0.013	0.013	0.013	0.013	0.014	0.015	0.013
DRV	0.019	0.020	0.024	0.023	0.024	0.022	0.024	0.021	0.021
FPV	0.010	0.014	0.010	0.011	0.010	0.012	0.011	0.011	0.011
IDV	0.009	0.010	0.009	0.008	0.008	0.008	0.010	0.009	0.009
LPV	0.008	0.008	0.007	0.007	0.006	0.008	0.007	0.007	0.006
NFV	0.010	0.013	0.008	0.008	0.009	0.008	0.008	0.008	0.008
SQV	0.013	0.013	0.012	0.012	0.011	0.011	0.010	0.012	0.012
TPV	0.035	0.027	0.042	0.036	0.038	0.046	0.054	0.049	0.046
3TC	0.011	0.015	0.014	0.012	0.013	0.014	0.015	0.012	0.012
ABC	0.023	0.030	0.025	0.024	0.023	0.023	0.022	0.022	0.022
AZT	0.013	0.016	0.016	0.016	0.015	0.018	0.014	0.014	0.013
D4T	0.042	0.037	0.035	0.038	0.040	0.036	0.038	0.039	0.037
DDI	0.032	0.041	0.040	0.035	0.033	0.035	0.035	0.033	0.033
TDF	0.028	0.037	0.029	0.028	0.028	0.031	0.030	0.025	0.029
EFV	0.014	0.013	0.012	0.009	0.011	0.011	0.011	0.011	0.010
ETR	0.057	0.037	0.047	0.043	0.041	0.050	0.048	0.047	0.044
NVP	0.011	0.012	0.009	0.008	0.009	0.014	0.007	0.008	0.008
RPV	0.060	0.121	0.107	0.109	0.095	0.102	0.081	0.088	0.082

than Overlap but worse than Jaccard. Regarding the categorical kernels, the Jaccard kernel performed better than or equivalent to Overlap in all cases.

Kernel Methods vs Non-kernel Methods. Predictive performances of unweighted kernels and RF were markedly different between protease and transcriptase inhibitors. RF was consistently worse than kernel methods for PI database, especially when compared to nonlinear kernel methods (Fig. 1). In the case of reverse transcriptase inhibitors, RF performance overlapped with those of kernel methods. In 3TC, DDI (Fig. 2) and, notably, NVP (Fig. 3), RF performed better than all the unweighted kernels, including Jaccard.

3.3   Comparison Between Methods (Weighted Case)

Figure 4 shows two representative examples of the relative importance of each protein position obtained from RF, which we used to weight the kernels. RF detected as important most of the major resistance protein positions described in [2]. Protease is shorter than reverse transcriptase, but as a general rule comparatively more positions were detected as important. To evaluate this numerically, we computed the Gini index (a measure of inequality) of the weight distributions for each one of the drugs. Reverse transcriptase inhibitors had, in general,

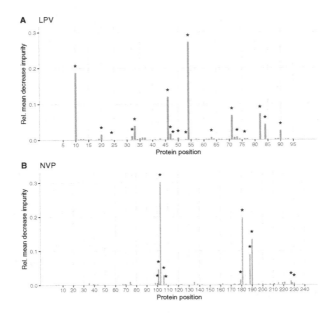

Fig. 4. RF relative importance of each protein position, averaged over the 40 replicates, for two drugs: a protease inhibitor (*A*) and a reverse transcriptase inhibitor (*B*). The stars mark the major drug-related positions reported in the literature.

higher Gini indexes than protease inhibitors. Then, we performed a one-tailed, two-sample Kolmogorov–Smirnov test to test if this difference was statistically significant. We obtained a p = 0.003, thus supporting this hypothesis.

Weighted vs Unweighted Kernels. Generally speaking, weighting was more effective in reverse transcriptase inhibitors, decreasing the error by ≈14% on average (Table 2). In these databases, weighted kernels outperformed RF in all cases, whereas their unweighted counterparts did not. This was particularly the case for 3TC, DDI, ETR, RPV, EFV and especially NVP, where the weighting decreased the Jaccard kernel error around a 50%. In contrast, the effect of weighting was less clear in the PI database: weighting made the error worse in two drugs (ATV and TPV), improved the error in also two drugs (NFV and SQV) and delivered a very similar error in the four remaining drugs. In general, the weighted kernels preserved the pattern observed in unweighted kernels, where the linear kernel had worse performance than RBF, RBF than Overlap, and Overlap than Jaccard. However, there were a few cases where weighted RBF had higher error than weighted Linear (IDV, for example), and one case where weighted RBF was slightly better than the weighted categorical kernels (DDI, Fig. 2).

4   Discussion

The particularity of this work lies in the data treatment. Even in the kernel field, there is a gap in the study of qualitative kernels [8]. We show that even using the simplest categorical kernel possible (as is the Overlap kernel) we improve RBF kernel results in most of all drugs tested, although the extent of the improvement depends on the specific drug. The Jaccard kernel, which is also categorical, improves the Overlap results slightly and has the advantage of handling directly the amino acid mixtures, which are present in ≈55% of the HIV sequences in this dataset. Both Overlap and Jaccard are basic kernel functions, but our kernel definition (6) is general enough to replace them for more sophisticated categorical kernels that may improve even more the resistance prediction.

As mentioned before, the study of major drug resistance associated residues is the base of the rule-based algorithms approach. Weighting the kernels makes possible to take advantage of the fact that not all positions are equally important for the development of resistance. In our results, we observe a distinct distribution of the importances in protease inhibitors and transcriptase reverse inhibitors (both NRTIs and NNRTIs). It is known that there exist differences in the mutational pattern between the two enzymes in regards to the drug resistance [2]. In the reverse transcriptase, the major resistance mutations tend to be located in specific places, i.e. the drug binding site. There are almost no relevant mutations described in the C-terminal ending; for that reason, Stanford HIV Drug Resistance Database only stores the first 240 residues (out of 560) of the sequence. Instead, in the protease case, the virus acquires resistance by accumulating mutations. There are at least 36 important residues (of a total of 99) involved in protease drug resistance mutations and, unlike reverse transcriptase, they are distributed along the whole sequence. In line with this, we observed higher proportion of polymorphic positions and higher number of different alleles in the protease data. These differences may explain why RF and, therefore, the weighted categorical kernels work better at the NRTI and NNRTI databases. If less positions of the protein are relevant to the drug resistance, applying the mean across all positions dilutes this relevant information within the non-relevant positions. In contrast, the compensatory secondary mutations of the protease probably introduce some degree of correlation between protein positions, which may explain why weighting in PI database does not result in a clear improvement of performance.

5   Conclusions

Machine learning is an effective approach to predict HIV drug resistance, and a straightforward alternative to the much slower and expensive in vitro assay. Here, we propose the use of categorical kernel functions adapted to HIV data particularities, as are the allele mixtures, and that integrate additional knowledge about the major resistance associated protein positions. Our results show that kernels that take into account both the categorical nature of the data and the presence of mixtures consistently result in the best prediction model. Using more sophisticated categorical kernels may improve even more the resistance prediction. As for the introduction of the protein position importance, we notice differences depending on the protein targeted by the drug. In the case of reverse transcriptase, weights based in the relative importance of each position greatly increase the prediction performance, while in protease results are mixed. These differences may be related to a distinct mutational profile regarding drug resistance in the two proteins.

Acknowledgments. Work funded by project AGL2016-78709-R (Ministerio de Economía y Competitividad, Spain) to MPE. ER has funding from a FI-AGAUR grant (Generalitat de Catalunya). CRAG center receives the support of "Centro de Excelencia Severo Ochoa 2016–2019" award SEV-2015-0533 (Ministerio de Economía y Competitividad, Spain).

References

1. Global HIV & AIDS statistics. http://www.unaids.org/en/resources/fact-sheet. Accessed 13 Nov 2018
2. Iyidogan, P., Anderson, K.S.: Current perspectives on HIV-1 antiretroviral drug resistance. Viruses **6**, 4095–4139 (2014)
3. Seitz, R.: Human immunodeficiency virus (HIV). Transfus. Med. Hemother. **43**, 203–222 (2016)
4. Shafer, R.W., Dupnik, K., Winters, M.A., Eshleman, S.H.: A guide to HIV-1 reverse transcriptase and protease sequencing for drug resistance studies. HIV Seq. Compend. **2001**, 1–51 (2001)
5. Bonet, I.: Machine learning for prediction of HIV drug resistance: a review. Curr. Bioinform. **10**(5), 579–585 (2015)
6. The R project for statistical computing. https://www.R-project.org/. Accessed 15 Nov 2018
7. Scholkopf, B., Vert, J.P., Tsuda, K.: Kernel Methods in Computational Biology, 1st edn. The MIT Press/A Bradford Book, Massachusetts (2004)
8. Belanche, L.A., Villegas, M.A.: Kernel functions for categorical variables with application to problems in the life sciences. Front. Artif. Intell. Appl. **256**, 171–180 (2013)
9. Genotype-Phenotype Stanford University HIV Drug Resistance Database. https://hivdb.stanford.edu/pages/genopheno.dataset.html. Accessed 13 Nov 2018
10. Bouchard, M., Jousselme, A.L., Doré, P.E.: A proof for the positive definiteness of the Jaccard index matrix. Int. J. Approx. Reason. **54**, 615–626 (2013)

Deciphering General Characteristics of Residues Constituting Allosteric Communication Paths

Girik Malik[1,2], Anirban Banerji[1], Maksim Kouza[1,3],
Irina A. Buhimschi[2,4], and Andrzej Kloczkowski[1,4,5(✉)]

[1] Battelle Center for Mathematical Medicine,
The Research Institute at Nationwide Children's Hospital, Columbus, OH, USA
Andrzej.Kloczkowski@nationwidechildrens.org
[2] Center for Perinatal Research,
The Research Institute at Nationwide Children's Hospital, Columbus, OH, USA
[3] Faculty of Chemistry, University of Warsaw, Pasteura 1,
02-093 Warsaw, Poland
[4] Department of Pediatrics, The Ohio State University College of Medicine,
Columbus, OH, USA
[5] Future Value Creation Research Center, Nagoya University, Nagoya, Japan

Abstract. Allostery is one of most important processes in molecular biology by which proteins transmit the information from one functional site to another, frequently distant site. The information on ligand binding or on posttranslational modification at one site is transmitted along allosteric communication path to another functional site allowing for regulation of protein activity. The detailed analysis of the general character of allosteric communication paths is therefore extremely important. It enables to better understand the mechanism of allostery and can be used in for the design of new generations of drugs.

Considering all the PDB annotated allosteric proteins (from ASD - AlloSteric Database) belonging to four different classes (kinases, nuclear receptors, peptidases and transcription factors), this work has attempted to decipher certain consistent patterns present in the residues constituting the allosteric communication sub-system (ACSS). The thermal fluctuations of hydrophobic residues in ACSSs were found to be significantly higher than those present in the non-ACSS part of the same proteins, while polar residues showed the opposite trend.

The basic residues and hydroxyl residues were found to be slightly more predominant than the acidic residues and amide residues in ACSSs, hydrophobic residues were found extremely frequently in kinase ACSSs. Despite having different sequences and different lengths of ACSS, they were found to be structurally quite similar to each other – suggesting a preferred structural template for communication. ACSS structures recorded low RMSD and high Akaike Information Criterion (AIC) scores among themselves. While the ACSS networks for all the groups of allosteric proteins showed low degree centrality and closeness centrality, the betweenness centrality magnitudes revealed nonuniform behavior. Though cliques and communities could be identified within the ACSS, maximal-common-subgraph considering all the ACSS could not be generated, primarily due to the diversity in the dataset. Barring one particular case, the entire ACSS for any class of allosteric proteins did not

© Springer Nature Switzerland AG 2019
I. Rojas et al. (Eds.): IWBBIO 2019, LNBI 11466, pp. 245–258, 2019.
https://doi.org/10.1007/978-3-030-17935-9_23

demonstrate "small world" behavior, though the sub-graphs of the ACSSs, in certain cases, were found to form small-world networks.

Keywords: Allosteric communication sub-system ·
B-factor of allosteric residues · Cliques and communities · Closeness centrality ·
Betweenness centrality · Maximum-common-subgraph · Small-world network

1 Introduction

Starting from Monod−Wyman−Changeux [1] and Koshland−Némethy−Filmer [2] models, investigations of allosteric regulation of protein function have over half-a-century long, rich and multifaceted history. There are so many excellent reviews that have attempted to capture the essence of various aspects of research [3–8]. To summarize these efforts, one can merely observe that while a lot has been unearthed about the physicochemical nature of allosteric signal transduction, the various modes through which the long-distant communication is achieved, the structural details of cooperativity revealed during this process, there are still significant aspects of allosteric regulation, especially in the context of generalized characterization of the process, that need to be better understood. The present work reports a few generalized findings about the allosteric communication.

Because the allosteric communication paths are constituted by a certain subset of residues, we attempted in the present work, to provide a quantifiable difference between the residues involved in allosteric communications and those which are not involved. Because of that, our study revolved principally around identifying the statistical and graph-theoretical differences between the two aforementioned set of residues. We tried to decipher some consistent patterns embedded latently in structural, biophysical and topological nature of allosteric communication sub-structures (ACSS).

We focused also on the analysis of mobilities of residues forming ACSS. We compared protein fluctuations derived from crystallographic Debye-Waller B-factors of experimentally solved crystal structures with those obtained from the root mean square fluctuations (RMSF) profile from computational modeling.

We were interested to know whether the sub-structures of the allosteric communication paths have structural similarities among themselves, so that the kinase's allosteric communication paths will be characterized by a certain set of canonical parameters, while the nuclear receptor's allosteric communication paths will be different by certain (structural) degrees, etc.

2 Materials

The curated database ASD (Allosteric Database) [9] was used to retrieve protein structures with information about the identified allosteric communication paths. Cases with differences in the description of protein structures provided by the ASD and PDB were not considered for the study. Retaining the typification scheme provided by the ASD, the finally selected set of 30 proteins were further divided in four groups:

kinases, nuclear receptors, peptidases and transcription factors. The PDB IDs of these 30 proteins are: 1CZA, 1DKU, 1E0T, 1PFK, 1S9I, 1SQ5, 2BTZ, 2JJX, 2OI2, 2VTT, 2XRW, 3BQC, 3EQC, 3F9M, 3MK6, 4AW0 (kinases); 1IE9, 1XNX, 2AX6 3S79 (nuclear receptors); 1SC3, 2QL9, 4AF8 (peptidases); and 1JYE, 1Q5Y, 1R1U, 1XXA, 2HH7, 2HSG, 3GZ5 (transcription factors).

3 Methodology

Resorting to a coarse-grained representation of residues, and a reduced amino acid alphabet is more likely to lead to generalized ideas from the investigation of the ACSS of 30 proteins. A mere two-letter hydrophobic-polar classification of the residues would have been too broad to reveal the complexity of the problem. Thus we resorted to a scheme [10, 11] which has been found to be extremely successful in protein structure prediction studies [12, 13]. Here the 20 amino acids are expressed with a reduced 8-letter alphabet scheme; that is: GLU and ASP - as acidic, ARG, LYS and HIS - as basic, GLN and ASN - as amides, SER and THR - as hydroxyls, TRP, TYR, PHE, MET, LEU, ILE and VAL - as hydrophobic, and GLY and ALA - as small residues. PRO and CYS are special among the 20 amino acids because of their special status; each one of them are placed as singleton groups. This coarse-grained description was used to study both population characteristics of the ACSS constituents and to undertake the network-based investigations of ACSS.

Various tests were conducted throughout the study to compare residues constituting ACSS with non-ACSS ones, to measure the extent by which residues involved in allosteric communication differ from all remaining residues. Atoms of the residues not identified (viz., color-coded) by the ASD as part of allosteric communication paths, were considered to be non-ACSS residues and atoms.

We used THESEUS 2.0 software [14] that superposes multiple protein structures without throwing away gaps in them and without causing significant information loss.

For network-based and complex network studies, Python's NetworkX was used as the graphing library, while matplotlib was used for image generation. Apart from these, Python's igraph was used to investigate cliques and communities.

Because statistical tests are necessary to categorically establish the general traits in the allosteric proteins and yet, because the present study considers a limited set of allosteric proteins as belonging to different four classes two non-parametric tests (Wilcoxon signed-rank test and Friedman's non-parametric test) [15–17] were employed to ascertain the traits of the obtained results.

In investigating the "small world network" characteristics, methodologies elaborated in [18] were implemented by us; details about the theoretical basis of the methodology, thus, can be found there. To gather the answer to the question of whether or not the ACSSs are SWNs or not, at multiple resolutions, we studied the problem by generating the Erdös-Rényi (E–R) random graph at three probabilities: 0.3, 0.5 and 0.7.

In our work, the CABS-flex method [19] was used for predicting protein fluctuations. CABS-flex employs a coarse grained CABS model [20] - efficient and versatile tool for modeling protein structure, dynamics and interactions [21–24]. Conformations obtained by CABS-flex simulations further can be reconstructed to physically sound atomistic systems using coarse-grained to atomistic mapping methods [21, 25]. The interactions between atoms are described by a realistic knowledge-based potential, while protein-solvent interactions are approximated using implicit solvent model [20, 26].

4 Results

4.1 The Thermal Fluctuation of Residues in Allosteric Communication Paths

Alongside the dynamics needed to ensure the propagation of the structural signal through protein, the ACSS residues possess their inherent thermal fluctuational dynamic. The construction of the residual-interaction networks depends on the value of the cutoff distance, that may be different than the commonly used value of 6.5 Å [27]. To quantify the extent of fluctuations of the ACSS residues, versus fluctuations of non-ACSS residues, we extracted B-factors from the coordinate files of the protein structures in protein data bank (PDB) [28] for all 30 proteins. Table 1 contains the details of this investigation.

To assess whether and by what extent the B-factors of different families of allosteric proteins differ from each other, we subjected the mean values to Friedman's non-parametric test (alternatively referred to as 'non-parametric randomized block analysis of variance') [15, 16]. We chose to employ Friedman's test because, ANOVA requires the assumptions of a normal distribution and equal variances (of the residuals) to hold, none of which is found to be existing in our case (viz., that in Table 1), while Friedman test is free from the aforementioned restrictions. The null hypothesis for the test was that the B-factors of the four types of ACSS are the same across repeated measures. Result obtained from the test categorically demonstrates that there indeed exists a substantial difference in the B-factors of these four classes of ACSSs. Results obtained from B-factors of four types of ACSS was Freidman $X^2 = 20.4 > 16.266$ (P value at 0.001, with 3 degrees of freedom), whereby the null hypothesis was rejected comprehensively.

To ascertain the degree to which the B-factors of ACSS residues in each of the four classes of allosteric proteins differ from the B-factors of the non-ACSS residues, each of the classes were subjected to Wilcoxon signed rank test (36), which is a non-parametric analogue of paired t-test for correlated samples, without assuming that the population is normally distributed. The null hypothesis for each of the comparisons was that the median difference between pairs of observations is zero. Result obtained from the tests revealed that the B-factors of ACSS residues in each of the classes differed significantly than the B-factors of the non-ACSS residues. For the kinase class of allosteric proteins we found, $W_{kinase} = 87 \gg 23$ ($[W(\alpha = 0.01, 17) = 23]$); for the peptidase class $W_{peptidase} = 8 > 2$ ($[W(\alpha = 0.05, 7) = 2]$) (we note that W is not defined in 0.01 at degrees of freedom 7 (though $W(0.01, 8) = 0$), whereby, the critical value comparison is being reported at the weaker 0.05 level); for the Nuclear Receptors, $W_{NR} = 15 > 5$ ($[W(\alpha = 0.01, 11) = 5]$); and for the transcription factors, $W_{TF} = 6 > 5$ ($[W(\alpha = 0.01, 11) = 5]$). Thus, the null hypothesis was rejected in each of the four cases with extremely high confidence.

Table 1. B-factor of residues constituting ACSS and those constituting non-ACSS

Kin. ACSS		Kin. non-ACSS		Pept. ACSS		Pept. non-ACSS		N.R. ACSS		N.R. non-ACSS		T.F. ACSS		T.F. non-ACSS	
Res.	B-factor	Res.	B-factor	Res.	B-factor	Res.	B-factor	Res.	B-factor	Res.	B-factor	Res.	B-factor	Res.	B-fact.
ALA	36.53(24.21)	ALA	30.34(17.73)			ALA	20.89(14.36)			ALA	32.89(22.47)	ALA	25.94(4.15)	ALA	37.71(19.94)
ARG	35.21(21.93)	ARG	37.69(21.34)	ARG	25.69(13.14)	ARG	23.68(16.86)	ARG	27.79(17.02)	ARG	46.01(24.03)	ARG	46.38(17.76)	ARG	46.82(22.25)
ASN	59.04(30.24)	ASN	36.84(19.79)			ASN	24.04(16.16)	ASN	13.86(5.30)	ASN	44.61(23.59)	ASN	28.83(19.51)	ASN	52.78(24.03)
ASP	35.64(13.31)	ASP	38.09(21.57)	ASP	14.62(3.92)	ASP	26.64(19.43)			ASP	40.96(24.86)	ASP	38.26(12.27)	ASP	42.54(21.98)
CYS		CYS	29.03(16.94)	CYS	35.69(1.38)	CYS	20.16(14.56)			CYS	38.71(27.67)	CYS	52.52(26.14)	CYS	28.70(13.13)
GLN	28.44(9.25)	GLN	39.92(22.96)			GLN	25.74(21.61)	GLN	54.70(5.09)	GLN	35.91(20.80)	GLN	36.90(4.65)	GLN	39.24(20.99)
GLU	39.59(17.93)	GLU	38.99(20.70)	GLU	15.95(4.99)	GLU	32.80(18.88)	GLU	51.66(18.04)	GLU	47.97(26.94)			GLU	53.41(24.19)
GLY	27.87(9.56)	GLY	32.86(17.22)			GLY	22.80(16.68)			GLY	47.41(27.43)			GLY	39.62(22.36)
HIS	44.97(20.35)	HIS	35.71(20.97)			HIS	27.35(20.58)	HIS	13.88(2.28)	HIS	37.59(23.26)	HIS	37.62(18.24)	HIS	34.54(21.18)
ILE	42.26(16.44)	ILE	29.78(15.62)			ILE	21.97(15.70)			ILE	44.0(28.05)			ILE	38.45(19.11)
LEU	40.69(19.53)	LEU	29.97(16.05)			LEU	19.34(13.89)	LEU	54.75(5.83)	LEU	34.14(23.81)			LEU	40.07(20.85)
LYS	38.72(11.67)	LYS	38.81(20.22)			LYS	31.09(18.67)			LYS	46.38(27.14)	LYS	47.32(3.57)	LYS	53.42(23.40)
MET	43.12(22.56)	MET	31.10(15.56)			MET	18.20(14.14)	MET	59.16(0.80)	MET	41.21(24.96)			MET	42.64(23.60)
PHE	28.48(6.08)	PHE	29.68(16.62)			PHE	22.49(15.94)	PHE	16.25(2.38)	PHE	37.18(23.42)			PHE	41.84(23.56)
		PRO	33.25(17.87)			PRO	21.53(16.09)			PRO	39.67(25.50)			PRO	41.43(19.17)
SER	34.86(13.03)	SER	33.41(19.38)	SER	11.33(0.60)	SER	23.36(17.02)	SER	8.37(0.82)	SER	38.91(26.37)	SER	17.47(1.97)	SER	41.46(21.70)
THR	45.03(27.45)	THR	32.29(18.18)	THR	12.35(0.40)	THR	20.91(15.51)			THR	42.18(25.24)	THR	51.90(20.00)	THR	40.01(19.26)
		TRP	31.22(23.75)			TRP	27.45(22.24)			TRP	44.96(32.35)			TRP	32.68(13.28)
TYR	52.48(24.13)	TYR	31.12(16.42)	TYR	36.82(1.73)	TYR	22.70(18.29)	TYR	11.47(3.65)	TYR	43.32(28.58)	TYR	67.23(2.49)	TYR	39.88(18.00)
VAL	38.10(23.12)	VAL	28.78(14.69)			VAL	19.41(15.28)	VAL	57.18(0.37)	VAL	38.22(24.76)			VAL	37.22(21.24)

B-factors were calculated at the residual level in ACSS and non-ACSS, they are presented as Mean (Std. Dev.)

4.2   Robustness of the Results Against CABS-Flex Simulations

The computational modelling is a key to solving many fundamental problems of molecular biology. Prediction of protein structures and interactions [29] as well as structural transformations taking places during unfolding, folding and aggregation processes have been studied by computer simulations at different levels of resolution and timescales [30–38]. For more efficient simulations one uses coarse-grained (CG) models which reduce the complexity of each amino acid by representing it by a single node or group of pseudo atoms [20, 29, 39, 40].

In order to address the question whether B-factors extracted from PDB file are consistent with root mean square fluctuations (RMSF) of atoms from simulations, limited simulations have been carried out with the help of CABS-flex simulations [19]. We computed the values of RMSF for three conceptually different proteins, 1Q5Y from transcription factors group, 1SC3 from the peptidases and 2JXX from the kinase group of allosteric proteins. RMSF profiles are shown as red curves (on right Y-axis) on middle plots in Figs. 1a, b and c. The values of B-factors are shown as black curves (on left Y-axis).

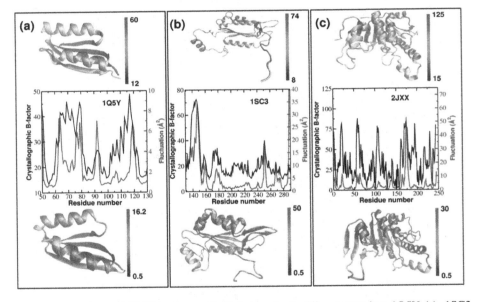

Fig. 1. Comparison of RMSF values with B factors for 3 different proteins: 1Q5Y (a), 1SC3 (b) and 2JXX (c). (Color figure online)

Although quantitative comparison between B-factors and RMSFs is not possible due to different temperatures and environmental factors used in simulations and experiments, qualitatively the data agree. The most fluctuating protein residues during near-native state simulations result in a series of peaks in RMSF profile (red curve, right X-axis) which correlate with experimentally measured B-factor values (black curve, left Y-axis).

Upper and bottom snapshots in Fig. 1 correspond to protein representation colored by crystallographic B-factor values from PDB and by RMSF values from CABS-flex simulation, respectively. The overall trend is that the mobility of atoms obtained from simulations are in reasonable agreement with the crystallographic B-factors.

4.3   Composition of the ACSS Population

Allosteric signalling achieved at the structural level show certain differences for various proteins [41–43]. Thus, we expect to observe differences in composition of ACSS residues for four different classes of proteins. We found that basic residues are more frequent in ACSS than the acidic ones. To demonstrate this prevalence let us take a closer look at the composition of ACSSs for kinases: the acidic residues were found in 12/133 cases, while the basic residues occurred in 32/133 cases. For the class of transcription factors the basic residues in ACSSs occurred in 13/35 cases, whereas the acidic residues occurred in 5/35 cases. The hydrophobic residues were found to occur in ACSSs of kinases with significant frequency (59/133 cases), but were be notably small in ACSS of transcription factors (2/35) and in peptidases (1/11).

Hydroxyl residues were found to be more common in ACSSs than the amide residues, for kinase ACSS: amide residues 6/133, and hydroxyl residues 14/133 cases. PRO and CYS populations although are extremely small in ACSS, show that CYS occurs slightly more frequently than PRO. The small amino acids (GLY and ALA) were found in very small frequency in ACSSs, while TRP was not found as part of any of the ACSSs.

4.4   Structural Superimposition of Multiple Allosteric Communication Paths

Results obtained from the structural superimposition of multiple ACSSs demonstrated clearly that the allosteric communication paths, for any type of allosteric protein, match closely each other in their structures. We superimposed the PDB-coordinates of ACSSs of all proteins for each of the four classes using 'Theseus' software. Here we report the two most prominent results, a: RMSD for the superposition, and b: the Akaike Information Criterion (AIC). AIC proposed by Akaike [44] has become commonly used tool for statistical comparison of multiple theoretical models characterized by different numbers of parameters. Because the RMSD of two superposed structures indicates their divergence from one another a small value is interpreted as a good superposition. In contrast, the higher magnitude of AIC indicates better superposition. We found that the ACSS paths, despite belonging to different proteins and corresponding to sequences of varying lengths, consistently demonstrated lower RMSD values and significantly higher AIC values in comparison to non-ACSSs parts of the structures.

4.5 Network Analyses of Allosteric Communication Paths

4.5.1 Centrality of ACSS

The process of allosteric signal communication is directional, but the richness of the constructs available to study networks becomes apparent by using non-directional graph-theoretical framework. Thus, instead of asking 'what is the route of allosteric signal propagation for a specific protein?', which is already provided by ADB, we asked questions like: 'how robust the ACSSs are, compared to non-ACSS parts of the proteins?', or, 'how does the fluctuation of one arbitrarily-chosen residue influence the spread of allosteric signal through ACSS?', or, 'how probable is it that allosteric communication occurs through a randomly chosen shortest path between two residues belonging to ACSS?', etc.

To answer these and similar questions of general nature, we started our investigation by studying the centrality aspects of the ACSS network. The centrality metrics quantify the relative importance of a protein residue (viz. the vertex) or a residue-to-residue communication path (viz., an edge) in the network description of ACSS. There are many centrality measures, we chose to concentrate upon three fundamental measures outlined in Freeman's classic works [45, 46], namely: degree centrality, betweenness centrality and closeness centrality.

4.5.2 Degree Centrality

Degree centrality for any protein residue in an ACSS network is calculated in a straightforward way, by counting the number of residue-residue communication links connecting that residue (implementing the classical definition [45] to the context of ACSS). Degree centrality of any ACSS residue provides an idea about the local structure around that residue, by measuring the number of other residues connected to it. We note that degree centrality is a local measure that does not provide any information about the network's global structure. We have found that the average degree centrality of ACSS residues, irrespective of the type of allosteric proteins, is lower than the average degree centrality of the non-ACSS residues. This result, alongside that obtained from the other centrality measures are presented in Table 2.

Table 2. The centrality indices for ACSS and non-ACSS for four groups of allosteric proteins.

	Kinase ACSS residues	Kinase non-ACSS residues	Peptidase ACSS residues	Peptidase non-ACSS residues	Nuclear receptor ACSS residues	Nuclear receptor non-ACSS residues	Transcription factor ACSS residues	Transcription factor non-ACSS residues
Average degree centrality	0.411	0.518	0.722	0.814	0.613	0.690	0.426	0.659
Average closeness centrality	0.488	0.596	0.809	0.843	0.707	0.754	0.474	0.719
Average betweenness centrality	0.075	0.122	0.194	0.038	0.137	0.141	0.062	0.119

The three types of major centrality measures calculated on the ACSS and non-ACSS graphs of the same size.

We note that the average degree centrality of ACSS fragments consistently show lower values than similar non-ACSS fragments. We note also that the typical differences between average degree centrality of ACSS and non-ACSS fragments are: ~ 0.10 for kinases, ~ 0.8 for nuclear receptors, ~ 0.9 for peptidases, and ~ 0.13 for transcription factors. Thus, the consistently lower values of average degree centrality observed in ACSS fragments suggests that nature attempts to shield them from perturbations which may destabilize allosteric communication.

4.5.3   The Global Centrality Measures

While the degree centrality provides a measure to assess the possibility of immediate involvement of a residue in influencing the signal communication in residue interaction network of a protein, the concepts of closeness centrality and betweenness centrality provide ideas of how the global topology of the network influences the signal propagation. Closeness centrality of any connected graph measures how "close" a vertex is to other vertices in a network; this is computed by summing up the lengths of the shortest paths between that vertex and other vertices in the network. Closeness of a vertex, thus, can be interpreted as a predictor of how long it may take for that vertex to communicate with all other vertices. In the framework of protein residue connectivity network, the residues with low closeness score can be identified as ones that are separated by short distances from other residues. It can be expected that they receive the structural signal (i.e. instantaneous fluctuation or perturbation) faster, being well-positioned to receive this information early. We indeed found that the average closeness centrality of the ACSS network is lower in comparison to the non-ACSS fragments, for all the types of allosteric proteins. However, the difference between the extent of average closeness centrality between ACSS and non-ACSS fragments was found to vary over a larger scale than what was observed for average degree centrality (see Table 2).

The betweenness centrality provides more idea about the global network structure; for every vertex of the network the betweenness centrality specifies the fraction of the shortest paths (geodesics) that pass through that vertex. In this sense, such measure assesses the influence that a given vertex (residue) has over the transmission of a structural signal. A residue with large betweenness centrality score can be expected to have a large influence on the allosteric signal propagating through the ACSS network. Results obtained by us shown in Table 2.

4.6   Cliques and Communities in ACSS

Cliques are the complete subgraphs, where every vertex is connected to every other vertex. A clique is considered maximal only if it is not found to be a subgraph of some other clique. Communities are identified through partitioning the set of vertices, whereby each vertex is made a member of one and only one community. Because of their higher order connectivity, the cliques detected in protein structures are considered to indicate regions of higher cohesion (in some cases, rigid modules). Do the ACSSs embody certain common characteristics in their connectivities which can be revealed through the cliques and communities? To answer this question, we subjected the ACSSs of each of the four classes of proteins to investigation, which implemented [47] and [48] algorithms through the Python-igraph package. We found that indeed the ACSS modules can be partitioned into cliques and communities.

4.7   Maximum Common Subgraphs to Describe the ACSS

A maximal common subgraph of a set of graphs is the common subgraph having the maximum number of edges. Many attempts have been made for the last two decades to apply this methodology in protein science [49–51]. Finding the maximal common subgraph is a NP-complete problem [52]. To solve this difficult problem a backtrack search algorithm proposed by McGregor [53] and a clique detection algorithm of Koch [52], are traditionally used. However, for our ACSSs, some of which are quite large in size, neither McGregor's nor Koch's algorithm was found to be applicable; primarily because of the huge computational costs incurred by the exponential growth of inter-mediary graphs of varying sizes. Thus, upon generating the subgraphs for each of ACSSs (using Python's NetworkX), we had to resort to the brute-force method to identify the maximum common subgraph for each of the ACSS classes. In some cases, the number of cliques was found to be large; e.g. for 2BTZ while in some other cases only one clique was found (e.g. for 2VTT or for 2XRW).

4.8   How Frequently Do the Allosteric Communication Paths Form Small World Network?

Investigating whether in general the ACSS residues belonging to the four different classes of allosteric proteins constitute 'small world' networks (SWN) or not is important; because SWNs are more robust to perturbations, and may reflect an evolutionary advantage of such an architecture [54, 55]. The SWN [56], constitute a compromise between the regular and the random networks, because on one hand they are characterized by large extent of local clustering of nodes, like in regular networks, and on the other hand they embody smaller path lengths between nodes, something that is distinctive for ran-dom networks. Because of the ability to combine these two disparate properties, not surprisingly, it has been shown that networks demonstrating the 'small-world' charac-teristics tend to describe systems that are characterized by dynamic properties different from those demonstrated by equivalent random or regular networks [56–61]. We have found that whether ACSSs exhibit SWN nature or not - is a complex problem; while the complete ACSS of a protein may not always demonstrate SWN characteristics, many sub-graphs of non-trivial lengths of the same ACSS reveal SWN character.

5   Conclusions

The aim of the present work was to decipher some general patterns of residues forming the ACSS of 30 allosteric proteins, and compare them with non-ACSS residues in the same proteins. Our aim was to report the general quantifiable differences between these two (aforementioned) sets of residues and not to study the general mechanism of allosteric communication. By performing the CABS-based simulations of proteins around their native conformations we demonstrated that protein fluctuations depicted by RMSF profiles can be mapped to B-factors and show satisfactory degree of agreement with experimental data.

Our results may benefit the protein engineering community and those studying the general mechanism of allosteric communication or in general, long-distance communication in proteins. The knowledge of the topological invariants of communication paths and the biophysical, biochemical and structural patterns may help in a better understanding of allostery. As many recent papers [62–66] have pointed out, the long-distance communication features within proteins involve several types of non-linear characteristics that may often be dependent on transient fluctuations, making it difficult to arrive at a generalized dynamic picture. However a generalized static picture of the long-distance communication route can be obtained, which may help to better understand such communication schemes, especially those related to allostery. The present work attempted to report such generalized findings. While certain yet-unknown (to the best of our knowledge) patterns regarding the thermal fluctuation profile of ACSS atoms, the structural and topological nature of the ACSS have come to light, incongruities of our findings regarding the extent of betweenness centrality in ACSS network and their small-world nature indicates the need for more focused studies directed at these issues, which in turn, may shed new light on allosteric signal communication. For example, proteins, in general, are fractal objects with known characteristics of trapping energy [67–70]. Do the findings on betweenness and on small-world network nature reported in this work indicate the possibility of energy traps in ACSSs? - We plan to probe into many such questions in future.

Acknowledgements. The second author, Dr. Anirban Banerji, passed away in Columbus, OH on Aug. 12, 2015 at the age of 39. A.K. acknowledges support from The Research Institute at Nationwide Children's Hospital, from the National Science Foundation (DBI 1661391) and from National Institutes of Health (R01GM127701 and R01GM127701-01S1). M.K. acknowledges the Polish Ministry of Science and Higher Education for financial support through "Mobility Plus" Program No. 1287/MOB/IV/2015/0. I.A.B. acknowledges support from the Eunice Kennedy Shriver National Institute of Child Health and Human Development (NICHD) R01HD084628 and The Research Institute at Nationwide Children's Hospital's John E. Fisher Endowed Chair for Neonatal and Perinatal Research.

References

1. Monod, J., Wyman, J., Changeux, J.P.: On the nature of allosteric transitions: a plausible model. J. Mol. Biol. **12**, 88–118 (1965)
2. Koshland Jr., D.E., Némethy, G., Filmer, D.: Comparison of experimental binding data and theoretical models in proteins containing subunits. Biochemistry **5**, 365–385 (1966)
3. Nussinov, R.: Introduction to protein ensembles and allostery. Chem. Rev. **116**, 6263–6266 (2016)
4. Ribeiro, A.A., Ortiz, V.: A chemical perspective on allostery. Chem. Rev. **116**, 6488–6502 (2016)
5. Dokholyan, N.V.: Controlling allosteric networks in proteins. Chem. Rev. **116**, 6463–6487 (2016)
6. Guo, J., Zhou, H.X.: Protein allostery and conformational dynamics. Chem. Rev. **116**, 6503–6515 (2016)

7. Papaleo, E., Saladino, G., Lambrughi, M., Lindorff-Larsen, K., Gervasio, F.L., Nussinov, R.: The role of protein loops and linkers in conformational dynamics and allostery. Chem. Rev. **116**, 6391–6423 (2016)

8. Wei, G.H., Xi, W.H., Nussinov, R., Ma, B.Y.: Protein ensembles: how does nature harness thermodynamic fluctuations for life? The diverse functional roles of conformational ensembles in the cell. Chem. Rev. **116**, 6516–6551 (2016)

9. Huang, Z.M., Mou, L.K., Shen, Q.C., Lu, S.Y., Li, C.G., Liu, X.Y., et al.: ASD v2.0: updated content and novel features focusing on allosteric regulation. Nucleic Acids Res. **42**, D510–D516 (2014)

10. Feng, Y.P., Kloczkowski, A., Jernigan, R.L.: Four-body contact potentials derived from two protein datasets to discriminate native structures from decoys. Proteins **68**, 57–66 (2007)

11. Feng, Y., Jernigan, R.L., Kloczkowski, A.: Orientational distributions of contact clusters in proteins closely resemble those of an icosahedron. Proteins **73**, 730–741 (2008)

12. Faraggi, E., Kloczkowski, A.: A global machine learning based scoring function for protein structure prediction. Proteins **82**, 752–759 (2014)

13. Gniewek, P., Kolinski, A., Kloczkowski, A., Gront, D.: BioShell-threading: versatile Monte Carlo package for protein 3D threading. BMC Bioinform. **15**, 22 (2014)

14. Theobald, D.L., Steindel, P.A.: Optimal simultaneous superpositioning of multiple structures with missing data. Bioinformatics **28**, 1972–1979 (2012)

15. Friedman, M.: The use of ranks to avoid the assumption of normality implicit in the analysis of variance. J. Am. Stat. Assoc. **32**, 675–701 (1937). J. Am. Stat. Assoc. **34**, 109 (1939)

16. Wilcoxon, F.: Individual comparisons by ranking methods. Biometrics Bull. **1**, 80–83 (1945)

17. Friedman, M.: A comparison of alternative tests of significance for the problem of m rankings. Ann. Math. Stat. **11**, 86–92 (1940)

18. Humphries, M.D., Gurney, K.: Network 'small-world-ness': a quantitative method for determining canonical network equivalence. PLOS One **3**, e0002051 (2008)

19. Jamroz, M., Kolinski, A., Kmiecik, S.: CABS-flex: server for fast simulation of protein structure fluctuations. Nucleic Acids Res. **41**, W427–W431 (2013)

20. Kolinski, A.: Protein modeling and structure prediction with a reduced representation. Acta Biochim. Pol. **51**, 349–371 (2004)

21. Kmiecik, S., Gront, D., Kouza, M., Kolinski, A.: From coarse-grained to atomic-level characterization of protein dynamics: transition state for the folding of B domain of protein A. J. Phys. Chem. B **116**, 7026–7032 (2012)

22. Wabik, J., Kmiecik, S., Gront, D., Kouza, M., Kolinski, A.: Combining coarse-grained protein models with replica-exchange all-atom molecular dynamics. Int. J. Mol. Sci. **14**, 9893–9905 (2013)

23. Blaszczyk, M., Kurcinski, M., Kouza, M., Wieteska, L., Debinski, A., Kolinski, A., et al.: Modeling of protein-peptide interactions using the CABS-dock web server for binding site search and flexible docking. Methods **93**, 72–83 (2016)

24. Jamroz, M., Orozco, M., Kolinski, A., Kmiecik, S.: Consistent view of protein fluctuations from all-atom molecular dynamics and coarse-grained dynamics with knowledge-based force-field. J. Chem. Theory Comput. **9**, 119–125 (2013)

25. Gront, D., Kmiecik, S., Kolinski, A.: Backbone building from quadrilaterals: a fast and accurate algorithm for protein backbone reconstruction from alpha carbon coordinates. J. Comput. Chem. **28**, 1593–1597 (2007)

26. Jamroz, M., Kolinski, A., Kmiecik, S.: Protocols for efficient simulations of long-time protein dynamics using coarse-grained CABS model. Methods Mol. Biol. **1137**, 235–250 (2014)

27. Sun, W.T., He, J.: From isotropic to anisotropic side chain representations: comparison of three models for residue contact estimation. PLOS One **6**, e19238 (2011)

28. Berman, H.M., Westbrook, J., Feng, Z., Gilliland, G., Bhat, T.N., Weissig, H., et al.: The protein data bank. Nucleic Acids Res. **28**, 235–242 (2000)
29. Kmiecik, S., Gront, D., Kolinski, M., Wieteska, L., Dawid, A.E., Kolinski, A.: Coarse-grained protein models and their applications. Chem. Rev. **116**, 7898–7936 (2016)
30. Sulkowska, J.I., Kloczkowski, A., Sen, T.Z., Cieplak, M., Jernigan, R.L.: Predicting the order in which contacts are broken during single molecule protein stretching experiments. Proteins-Struct. Funct. Bioinform. **71**, 45–60 (2008)
31. Scheraga, H.A., Khalili, M., Liwo, A.: Protein-folding dynamics: overview of molecular simulation techniques. Annu. Rev. Phys. Chem. **58**, 57–83 (2007)
32. Nasica-Labouze, J., Nguyen, P.H., Sterpone, F., Berthoumieu, O., Buchete, N.V., Cote, S., et al.: Amyloid beta protein and Alzheimer's disease: when computer simulations complement experimental studies. Chem. Rev. **115**, 3518–3563 (2015)
33. Kouza, M., Co, N.T., Nguyen, P.H., Kolinski, A., Li, M.S.: Preformed template fluctuations promote fibril formation: insights from lattice and all-atom models. J. Chem. Phys. **142**, 145104 (2015)
34. Kouza, M., Banerji, A., Kolinski, A., Buhimschi, I.A., Kloczkowski, A.: Oligomerization of FVFLM peptides and their ability to inhibit beta amyloid peptides aggregation: consideration as a possible model. Phys. Chem. Chem. Phys. **19**, 2990–2999 (2017)
35. Kmiecik, S., Kouza, M., Badaczewska-Dawid, A.E., Kloczkowski, A., Kolinski, A.: Modeling of protein structural flexibility and large-scale dynamics: coarse-grained simulations and elastic network models. Int. J. Mol. Sci. **19**, 3496 (2018)
36. Kouza, M., Banerji, A., Kolinski, A., Buhimschi, I., Kloczkowski, A.: Role of resultant dipole moment in mechanical dissociation of biological complexes. Molecules **23**, 1995 (2018)
37. Kouza, M., Co, N.T., Li, M.S., Kmiecik, S., Kolinski, A., Kloczkowski, A., et al.: Kinetics and mechanical stability of the fibril state control fibril formation time of polypeptide chains: a computational study. J. Chem. Phys. **148**, 215106 (2018)
38. Lan, P.D., Kouza, M., Kloczkowski, A., Li, M.S.: A topological order parameter for describing folding free energy landscapes of proteins. J. Chem. Phys. **149**, 175101 (2018)
39. Shakhnovich, E.: Protein folding thermodynamics and dynamics: where physics, chemistry, and biology meet. Chem. Rev. **106**, 1559–1588 (2006)
40. Liwo, A., He, Y., Scheraga, H.A.: Coarse-grained force field: general folding theory. Phys. Chem. Chem. Phys. **13**, 16890–16901 (2011)
41. Banerji, A.: An attempt to construct a (general) mathematical framework to model biological "context-dependence". Syst. Synth. Biol. **7**, 221–227 (2013)
42. Tuncbag, N., Gursoy, A., Nussinov, R., Keskin, O.: Predicting protein-protein interactions on a proteome scale by matching evolutionary and structural similarities at interfaces using PRISM. Nat. Protoc. **6**, 1341–1354 (2011)
43. Ozbabacan, S.E.A., Gursoy, A., Keskin, O., Nussinov, R.: Conformational ensembles, signal transduction and residue hot spots: application to drug discovery. Curr. Opin. Drug Disc. **13**, 527–537 (2010)
44. Akaike, H.: A new look at the statistical-model identification. IEEE Trans. Autom. Control **19**, 716–723 (1974)
45. Freeman, L.C.: Centrality in social networks conceptual clarification. Soc. Netw. **1**, 215–239 (1979)
46. Freeman, L.C., Borgatti, S.P., White, D.R.: Centrality in valued graphs - a measure of betweenness based on network flow. Soc. Netw. **13**, 141–154 (1991)
47. Reichardt, J., Bornholdt, S.: Statistical mechanics of community detection. Phys. Rev. E **74**, 016110 (2006)

48. Traag, V.A., Bruggeman, J.: Community detection in networks with positive and negative links. Phys. Rev. E **80**, 036115 (2009)

49. Grindley, H.M., Artymiuk, P.J., Rice, D.W., Willett, P.: Identification of tertiary structure resemblance in proteins using a maximal common subgraph isomorphism algorithm. J. Mol. Biol. **229**, 707–721 (1993)

50. Koch, I., Lengauer, T., Wanke, E.: An algorithm for finding maximal common subtopologies in a set of protein structures. J. Comput. Biol. **3**, 289–306 (1996)

51. Raymond, J.W., Willett, P.: Maximum common subgraph isomorphism algorithms for the matching of chemical structures. J. Comput. Aid. Mol. Des. **16**, 521–533 (2002)

52. Koch, I.: Enumerating all connected maximal common subgraphs in two graphs. Theor. Comput. Sci. **250**, 1–30 (2001)

53. McGregor, J.J.: Backtrack search algorithms and the maximal common subgraph problem. Softw. Pract. Exp. **12**, 23–34 (1982)

54. Barabasi, A.L., Albert, R.: Emergence of scaling in random networks. Science **286**, 509–512 (1999)

55. Barabasi, A.L., Oltvai, Z.N.: Network biology: understanding the cell's functional organization. Nat. Rev. Genet. **5**, 101–113 (2004)

56. Watts, D.J., Strogatz, S.H.: Collective dynamics of 'small-world' networks. Nature **393**, 440–442 (1998)

57. Barahona, M., Pecora, L.M.: Synchronization in small-world systems. Phys. Rev. Lett. **89**, 054101 (2002)

58. Nishikawa, T., Motter, A.E., Lai, Y.C., Hoppensteadt, F.C.: Heterogeneity in oscillator networks: are smaller worlds easier to synchronize? Phys. Rev. Lett. **91**, 014101 (2003)

59. Roxin, A., Riecke, H., Solla, S.A.: Self-sustained activity in a small-world network of excitable neurons. Phys. Rev. Lett. **92**, 198101 (2004)

60. Lago-Fernandez, L.F., Huerta, R., Corbacho, F., Siguenza, J.A.: Fast response and temporal coherent oscillations in small-world networks. Phys. Rev. Lett. **84**, 2758–2761 (2000)

61. del Sol, A., O'Meara, P.: Small-world network approach to identify key residues in protein-protein interaction. Proteins **58**, 672–682 (2005)

62. Kim, H., Zou, T.S., Modi, C., Dorner, K., Grunkemeyer, T.J., Chen, L.Q., et al.: A hinge migration mechanism unlocks the evolution of green-to-red photoconversion in GFP-like proteins. Structure **23**, 34–43 (2015)

63. Na, H., Lin, T.L., Song, G.: Generalized spring tensor models for protein fluctuation dynamics and conformation changes. Adv. Exp. Med. Biol. **805**, 107–135 (2014)

64. Song, G., Jernigan, R.L.: An enhanced elastic network model to represent the motions of domain-swapped proteins. Proteins **63**, 197–209 (2006)

65. Jamroz, M., Kolinski, A., Kihara, D.: Structural features that predict real-value fluctuations of globular proteins. Proteins **80**, 1425–1435 (2012)

66. Yang, Y.D., Park, C., Kihara, D.: Threading without optimizing weighting factors for scoring function. Proteins **73**, 581–596 (2008)

67. Enright, M.B., Leitner, D.M.: Mass fractal dimension and the compactness of proteins. Phys. Rev. E **71**, 011912 (2005)

68. Banerji, A., Ghosh, I.: Revisiting the myths of protein interior: studying proteins with mass-fractal hydrophobicity-fractal and polarizability-fractal dimensions. PLOS One **4**, e7361 (2009)

69. Leitner, D.M.: Energy flow in proteins. Annu. Rev. Phys. Chem. **59**, 233–259 (2008)

70. Reuveni, S., Granek, R., Klafter, J.: Anomalies in the vibrational dynamics of proteins are a consequence of fractal-like structure. Proc. Natl. Acad. Sci. U.S.A. **107**, 13696–13700 (2010)

Feature (Gene) Selection in Linear Homogeneous Cuts

Leon Bobrowski[1,2] and Tomasz Łukaszuk[1(✉)]

[1] Faculty of Computer Science, Bialystok University of Technology,
ul. Wiejska 45A, Bialystok, Poland
{l.bobrowski, t.lukaszuk}@pb.edu.pl
[2] Institute of Biocybernetics and Biomedical Engineering, PAS, Warsaw, Poland

Abstract. A layer of formal neurons ranked based on given learning data sets can linearize these sets. This means that such sets become linearly separable as a result of transforming feature vectors forming these sets through the ranked layer. After the transformation by the ranked layer, each learning set can be separated by a hyperplane from the sum of other learning sets.

A ranked layer can be designed from formal neurons as a result of multiple homogenous cuts of the learning sets by separating hyperplanes. Each separating hyperplane should cut off a large number of feature vectors from only one learning set. Successive separating hyperplanes can be found through the minimization of the convex and piecewise-linear (*CPL*) criterion functions. The regularized *CPL* criterion functions can be also involved in the feature selection tasks during successive cuts.

Keywords: Labeled data · Formal neurons · Ranked layers ·
Data sets linearization · *CPL* criterion function · Feature selection

1 Introduction

Learning data sets in the classification problem represent objects assigned to particular categories (classes). It is assumed that these objects are represented in a standardized manner by multivariate feature vectors of the same dimension or as points in the same feature space [1].

Multivariate feature vectors can be transformed by layers of binary classifiers. Binary classifiers transform feature vectors into numbers equal to one or to zero depending on values of particular features and the classifier parameters. Classifiers can be designed on the basis learning data sets according to a variety of pattern recognition methods [2, 3].

A formal neuron is an important example of a binary classifier. A formal neuron and the error-correction learning algorithm were used in the fundamental *Perceptron* model of the neural system plasticity [4, 5]. One of the key assumptions of the *Perceptron* model was the linear separability of the learning sets.

The perceptron criterion function is linked to the error-correction algorithm and belongs to the family of *convex and piecewise linear (CPL)* criterion functions [6]. The perceptron criterion function is linked to the error-correction algorithm and belongs to

© Springer Nature Switzerland AG 2019
I. Rojas et al. (Eds.): IWBBIO 2019, LNBI 11466, pp. 259–270, 2019.
https://doi.org/10.1007/978-3-030-17935-9_24

the family of *convex and piecewise linear (CPL)* criterion functions [6]. Applications of the CPL criterion functions in exploration of large, multidimensional data sets are based on an efficient and precise computational technique, which is called the *basis exchange algorithm* [7]. The basis exchange algorithm uses the Gauss – Jordan transformation and for this reason is similar to the *Simplex* algorithm used with spectacular successes in linear programming. The minimization of the *CPL* criterion functions allows, among others, for effective designing linear classifiers combined with a selection of feature subsets in accordance with the *relaxed linear separability (RLS)* method [8].

The minimization of the *CPL* criterion functions allows for designing ranked layers of binary classifiers [8]. Ranked layers of binary classifiers have an important property of data sets linearization [9, 10]. The possibility of inclusion of the feature selection tasks in the process of the ranked layer designing from formal neurons is examined for the first time in the presented paper.

2   Separable Learning Sets

We consider m objects O_j ($j = 1, \ldots, m$) represented as feature vectors $\mathbf{x}_j = [x_{j,1}, \ldots, x_{j,n}]^T$, or as points $\mathbf{x}_j$ in the n-dimensional feature space $F[n]$ ($\mathbf{x}_j \in F[n]$). Components $x_{j,i}$ of the feature vector $\mathbf{x}$ represent particular features x_i as numerical results of different measurements on a given object O_j ($x_{j,i} \in \{0,1\}$ or $x_{j,i} \in R$).

The labelled feature vector $\mathbf{x}_j(k)$ ($j = 1, \ldots, m$) contains additional information about the *category (class)* ω_k ($k = 1, \ldots, K$) of j-th object $O_j(k)$. The k-th learning set C_k contains m_k examples of the feature vectors $\mathbf{x}_j(k)$ assigned to the k-th category ω_k

$$C_k = \{\mathbf{x}_j(k)\} \quad (j \in J_k) \tag{1}$$

where J_k is the set of indices j of the feature vectors $\mathbf{x}_j(k)$ assigned to the class ω_k.

Definition 1. The learning sets C_k (1) are *separable* in the feature space $F[n]$, if they are disjoined in this space ($C_k \cap C_{k'} = \varnothing$, if $k \neq k'$). This means that the feature vectors $\mathbf{x}_j(k)$ and $\mathbf{x}_{j'}(k')$ belonging to different learning sets C_k and $C_{k'}$ cannot be equal:

$$(k \neq k') \Rightarrow (\forall j \in J_k) \; and \; (\forall j' \in J_{k'}) \; \mathbf{x}_j(k) \neq \mathbf{x}_{j'}(k') \tag{2}$$

We are also considering separation of the sets C_k (1) by the hyperplanes $H(\mathbf{w}_k, \theta_k)$ in the feature space $F[n]$:

$$H(\mathbf{w}_k, \theta_k) = \{\mathbf{x} : \mathbf{w}_k^T \mathbf{x} = \theta_k\}. \tag{3}$$

where $\mathbf{w}_k = [w_{k,1}, \ldots, w_{k,n}]^T \in R^n$ is the k-th weight vector, $\theta_k \in R^1$ is the threshold, and $\mathbf{w}_k^T \mathbf{x}$ is the inner product.

Definition 2. The feature vector $\mathbf{x}_j$ is situated on the *positive side* of the hyperplane $H(\mathbf{w}_k, \theta_k)$ (3) if and only if $\mathbf{w}_k^T \mathbf{x}_j > \theta_k$. Similarly, vector $\mathbf{x}_j$ is situated on the *negative side* of $H(\mathbf{w}_k, \theta_k)$ if and only if $\mathbf{w}_k^T \mathbf{x}_j < \theta_k$.

Definition 3. The learning sets (1) are *linearly separable* in the n-dimensional feature space $F[n]$ if each of the sets C_k can be fully separated from the sum of the remaining sets C_i by some hyperplane $H(\mathbf{w}_k, \theta_k)$ (3):

$$(\forall k \in \{1, \ldots, K\}) \, (\exists \mathbf{w}_k, \theta_k) (\forall \mathbf{x}_j(k) \in C_k) \quad \mathbf{w}_k^T \mathbf{x}_j(k) > \theta_k$$
$$\textbf{and } (\forall \mathbf{x}_j(k') \in C_{k'}, k' \neq k) \quad \mathbf{w}_k^T \mathbf{x}_j(k') < \theta_k \tag{4}$$

In accordance with the inequalities (4), all the vectors $\mathbf{x}_j(k)$ from the learning set C_k are situated on the positive side of the hyperplane $H(\mathbf{w}_k, \theta_k)$ (3) and all the vectors $\mathbf{x}_j(k')$ from the remaining sets $C_{k'}$ are situated on the negative side of this hyperplane.

3   Ranked Layers of Formal Neurons

The decision rule $r(\mathbf{w}, \theta; \mathbf{x})$ of the formal neuron $FN(\mathbf{w}, \theta)$ can be given as [2]:

$$\textit{if } \mathbf{w}^T \mathbf{x} \geq \theta, \textit{then } r(\mathbf{w}, \theta; \mathbf{x}) = 1, \textit{ otherwise } r(\mathbf{w}, \theta; \mathbf{x}) = 0 \tag{5}$$

where $\mathbf{w} = [w_1, \ldots, w_n]^T \in R^n$ is the weight vector, and $\theta \in R^1$ is the threshold.

The formal neuron $FN(\mathbf{w}, \theta)$ is activated ($r(\mathbf{w}, \theta; \mathbf{x}) = 1$) if and only if the weighed sum $w_1 x_1 + \ldots + w_n x_n$ of n inputs x_i ($x_i \in R$) is greater than the threshold θ. The decision rule $r(\mathbf{w}, \theta; \mathbf{x})$ (5) of the formal neuron $FN(\mathbf{w}, \theta)$ depends on the $n + 1$ parameters w_i ($i = 1, \ldots, n$) and θ.

The activation field $A(\mathbf{w}, \theta)$ of the formal neuron $FN(\mathbf{w}, \theta)$ is defined as the below half space:

$$A(\mathbf{w}, \theta) = \{\mathbf{x}: r(\mathbf{w}, \theta; \mathbf{x}) = 1\} = \{\mathbf{x}: \mathbf{w}^T \mathbf{x} \geq \theta\} \tag{6}$$

The activation field $A(\mathbf{w}, \theta)$ is the set of such vectors (points) $\mathbf{x}$ which are located on the positive side of the hyperplane $H(\mathbf{w}, \theta)$ (3). Each feature vector $\mathbf{x}$ located in the field $A(\mathbf{w}, \theta)$ activates the formal neuron $FN(\mathbf{w}, \theta)$ ($r(\mathbf{w}, \theta; \mathbf{x}) = 1$).

The ranked layer of L formal neurons $FN(\mathbf{w}_i, \theta_i)$ (5) transforms feature vectors $\mathbf{x}_j(k)$ from the learning sets C_k (1) into the sets D_k of the output vectors $\mathbf{r}_j(k)$:

$$D_k = \{\mathbf{r}_j(k)\} \quad (j \in I_k) \tag{7}$$

where each of the m output vector $\mathbf{r}_j(k) = [r_{j,1}, \ldots, r_{j,n}]^T$ has L binary components $r_{j,i} = r(\mathbf{w}_{,i}, \theta_{,i}; \mathbf{x}_j)$ (5).

The sets D_k (7) of the vectors $\mathbf{r}_j(k)$ transformed by the ranked layer become linearly separable (4) with the thresholds θ_k equal to zero:

$$(\forall k \in \{1, \ldots, K\}) \, (\exists \mathbf{v}_k) \, (\forall \mathbf{r}_j(k) \in D_k) \quad \mathbf{v}_k^T \mathbf{r}_j(k) > 0$$
$$\textbf{and } (\forall \mathbf{r}_j(k') \in D_{k'}, k' \neq k) \quad \mathbf{v}_k^T \mathbf{r}_j(k') < 0 \tag{8}$$

The proof of the above property can be found in the earlier paper [9].

Definition 4. The k-th *activation field* $A_k[(\mathbf{w}_1, \theta_1), \ldots, (\mathbf{w}_L, \theta_L)]$ of the layer of L formal neurons $FN(\mathbf{w}_i, \theta_i)$ is the set of such feature vectors $\mathbf{x}$ which give at the layer's output the k-th binary vector (*pattern*) $\mathbf{r}_k = [r_{k,1}, \ldots, r_{k,L}]^T$, where $r_{k,i} \in \{0,1\}$:

$$A_k[(\mathbf{w}_1, \theta_1), \ldots, (\mathbf{w}_L, \theta_L)] = \{\mathbf{x}: r(\mathbf{w}_1, \theta_1; \mathbf{x}) = r_{k,1}, \ldots, r_L(\mathbf{w}_L, \theta_L; \mathbf{x}) = r_{k,L}\} \quad (9)$$

Each non-empty activation field $A_k[(\mathbf{w}_1, \theta_1), \ldots, (\mathbf{w}_L, \theta_L)]$ (9) of the layer of L formal neurons $FN(\mathbf{w}_k, \theta_k)$ is a convex polyhedron in the feature space $F[n]$ with the walls formed by the hyperplanes $H(\mathbf{w}_k, \theta_k)$ (3).

Definition 5. The k-th formal neuron $FN(\mathbf{w}_k, \theta_k)$ with the decision rule $r(\mathbf{w}_k, \theta_k; \mathbf{x})$ (5) is *admissible* in respect to the i-th data set C_i (1) if and only if the activation field $A_{FN}(\mathbf{w}_k, \theta_k)$ (6) contains the labeled vectors $\mathbf{x}_j(i)$ from only this data set C_i (1).

The multistage procedure of the ranked layer designing is based on a sequence of the *homogeneous cuts* by the formal $FN(\mathbf{w}_k, \theta_k)$ (5) with the *admissible* decision rules $r(\mathbf{w}_k, \theta_k; \mathbf{x})$ Such formal neuron $FN(\mathbf{w}_k, \theta_k)$ (5) which is admissible to the i-th data set C_i allows for a homogeneous cut of this data set. In result of the first homogeneous cut, the i-th data set C_i (1) is reduced to the data set $C_i{}'$ by the set $R_{i,k}$ of such labeled vectors $\mathbf{x}_j(i)$ ($\mathbf{x}_j(i) \in C_i$) which activate only the k-th neuron $FN(\mathbf{w}_k, \theta_k)$:

$$C'_{i,k} = C_i / R_{i,k} \quad (10)$$

where

$$R_{i,k} = \{\mathbf{x}_j(i): r(\mathbf{w}_k, \theta_k, \mathbf{x}_j(i)) = 1\} = \{\mathbf{x}_j(i): \mathbf{x}_j(i) \in A_{FN}(\mathbf{w}_k, \theta_k)(6)\} \quad (11)$$

and

$$(\forall i' \neq i) \, R_{i',k} = \emptyset \quad (12)$$

In result of the homogeneous cut of the set C_i (10) the family of the data sets C_k (1) takes the below form:

$$C_1, \ldots, C_{i-1}, C'_{i,k}, C_{i+1}, \ldots, C_i \quad (13)$$

During the next stages of the ranked procedure one of the above sets C_i or $C_i{}'$ (13) will be reduced by the homogeneous cut (10). The sequence of the homogeneous cuts (10) is stopped if each of the learning sets C_i (1) is reduced to the empty set $\emptyset$ [7].

The procedure outlined above allows to generate the sequence of the parameters $(\mathbf{w}_k, \theta_k)$ of the admissible formal neurons $FN(\mathbf{w}_k, \theta_k)$ (*Definition 5*):

$$(\mathbf{w}_1, \theta_1), \ldots, (\mathbf{w}_L, \theta_L) \quad (14)$$

Each pair of the parameters $(\mathbf{w}_k, \theta_k)$ (12) defines the admissible formal neuron $FN(\mathbf{w}_k, \theta_k)$ (5).

Definition 6. The layer of L formal neurons $FN(\mathbf{w}_k, \theta_k)$ (5) defined by the sequence of the parameters $(\mathbf{w}_k, \theta_k)$ (14) is ranked in respect to the family of the data sets C_k (1).

It has been proved that the ranked layer of L formal neurons $FN(\mathbf{w}_k, \theta_k)$ (5) defined by the sequence of the parameters $(\mathbf{w}_k, \theta_k)$ (12) linearizes the learning sets C_k (1) [9].

4   The Perceptron Penalty and Criterion Functions

The minimization of the *CPL* criterion functions allows for an effective designing formal neurons $FN(\mathbf{w}_k, \theta_k)$ (5) with the admissible decision rules $r(\mathbf{w}_k, \theta_k; \mathbf{x})$ (*Definition 5*). In particularly, such technique allows to find the admissible decision rules $r(\mathbf{w}_k, \theta_k; \mathbf{x})$ (55) with a large numbers m_k of the feature vectors $\mathbf{x}_j$ from the k-th learning set $(\mathbf{x}_j \in C_k$ (1)) in the activation field $A(\mathbf{w}_k, \theta_k)$ (6).

The perceptron criterion function is linked to the error-correction algorithm used in the neural networks and belongs to the family of the *convex and piecewise linear (CPL)* criterion functions [6]. Let us introduce the positive set G_k^+ with m_k^+ vectors $\mathbf{x}_j$ from the k-th learning sets C_k ($\mathbf{x}_j \in C_k$) and the negative sets G_k^- with m_k^- vectors $\mathbf{x}_j$ from the remaining learning sets C_i (1):

$$G_k^+ = \{\mathbf{x}_j : j \in J_k^+\} \; and \; G_k^- = \{\mathbf{x}_j : j \in J_k^-\}, \tag{15}$$

where J_k^+ and J_k^- are disjoined sets of the indices j ($J_k^+ \cap J_k^- = \varnothing$).

If the sets G_k^+ and G_k^- are linearly separable (4), then all the elements $\mathbf{x}_j$ of the set G_k^+ can be located on the *positive side* ($\mathbf{w}_k^T \mathbf{x}_j > \theta_k$) of some hyperplane $H(\mathbf{w}_k, \theta_k)$ (3) in the feature space and all the elements $\mathbf{x}_j$ of the set G_k^- can be situated on the *negative side* of this hyperplane ($\mathbf{w}_k^T \mathbf{x}_j < \theta_k$).

The perceptron penalty functions $\varphi_j^+(\mathbf{w}, \theta)$ are defined for all the feature vectors $\mathbf{x}_j$ from the set G_k^+ (15) [6]:

$$\begin{aligned} &\left(\forall \mathbf{x}_j \in G_k^+\right) \\ &\varphi_j^+(\mathbf{w}, \theta) = 1 + \theta - \mathbf{w}^T \mathbf{x}_j \quad if \; \mathbf{w}^T \mathbf{x}_j < \theta + 1 \\ &\varphi_j^+(\mathbf{w}, \theta) = 0 \qquad\qquad\quad if \; \mathbf{w}^T \mathbf{x}_j \geq \theta + 1 \end{aligned} \tag{16}$$

Similarly, the perceptron penalty functions $\varphi_j^-(\mathbf{w}, \theta)$ are defined for all the feature vectors $\mathbf{x}_j$ from the set G_k^- (15) (Fig. 1):

$$\begin{aligned} &\left(\forall \mathbf{x}_j \in G_k^-\right) \\ &\varphi_j^-(\mathbf{w}, \theta) = 1 - \theta + \mathbf{w}^T \mathbf{x}_j \quad if \; \mathbf{w}^T \mathbf{x}_j > \theta - 1 \\ &\varphi_j^-(\mathbf{w}, \theta) = 0 \qquad\qquad\quad if \; \mathbf{w}^T \mathbf{x}_j \leq \theta - 1 \end{aligned} \tag{17}$$

The function $\varphi_j^+(\mathbf{w}, \theta)$ (16) is equal to zero if the feature vector $\mathbf{x}_j \in G_k^+$ is situated on the positive side of the hyperplane $H(\mathbf{w}_k, \theta_k)$ (3) and is not too close to it. Similarly, $\varphi_j^+(\mathbf{w}, \theta)$ (17) is equal to zero if the vector $\mathbf{x}_j \in G_k^-$ is situated on the negative side of the hyperplane $H(\mathbf{w}_k, \theta_k)$ and is not too close to it.

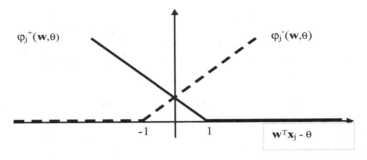

Fig. 1. The perceptron penalty functions φ_j^+ $(\mathbf{w}, \theta)$ (—) (16), and $\varphi_j^-(\mathbf{w}, \theta)$ (- - - -) (17).

The *perceptron criterion function* $\Phi_k(\mathbf{w}, \theta)$ is defined on elements $\mathbf{x}_j$ of the data sets G_k^+ and G_k^- (15) as the weighted sum of the penalty functions φ_j^+ $(\mathbf{w}, \theta)$ and φ_j^+ $(\mathbf{w}, \theta)$ [7]:

$$\Phi_k(\mathbf{w}, \theta) = \sum_{j \in J_k^+} \alpha_j \varphi_j^+ (\mathbf{w}, \theta) + \sum_{j \in J_k^-} \alpha_j \varphi_j^+ (\mathbf{w}, \theta) \tag{18}$$

where the positive parameters α_j can represent *prices* of particular feature vectors $\mathbf{x}_j$.

The standard values of the parameters α_j are equal to $\alpha_j = 1/(2\ m_k^+)$ for $\mathbf{x}_j \in G^+$ and $\alpha_j = 1/(2\ m_k^+)$ for $\mathbf{x}_j \in G$ [6].

The optimal parameters $(\mathbf{w}_k^*, \theta_k^*)$ define the minimal value $\Phi_k(\mathbf{w}_k^*, \theta_k^*)$ of the convex criterion function $\Phi_k(\mathbf{w}, \theta)$ (18):

$$(\forall (\mathbf{w}, \theta) \in R^{n+1})\ \ \Phi_k(\mathbf{w}, \theta) \geq \Phi_k(\mathbf{w}_k^*, \theta_k^*) \tag{19}$$

Remark 1. The minimal value $\Phi_k(\mathbf{w}_k^*, \theta_k^*)$ (19) of the perceptron criterion function $\Phi_k(\mathbf{w}, \theta)$ (18) is equal to zero if and only if the learning sets G_k^+ and G_k^- (15) are linearly separable (4) [6]:

$$(\forall \mathbf{x}_j \in G_k^+)\left(\mathbf{w}_k^*\right)^{\mathrm{T}} \mathbf{x}_j > \theta_k^*,\ and$$
$$(\forall \mathbf{x}_j \in G_k^-)\left(\mathbf{w}_k^*\right)^{\mathrm{T}} \mathbf{x}_j < \theta_k^* \tag{20}$$

5   The Regularized *CPL* Criterion Functions $\Psi_{K,\lambda}(\mathbf{w}, \theta)$

The regularized criterion function $\Psi_{k,\lambda}(\mathbf{w}, \theta)$ is used for the purpose of feature selection [8]. The regularized function $\Psi_{k,\lambda}(\mathbf{w}, \theta)$ is the sum of the perceptron criterion function $\Phi_k(\mathbf{w}, \theta)$ (18) and the additional *CPL* penalty functions equal to the absolute values of components $|w_i|$ of the weight vector $\mathbf{w}$ multiply by the *costs* γ_i ($\gamma_i > 0$) of particular features:

$$\Psi_{k,\lambda}(\mathbf{w}, \theta) = \Phi_k(\mathbf{w}, \theta) + \lambda \sum \gamma_i |w_i|$$
$$i \in \{1, \ldots, n\} \tag{21}$$

where $\lambda \geq 0$ is the *cost level*. The standard values of the parameters γ_i are equal to one.

The regularized criterion function $\Psi_{k,\lambda}(\mathbf{w}, \theta)$ (21) is used in the framework *relaxed linear separability* (*RLS*) method of feature subset selection [8]. The regularization components $\lambda \Sigma \gamma_i |w_i|$ of the criterion function $\Psi_{k,\lambda}(\mathbf{w}, \theta)$ (21) are similar to those used in the *Lasso* method of feature selection [11].

The main difference between the *Lasso* and the *RLS* methods is in the types of the basic criterion functions. The basic criterion function typically used in the *Lasso* method is the *residual sum of squares*, whereas the perceptron criterion function $\Phi_k(\mathbf{w}, \theta)$ (18) is used in the *RLS* method. This difference effects the computational techniques applied to minimize the criterion functions. The criterion function $\Psi_{k,\lambda}(\mathbf{w}, \theta)$ (21), similarly to the perceptron function $\Phi_k(\mathbf{w}, \theta)$ (18), is convex and piecewise-linear (*CPL*). The basis exchange algorithms, which are similar to linear programming, allow the identification of the minimum of the functions $\Psi_{k,\lambda}(\mathbf{w},\theta)$ (21) or $\Phi_k(\mathbf{w}, \theta)$ (18) even in the case when the learning sets G_k^+ and G_k^- (15) are big, formed by many high dimensional feature vectors $\mathbf{x}_j$.

The minimal value of the regularized function $\Psi_{k,\lambda}(\mathbf{w}, \theta)$ (21) defines the optimal weight vector $\mathbf{w}_\lambda^* = [w_{\lambda,1}^*, \ldots., w_{\lambda,n}^*]^T$ and the threshold θ_λ^*:

$$\left(\forall (\mathbf{w}, \theta) \in R^{n+1}\right) \ \Psi_{k,\lambda}(\mathbf{w}, \theta) \geq \Psi_{k,\lambda}\left(\mathbf{w}_\lambda^*, \theta_\lambda^*\right) \tag{22}$$

The optimal parameters $\mathbf{w}_\lambda^* = [w_{\lambda,1}^*, \ldots., w_{\lambda,n}^*]^T$ are used in the following *feature selection rule* [8]:

$$\left(w_{\lambda,i}^* = 0\right) \Rightarrow \text{(the } i\text{-th feature } x_i \text{ is reduced (neglected)}$$
$$\text{in the feature space } F[n]) \tag{23}$$

The n-dimensional *feature space* $F[n]$ is composed of all the n features x_i from the set $\{x_1, \ldots, x_n\}$. Feature reduction based on the rule (23) results in appearance of the reduced feature subspaces F_k. The symbol F_k denotes a feature subspace that is composed of k features x_i from the set $\{x_1, \ldots, x_n\}$. The reduced learning sets G_k^+ and G_k^- (15) can be composed from the reduced feature vectors $\mathbf{x}_j$ with the same k features x_i.

Successive increase of the value of the cost level λ in the criterion function $\Psi_\lambda(\mathbf{w}, \theta)$ (21) causes reduction of additional features x_i and, as the result, allows to generate the descended sequence of feature subspaces F_k with the decreased dimensionality n_k [8]:

$$F[n] = F_n \rightarrow F_{n-1} \rightarrow \ldots \rightarrow F_k \rightarrow \ldots \rightarrow F_1 \tag{24}$$

where $(\forall k \in \{2, \ldots., n\})F_k \supset F_{k-1}$.

The reduced feature subspaces F_k (24) are formed from the feature space $F[n]$ as a result of certain features x_i omission according to rule (6). The reduced feature vectors $\mathbf{x}_j'$ $(\mathbf{x}_j' \in F_k)$ are obtained from the feature vectors $\mathbf{x}_j$ in a similar manner, by reducing selected components $x_{j,i}$. The reduction of some features x_i may result in the loss of linear separability (20) of the learning sets G_k^+ and G_k^- (15).

In accordance with the relaxed linear separability (*RLS*) method, the generation of sequence (24) of the feature subspaces F_k is done deterministically [8]. Each step $F_k \rightarrow F_{k-1}$ can be realized by an increase the cost λ value: $\lambda_k \rightarrow \lambda_{k-1} = \lambda_k + \Delta_k$ ($\Delta_k > 0$) in the criterion function $\Psi_\lambda(\mathbf{w}, \theta)$ (21). The increasing of the parameter λ allows to reduce successive features x_i from the subspace F_k.

The feature reduction process should be stopped by fixing the cost λ at a specific level. The choice of the *stop criterion* of the sequence (21) is a crucial issue in the *RLS* method [8].

6 Linear Homogeneous Cuts with Feature Selection

The ranked layer designing from formal neurons $FN(\mathbf{w}_k, \theta_k)$ (5) can start on the basis of homogenous truncations (cuts) of feature vectors $\mathbf{x}_j$ belonging to a single learning sets C_k (1). The minimization of the k-th criterion function $\Phi_k(\mathbf{w}, \theta)$ (18) can be used for the separation of a large number of elements $\mathbf{x}_j$ of the k-th set C_k from all the elements $\mathbf{x}_j$ of the remaining sets C_i ($i \neq k$). Let us define for this purpose the positive set G_k^+ and the negative set G_k^- (15) in the below manner:

$$G_k^+ = \{\mathbf{x}_j : \mathbf{x}_j \in C_k\}, and \ G_k^- = \{\mathbf{x}_j : \mathbf{x}_i \in \cup_{i \neq k} C_i\} \tag{25}$$

The positive set G_k^+ is formed by m_k^+ feature vectors $\mathbf{x}_j$ from the k-th learning set C_k and the negative set G_k^- is formed by m_k^- feature vectors $\mathbf{x}_j$ from the remaining learning sets C_i ($i \neq k$) (1).

The k-th perceptron criterion function $\Phi_k(\mathbf{w}, \theta)$ (18) is defined on elements $\mathbf{x}_j$ of the sets G_k^+ and G_k^- (25). The regularized function $\Psi_{k,\lambda}(\mathbf{w}, \theta)$ (21) is the sum of the k-th perceptron criterion function $\Phi_k(\mathbf{w}, \theta)$ (18) and the additional *CPL* penalty functions $\gamma_i |w_i|$. Such optimal parameters $\mathbf{w}_\lambda^* = [w_{\lambda,1}^*, \ldots, w_{\lambda,n}^*]^T$ and θ_λ^* which constitute the

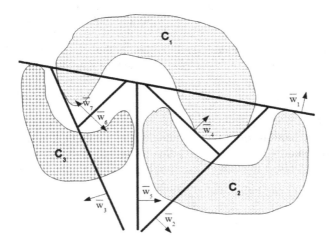

Fig. 2. Example of homogenous cuts of the three learning sets C_i (i = 1, 2, 3).

minimal value $\Psi_{k,\lambda}(\mathbf{w}_\lambda^*, \theta_\lambda^*)$ (22) of the k-th regularized function $\Psi_{k,\lambda}(\mathbf{w}, \theta)$ (21) can be used in searching for the homogeneous cut of the k-th learning set C_k (1) (Fig. 2).

Definition 7. The hyperplane $H(\mathbf{w}_\lambda^*, \theta_\lambda^*)$ (3) defined by the optimal parameters $\mathbf{w}_\lambda^*$ and θ_λ^* (22) allows to reduce (*cut*) some elements of $\mathbf{x}_j$ of the learning sets C_i (1) in accordance with the below rule:

$$if\left(\mathbf{w}_\lambda^*\right)^T \mathbf{x}_j > \theta_\lambda^* \tag{26}$$

then the j-th feature vector $\mathbf{x}_j$ ($\mathbf{x}_j \in C_i$) is reduced from the i-th set C_i (1)

Remark 2. The parameters α_j (18) and λ (21) of the k-th regularized function $\Psi_{k,\lambda}(\mathbf{w}, \theta)$ (21) can be selected in such a way that all reduced feature vector $\mathbf{x}_j$ (26) belongs to only learning set C_k ($\mathbf{x}_j \in C_k$). In this case, the k-th formal neuron $FN(\mathbf{w}_k, \theta_k)$ (5) cuts (26) the learning sets C_i (1) in the homogeneous manner.

The *Remark 2* has been proved theoretically and can be demonstrated experimentally. In order to obtain the ranked layer of formal neurons $FN(\mathbf{w}_k, \theta_k)$ (5) with a large generalization power, the following postulate is introduced:

Designing Postulate: The formal neurons $FN(\mathbf{w}_k, \theta_k)$ with the admissible decision rules $r(\mathbf{w}_k, \theta_k; \mathbf{x})$ (5) should be designed for the ranked layer in such a way, that each neuron operates in a low-dimensional feature space F_k (24) and reduces (23) a large number of feature vectors $\mathbf{x}_j$ (1) in the homogeneous manner.

7 Experimental Results

The data set collected by The Pan-Cancer Atlas project [13] were used in the experiment. This *RPPA (Final)* data set were available at the address: http://api.gdc.cancer.gov/data/fcbb373e-28d4-4818-92f3-601ede3da5e1. The original set contains 7790 objects representing the examined patients, each object is described by 198 features, expressions of selected genes. Moreover, each object is labeled with the abbreviated name of the type of cancer diagnosed in the patient. The set collects data from patients with 31 types of cancer.

For the purpose of the experiment, objects marked with BRCA, CORE, KIRC, LUAD and PRAD labels as well as features without missing values in the selected objects were selected from the *RPPA (Final)* dataset. The result is a new dataset containing 2401 objects described by 190 features. The objects of the new dataset were then divided randomly into a training and testing part in the ratio 80/20.

Particular learning sets C_k of feature vectors $\mathbf{x}_j = \mathbf{x}_j(k)$ (1) were cut in the homogeneous manner by the hyperplanes $H(\mathbf{w}_k, \theta_k)$ (3) with $\theta_k = 1$ during the experiment. The k-th hyperplane $H(\mathbf{w}_k, \theta_k)$ (3) truncated (4) a large number of feature vectors $\mathbf{x}_j$ from only one learning set C_k (1). The parameters $\mathbf{w}_k$ and θ_k of particular admissible

Table 1. The parameters w_{ki} and θ_k together with names of features for successive homogenous cuts made in the experiment.

$H(w_1, \theta_1)$ (BRCA)		$H(w_2, \theta_2)$ (CORE)		$H(w_3, \theta_3)$ (KIRC)		$H(w_4, \theta_4)$ (PRAD)		$H(w_5, \theta_5)$ (LUAD)	
2,962	AMPKALPHA	1,254	BAK	−7,654	MYOSINIIA_pS1943	9,383	AR	2,839	PEA15_pS116
1,393	PKCALPHA_pS657	−0,962	CHK2	5,843	FASN	−8,914	RAPTOR	0,996	PDCD4
−1,364	GATA3	−0,745	NOTCH1	−2,828	PEA15_pS116	1	θ_4	−0,891	TRANSGLUTAMINASE
−1,129	COLLAGENVI	−0,735	P70S6K_pT389	−2,184	NDRG1_pT346			0,799	FIBRONECTIN
1,035	SMAD3	−0,716	TIGAR	1,075	MYH11			0,724	GAPDH
1,021	SMAC	0,628	AR	0,911	CLAUDIN7			0,549	RICTOR
1,018	YAP	−0,497	ETS1	1	θ_3			1	θ_5
−0,679	ERALPHA	1	θ_2						
0,377	MYH11								
0,286	PAI1								
1	θ_1								

hyperplanes $H(w_k, \theta_k)$ are presented in following Table 1 (*Definition* 5). Table 1 contains also the names of the selected genes in the each homogeneous cut.

The designed layer of the five formal neurons $FN(w_k, \theta_k)$ (5) ($k = 1, \ldots, 5$) based on the admissible hyperplanes $H(w_k, \theta_k)$ (*Definition* 5) were used as the classifier of the feature vectors x_j of unknown origin ($x_j \in C_k$ (1)) with the following decision rule:

$$if \ w_k^T x_j > 1, \tag{27}$$

then the feature vector x_j is assigned to the k-th *category* ω_k ($k = 1, \ldots, 5$).

Table 2. Confusion matrix with detailed classification results obtained for feature vectors x_j from the training data

		Predicted class					
		BRCA	CORE	KIRC	LUAD	PRAD	No class
Actual class	BRCA	699	0	0	0	0	21
	CORE	0	325	0	0	0	39
	KIRC	0	0	356	0	0	7
	LUAD	0	0	0	187	0	2
	PRAD	0	0	0	0	272	8

Table 3. Confusion matrix with detailed classification results obtained for feature vectors x_j from the testing data

		Predicted class					
		BRCA	CORE	KIRC	LUAD	PRAD	No class
Actual class	BRCA	161	0	1	0	1	9
	CORE	0	91	0	0	2	9
	KIRC	0	0	86	0	1	4
	LUAD	0	1	0	47	0	0
	PRAD	0	0	0	0	70	2

The classifier (27) correctly assigns 96% of feature vectors x_j from the training data and 93.8% of objects from the testing data. Detailed classification results are presented in confusion matrices presented in Tables 2 and 3.

8 Concluding Remarks

Designing ranked layers of formal neurons $FN(\mathbf{w}, \theta)$ (5) is described in the paper. The ranked layer can be built as a result of homogenous cuts of selected feature vectors x_j from various learning sets C_k (1). During each homogenous cut some feature vectors x_j from only one learning set are eliminated.

The ranked layer transforms feature vectors x_j belonging to particular learning sets C_k (1) into linearly separable (8) sets D_k (7) of vectors $\mathbf{r}_j(k)$ with binary components $r_{j,i}$ ($r_{j,i} \in \{0, 1\}$). The hierarchical network of two ranked layers allows to transform all feature vectors x_j from particular data sets C_k (1) into only one vector with K binary components.

The ranked layer can be designed in result of a series of minimizations of the perceptron criterion functions $\Phi_k(\mathbf{w}, \theta)$ (18) which belong to the family of the convex and piecewise linear (*CPL*) functions. The minimizations of the regularized criterion functions $\Psi_{k,\lambda}(\mathbf{w}, \theta)$ (21) allowed to involve feature selection process into designing ranked layers of formal neurons $FN(\mathbf{w}, \theta)$ (5). Designing ranked layers combined with a selection of feature subsets in accordance with the relaxed linear separability (*RLS*) method could open a new research area aimed at exploration of large data sets [12].

Acknowledgments. The presented study was supported by the grant S/WI/2/2018 from Bialystok University of Technology and funded from the resources for research by Polish Ministry of Science and Higher Education.

References

1. Fukunaga, K.: Introduction to Statistical Pattern Recognition. Academic Press, Norwell (1972)
2. Duda, R.O., Hart, P.E., Stork, D.G.: Pattern Classification. Wiley, Hoboken (2003)
3. Bishop, C.M.: Pattern Recognition and Machine Learning. Springer, Boston (2006). https://doi.org/10.1007/978-1-4615-7566-5
4. Rosenblatt, F.: Principles of Neurodynamics. Spartan Books, Washington (1962)
5. Minsky, M.L., Papert, S.A.: Perceptrons. MIT Press, Cambridge (1969)
6. Bobrowski, L.: Data Mining Based on Convex and Piecewise Linear (*CPL*) Criterion Functions (in Polish). Bialystok University of Technology Press, New York (2005)
7. Bobrowski, L.: Design of piecewise linear classifiers from formal neurons by some basis exchange technique. Pattern Recogn. **24**(9), 863–870 (1991)
8. Bobrowski, L., Łukaszuk, T.: Relaxed linear separability (RLS) approach to feature (Gene) subset selection. In: Xia, X. (ed.) Selected Works in Bioinformatics, pp. 103–118. INTECH (2011)

9. Bobrowski, L.: Induction of linear separability through the ranked layers of binary classifiers. In: Iliadis, L., Jayne, C. (eds.) AIAI/EANN -2011. IAICT, vol. 363, pp. 69–77. Springer, Heidelberg (2011). https://doi.org/10.1007/978-3-642-23957-1_8

10. Bobrowski, L., Topczewska, M.: Linearizing layers of radial binary classifiers with movable centers. Pattern Anal. Appl. **18**(4), 771–781 (2015)

11. Tibshirani, R.: Regression shrinkage and selection via the lasso. J. Roy. Stat. Soc. B **58**(1), 267–288 (1996)

12. Bobrowski, L.: Data Exploration and Linear Separability, pp. 1–172. Academic Publishing, Germany, Lambert (2019)

13. Pancancer Atlas publication page. https://gdc.cancer.gov/about-data/publications/pancanatlas. Accessed 5 Dec 2018

DNA Sequence Alignment Method Based on Trilateration

Veska Gancheva[✉] and Hristo Stoev

Faculty of Computer Systems and Control, Technical University of Sofia,
8 Kliment Ohridski, 1000 Sofia, Bulgaria
vgan@tu-sofia.bg, hristomihaylovstoev@gmail.com

Abstract. The effective comparison of biological data sequences is an important and a challenging task in bioinformatics. The sequence alignment process itself is a way of arranging DNA sequences in order to identify similar areas that may have a consequence of functional, structural or evolutionary relations between them. A new effective and unified method for sequence alignment on the basic of trilateration, called CAT method, and using C (cytosine), A (adenine) and T (thymine) benchmarks is presented in this paper. This method suggests solutions to three major problems in sequence alignment: creating a constant favorite sequence, reducing the number of comparisons with the favorite sequence, and unifying/standardizing the favorite sequence by defining benchmark sequences.

Keywords: Bioinformatics · DNA · Sequence alignment · Trilateration

1 Introduction

Main task in biological data processing is searching of a similar sequence in database [1]. Algorithms such Needleman-Wunsch [2] and Smith-Waterman [3], which accurately determine the level of similarity of two sequences, are very time consuming applying them on large dataset. In order to increase the searching in large database, scientists apply heuristic methods, which significantly accelerate the searching time, but quality of the results decreases. FASTA is a DNA and protein sequence alignment software package introducing heuristic methods for aligning a query sequence to entire database [4]. BLAST is one of the most widely used sequence searching tool [5]. The heuristic algorithm it uses is much faster than other approaches, such as calculating an optimal alignment. This emphasis on speed is vital to turning the algorithm into the practice of the vast genomic databases currently available, although the following algorithms may be even faster. BLAST is more time efficient comparing with FASTA, searching only for the more significant sequences but with comparative sensitivity. Even parallel implementation of above algorithms is limited by the hardware systems [6–9]. A metaheuristics method for multiple sequence alignment, currently used for increasing the performance, have adopted the idea of generating of sequence favorite, after that all other sequences from the database are comparing to the favorite sequence [10]. On this way the sequence favorite turns into a benchmark for the other sequences in the database. Using this approach some problems occurs like the case of entering

I. Rojas et al. (Eds.): IWBBIO 2019, LNBI 11466, pp. 271–283, 2019.
https://doi.org/10.1007/978-3-030-17935-9_25

new data into the database or deleting some of the existing records. Since the sequence favorite is generated based on the records: (1) Changing the data leads to recalculation the sequence favorite. (2) Each of sequences at the database have to be compared again with the new sequence favorite in order to get new result, and this takes time and resources for the calculation. (3) There is a different favorite sequence for each database, and it can lead problems in merging different databases, especially in big data – collection of multiple different database structures and access.

In order to enhance existing heuristics algorithms' idea, improvements are proposed in each of the following fields:

1. Constant sequence favorite – i.e. it should not depend on the data set and should remain unchanged in case data set is modified.
2. Avoid comparisons or reduce the number of comparisons with the favorite sequence in the database searching (for each sequence we apply complicated algorithm for comparison with sequence favorite).
3. Unify/standardize of sequence favorite for all databases.

The goal of this paper is to present a new effective and unified method for sequence alignment on the basic of trilateration method. This method suggests solutions to three major problems in sequence alignment: (1) creating a constant favorite sequence, (2) reducing the number of comparisons with the favorite sequence, and (3) unifying/standardizing the favorite sequence by defining benchmark sequences.

2   An Effective and Unified Method for DNA Sequence Alignment Based on Trilateration

If we introduce some kind of coordinate system or find 3 or more benchmarks, then with the help of analytic geometry or in particular trilateration, we could fix the position of the points one against another, which actually represents the similarity between the database records. Also, the necessity of calculation of the sequence favorite will disappear. The idea of a sequence favorite is to find a starting point – benchmark, based on which to compare and analyze the rest of the records in the database. From a mathematical point of view, the favorite sequence can be represented as a function of N variables (in the case of DNA, the variables are the 4 bases: adenine, thymine, guanine and cytosine). Then we can represent the rest of the records in the database again as a function of the same variables. In that case, similarity comparison would be represented by the distance between each of the sequence to the sequence favorite. In other words, the location of the point representation, described by the function of the sequence, is localized against another point defined by the function of the favorite sequence. Representing each record from the data base with a point, will form a cloud of points, and point representation of sequence favorite, must be somewhere in the middle of this cloud. Since we do not have a specified coordinate system, each data base would form its own could of points with own center and merging of those data base will be very hard. If we introduce standardized coordinate system, applicable for all data basses, it would be possible, with the aid of elementary analytical geometry or trilateration, to determine the positions of the points relative to each other, which will reflect the degree of similarity between the records in the database and will be the same across database. Also, the need of calculate the favorite sequence will be dropped. Since DNA is built of only 4 bases, we can compose coordinate system with benchmark of endless sequence generated from each of the base. This will allow us to apply principals of trilateration, in order to fix the position of each sequence of the database against the coordinate system. Assess faster similarity of two or more sequence, and if we should process them further with more accurate algorithm (Similar method is used in Global Positioning System GPS [11]).

Trilateration is a method for positioning of objects using circle geometry. This method uses the known position of two or more reference points and measure the distance between the object and each of the reference points [12, 13]. In order to use trilateration, at least three reference points are required to determine accurately the point position in the plane. Trilateration is a method similar to triangulation, using angular measurements and a certain distance for position determining (Fig. 1).

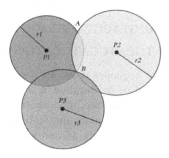

Fig. 1. Schematic overview of trilateration method

The relative position of point B relative to the points P1, P2 and P3 is determined by measuring the distance r1, r2 and r3. The distance r1 determines the position as a point of a circle with a radius r1 and center P1. r2 reduces opportunities position to be one of the crossed letters A or B. The third measure r3 determine the exact coordinates of the point. More than three measurements can be made in order to reduce the error. If applied similar reasoning to compare DNA data, points P1, P2 and P3 will be the benchmark sequences on each of the bases. Here a question arises "Once the bases are 4 why do we have 3 benchmark sequences?". In the structure of DNA there are certain rules that bases can connect: A + T; T + A; G + C; C + G, so that an "A" on one string of DNA will "connect" successfully only with "T" on the other string. It should be noted that the order is important: A + T is not equivalent to T + A and C + G is not the same as G + C. As a consequence of this rule, it can be stated that knowing one half of the helix of DNA, automatically can be reproduced the other. This means that from the mathematics viewpoint is not necessary to examine the entire helix. Or only benchmark of two bases would be enough for the accuracy of the results from a biological viewpoint, but we also introducing the third benchmark for greater precision.

Paying attention to the notes exposed a few lines above ("order is important: A + T is not equivalent to T + A") means that the generating of basic benchmarks as an infinite number of single basis is not good idea, because in fact position of the basis is matter and if the benchmark is formed form single basis we will result of simple counting of that basis of the processed sequence. If we write every base with the letter of the Latin alphabet – ACGT and repeat this string endlessly, we will get first proposed a benchmark sequence. It will look as following:

ACGTACGTACGTACGTACGTACGTACGTACGTACGTACGTACGT......

It can be called A-benchmark. In order to obtain good results by trilateration, it is advisable to have a uniform distribution of the benchmarks. Considering this recommendation and the way bases bound, the next benchmark sequence can be derived. It will be the second part of the hypothetical DNA created from A – A-benchmark or the benchmark should be supplemented to create a full DNA helix. Thus it is introduced exactly mirror allocation of A-benchmark, which provide more accurate results for the implementation of trilateration. Obtained following a benchmark sequence:

ACGTACGTACGTACGTACGTACGTACGTACGT... – A-Benchmark

TGCATGCATGCATGCATGCATGCATGCATGCA... – T-Benchmark

Thus, the second benchmark sequence could be named T-Benchmark. For a more precise application of the method of trilateration it is necessary to introduce a third benchmark, preferably equidistant from the previous two. Since two of the bases are already in use, to generate the next benchmark should be chosen between the other two, but at the same time the third benchmark have to be equal distant from both A-benchmark and T- benchmark.

Ideally, it should be 50% identical to the A-benchmark and 50% identical to the T-benchmark. To achieve this requirement, there is provided a sequence of bases where the two coincide in position with the respective bases of the A-benchmark and two bases coincide in position with T-benchmark. Or third benchmark would type:

ACGTACGTACGTACGTACGTACGTACGTACGT...– A-Benchmark

TGCATGCATGCATGCATGCATGCATGCATGCA... – T-Benchmark

TGCATGCATGCATGCATGCATGCATGCATGCA...– medial Benchmark 1

But if applied similar reasoning T-Benchmark against A-Benchmark is obtained:

ACGTACGTACGTACGTACGTACGTACGTACGT...– A-Benchmark

TGCATGCATGCATGCATGCATGCATGCATGCA... – T-Benchmark

TCGATCGATCGATCGATCGATCGATCGATCGA... – medial Benchmark 2

Where and how to fit median benchmarks in the method of trilateration?

In the process of trilateration than centers of circles, which we will attempt to represent with reference sequences, we should introduce equivalent radiuses. Since our task is to identify similarity between sequences, it is reasonable, radius to be equated with the grade of similarity. For the degree of similarity, a percentage of complete alignment of the bases from the benchmark sequence to the test sequence is selected. In order to calculate this rate, the benchmark sequence is limited to the size of the test sequence and thus calculates the percentage of matches. So, the number of matches to the full length of the test sequence will give wanted percentage of matches. The algorithm is with linear complexity and will always give the same result for the same sequence. This means that these calculations can be done once for all sequences in the database, then can be stored as sequences meta data and to be used in the study, which is a good solution for the problem (2).

Let's go back to the radius and how the degree of similarity must be applied in this case. The closer to the center of the circle is the benchmark sequence, the degree of matching is greater. When the match is 100%, the examined sequence is in the center of the circle. At this way the radius in absolute value is 1, as in the center of the circle, the distance is 1 and in the periphery is 0. The center of the circle of A-benchmark should lie on the periphery of the circle of T-benchmark and the center of the circle of T-benchmark should lie on the periphery of the circle of A-benchmark (Fig. 2).

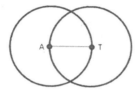

Fig. 2. Intersection between A-benchmark and T-benchmark area of sameness

This definition is used to determine the location of the medial benchmarks to A-benchmark and T-benchmark. In essence the medial benchmarks are 50% identical with A-benchmark and 50% identical with T-benchmark, which means that the benchmark should lie in the middle of the radius of the circle describing the similarity with A-benchmark and T-benchmark. That is, the middle benchmark is lying on the line between the centers of the two circles. This fact actually does not help us locating by the trilateration method, because the angle between the intersection point and the center of any of the circles is 0 degrees, but it raises an interesting question.

There are two middle benchmarks, each of which has an intersection point with the line created by the centers. These points of intersection are the middle of the intercept between the two centers, and they are perpendicular to the line passing through the intercept. Mathematically, this should mean that both middle benchmarks coincide, but this is not true. Then what creates this paradox in mathematics? The answer lies in biology. At the beginning of this article, we noted that the order of base pairing is important. This characteristic is expressed in the mathematical models applied to the biological data. In the first case there is alignment - A-benchmark against T-benchmark (left to right) with middle benchmark 1, in the other case the order is T-benchmark against A-benchmark (left to right) with middle benchmark 2.

Regarding trilateration it is not good to select three points lying on the same line. Therefore we continue to seek a third point, which will help to implement trilateration with certain accuracy. In this case, would it be possible to seek "equation" of the intersections of the circles described by A-benchmark and T-benchmark?

The radius has been defined such that the periphery of the circle is equal to 0% match. This means that the intersections in the circles should be 0% similar to the A- and T-benchmarks. Since the bases are only 4 and already two specific layouts were used and in the same time we seek for arrangements that do not match with the positions of the existing ones, we can easily write the following arrangements:

CTAGCTAGCTAGCTAGCTAGCTAGCTAGCTAGCTAGCTAGCTAGCTAG

GATCGATCGATCGATCGATCGATCGATCGATCGATCGATCGATCGATC

CATGCATGCATGCATGCATGCATGCATGCATGCATGCATGCATGCATG

GTACGTACGTACGTACGTACGTACGTACGTACGTACGTACGTACGTAC

......

Which of the above sequences are the targeted ones? The targeted sequences should meet another one condition. Let's try to find conditions that point M from Fig. 3 must satisfy. Pint M point is intersection of A-benchmark and T-benchmark circles, which meant that it should not match at all any of A and T benchmark. It is a peak of the ATM triangle, and this triangle is equilateral triangle, all sides are equal to the radius. The height from peak M to line AT, is a median of that triangle. And this is a point, which sequence we already know, or we can easily form. So, if we find the value of h, we will know the percentage of similarity of M to the medial sequence and we can find suitable sequence for point M.

At Fig. 3 we made de calculations (1), (2) and (3) and the similarity between medial sequence and sequence represented by point M should be around $\sim 13\%$ (Fig. 3).

$$h^2 = r^2 - \left(\frac{r}{2}\right)^2 \tag{1}$$

$$h^2 = \frac{3r^2}{4} \tag{2}$$

$$h = \frac{\sqrt{3}}{2}r \tag{3}$$

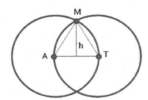

Fig. 3. Calculation of h – distance with Pythagoras theorem

The formed triangle by the centers of the circles and the intersection of triangle is equilateral. The distance from the edge of the triangle to the angle is 0.866025403 7844386467637231707075294. But the further away from the center, so the coincidence decreases, which means that the middle point is $\sim 13\%$ match.

It would be quite difficult to find the right "ingredients" of this sequence to satisfy exact match of the middle point of 13% and it would be fairly long sequence. Each match from M to the middle point automatically means matching A-benchmark or T-benchmark. Forming of sequence satisfying exact match of 13%, will require diving deep in the deeps of mathematics and will complicate further processing of the algorithm. Therefore, it's better to seek for a simpler solution of the problem and if possible to limit the sequence with a string of 4 bases which would be repeated endlessly.

Equilateral triangle formed by the centers A, T and the intersection of the circles will be used again. But this time an intercept with the length of 0.5 will be taken from the medial point. That's where would be point C and the Pythagorean Theorem will be applied in order to find the length of the hypotenuse, which should give the percentage of match with A-benchmark. Or (Fig. 4):

$$x^2 = \left(\frac{r}{2}\right)^2 + \left(\frac{r}{2}\right)^2 \tag{4}$$

$$x^2 = \frac{2r^2}{4} \tag{5}$$

$$x = \frac{1}{\sqrt{2}} r \tag{6}$$

$$x \approx 0.71 \tag{7}$$

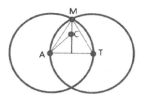

Fig. 4. Calculation of C-Distance with Pythagoras theorem

The further away from the center, the more similarity is reduced. This means that the point C is below 30% match similar. At 4 numbers of bases 25% match means that a base from the ideal A-benchmark must match a base of C-benchmark, and two bases of C-benchmark should match with two bases of medial benchmark (the distance from the tip to C is 0.5 radiuses, a 50% match). If the same reasoning is applied to T-benchmark, this means that one base of T-benchmark also must match with one of C-benchmark. After the present, it is much easier to find the components of the C-benchmark.

ACGTACGTACGTACGTACGTACGTACGTACGTACGTACGT... – A-Benchmark

TGCATGCATGCATGCATGCATGCATGCATGCATGCATGCA... – T-Benchmark

CGATCGATCGATCGATCGATCGATCGATCGATCGAT... – C-Benchmark

GCTAGCTAGCTAGCTAGCTAGCTA...– G-Benchmark fullfit Benchmark C

By drawing all the circles for each of the found benchmarks (A, C, G and T), the area in which the four circles intersect contains all the possible profiles comprised of the four bases (Fig. 5). For the purpose of trilateration are required only 3 circles. Therefore, only benchmarks C, A and T (CAT) will be considered.

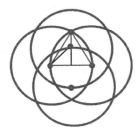

Fig. 5. Intersection between A, T, C and G benchmarks area of sameness. All possible sequences are locked this area

Once the three constant benchmarks have been established to implement trilateration, a problem (1) is solved. Constant sequence favorite does not depend on the data in the database and remain the same when data set changed. In fact, there are three constant sequences as a favorite sequence, which, if followed by standard multiple sequence alignment algorithms with sequence favorite, means there are three times more comparisons. Triples problem (2) - avoid comparisons or reduce the number of comparisons with sequences favorite during search in the database (for each sequence applies a complex algorithm to compare against the sequence favorite).

As founded benchmarks sequences are constant (i.e. not depend on either the data or the number), this allows to make comparisons at the outset - the introduction of the sequences in the database and this is description information (meta data) accompanying each sequence. In this way it is not needed comparison of the sequences during the searching process (which is the slowest operation), but instead will be compared only description information generated at the data input process.

By establishing benchmark sequences is solved problem (3) - Unification/ standardization of sequences favorites for all database. There is now a uniform sequences that are standardized for all the bases, using the described algorithm for comparison.

When two sequences have same profiles, this means that they have sections with same alignment and can be expected complete coincidence of one sequence to the other. But how to evaluate sequences that does not have identical profiles?

For the evaluation of random sequences, it is necessary to calculate the distance of the S1S2 segment in Fig. 6.

$$\sqrt{|AD_1 - AD_2|^2 + |h_1 - h_2|^2} \tag{8}$$

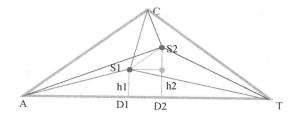

Fig. 6. Calculation of the distance between two profiles with cosine theorem

Currently regarded triangle AS1T, then carry out similar calculations and reasoning for AS2T. What is known about the upper triangle are the sides AT = |1|, AS1 = distance from S1 to A-benchmark (it is known), S1T = distance from S1 to T-benchmark. Used cosine theorem to find angle TAC, after which the side AD1:

$$S_1T^2 = AS_1^2 + AT^2 - 2.AS_1.AT.\cos(\alpha_1) \tag{9}$$

$$\cos(\alpha_1) = \frac{AS_1^2 + AT^2 - S_1T^2}{2.AS_1.AT} \tag{10}$$

$$AD_1 = AS_1.\cos(\alpha_1) \tag{11}$$

$$h_1 = \sqrt{AS_1^2 - (AS_1.\cos(\alpha_1))^2} \tag{12}$$

Similarly for triangle AS1T:

$$S_2T^2 = AS_2^2 + AT^2 - 2.AS_2.AT.\cos(\alpha_2) \tag{13}$$

$$\cos(\alpha_2) = \frac{AS_2^2 + AT^2 - S_2T^2}{2.AS_2.AT} \tag{14}$$

$$AD_2 = AS_2.\cos(\alpha_2) \tag{15}$$

$$h_2 = \sqrt{AS_2^2 - (AS_2.\cos(\alpha_2))^2} \tag{16}$$

After replacing the values found, a value for S1S2 is obtained as distance between two points on the formula:

$$S_1S_2 = \sqrt{(S_1D_1 - S_1D_2)^2 + (AD_1 - AD_2)^2} \tag{17}$$

The lower value for the segment means the greater probability of complete match, expressed in percentages. This allows quicker and accurate enough sequence database search, with a certain percentage of similarity, that later will be aligned and compared with more accurate algorithms such as Needleman-Wunsch or Smith-Waterman.

3 Experimental Results and Analysis

The objective of the experiments is to estimate experimentally efficiency of the designed CAT method for DNA sequence alignment. For this purpose a program implementation is developed. Let us examine these two sequences and their optimal alignment using Needleman-Wunsch algorithm:

```
G A A T T C A G T T A
|   | |   |   |       |
G G A T - C - G - - A
```

After the alignment there are six full matches out of a possible seven or 85.71% matched according Needleman-Wunsch algorithm. The same data have been used for experiments with both methods. Experimental results in the case of proposed CAT method are shown in Fig. 7.

```
GAATTCAGTTA
C-benchmark - 0% off 1
A-benchmark - 18.18% off 0.8181
T-benchmark - 18.18% off 0.8181
cosA = 0.6111111;
AD = 0.5;
h = 0.64762758403522769;
GGATCGA
C-benchmark - 42.85% off .5714285714285714
A-benchmark - 14.28% off .8571428571428571
T-benchmark - 28.57% off .7142857142857143
cosA = 0.71428571428571;
AD = 0.612244897959183;
h = 0.599875039048941;
S1S2 = 0 = 12198041920953887
=> sameness % 87.80%
```

Fig. 7. Experimental results for DNA sequence alignment method based on the trilateration

After finding suitable sequences in the database, such as the minimum value for S1S2, a more accurate alignment algorithm could be applied. Calculations obtained through the proposed CAT method are relatively simple and quick for implement; make it suitable to be applied as a first step in more accurate algorithm such as FASTA and to perform experimental studies utilizing big data sets.

A series of sequence alignment experiments have been carried out through different combinations of DNA sequences comparing both methods. The results for the calculated similarity of the fixed-length sequence alignment are shown in Table 1. DNA1 consists of 100 nucleotides and DNA2 consists of 30 nucleotides respectively. Results of the calculated similarity after alignment of random generated DNA sequences with random length are presented in Table 2. Table 3 presents the results of a benchmark sequence alignment, wherein DNA2 sequence is a subsequence of DNA1 sequence, i.e. there is a 100% match of the second over the first, which is obvious from the column with results of Needleman-Wunsch algorithm. Column Delta in all tables represents deviation between calculations based on CAT algorithm with respect to the Needleman-Wunsch algorithm.

The analysis of experimental results obtained by the sequence alignment shows little deviation of CAT method that could be ignored if this deviation is permissible at the expense of performance. The execution time of Needleman-Wunsch algorithm increases with increasing the sequences length. Time performance of the CAT method remains constant regardless of the sequences length. Therefore, the advantage of the proposed method is the rapid operation of large sequence alignment for which the accurate algorithms execution takes a long time.

Table 1. Experimental results of sequence alignment in case of fixed length sequences

Experiment no	DNA1 length	DNA2 length	CAT similarity	Exact match	Needleman Wunsch	Delta
1	110	30	0.668481925	24	0.8	0.131518075
2	110	30	0.901684169	23	0.766666667	0.135017502
3	110	30	0.709322998	26	0.866666667	0.157343669
4	110	30	0.949633278	24	0.8	0.149633278
5	110	30	0.844141279	25	0.833333333	0.010807946
6	110	30	0.845657014	24	0.8	0.045657014
7	110	30	0.868962098	26	0.866666667	0.002295431
8	110	30	0.861974197	22	0.733333333	0.128640864
9	110	30	0.825153668	26	0.866666667	0.041512999
10	110	30	0.794094919	21	0.7	0.094094919

Table 2. Experimental results of sequence alignment in case of various length sequences

Experiment no	DNA1 length	DNA2 length	CAT similarity	Exact match	Needleman Wunsch	Delta
1	978	149	0.92892273	133	0.89261745	0.03630528
2	945	194	0.958308962	171	0.881443299	0.076865663
3	784	257	0.956673586	210	0.817120623	0.139552964
4	877	280	0.982422073	226	0.807142857	0.175279216
5	907	182	0.886614319	154	0.846153846	0.040460473
6	503	283	0.926791731	197	0.696113074	0.230678657
7	542	136	0.947658604	117	0.860294118	0.087364487
8	723	203	0.951157669	170	0.837438424	0.113719245
9	742	346	0.98153563	263	0.760115607	0.221420023
10	667	379	0.945220218	265	0.701058201	0.244162017

Table 3. Experimental results of sequence alignment in case of exact matching

Experiment no	DNA1 length	DNA2 length	CAT similarity	Exact match	Needleman Wunsch	Delta
1	110	30	0.797937803	30	1	0.202062197
2	110	30	0.878802211	30	1	0.121197789
3	110	30	0.780298459	30	1	0.219701541
4	110	30	0.83460827	30	1	0.16539173
5	110	30	0.790112287	30	1	0.209887713
6	110	30	0.805192591	30	1	0.194807409
7	110	30	0.791356336	30	1	0.208643664
8	110	30	0.902574547	30	1	0.097425453
9	110	30	0.893694124	30	1	0.106305876
10	110	30	0.779823187	30	1	0.220176813

4   Conclusion

An innovative method for DNA sequences alignment based on the trilateration method has been proposed in this paper. Three constant benchmarks for the trilateration implementation have been defined, which create a constant favorite sequence, i.e. it does not depend on the data in the database. This allows making comparisons at the outset – during input of the sequences in the database and it can be stored as meta data to each sequence. Thus, there is no need to make a comparison of the sequences during the search, but instead will only compare the meta data. By establishing benchmark sequences have been solved the problem of unification/standardization of favorite sequence for all facilities using the described algorithm to compare. Calculations obtained in the proposed CAT method are relatively simple and quick to implement, making it suitable for application as a first step in more accurate algorithm. Future work is to apply the proposed algorithm and to perform experimental studies with big data sets in order to investigate the accuracy.

Acknowledgment. This work is supported by Grant DN07/24.

References

1. Mount, D.: Bioinformatics: Sequence and Genome Analysis, 2nd edn. Cold Spring Harbor Laboratory Press, New York (2009)
2. Needleman, S., Wunsch, C.: A general method applicable to the search for similarities in the amino acid sequence of two proteins. J. Mol. Biol. **48**, 443–453 (1970)
3. Smith, T., Waterman, M.: Identification of common molecular subsequences. J. Mol. Biol. **147**, 195–197 (1981)
4. Lipman, D.J., Pearson, W.R.: Rapid and sensitive protein similarity searches. Science **227** (4693), 1435–1441 (1985). https://doi.org/10.1126/science.2983426. PMID 2983426
5. Altschul, S., et al.: Basic local alignment search tool. J. Mol. Biol. **215**(3), 403–410 (1990)

6. Altschul, S., et al.: Gapped BLAST and PSIBLAST: a new generation of protein database search programs. Nucleic Acids Res. **25**, 3389–3402 (1997)
7. Lin, H., et al.: Efficient data access for parallel BLAST. In: 19th International Parallel & Distributed Processing Symposium (2005)
8. Zhang, F., Qiao, X.Z., Liu, Z.Y.: A parallel Smith-Waterman algorithm based on divide and conquer. In: Proceedings of the Fifth International Conference on Algorithms and Architectures for Parallel Processing ICA3PP 2002 (2002)
9. Farrar, M.: Striped Smith-Waterman speeds database searches six times over other SIMD implementations. Bioinformatics **23**(2), 156–161 (2007)
10. Borovska, P., Gancheva, V., Landzhev, N.: Massively parallel algorithm for multiple biological sequences alignment. In: Proceeding of 36th IEEE International Conference on Telecommunications and Signal Processing (2013). https://doi.org/10.1109/TSP.2013.6614014
11. Motlagh, O., Tang, S.H., Ismail, H., Ramli, A.R.: A review on positioning techniques and technologies: a novel AI approach. J. Appl. Sci. **9**, 1601–1614 (2019). https://doi.org/10.3923/jas.2009.1601.1614
12. Thapa, K., Case, S.: An indoor positioning service for bluetooth ad hoc networks. In: Proceedings of the Midwest Instruction and Computing Symposium, 11–12 April 2003, Duluth, MN, USA, pp. 1–11 (2003)
13. Ciurana, M., Barcelo-Arroyo, F., Izquierdo, F.: A ranging system with IEEE 802.11 data frames. In: 2007 IEEE Radio and Wireless Symposium (2007). https://doi.org/10.1109/rws.2007.351785

Convolutional Neural Networks for Red Blood Cell Trajectory Prediction in Simulation of Blood Flow

Michal Chovanec[1], Hynek Bachratý[2], Katarína Jasenčáková[2],
and Katarína Bachratá[2(✉)]

[1] Department of Technical Cybernetics,
Faculty of Management Science and Informatics,
University of Žilina, Žilina, Slovakia
michal.chovanec@fri.uniza.sk
[2] Department of Software Technology,
Faculty of Management Science and Informatics,
University of Žilina, Žilina, Slovakia
{hynek.bachraty,katarina.jasencakova,katarina.bachrata}@fri.uniza.sk
http://cell-in-fluid.kst.fri.uniza.sk

Abstract. Computer simulations of a blood flow in microfluidic devices are an important tool to make their development and optimization more efficient. These simulations quickly become limited by their computational complexity. Analysis of large output data by machine learning methods is a possible solution of this problem. We apply deep learning methods in this paper, namely we use convolutional neural networks (CNNs) for description and prediction of the red blood cells' trajectory, which is crucial in modeling of a blood flow. We evaluated several types of CNN architectures, formats of theirs input data and the learning methods on simulation data inspired by a real experiment. The results we obtained establish a starting point for further use of deep learning methods in reducing computational demand of microfluid device simulations.

Keywords: Microfluidic devices · Simulation experiment ·
Deep learning · Convolutional neural networks · Trajectory prediction

M. Chovanec—Author of this work was supported by the Ministry of Education, Science, Research and Sport of the Slovak Republic under the contract No. VEGA 1/0643/17.
K. Bachratá—Authors of this work were supported by the Slovak Research and Development Agency under the contract No. APVV-15-0751.

I. Rojas et al. (Eds.): IWBBIO 2019, LNBI 11466, pp. 284–296, 2019.
https://doi.org/10.1007/978-3-030-17935-9_26

1   Introduction

Computer Based Simulations of Blood Flow

Microfluidic devices have been investigated for multiple medical applications. For instance, these devices can be used for capturing of circulating tumor cells (CTCs) and thus for the early diagnostics and the targeted treatment of cancer. Since CTC are greatly outnumbered by other cells, it is highly desirable that the microfluidic devices are very efficient. The efficiency can be optimized by altering the shape, dimensions and other parameters of these devices. However, testing, verification and comparison of their properties on real prototypes of microfluidic devices is labourious and technically demanding. We have proposed replacing it with numerical simulation, which needs to be sufficiently authentic [1,5].

The model we use considers blood as a suspension. The suspensions' model consists of plasma and of elastic models of its solid components. The dominant part of solid components is formed by red blood cells (RBCs). Thus, the correct setup of elastic parameters of the RBCs' behavior and their interactions with the liquid, surface of device and other objects, are crucial for correctness and accuracy of used simulation model. Our research group uses the software tool ESPResSo for these simulations. We expanded it with a module of immersed elastic objects, which is still under development [6].

Nowadays, there is increasingly obvious need to augment numerical simulations. The need arises from the high modeling complexity together with the scope of simulations. The complexity involves the size and the topology of the simulated device, the amount of the modeled RBCs, the duration of the simulated experiment as well as a range of the monitored and recorded measurements. For these reasons, the simulation experiments performed at present have very high requirements on hardware efficiency as well as on the required computation time. Often times simulations run for several days or even weeks. Hence it happens to be a limitation for rerunning or extension of the simulation experiments with the same or slightly modified parameters. Even though a typical simulation provides a large amount of the observation data, we usually need to use only a fraction of it for a given end goal. Several analyses [2–4] suggest the experiments can be universally described and characterized by miscellaneous statistics on files from the output data. This inspired us to apply machine learning methods for complex processing of all output data from the simulation experiments. This approach promises to extract relationships from the simulation data which would one allow to obtain more results without the need for new simulation experiments.

The movement and the behavior of RBCs in general play the key role in our simulation experiments. Therefore, we decided to use machine learning methods firstly on trajectory prediction of RBCs immersed in the blood flow in the artificial microfluidic channel (Fig. 1). Our first experience in this topic can be found in [3]. There, the data from the simulation experiments were used as the input for the basis functions of Kohonen neural network. These basis functions allowed miscellaneous types of RBCs' trajectory prediction in the particular device. We obtained acceptable accuracy for the predictions, even with the relatively unsophisticated machine learning method.

In this paper, we decided to verify the possibilities of applications of the newest methods in deep neural networks learning. We used it on the data from a different type of the simulation experiment. To our best knowledge this is the first such application of CNN for prediction of trajectories of RBCs. Therefore, the aim of this work is to test various approaches to data preprocessing, different network networks architectures, various learning methods and utilization.

2   Design of a Simulation Experiment

We used data from the simulation experiment, which was inspired by the real experiment described in [8]. We simulated the movement of 38 RBCs in the micro channel depicted in the Fig. 2. The amount of the RBCs corresponds to the 1% hematocrit used in the laboratory experiment. Initial seeding of the simulated cells was in the left part of the channel, free from any obstacles. The main task of the experiment was to monitor characteristics like cells' deformation and velocity in the special narrow slits in the channel. The channel was periodic in direction of its main horizontal x-axis in the computer simulation. This direction also corresponds to the blood flow. We used the standard and calibrated parameters of the simulation model (see the Table 1). For the detailed description of the simulation model, see [9]. In terms of using the simulations results for a design and a validation of the CNNs architecture and their learning methods, the more important is the size and the quality of the data observed from the simulation described in the following section.

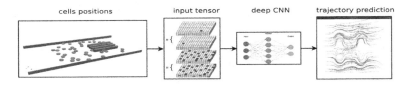

Fig. 1. RBCs' trajectory prediction using CNN from the simulation experiment.

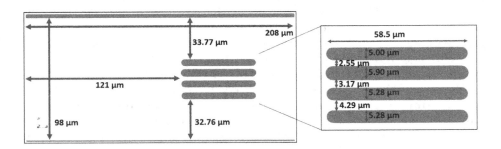

Fig. 2. Schema of the simulation channel with its dimensions.

3   Dataset Description

3.1   Data Processing

Data for the neural network comes from a simulation experiment. Its output includes information about trajectory properties of each cell. Before giving more in depth description of these data, it is good to acknowledge that the input of a neural network consists of several cell positions and output is the velocity of that cell at a particular time. There is record about movement of each cell during the simulation. Output files include information about the position and velocity of cell centers and also the velocity of border points, which are important when determining cell's rotation. The user can choose how often to record this data, since the time step is usually around tenths of microseconds. Data for this study were recorded every 0.001 s (once in 1000 simulation steps), which results in almost 9 200 records for each cell.

Table 1. First seven parameters are the elastic parameters of the cell model used in our experiments. In the last three rows are numerical parameters of simulation liquid.

Parameter	SI unit
Radius	$3.91 * 10^{-6}$ m
Stretching coefficient k_s	$6 * 10^{-6}$ N/m
Bending coefficient k_b	$8 * 10^{-18}$ Nm
Coefficient of local area conservation k_{al}	$1 * 10^{-6}$ N/m
Coefficient of global area conservation k_{ag}	$9 * 10^{-4}$ N/m
Coefficient of volume conservation k_v	$5 * 10^2$ N/m^2
Membrane viscosity	0 m^2/s
Density	$1 * 10^3$ kg/m^3
Kinematic viscosity	$1 * 10^{-6}$ m^2/s
Friction coefficient	$1.15 (-)$

Training and testing sets used for neural network are extracted from simulations which differ only in initial seeding of the cells. Therefore trajectories of the cells in these experiments are different. However, there should be similar behavior of cells' trajectories, since everything, apart from the initial seeding of the cells, is fixed. This includes the geometry of the channel, the elastic parameters of the cells and fluid parameters. In later work, we would like to apply our framework also on simulations with various geometries and parameter settings.

As was already mentioned, the input for the neural network are the positions of the cell center and the output is the velocity of this cell at given position. The learning set for our neural network consist of the simulation output which also includes redundant data. We processed the data in the following steps:

1. We loaded the data from the experiments and checked their correctness. Afterwards, we extracted data, which were relevant for the neural network and found the extreme values.

2. Next step was the normalization of the data. Using the extreme values from the previous step, we obtained data in interval $\langle 0, 1 \rangle$. Normalization of data is common in neural networks. It is needed for instance, when channels used in training and testing have different lengths or if the duration of training and testing simulation is different for some reason.
3. We made the desired pairs of the inputs and the outputs from the normalized data.
4. We created the desired input for the neural network. It is one of the tensors specified in the next section.

3.2   Neural Network Input

Correct type of input is very important for the results of the neural network. We work with two types of the input tensor in the form of a 3−dimensional array. Notice, that these tensors have some adjustable functions.

Input Tensor Based on Coordinates of the Cell Centres

This input type is set by algorithm parameter called *no_spatial_tensor*. Each row represents x, y and z coordinates of the cell centre at given time. Series of rows than represents the movement of the cell centre. To get an ample information about the cell's movement it is sufficient to use every $800th$ row from simulation output. (In the following text we will call this time step.) For this kind of two consecutive rows, the cell usually moves a little bit less than radius of its largest dimension. Which means, that cell at two consecutive positions probably has a small intersection with itself. The distance is computed using Euclidean norm. For our tensor we used zero padding of width p. The horizontal layers of the tensor represent different cells, where the cell we are focused on is on the top of the tensor. By setting boolean variable *use_depth* true or false, we can choose if we want to consider all cells or only the one cell we are focused on. Thus the dimension of this tensor is:

$$(3 + 2p) \times (n + 2p) \times 1, \qquad \text{if } use\_depth = false,$$
$$(3 + 2p) \times (n + 2p) \times c, \qquad \text{if } use\_depth = true,$$

where n is the number of time steps and c is the number of the cells.

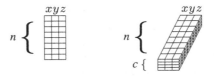

Fig. 3. Schemes of the *no_spatial_tensor* inputs (without padding). The input tensor for *use_depth* = *false* is on the left side. The input tensor for *use_depth* = *true* is on the right side.

In the experiment we used $n = 8$, number of the cells was $c = 38$ and we used padding $p = 1$. Due to our processing limitations we chose to use_depth false and therefore, the dimension of the input was $5 \times 10 \times 1$ (Fig. 3).

Input Tensor Based on Spatial Discretization

This input type for the neural network set by parameter called *spatial_tensor* is inspired by the image processing. The spatial tensor consists of n or $2n$ smaller tensors $A^{(i)}$, $i = 1, 2, \ldots, n$ which describe the position of the cell centres. Where n represents the number of the time steps.

We split the channel into $disc\_x \times disc\_y \times disc\_z$ identical rectangular areas. We obtain a small tensor A by determining if the cell is present or not, in each of the $disc\_x \times disc\_y \times disc\_z$ channel areas. Where

$$A_{ijk} = \begin{cases} 1, & \text{if there is a centre of some cell,} \\ 0, & \text{if there is no centre of any cell.} \end{cases} \tag{1}$$

Since this input is orthogonal the neuron networks gives us more accurate output values.

We also considered another approach, where the value 1, at the particular position A_{xyz}, is replaced by values of a Gauss function. More precisely, we replace all values in the s^3 neighborhood of the element A_{xyz}. We set s to be an odd number in order for A_{xyz} to be in the centre with the maximum value of Gauss function. To set up this approach, we set parameter use_gaussian_kernel.

We make n tensors, $A^{(1)}, \ldots, A^{(n)}$, for each time step. Each tensor records the position of the centre of one cell. Thus, if we don't use the gaussian kernel, $A^{(i)}$ is a sparse matrix with only one value 1. The input tensor consists of smaller tensors $A^{(n)}, \ldots, A^{(1)}$ in that order.

If we want to incorporate all the cells, we set $use\_depth = true$ to create another n tensors $B^{(1)}, \ldots, B^{(n)}$. Therefore each of these tensors includes the information about all of the cell centres. Then the input tensor consists of small tensors $A^{(n)}, \ldots, A^{(1)}, B^{(n)}, \ldots, B^{(1)}$ in this order. The small tensor $A^{(n)}$ is at the top horizontal layer of the input tensor and $B^{(1)}$ is at the bottom layer of the input tensor. Therefore the dimension of the input tensor is:

$$disc\_x \times disc\_y \times disc\_z * n, \qquad \text{if } use\_depth = false,$$
$$disc\_x \times disc\_y \times disc\_z * 2n, \qquad \text{if } use\_depth = true,$$

We can see the schemes for these inputs in the Fig. 4. Until now, we did not use the inputs for the neural network, where all cells are included. In our computations, we used $disc\_x = 16, disc\_y = 16, disc\_z = 3$ and $n = 8$. We also used the gaussian kernel. Hence, the dimension of the input tensor for the neural network was $16 \times 16 \times 24$.

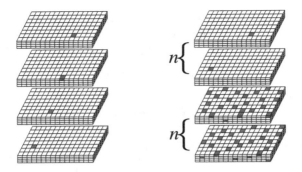

Fig. 4. Schemes of the *spatial_tensor* inputs, for $disc\_x = 16$, $disc\_y = 8$, $disc\_z = 4$. Marked areas represent positions of the cell centres. There are tensors $A^{(i)}$ illustrated on the left side, where position of the one cell centre is recorded. If we join these tensors together as is described above, we obtain *spatial_tensor* for *use_depth* = *false* with $n = 4$. There is *spatial_tensor* for *use_depth* = *true* depicted on the right side.

3.3 Trajectory Prediction Based on Network's Output

At the beginning we choose time $T > 0$, from which we want to predict cell's trajectories. In order to do so, our framework first predicts the velocity of the cell from past n positions of this cell recorded in the input tensor. Then it predicts positions from these predicted velocities. This is computed for all of the cells at given time. And then it is repeated with the data in the input tensor shifted by one simulation step. Data for the input tensors are selected as follows. The data from *simulation time* $< T$ are from the simulation output and data from *simulation time* $\geq T$ are predicted by the algorithm.

The next scheme (Fig. 5) illustrates the prediction of RBCs' trajectories from theirs velocities.

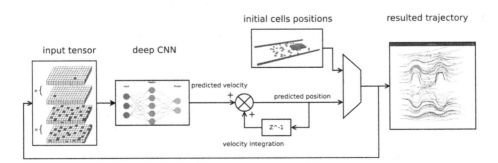

Fig. 5. RBCs' trajectory prediction from predicted velocities using our CNN from the simulation experiment.

4   Different Types of Network Architectures

In 2018, Bachratý et al. [3], trained and then used network for trajectory prediction, inspired by the radial basis function network and Kohonens' self-organized maps. Our research advanced to framework using CNNs. This type of network is mostly applied to image and video recognition. It has quite little pre-processing in comparison to other image classification algorithms. Since spatial discretization based input is derived from the image of the channel, we expect the use of framework based on CNN as appropriate. Nevertheless, it doesn't mean that this approach cannot be used with the coordinates based input. We confirmed our hypothesis, that framework works better with the *spatial_tensor* parameter. Besides using convolution or fully connected layers, we also used a relatively new type of layers, the dense convolution layers, introduced in [7].

Here we present results for the 8 neural network experiments of trajectory predictions. There were 6 different architectures of neural networks, see the Table 2. First four experiments, named Experiment_0 to Experiment_3 have *no_spatial_tensor* input with the same parameters, but mutually different

Table 2. Network architectures.

layer	net 0	net 1	net 2	net 3	net 4	net 5	net 6	net 7
0	fc 256	conv 3x3x32	dense conv 3x3x8	dense conv 3x3x8	dense conv 3x3x8	dense conv 3x3x8	dense conv 3x3x8	dense conv 3x3x8
1	fc 64	fc 64	dense conv 3x3x8	dense conv 3x3x8	dense conv 3x3x8	dense conv 3x3x8	dense conv 3x3x8	dense conv 3x3x8
2	fc 32	fc 32	dense conv 3x3x8	dense conv 3x3x8	dense conv 3x3x8	dense conv 3x3x8	dense conv 3x3x8	dense conv 3x3x8
3	fc 3	fc 3	dense conv 3x3x8	dense conv 3x3x8	dense conv 3x3x8	dense conv 3x3x8	dense conv 3x3x8	dense conv 3x3x8
4			conv 1x1x32	conv 1x1x16	conv 1x1x32	conv 1x1x16	conv 1x1x16	conv 1x1x32
5			fc 3	dense conv 3x3x8	fc 3	dense conv 3x3x8	dense conv 3x3x8	dense conv 3x3x8
6				dense conv 3x3x8		dense conv 3x3x8	dense conv 3x3x8	dense conv 3x3x8
7				dense conv 3x3x8		dense conv 3x3x8	dense conv 3x3x8	dense conv 3x3x8
8				dense conv 3x3x8		dense conv 3x3x8	dense conv 3x3x8	dense conv 3x3x8
9				conv 1x1x32		conv 1x1x32	conv 1x1x16	conv 1x1x32
10				fc 3		fc 3	dense conv 3x3x8	dense conv 3x3x8
11							dense conv 3x3x8	dense conv 3x3x8
12							dense conv 3x3x8	dense conv 3x3x8
13							dense conv 3x3x8	dense conv 3x3x8
14							conv 1x1x32	conv 1x1x64
15							fc 3	fc 3

networks architectures. Similarly for the Experiment_4, to Experiment_7, but with the *spatial_tensor* parameter. The network architectures for experiment pairs 2 and 4 and for 3 and 5 are the same.

Hyperparameters

Used neural network architecture hyperparameters are these: Weights are initialized in the range *xavier*, bias is set to 0. Learning rate is 0.0002, $\lambda_1 = \lambda_2 = 0.000001$, dropout $= 0.02$ and minibatch size is 32.

Computations ran on CUDA, graphics card GeForce GTX 1080 Ti. Our algorithm is written in C++ and Python. The training phase lasts about 6 hours for more complicated network architectures. The trajectory predictions calculated using already trained network took about half an hour on the graphics card GTX 960.

5   Results

Comparing the difference between the target and resulted values given from neural networks, we observed that networks with *spatial_tensor* parameter (Experiments 4 to 7) give better results than for *no_spatial_tensor* parameter. The Experiment_7, where the network has the most layers gave us the best result. In Fig. 6

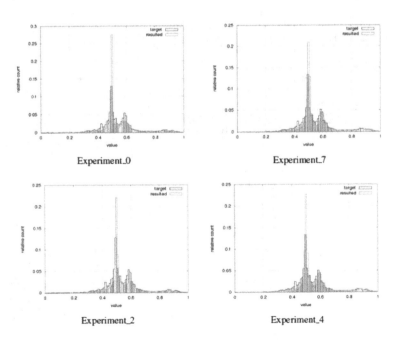

Fig. 6. Histograms of target (red) and resulted (green) values. The experiments with *no_spatial_tensor* input are on the left side and experiments with *spatial_tensor* input are on the right side. Note, that Experiment 0, 2 and 7 have different architecture and 4 has the same architecture as 2. (Color figure online)

the histograms of target and resulted output values of velocities are depicted. Note, that these values are normalized, but it doesn't qualitatively affect the information which it provides. These histograms are necessary verification of the framework we used, but not sufficient. The Fig. 7 shows relative error of predicted velocities summed in all three dimensions. Note, that the ranges on the vertical axis are different. We can immediately see, that the experiments with *spatial_tensor* parameter have smaller error. This holds for all of the 8 experiments.

We can see a considerable difference between testing progress of the first four experiments and the other four experiments on the left side of the Fig. 8. The Figure on the right shows, in detail, the testing progress from Experiment_4 to Experiment_7.

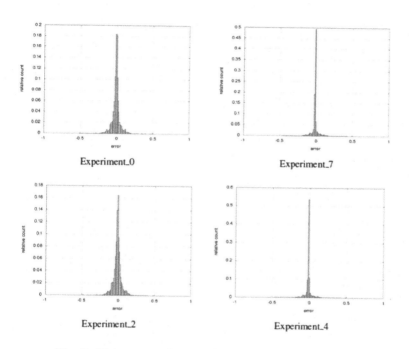

Fig. 7. Histograms of sums of errors in x, y and z axis.

The Fig. 9 shows the trajectory prediction of RBCs. Red trajectories are target and blue trajectories are resulted from neural networks experiments. The experiments on right side with *spatial_tensor* parameter give better results.

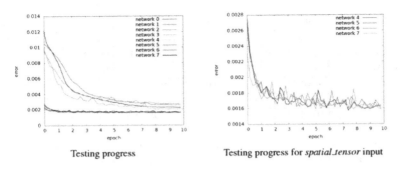

Fig. 8. Testing progress of networks.

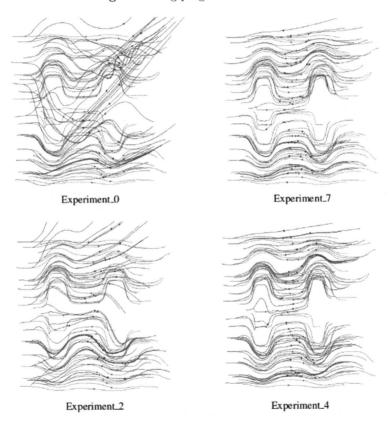

Fig. 9. RBCs' trajectory prediction. Red trajectories are target and blue trajectories are resulted. (Color figure online)

6   Conclusion

We introduced our approach on trajectory prediction by neural networks. We described data processing and our first results. From these we can determine, that spatial discretization based input and CNN architectures with the largest number of hidden layers are the best the trajectory prediction. There seems to be more applications of this prediction, e.g.:

- prediction of RBCs' trajectory prolongation from executed simulation experiment,
- prediction of RBCs' local behavior in a random position of a microfluidic channel. Including such positions, where cell never occurred during the simulation experiment,
- prediction and construction of fictive RBCs' trajectory with an arbitrary starting position in a channel,
- construction of completely virtual simulation experiment with given initial seeding of a larger number of RBCs,
- prediction of RBCs' trajectory on video records, useful for tracking algorithms.

We would like to apply our framework on other types of blood flow simulation experiments based on real laboratory experiments. Because it is not easy to find suitable examples of precisely described experiments, we welcome cooperation in this area.

References

1. Bachratá, K., Bachratý, H.: On modeling blood flow in microfluidic devices. In: ELEKTRO 2014: 10th International Conference, pp. 518–521. IEEE (2014). ISBN 978-4799-3720-2
2. Bachratá, K., Bachratý, H., Slavík, M.: Statistics for comparison of simulations and experiments of flow of blood cells, EPJ Web of Conferences, vol. 143 (2017). Art. no. 02002
3. Bachratý, H., Bachratá, K., Chovanec, M., Kajánek, F., Smiešková, M., Slavík, M.: Simulation of blood flow in microfluidic devices for analysing of video from real experiments. In: Rojas, I., Ortuño, F. (eds.) IWBBIO 2018. LNCS, vol. 10813, pp. 279–289. Springer, Cham (2018). https://doi.org/10.1007/978-3-319-78723-7_24
4. Bachratý, H., Kovalčíková, K., Bachratá, K., Slavík, M.: Methods of exploring the red blood cells rotation during the simulations in devices with periodic topology. In: 2017 International Conference on Information and Digital Technologies (IDT), Zilina, pp. 36–46 (2017)
5. Cimrák, I., et al.: Object-in-fluid framework in modeling of blood flow in microfluidic channels. Comun. Sci. Lett. Univ. Zilina 18(1a), 13–20 (2016)
6. Cimrák, I., Gusenbauer, M., Jančigová, I.: An ESPResSo implementation of elastic objects immersed in a fluid. Comput. Phys. Commun. 185, 900–907 (2014)

7. Huang, G., Liu, Z., Maaten, L.V., Weinberger, K.Q.: Densely connected convolutional networks. In: 2017 IEEE Conference on Computer Vision and Pattern Recognition (CVPR), pp. 2261–2269 (2017)

8. Tsai, C.H.D., et al.: An on-chip RBC deformability checker significantly improves velocity-deformation correlation. Micromachines **7**, 176 (2016)

9. Kovalčíková, K., Bachratý, H., Bachratá, K., Jasenčáková, K.: Influence of the red blood cell model on characteristics of a numerical experiment. In: Experimental Fluid Mechanics conference, Prague (2018, in press)

Platform for Adaptive Knowledge Discovery and Decision Making Based on Big Genomics Data Analytics

Plamenka Borovska[1], Veska Gancheva[1(✉)], and Ivailo Georgiev[2]

[1] Technical University of Sofia, Kliment Ohridski 8, 1000 Sofia, Bulgaria
`{pborovska, vgan}@tu-sofia.bg`
[2] The Stephan Angeloff Institute of Microbiology,
Georgi Bonchev 26, 1113 Sofia, Bulgaria
`ivailo@microbio.bas.bg`

Abstract. In the past years, researchers and analysts worldwide determine big data as a revolution in scientific research and one of the most promising trends that has given impetus to the intensive development of methods and technologies for their investigation and has resulted in the emergence of a new paradigm for scientific research Data-Intensive Scientific Discovery (DISD). The paper presents a platform for adaptive knowledge discovery and decision making tailored to the target of scientific research. The major advantage is the automatic generation of hypotheses and options for decisions, as well as verification and validation utilizing standard data sets and expertise of scientists. The platform is implemented on the basis of scalable framework and scientific portal to access the knowledge base and the software tools, as well as opportunities to share knowledge and technology transfer.

Keywords: Big genomic data analytics · Data integration ·
Knowledge discovery from data

1 Introduction

During the last years leading scientists, researchers and analysts determine big data as revolution in scientific studies and one of the most challenging trends in technological innovations. As a result of the computer simulations a huge amount of data was generated during the in silico experiments [1]. One of the fundamental scientific areas, strongly dependent on big data technologies, is molecular and computational biology [2]. The technological progress, as well as next generation sequencing yielded exponential growth of experimental genomic data, and as a result the well-known methods and technologies became not applicable to the new challenges of big data. This has stimulated the development of methods and technologies for processing of large amount of data and radical changes in the scientific research paradigms.

In the biological sciences there are very well established practices of collecting data in the public and generally accessible data bases, which are used by scientists all over the world, in building up solutions for specific problems. The advance in bioinformatics stimulates innovative methods for processing and analyzing the collected data.

© Springer Nature Switzerland AG 2019
I. Rojas et al. (Eds.): IWBBIO 2019, LNBI 11466, pp. 297–308, 2019.
https://doi.org/10.1007/978-3-030-17935-9_27

With the advances in bioinformatics, the volume of collected biological data and the number of databases in which they are stored have been steadily growing. They are supported by various organizations and institutions dealing with human genome research, virus research and their mutations, protein research, drug synthesis, and so on. A major problem for the biological data integration is the data structure and format. They are not always standardized and data access is not centralized, making it extremely difficult and time-consuming to search throughout all databases.

Biological knowledge is distributed in specialized databases data sources. Each database has its own complex data structures reflecting the scientific concept of the model [3]. Many data sources have overlapping data elements with conflicting definitions. Data integration from heterogeneous sources is very important for the effective use of biological information. It is important to interpret the different data formats, download data from different sources and convert them to integrated information. Biological data sources are characterized by an extremely high degree of heterogeneity with regard to the type of data model and the relevant data pattern, as well as the incompatible formats and nomenclatures of values [4].

Biological databases are highly decentralized, with high levels of terminology, record specificity, data representation, and request formats [5]. This in turn is associated with problems with manually executing queries from multiple databases. Therefore, there is a need to automate the integration of biological databases, with much more than simply extracting and modifying the data [6, 7]. Integration requires the use of binding formats in different databases, but large scale and redundancies make such integration impossible.

The next generation of methods for data analysis has to manage huge amounts of data from various types of sources with differentiated characteristics, levels of trust, and frequency of updates. Data analyses have to acquire knowledge in effective and sustainable way. To achieve this, it is necessary to build up complicated predictive models and methods for heterogeneous and big data analytics. On the other hand, these models and methods have to be implemented in real time for big data streams. This is a great challenge, because big data, besides its huge volume, are strongly heterogeneous and dynamic, requiring high performance and scalability.

The vast amount of accumulated data contains valuable hidden knowledge that can be useful to facilitate and improve the decision making process. That is why there is an objective need to create automated methods for extracting knowledge from data. The extracted knowledge have to meet the following requirements: be accurate, understandable and useful. In addition, knowledge should have the potential to predict. Data acquisition tasks include classification, dependency modeling, clustering, function retrieval, and associative rules detection.

Knowledge discovery and data mining is an area focused on methodologies for extracting useful knowledge from data. The ever increasing growth of data and the pervasive use of databases have demanded challenges for innovative knowledge data discovery methodologies. Data acquisition skills are based on research in statistics, databases, modeling, machine learning, data visualization, optimization, and high performance computing to deliver automated sophisticated and intelligent solutions.

The work presented in this paper is a part of a project that offers a scientific platform for adaptive in silico knowledge data discovery based on big genomic data analytics. The focus is on advanced information technologies and the fourth scientific research paradigm Data Intensive Scientific Discovery (DISD) in support of precision medicine, specifically, for the case study of fighting breast cancer.

An intelligent method for adaptive in silico knowledge discovery based on big genomic data analytics which is adaptable to important biological, medical and computational aspects has been suggested in [8]. The method is built upon the parallel phase paradigm comprising two overlapping and correlated phases – machine learning phase and operational phase.

The goal of this paper is to suggest platform for adaptive knowledge discovery and decision making based on big genomic data analytics based on the designed method for knowledge discovery stated above. The platform integrates scalable framework and scientific portal to access the knowledge base and the software tools, as well as opportunities to share knowledge and technology transfer.

The paper is structured as follows: Knowledge discovery from data pipeline is discussed in Sect. 2. Scalable adaptive method for knowledge discovery and decision making based on big data analytics is presented in Sect. 3. Conceptual architecture of data retrieval and integration system is presented in Sect. 4. Section 5 is focused on the scientific application specific platform.

2 Knowledge Discovery Based on Data Analytics

The objectives of knowledge discovery based on data analytics are predictive and descriptive as follows.

- The predictive includes the use of some variables or fields in a database to predict unknown or future values of other variables of interest.
- The description focuses on finding interpretative models describing the data.

Predictive and description objectives are achieved by using the following data retrieval tasks:

- Classification is a function that classifies a data element in one of several predefined classes.
- Regression is a function that classifies a data element of a predictive variable with a real value.
- Clustering is a general descriptive task that identifies a limited set of categories or clusters to describe the data.
- Grouping includes methods for finding a compact description for a subset of data.
- Modeling the relationship is finding a model that describes significant dependencies between variables.
- The change and detection of deviation is focused on detecting the most significant changes in data from previous measured or normative values.

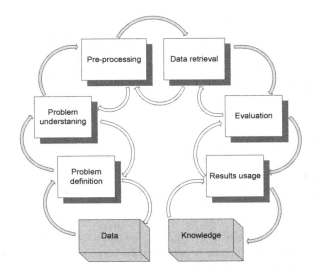

Fig. 1. The knowledge discovery based on data analytics process

Knowledge discovery and data mining process (Fig. 1) consists of six main stages:

1. Understanding the problem area - this is the initial stage that focuses on defining research goals and relevant requirements from the user's point of view. Once this stage has been completed, this knowledge has to be translated into definitions of data retrieval tasks and a preliminary plan for how these goals can be achieved.
2. Understanding the data - begins with initial data collection and continues with activities aimed at deepening the knowledge in respect to the data nature. At this stage, it is necessary to identify problems related to the data quality, to get an initial opinion on the data nature, to find the relevant subsets of data in order to form initial hypotheses about the hidden knowledge.
3. Data preparation - covers all the activities of creating raw data out of the final set of raw data. The stage of data preparation often has to be repeated many times at different stages of the computational pipeline. The tasks of data preparation include data selection, determining of attributes, exploring individual records, as well as data transformation and clearing data.
4. Modeling - this stage consists of selecting and applying various modeling techniques to derive data dependencies. The model parameters are adjusted to their optimal values. Since some models have their own specific data format requirements, it is often necessary to return to the data preparation stage.
5. Model evaluation - consists of carefully reviewing all the steps implemented in building up this model to ensure that they achieve the specific goals. At the end of this stage, a decision is made to use the results obtained during the drilling process.
6. Exploitation of the model - related to the monitoring and exploitation strategy applied. At this stage it should be determined whether and when to resume the procedure of data mining and under what conditions.

Machine learning is artificial intelligence technique for data analytics and its algorithms allow software applications to become more accurate in predicting outcomes without being explicitly programmed. The basic premise of machine learning is to design algorithms that can process input data and use statistical analysis to predict an output while updating outputs as new data becomes available.

Classification is the most intensively studied task. Data subjected to analysis are divided in two groups: training set and validation set. The algorithm for knowledge extraction should build up rules, applying the training set. After completing the process of learning and establishing the classification set of rules, the effectiveness of the rules is evaluated on the basis of the validation set. The task of dependency modeling can be regarded as generalization of the classification task. In this case the goal is to predict the values of several attributes.

The conceptual model of knowledge discovery and decision making based on big data analytics is shown in Fig. 2.

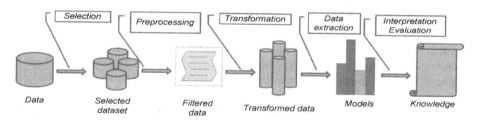

Fig. 2. Conceptual model for knowledge discovery and decision making based on big data analytics

The complete process of extracting and interpreting data models involves re-implementing the following steps:

- Determine the purpose of the knowledge discovery process - define the task and the relevant advance knowledge and its application aspects.
- Define application domain, relevant knowledge and end-user's goals.
- Create target data set: selecting a data set or focusing on a subset of variables or data samples.
- Filter and pre-process: remove noise or negative values; collect the necessary information for modeling or noise reporting; establish strategies for processing the missing data fields.
- Simplify the dataset by removing undesirable variables: find useful functions for presenting data in respect to the goal of to the task; apply dimensional or transformation methods to reduce the effective number of variables under consideration or to find invariant data representations.
- Combine the objectives of the data discovery process with data mining methods - determine whether the objective of the knowledge-based process is classification, regression, cauterization, etc.

- Select data mining algorithm. This process involves making decision which models and parameters may be appropriate for the overall process: select the method(s) to be used to search for model in data; determine which models and parameters may be appropriate; conformity of a particular data mining method with the common criteria of the knowledge discovery process.
- Data extraction - searching for models of interest in a specific presentation form or a set of such representations such as classification rules or trees, regression, clustering, etc.
- Interpret basic knowledge of extracted models.
- Utilize knowledge and its inclusion in another system for further action.

3 Scalable Adaptive Approach for Knowledge Discovery and Decision Making Based on Big Data Analytics

The aim is to suggest an integrated approach for supporting the knowledge discovery and decision making processes based on big data analytics, adaptive machine learning and adaptive procedures for generating rules according to the specific goal of the scientific research. The main advantage is the automated generation of hypotheses and solutions for the specific case under study. The interest of the knowledge discovered is estimated by the expertise of scientists in the relevant field. The adaptability of the approach is achieved by means of a synthesized collection of modules based on techniques such as data analysis, machine learning, and metaheuristics. Depending on the specific purpose of the study, the relevant modules are applied for pre-processing of data flows, for knowledge extraction and post-processing. Regarding the knowledge extraction, a hybrid approach of machine learning methods and procedures for rules generation is applied.

A scalable framework for adaptive knowledge discovery is based on big data streams analytics, providing a set of software tools for applying the method in research and experimental activities for wide spectrum of scientific areas. Streaming technology eliminates the need for significant resources of disc memory as raw data derived from databases are subjected directly to online processing. The scalability of the working framework reduces computational time by involving additional resources and parallel processing.

The advantage of the proposed framework is the automatic generation of the hypothesis and options for decisions making on the basis of the learning data set analysis, while the verification and validation is conducted via benchmark testing set and the expertise of the researchers of the relevant area. The interest of the discovered knowledge is estimated by hybrid approach – a combination of objective criteria and the expertise of scientists.

The research techniques follow the processing pipeline for discovering knowledge of interest out of a collection of data and imply the following:

- data preparation, cleansing and selection;
- knowledge discovery and decision making, and
- results visualization and interpretation.

Pre-processing of data in knowledge discovery covers: integration of data from different sources, data clearing in terms of accuracy, sampling, and selection of functions in terms of relevance. The selection of functions is done by applying algorithms for searching and optimization, and the feature set is optimized by iterative execution of machine learning algorithm. The combination of machine learning, data mining, knowledge discovery and decision making methods is applied to achieve high accuracy and precision of the extracted knowledge. Post-processing involves verification, validation, visualization and evaluation of the discovered knowledge.

The basic functional units for knowledge data discover are scientific analytics workflows being exceedingly useful for facilitating e-science. Actually, a workflow represents the computational pipeline for KDD based on specific analytics method (model and/or rules). The workflow provides a solution, supporting the design and conducting experiments using the available data and tools and high-level definition of the objectives of the experiment, modeled by workflow of scientific tasks. Different types of tasks can be performed within a workflow, especially when outputs from one task are used as input for the next task. The workflow process for knowledge discovery from big genomics data is shown in Fig. 3.

Fig. 3. Workflow process for knowledge discovery from big genomics data

The purpose of the machine learning phase is to build up repositories of synthesized collection of models and rules that will be used as components to build up an integrated KDD workflow. Machine learning (ML) phase is based on parallel workflows (computational pipelines), utilizing diverse ML models, including classifying and clustering methods. ML phase performs on the training and validation sets and the basic computational units are bundles of differentiated workflows, each differentiated workflow performing specific type of analytics. Differentiated analytics workflows are executed in parallel in order to accelerate the KDD process. Each analytics workflow builds up a model that is stored in models repository or a set of rules stored in rules repository. Once feature extraction and dimension reduction have been done, analytics model is generated utilizing the training set by configuring the respective differentiated workflow. Afterwards, the outcome of the differentiated workflow is subjected to validation by utilizing the validation data set.

The components of the big data analytics system architecture are shown in Fig. 4. The conceptual architecture of the scalable framework for adaptive knowledge

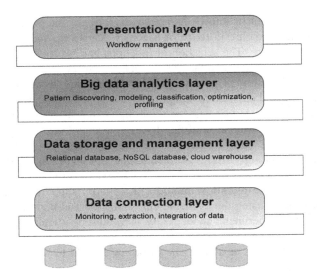

Fig. 4. Big data analytics system architecture

discovery comprises hardware and software resource reconfiguration. Software is developed utilizing streaming and parallel processing technologies (multithreading and clustering). The architecture consists of separate, independent components:

- access to progressively increasing amounts of data in multiple formats, extraction, interoperability, real time integration of various types of data and information;
- pre-processing of large data streams, including selection of attributes, filtering, discretization;
- knowledge discovery out of big data streams by applying methods for generating rules and machine learning and
- post-processing of data - knowledge interpretation and results visualization.

4   Data Integration

The requirements for scalable data integration systems for modern biology are indisputable due to the existence of very large, heterogeneous and complex datasets in the public database. Managing and merging these big data with local databases is a great challenge as it is the basis of computational analyzes and models that are then experimentally generated and validated through portal access to distributed modern high-performance infrastructure and software tools for big genomics data processing and visualization.

A conceptual architecture for an integrated and effective access to the exponentially growing volume of data in multiple formats is proposed (Fig. 5). The architecture allows the rapid management of large volumes of diverse data sets represented in different formats - relational, NoSQL, flat files. The integration system consists of services for transforming the common request into a specific language request for each local database, depending on its type. Additionally, the possibility of making a permanent access to the state of research in order to compare the results with the available information (access to a constantly updated representation of all the accumulated knowledge in the relevant field) is further explored.

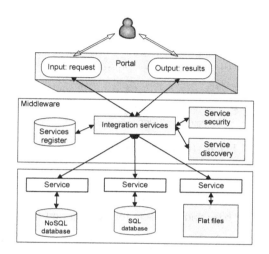

Fig. 5. Conceptual architecture of data retrieval and integration system

5   Scientific Specific Application Platform

The objective of an academic scientific portal gateway is to allow a large number of users to have transparent access to available distributed advanced computing infrastructure, software tools, visualization tools and resources.

The high-performance platform offers services for carrying out remote simulations and consists of HPC resources, resource database, knowledge discovery resources as services, software tools and web portal is presented in Fig. 6.

The portal provides user-centric view of all available services and distributed resources. The web-based environment is organized to provide a common user interfaces and services that securely access available heterogeneous computing resources, data and applications. It also allows many academic and scientific users to interact with these resources.

The experimental infrastructure is achieved through a customized application specific gateway, based on the platform and can be summarized as follows:

- identifying and clarifying the requirements of specific user communities, user scenarios and needs of the target groups of scientists;
- defining the specific user communities views in accordance with their requirements and develop custom scripts for specific applications;
- identification of methods and tools for user communities and their testing;

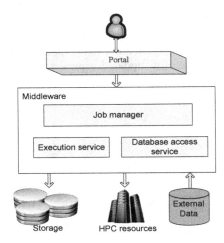

Fig. 6. High performance infrastructure

- development of workflow templates representing different user scenarios of low level and adapted to the specific requirements of customers;
- creating a knowledge base and data sets for training and testing.

The scientific platform will facilitate access and use of existing infrastructure in research and will provide:

- a central access point to the status of research in order to compare the results with the available information (access to constantly updated representation of the accumulated knowledge in the relevant area),
- interactive cooperation via various channels
- opportunities for knowledge and technology transfer, partnership and providing information and services.

The web-based environment is organized so as to provide a common user interfaces and services that provide secure access to currently available heterogeneous computing resources, data and applications.

The web portal provides also: user profile management, e.g., different views for different user; personalized access to information, software tools and processes; getting information from local or remote data sources, e.g., from databases, transaction systems, or remote web sites; aggregated the information into composite pages to provide information to users in a compact and easily consumable form. In addition, the portal also includes applications, software tools, etc.

The proposed platform is verified for the case studies of multiple sequence alignment based on social behavioral model, enhancer-promoter detection and early detection of breast cancer. An investigation for detection of enhancer-promoter interactions out of genomic big data based on machine learning is presented in [9]. A pipeline for detection of enhancer-promoter interactions using Decision Tree and Support Vector Machine classifiers is proposed. The experimental framework is based on Apache Spark environment that allows streaming and real time analysis of big data. The experimental results for detection of enhancer-promoter interactions have been performed with GM12878 and K562.

An innovative parallel method MSA_BG for multiple biological sequences alignment that is highly scalable and locality aware is designed [10]. The algorithm is iterative and is based on the concept of Artificial Bee Colony metaheuristics and the concept of algorithmic and architectural spaces correlation. The metaphor of the ABC metaheuristics has been constructed and the functionalities of the agents have been defined. The conceptual parallel model of computation has been designed. Parallelization and optimization of the multiple sequence alignment software MSA_BG in order to improve the performance, for the case study of the influenza virus sequences is proposed [11]. For this purpose a parallel MPI + OpenMP - based code optimization has been implemented and verified. The experimental results show that the hybrid parallel implementation provides considerably better performance.

6 Conclusion and Future Work

In this paper a platform for adaptive knowledge discovery and decision making based on big data analytics is proposed. The major advantage is the automatic generation of hypotheses and options for decisions, as verification and validation are performed using standard data sets and expertise of scientists. The tools for utilizing the platform are scalable framework and scientific portal to access the knowledge base and the software tools, as well as opportunities to share knowledge, and technology transfer. Web portal provides services to access and extract knowledge out of biological data and execute parallel software applications for big genomics data analysis.

An integrated approach to support knowledge discovery and decision making based on big data analytics, adaptive machine learning and adaptive procedures for generating rules according to the goal of scientific research is presented. A conceptual architecture for an integrated and effective access to the exponentially growing volume of data in multiple formats is proposed. The architecture allows the rapid management of large volumes of diverse data sets represented in different formats - relational, NoSQL, flat files. The integration system consists of services for transforming the common request into a specific language request for each local database, depending on its type.

The future work is to make in silico experiments on the platform based on big genomic data analytics for scientific research in the area of molecular biology. The spectrum of case studies under investigation comprises identifying regulatory elements in sequenced genomes, and prediction of the type and malignance of breast cancer. This will enable fast processing of clinical observations and laboratory analyzes data and comparison with the available data accumulated so far in support of precision medicine.

Acknowledgment. This paper presents the outcomes of research project "Intelligent Method for Adaptive In-silico Knowledge Discovery and Decision Making Based on Analysis of Big Data Streams for Scientific Research", contract DN07/24, financed by the National Science Fund, Competition for Financial Support for Fundamental Research, Ministry of Education and Science, Bulgaria.

References

1. Chen, P., Zhang, C.: Data-intensive applications, challenges, techniques and technologies: a survey on big data. J. Inf. Sci. **275**, 314–347 (2014). https://doi.org/10.1016/j.ins.2014.01.015
2. Roy, A.K.: Trends in computational biology and bioinformatics in the era of big data analytics. In: Conference International Workshop on Bioinformatics in Fisheries and Aquaculture' held at ICAR- CIFRI (2017). https://doi.org/10.13140/rg.2.2.21016.39680
3. Thiam Yui, C., Liang, L.J., Jik Soon, W., Husain, W.: A survey on data integration in bioinformatics. In: Abd Manaf, A., Sahibuddin, S., Ahmad, R., Mohd Daud, S., El-Qawasmeh, E. (eds.) ICIEIS 2011. CCIS, vol. 254, pp. 16–28. Springer, Heidelberg (2011). https://doi.org/10.1007/978-3-642-25483-3_2
4. Nguyen, H., Michel, L., Thompson, J.D., Poch, O.: Heterogeneous biological data integration with declarative query language. IBM J. Res. Dev. **58**(2/3), 1–12 (2014)
5. Rao, C.S., Somayajulu, D.V.L.N., Banka, H., Ro, S.: Feature binding technique for integration of biological databases with optimized search and retrieve. In: 2nd International Conference on Communication, Computing & Security [ICCCS-2012], pp. 622–629 (2012)
6. Paton, N.W., Missier, P., Hedeler, C. (eds.): DILS 2009. LNCS, vol. 5647. Springer, Heidelberg (2009). https://doi.org/10.1007/978-3-642-02879-3
7. Zhang, Z., Bajic, V.B., Yu, J., Cheung, K.-H., Townsend, J.P.: Data integration in bioinformatics: current efforts and challenges. In: Bioinformatics - Trends and Methodologies. pp. 41–56 (2011). ISBN 978-953-307-282-1. http://doi.org/10.5772/21654
8. Borovska, P., Ivanova, D.: Intelligent method for adaptive in silico knowledge discovery based on big genomic data analytics. In: AIP Conference Proceedings, vol. 2048, p. 060001 (2018). https://doi.org/10.1063/1.5082116
9. Ivanova, D., Borovska, P., Gancheva, V.: Experimental investigation of enhancer-promoter interactions out of genomic big data based on machine learning. Int. J. Comput. **3**, 58–62 (2018). ISSN: 2367-8895
10. Borovska, P., Gancheva, V., Landzhev, N.: Massively parallel algorithm for multiple biological sequences alignment. In: Proceedings of the IEEE International Conference on Telecommunications and Signal Processing (TSP), Rome, Italy, pp. 638–642. ISBN 978-1-4799-0402-0
11. Borovska, P., Gancheva, V.: Parallelization and optimization of multiple biological sequence alignment software based on social behavior model. Int. J. Comput. **3**, 69–74 (2018). ISSN: 2367-8895

Simulation Approaches to Stretching of Red Blood Cells

Alžbeta Bohiniková[(✉)] and Katarína Bachratá

Faculty of Management Science and Informatics, Department of Software Technology,
University of Žilina, Žilina, Slovakia
alzbeta.bohinikova@fri.uniza.sk
http://cell-in-fluid.fri.uniza.sk

Abstract. In this article we will give brief overview of elastic parameter of our red blood cell (RBC) model. Next we will discuss the importance of calibration of these parameters to get behaviour close to the behaviour of the actual RBC. For this purpose we used a widely known experiment where RBC are stretched by attaching silica beads to their membrane and applying force to these beads. The main focus of this article is how to model this stretching process and more precisely, how to choose between several options available to simulate the optical tweezers stretching. We compared three approaches - "points", "hat" and "ring", and based on simulation results selected the "ring" approach. Afterwards we computed working "rings" for different cell discretizations.

Keywords: Computational model · Red blood cell model ·
Stretching experiment · Numerical simulations

1 Introduction

Microfluidics is one of the highly sought after areas of research. There is increasing interest in this topic due to the many way in which a microfluidic device can be used. Just to mention a few interesting areas such as detection of specific cells in samples that can by run trough the channel for diagnostic purposes. There are microfluidic channels design to sort cells that can then be used for further analysis which might improve the treatments for patients.

Creating new designs and improving the existing ones of these devices can be a long and expensive process. The progress of accessible computational power allows us to use mathematical model to try and help with the process. Instead of testing 10 different geometries for microfluidic channel, one might need to maybe test only one in laboratory and rest can be done in computer simulations.

A. Bohiniková and K. Bachratá—This work was supported by the Slovak Research and Development Agency (contract number APVV-15-0751) and by the Ministry of Education, Science, Research and Sport of the Slovak Republic (contract No. VEGA 1/0643/17).

© Springer Nature Switzerland AG 2019
I. Rojas et al. (Eds.): IWBBIO 2019, LNBI 11466, pp. 309–317, 2019.
https://doi.org/10.1007/978-3-030-17935-9_28

One of the most studied liquids in these devices is blood. Red blood cells (RBC) account for about 40 to 45% of whole blood volume. Which means that working model of RBC can be quite helpful for developing microfluidic devices [1,2].

Our research group has mainly worked on this particular task. We model elastic objects, mainly RBC, immersed in fluid. All the computations are run on open-source software ESPResSo with object-in-fluid implementation added by our research group. Detailed information can be found in [3].

The fluid is modelled using lattice-Bolzmann method, where the space, trough which the fluid moves, is represented by lattice of discrete points. These points are fixed throughout the whole simulation. During simulation, fictitious fluid particles move and encounter other particles, which transfers information about their velocity and direction. Information about number and speed of the particles passing trough each of the fixed lattice is stored. In depth information can be found in [6].

2   Elastic Parameters

The elastic cells in the simulations are represented by triangulation of the cell membrane. Five elastic moduli are responsible for the elastic behaviour of the cell. In the next section we give a brief overview. More details are in [6].

2.1   Stretching

This parameter is responsible for rigidity of the cell. When the cell membrane is stretched, this model reacts and applies forces at given points in order to achieve the original relaxed state of the cell. We can imagine the edges of our triangulation as springs that try to stay in their original relaxed length.

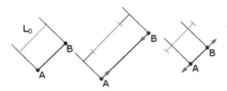

Fig. 1. Dynamics of stretching modulus. L_0 is the relaxed length. Adopted from [8]

2.2   Bending

This parameter is also responsible for rigidity of the cell. However, it is responsible for different kind of deformation. In the triangulation, each pair of the triangles has their relaxed angle. In case the cell deforms and some of the angles change, this parameter acts opposite to this deformation and tries to return the cell to the original relaxed angles.

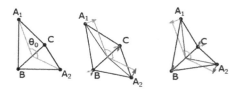

Fig. 2. Dynamics of bending modulus. θ_0 is the relaxed angle. Adopted from [8]

2.3   Local Area Conservation

Purpose of this parameter is clear from its name. Each triangle in the triangulation should ideally be equilateral with the same area. This is the relaxed area. In case cell deforms and these triangles change, this parameter is responsible for applying forces in order to achieve the original area. Its dynamics are similar to the stretching parameter.

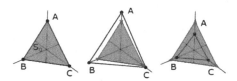

Fig. 3. Dynamics of local area conservation modulus. S_0 is the relaxed area. Adopted from [8]

2.4   Global Area Conservation

Similar to the previous parameter. The difference is that instead of local area, this parameter is responsible for the global area. In biological cells, when the changes are too large, the cell dies. This behaviour is implemented in the model by controlling these changes. If the changes in total area is larger then 3%, the cell stops existing, dies, for the simulation.

Fig. 4. Dynamics of global area conservation modulus. S_0 is the relaxed area. Adopted from [8]

2.5　Global Volume Conservation

The mechanics of this parameter are similar to global areal conservation. Instead of area, this parameter is responsible for volume. Also for large changes in volume, there is a control. If the change is larger then 3%, the cell stops existing, dies, for the simulation.

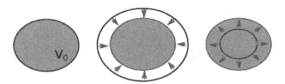

Fig. 5. Dynamics of global volume conservation modulus. V_0 is the relaxed volume. Adopted from [8]

3　Calibration and Stretching Experiment

In order to use the model of elastic cell for modelling RBC flowing trough different microfluidic devices we need to set up the elastic parameters so they resemble the behaviour of actual biological cells. General idea in these cases is to find experiment that can be reproduced in computational setting. Set everything up according to experiment and then change the values of elastic parameters of our cell in order to mimic the movement and deformation of the actual biological experiment. For this purpose we chose biological experiment, where RBC are stretched by optical tweezers [4].

In this experiment two silica beads are attached to the cell. They are used as handles. One is fixed, adhered to the glass surface, and the other is trapped using laser beam. By moving the laser beam, force can be applied to the cell. After the force is applied, the changes in the shape of the cell are measured. More specifically the change in the axial, direction of the stretch, and transverse, direction perpendicular to the stretch, diameter. Same steps are repeated for different values of the applied force.

4　Simulation of Stretching Experiment

The main focus of this article is to find how to apply the stretching force in the simulation model. First, we set up all the simulation parameters according to the biological experiment. Parameters can be found in Table 1. Second, we have several options how to deal with the stretching. The main purpose of the silica beads in the biological experiment was to protect the cell from laser and to serve as a grip for stretching.

However, in the simulation setting there is no need to model the silica beads, since we are able to apply forces to the points on the triangular mesh.

Fig. 6. Stretching of cell.

Table 1. Simulation parameters

Timestep	10^{-7} s
lbgrid	10^{-6} m
Stretching coefficient	$4 \cdot 10^{-6}$ N/m
Bending coefficient	$8 \cdot 10^{-12}$ N
Local area conservation coefficient	$3.9 \cdot 10^{-5}$ N/m
Global area conservation coefficient	$7 \cdot 10^{-4}$ N/m
Volume conservation coefficient	$9 \cdot 10^{2}$ N/m^2
Radius of RBC	$3,91 \cdot 10^{-6}$ m
Friction coefficient	$3,39 \cdot 10^{-9}$ N $\cdot$ s $\cdot$ m^{-1}
Fluid density	10^{3} kg $\cdot$ m^{-3}
Fluid viscosity	$1.5 \cdot 10^{-6}$ m^2/s

This spares us computational time, since modelling the beads would require more points and interactions in the simulation. On the other hand, we have to decide where precisely should we apply the stretching force to achieve same effect as in the biological experiment. We devised three possibilities for this aim.

1. **Point.**
   Probably the most simple way is to apply given stretching force to two points. One for each silica bead. One with the smallest x-coordinate (for the left bead) and one with the largest x-coordinate (for the left bead). Since we chose the stretching to be applied along the x-axes.

2. **Ring.**
   The second option is to create similar contact area as was done in the biological experiment. The contact diameter of silica beads is estimated to be 2 m. So we created non-circular ring from points in order to match the diameter. The points can be defined by setting interval of values for their x-coordinates. The contact diameter is then computed as an average of distances of the chosen points from the middle point of this ring.
   Lets have three points that have the value of x-coordinates in the correct interval: $A[x_a, y_a, z_a]$, $B[x_b, y_b, z_b]$ and $C[x_c, y_c, z_c]$. Than the "contact diameter" is computed as:

$$C_d = 2\frac{\sqrt{(y_a^2 + z_a^2)} + \sqrt{(y_b^2 + z_b^2)} + \sqrt{(y_c^2 + z_c^2)}}{3}. \tag{1}$$

So the task here was to find the correct interval of x-coordinate values to achieve the contact diameter of 2 m. Since we have discrete model, we allowed some leeway in order to have symmetry between left and right side. We marked as working those diameters that were in the margin of error from 1.7 m to 2.2 m.

Once we have the ring of point we apply the stretching force. In order to achieve the same effect as in the biological experiment we split equally given force between all of the points.

3. **Hat.**

The last option risen from the fact that finding the right interval for the x-coordinate values in the ring option was a bit intricate. And we wanted to test if there will be a difference between stretching the ring and stretching the hat, where we only need a threshold value for the x-coordinate of the points. Here the criterion was that the contact diameter was larger then 1.7 m. Again, the stretching force was applied among all the points of each hat (right and left) equally.

Fig. 7. Comparison of different approaches to stretching. From Left to right: Point, Ring, Hat.

4.1   Comparison of Different Stretching Approaches

The above three methods of stretching were compared by running stretching simulation for each of them. As a starting option we choose cell with 374 mesh points. For each option, set of five simulations was run, with parameters set at the values given in Table 1 and the applied stretching force was sequentially: $0.016, 0.047, 0.088, .130, 0.192$ nN.

For each simulations we recorded axial and transverse diameter of the cell. Leaving us with ten values for each approach. These values are plotted with the measurements from the biological experiment [4] in Figs. 8, 9 and 10.

From the Figures it is clear that the point approach is not very useful. Even at the very low values of stretching force the axial diameter of the cell is far away from the biological values. We can discard this approach.

When comparing the "ring" and the "hat" both seem to give quite good and similar results of the axial and travers diameters. To distinguish between their

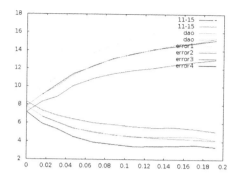

Fig. 8. Results of stretching simulations for the "ring" approach. Lines marked as "dao", denote the measurement from the biological experiment [4]. Also the lines marked as "error" are the errors from this experiment. The lines denoted as 11–15 are the results from our simulations.

quality we computed normalized mean square errors from the biological data. For each value first the normalized error was computed. The difference of the simulation and experimental values divided by the range of the error values:

$$err_{n_i} = \frac{\|d_{sim} - d_{bio}\|}{err_{max} - err_{min}}. \tag{2}$$

And the resulting error was computed as

$$err_{final} = \frac{\sum_{i=1}^{10} err_{n_i}^2}{10}. \tag{3}$$

For the "ring" approach the error was 1.7% and for "hat" it was 4%. Based on these result we concluded that "ring" approach is the best.

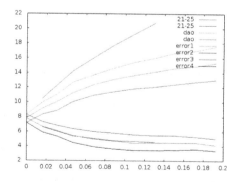

Fig. 9. Results of stretching simulations for the "point" approach. Lines marked as "dao", denote the measurement from the biological experiment [4]. Also the lines marked as "error" are the errors from this experiment. The lines denoted as 21–25 are the results from our simulations.

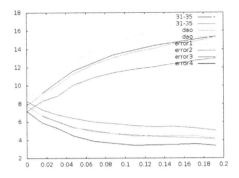

Fig. 10. Results of stretching simulations for the "hat" approach. Lines marked as "dao", denote the measurement from the biological experiment [4]. Also the lines marked as "error" are the errors from this experiment. The lines denoted as 31–35 are the results from our simulations.

4.2   Determining Intervals for x-coordinate

Once we chose the best approach we tested how precisely it is necessary to set the intervals for the x-coordinates. We tested with set of 5 intervals. All done for the cell with 374 points. The results are in Table 2.

Table 2. Results of stretching simulations for different intervals of x-coordinate

	Sim11–15	Sim41–45	Sim51–55	Sim61–65	Sim71–75
x_{min}	0.9	0.91	0.86	0.86	0.92
x_{max}	0.98	0.99	0.99	0.98	0.99
Diameter	1.87	1.79	2.11	2.21	1.70
No. points	10	12	18	16	10
Error	1.78%	2.60%	8.31%	10.38%	6.41%

Based on these results we can conclude that in order to achieve more accurate simulations it is worthwhile to set the ring diameter closely to the contact diameter of 2 m. Even small changes in the interval set up can lead to large differences in the measurements of the axial and travers diameter of the stretched cell.

In our simulations other discretizations of the cell are used. Since there were significant differences between the interval values for the 374 points we also calibrated the correct values for other discretizations used in our work. The results are in Table 3.

Table 3. Intervals of x-coordinate for different discretizations

Discretization	482	642	1002
x min	0.92	0.92	0.905
x max	0.98	0.97	0.98
Diameter	1.88	1.90	2.05
No. points	12	12	18

5 Conclusion

We have compared three different approaches how to simulate stretching of the cell by silica beads pulled by a laser. As we expected, the point approach did not produce useful results. Interesting were the difference between the hat and ring approach. At first glance, one might determine that these two approaches are equally efficient. However, the quantitative measurement of the normalized mean square error gave more in depth comparison.

Our conclusion is that for the most accurate results in optical tweezers stretching experiment the ring approach is the best from the three tested here. In this approach the forces are equally distributed and applied to the points on rings. These consist of number of points which create non-circular rings with average diameter close to the contact diameter from the biological experiment.

In the last part of the article we also give the correct values for the ring intervals for other discretizations of the RBC.

References

1. Cimrák, I., Gusenbauer, M., Schrefl, T.: Modelling and simulation of processes in microfluidic devices for biomedical applications. Comput. Math. Appl. **64**(3), 278–288 (2012)
2. Bachratá, K., Bachratý, H.: On modeling blood flow in microfluidic devices. In: ELEKTRO 2014: 10th International Conference, pp. 518–521. IEEE, Slovakia (2014)
3. Cimrák, I., Gusenbauer, M., Jančigová, I.: An ESPResSo implementation of elastic objects immersed in a fluid. Comput. Phys. Commun. **185**(3), 900–907 (2014)
4. Dao, M., Lim, C.T., Suresh, S.: Mechanics of the human red blood cell deformed by optical tweezers. J. Mech. Phys. Solids **51**(11), 2259–2280 (2003)
5. Dao, M., Li, J., Suresh, S.: Molecularly based analysis of deformation of spectrin network and human erythrocyte. Mate. Sci. Eng. C **26**(8), 1232–1244 (2005)
6. Cimrák, I., Jančigová, I.: Computational Blood Cell Mechanics: Road Towards Models and Biomedical Applications, p. 268. CRC Press, Boca Raton (2018)
7. Mills, J.P., Qie, L., Dao, M., Lim, C.T., Suresh, S.: Nonlinear elastic and viscoelastic deformation of the human red blood cell with optical tweezers. MCB **1**, 169–180 (2004)
8. Bušík, M.: Development and optimization model for flow cells in the fluid. [Dissertation thesis] - University of Žilina. Faculty of Management Science and Informatics. Department of Software Technology. - Supervisor: doc. Mgr. Ivan Cimrák, Dr. - Žilina, FRI ZU, 114 p (2017)

Describing Sequential Association Patterns from Longitudinal Microarray Data Sets in Humans

Augusto Anguita-Ruiz[1][✉], Alberto Segura-Delgado[2], Rafael Alcala[2],
Concepción Maria Aguilera[1], and Jesus Alcala-Fernandez[2]

[1] Department of Biochemistry and Molecular Biology II,
Institute of Nutrition and Food Technology "José Mataix",
Center of Biomedical Research, University of Granada,
Avda. del Conocimiento s/n, 18016 Armilla, Granada, Spain
augustoanguita@ugr.es
[2] DaSCI Research Institute, University of Granada, 18071 Granada, Spain

Abstract. DNA microarray technology provides a powerful vehicle for exploring biological processes on a genomic scale. Machine-learning approaches such as association rule mining (ARM) have been proven very effective in extracting biologically relevant associations among different genes. Despite of the usefulness of ARM, time relations among associated genes cannot be modeled with a standard ARM approach, though temporal information is critical for the understanding of regulatory mechanisms in biological processes. Sequential rule mining (SRM) methods have been proposed for mining temporal relations in temporal data instead. Although successful, existing SRM applications on temporal microarray data have been exclusively designed for *in vitro* experiments in yeast and none extension to *in vivo* data sets has been proposed to date. Contrary to what happen with *in vitro* experiments, when dealing with microarray data derived from humans or animals the "subject variability" is the main issue to address, so that databases include multiple sequences instead of a single one. A wide variety of SRM approaches could be used to handle with these particularities. In this study, we propose an adaptation of the particular SRM method "CMRules" to extract sequential association rules from temporal gene expression data derived from humans. In addition to the data mining process, we further propose the validation of extracted rules through the integration of results along with external resources of biological knowledge (functional and pathway annotation databases). The employed data set consists on temporal gene expression data collected in three different time points during the course of a dietary intervention in 57 subjects with obesity (data set available with identifier GSE77962 in the Gene Expression Omnibus repository). Published by Vink [1], the original clinical trial investigated the effects on weight loss of two different dietary interventions (a low-calorie diet or a very low-calorie diet). In conclusion, the proposed method demonstrated a good ability to extract sequential association rules with further biological relevance within the context of obesity. Thus, the application of this method could be successfully extended to other longitudinal microarray data sets derived from humans.

© Springer Nature Switzerland AG 2019
I. Rojas et al. (Eds.): IWBBIO 2019, LNBI 11466, pp. 318–329, 2019.
https://doi.org/10.1007/978-3-030-17935-9_29

Keywords: DNA microarray · Sequential rule mining · Association rules · Obesity · Human data · Weight loss

1 Introduction

Overweight and obesity are a public health problem that has raised concern worldwide. Characterized by an expansion of the adipose tissue (AT), obesity plays important roles in the development of cardiometabolic alterations which increase morbidity and mortality. It has been proposed that AT dysfunction is the main mechanism leading to these metabolic co-morbidities observed in obesity. Several investigations have confirmed that, on the other hand, intensive lifestyle interventions can increase weight loss (WL) as well as reduce the later risk of cardiometabolic disease in people with obesity [2]. The exact molecular mechanisms responsible for the beneficial effects of WL on AT remain not fully understood though.

DNA microarray technology provides a powerful vehicle for exploring biological processes on a genomic scale. Regarding dietary interventions, microarray experiments have revealed hundreds to thousands of genes differentially expressed (DE) during the course of WL in human AT [3]. Gene expression patterns have also been identified to vary strongly between the WL and the weight stabilization phase of a dietary intervention [3]. At the functional level, results point to the fact that WL might improve mitochondrial function and innate immunity in AT. Despite of all these findings, we still do not know how main WL-related genes interact each other in an orchestrated way to reach the outcome; which are the underlying regulatory gene networks for these biological processes? Is the upregulated expression of a certain gene, elicited by a dietary intervention, the trigger for later changes in the expression of other genes during the weight stabilization phase? Finding an answer to these questions is not an easy issue, especially due to the fact that classical statistical methods in molecular biology generally work on a 'one gene in one experiment' basis. Standards bioinformatic analyses currently employed in microarray experiments (such as clustering solutions) also present a limited throughput to get the overall picture of gene networks; particularly because genes can participate in more than one gene network all at once.

Machine-learning approaches such as association rule mining (ARM) have been proven very effective in extracting biologically relevant associations between different genes [4]. As one of the most popular knowledge discovery techniques, ARM methods detect sets of elements that frequently co-occur in a database and establish relationships between them of the form of X → Y, which means that when X occurs it is likely that Y also occurs. In genetics, these relationships can be translated into "how the expression of one (several) gene(s) may be linked or associated with the expression of a set of genes". Despite of the usefulness of ARM, time dependencies among associated genes cannot be modeled with a standard ARM approach, even though temporal information is critical for the understanding of regulatory mechanisms in biological processes. Various methods have been proposed for mining temporal relations in temporal data instead. Among the most popular techniques, it highlights sequential rule mining (SRM) [5], which further counts on a successful adaptation to time-series microarray data [6]. SRM methods extract gene association patterns taking into account temporal dependencies among items (genes) belonging to different time points (e.g. time delay between related expression profiles). An example of the sequential association rules that can be extracted from SRM on longitudinal microarray data has the form of [gene A↑, gene B↓] → (4 weeks) [gene C↑, gene D↑, gene E↑], which represents that upregulation of gene A and significant repression of gene B are followed by significant upregulation of genes C, D and E after 4 weeks of intervention.

In this study, we aimed to extract sequential association rules from human gene expression data through the application of SRM methods. For that purpose, a specific SRM algorithm named "CMRules" was adapted and applied to longitudinal microarray data derived from AT biopsies of 57 obese subjects participating in a dietary intervention [1]. We sought to describe the exact gene networks underlying WL-induced gene regulation in AT as well as to identify novel gene-gene interactions that aid knowledge into the functionality of WL-induced AT cellular responses. The article is organized as follows; in Sect. 2, main characteristics of the employed data set as well as primary analyses of raw expression data are detailed. In Sect. 3, we cover main concepts related to association rules and sequential rule mining. Next, Sect. 4 explains how two sequential databases were constructed from the original data and how a specific SRM algorithm named "CMRules" was adapted to the problem. The Sect. 5 explains how we further incorporate well-known biological data sources (such as functional and pathway annotation databases) to guide the method and the evaluation of extracted rules in a biological context. Section 6 describes codes and libraries employed for the analysis, which are readily available upon request. Finally, in Sect. 7 we present main discovered sequential association rules and discuss its possible meaning within the molecular framework of obesity. A brief discussion about the performance of the algorithm is also included.

2 Data Set Description and Preparation

The employed data set consists on temporal gene expression data derived from human AT samples collected in three different time points during the course of a dietary intervention. Published by Vink [1], the original clinical trial investigated the effects on WL of two different dietary interventions in 57 subjects with obesity. Subjects were randomly assigned to each experimental group: a low-calorie diet (LCD; 1250 kcal/day) for 12 weeks (slow weight loss) or a very-low-calorie diet (VLCD; 500 kcal/day) for 5 weeks (rapid weight loss). In both experimental conditions, the WL period was followed by a 4-week additional phase of weight stabilization. Abdominal subcutaneous AT biopsies were collected from each subject at each time point (baseline, after WL and after weight stabilization) and submitted to microarray analysis using the platform "*whole-genome Human Gene 1.1 ST*" from the Affymetrix company (one array per subject and time). A more detailed description of the study design can be found in the original publication of the data set [7].

The full data set was downloaded from the public repository *gene expression omnibus* (GEO) with identifier GSE77962[1]. Data were downloaded as raw fluorescence intensity files (one *.cel* file per subject and time point) and transformed into the form of an *N x M* matrix of expression levels, where the *N* rows correspond to subjects and the *M* columns correspond to probes under study. The data set presented valid fluorescence measures for 33.297 probes (*M* columns) mapping 19.654 unique genes distributed across the whole genome. The number of individuals presenting valid gene expression data was 24 on the VLCD group and 22 on the LCD group. Since all available time records were merged into a single database, each individual presented three consecutive entries in the database (long format), corresponding to gene expression levels at each temporal point (baseline, after WL and after the weight stabilization period). The total number of *N* rows was 138. Chip pseudo-images, histograms of *log2*(intensities) and MA-plots were generated from the data as part of the quality control check. Microarray fluorescence signals were normalized by means of the robust multichip average (RMA) method. Probes under study were annotated based on the latest released version of the "*org.Hs.eg.db*" database. All described primary analyses were performed in R environment.

3 Association Rule Mining (ARM): Concepts

The concept of association rules was first proposed by Agrawal [8] to discover what items are bought together within a transactional dataset (T). As previously mentioned, an association rule has the form of LHS (Left Hand Side) $\rightarrow$ RHS (Right Hand Side), where LHS and RHS are sets of items, and it represents that the RHS set being likely to occur whenever the LHS set occurs. Three main measures are used for evaluating the quality of an extracted rule in ARM: support (1), confidence (2) and lift (3). They are defined as:

$$\text{Support}(\text{LHS} \rightarrow \text{RHS}) = \frac{\text{Support}(\text{LHS} \cup \text{RHS})}{|T|} \tag{1}$$

$$\text{Confidence}(\text{LHS} \rightarrow \text{RHS}) = \frac{\text{Support}(\text{LHS} \rightarrow \text{RHS})}{\text{Support}(\text{LHS})} \tag{2}$$

$$\text{Lift}(\text{LHS} \rightarrow \text{RHS}) = \frac{\text{Support}(\text{LHS} \rightarrow \text{RHS})}{\text{Support}(\text{LHS}) \times \text{Support}(\text{RHS})} \tag{3}$$

Where |T| is a set of transactions of a given database and Support(LHS $\cup$ RHS) is computed as the proportion of transactions *t* in the dataset which contains the itemset (LHS $\cup$ RHS). Throughout the last decades, ARM methods have been employed in many different domains (including genetics or biomedicine) where they have shown a good ability to mine hidden patterns and relationships between items.

---

[1] https://www.ncbi.nlm.nih.gov/geo/query/acc.cgi?acc=GSE77962.

Notwithstanding, in real-world applications data is usually time-varying and most of existing ARM algorithms do not pay particular attention to temporal information. Especially in biology, temporal information is critical for the understanding of the regulatory mechanisms in biological processes.

Many new algorithms capable of handling with this kind of information have been proposed in recent years [5]. Known as sequential rule mining (SRM) methods, these approaches present a greater predictive and descriptive power and provide an additional degree of interestingness in those applications where the time dimension plays an important role. Under SRM, association rules are mined from a sequence database (SD) which is a generalization of a transaction database that contains time information about the occurrence of items [9]. As a result, we obtain sequential rules indicating that if some event(s) occurred, some other event(s) are likely to occur afterward with a given confidence or probability.

In a previous work, Nam [6] proposed an adaptation of SRM to time-series microarray data from yeast and demonstrated a good ability of the method to extract biologically relevant gene association patterns. This application, although interesting, was focused on *in vitro* gene expression data and none extension to *in vivo* temporal microarrays has been proposed to date. Contrary to what happen with *in vitro* gene expression data, in microarray experiments derived from humans or animals the "subject variability" is the main issue to address, so that databases include multiple sequences instead of a single one. Hopefully, many SRM approaches could be used to handle with these particularities. In next sections, we present how we adapt a particular SRM approach to mine sequential rules from temporal microarray data derived from human.

4   Sequential Database (SD) Construction and Adaptation of Sequential Rule Mining (SRM) Methods

Since SRM methods require a SD composed of items and sequences to work, continuous gene expression values from the gene expression matrix were transformed into discrete events (items). Particularly, two SDs were constructed from the original data (one per diet group) where each sequence corresponded to a subject and each event represented the change in gene expression of a certain probe during a particular time interval (WL period or WS period). Three possible states (no change, upregulation or downregulation) were allowed by probe and time interval following the "mean" criteria:

- For probes showing a positive $log2$(FoldChange)

  If $log2(FoldChange)_{ikj}$ > $Average(log2(FoldChange)_{ik})$ Then assign the label 'Upregulation' , Otherwise type 'No change'

- For probes showing negative $log2$(FoldChange)

If $log2\,(FoldChange)_{ikj} < Average\,(log2\,(FoldChange)_{ik})$ Then assign the label 'Downregulation', Otherwise type 'No change'

Where the "FoldChange" term refers to change in gene expression for the i probe, in the k time interval and the j subject. The term average refers to the mean "Fold-Change" for that particular ik probe in all subjects from the diet under study. An example of the general structure of constructed SDs is presented in Fig. 1. Of note is that we modeled time interval changes in gene expression rather than discrete gene expression levels at each time record. The main reason for that is that we were particularly interested in the long-term molecular consequences elicited by dietary intervention in AT. That is, which are the gene expression patterns at the end of the WS period elicited by WL-induced gene regulation?

	7923578/FMOD	7925929/AKR1C3	7926786/APBB1IP	7928872/SNCG	\<end-of-time-interval\>
AP06	1046	1050	NA	1056	-1
AP08	1046	NA	NA	NA	-1
AP09	NA	NA	NA	NA	-1
AP10	1046	1050	NA	1056	-1
AP11	1046	NA	NA	NA	-1
AP13	1046	NA	NA	1056	-1
AP14	1046	1050	NA	1056	-1
AP19	NA	NA	NA	NA	-1
AP20	1046	NA	NA	NA	-1
AP23	NA	NA	NA	1056	-1
AP24	NA	NA	NA	1056	-1
AP31	1046	NA	NA	1056	-1
AP33	NA	1050	NA	NA	-1
AP34	1046	1050	NA	1056	-1
AP35	1046	1050	NA	NA	-1
AP36	1046	NA	NA	NA	-1
AP39	NA	1050	NA	NA	-1
AP41	1046	1050	NA	NA	-1
AP48	NA	NA	NA	1056	-1
AP49	NA	NA	NA	1056	-1
AP50	NA	NA	NA	1056	-1
AP51	NA	NA	NA	1056	-1

7923578/FMOD	7925929/AKR1C3	7926786/APBB1IP	7928872/SNCG	\<end-of-time-interval\>	\<end-of-sequence\>
NA	NA	NA	2056	-1	-2
NA	NA	NA	NA	-1	-2
NA	1050	NA	NA	-1	-2
NA	NA	NA	2056	-1	-2
NA	NA	NA	NA	-1	-2
NA	NA	NA	2056	-1	-2
NA	NA	NA	2056	-1	-2
NA	NA	NA	2056	-1	-2
NA	1050	NA	NA	-1	-2
NA	NA	NA	2056	-1	-2
NA	1050	NA	2056	-1	-2
NA	1050	NA	2056	-1	-2
NA	NA	NA	NA	-1	-2
NA	NA	NA	2056	-1	-2
NA	1050	NA	NA	-1	-2
NA	1050	NA	NA	-1	-2
NA	1050	NA	NA	-1	-2
NA	NA	NA	NA	-1	-2
NA	1050	NA	NA	-1	-2
NA	1050	NA	NA	-1	-2
NA	NA	NA	2056	-1	-2
NA	1050	NA	2056	-1	-2

Fig. 1. Structure for the constructed LCD-SD containing discrete temporal information (events) for the change in gene expression. Events were represented using a 4-digit code (where the first digit represented the change in gene expression (1 = Downregulation, 2 = Upregulation) and the next three digits represented the identifier for the i probe).

Once the data were properly formatted, we proposed the use of "CMRules" for mining them [10]. Contrarily to other SRM methods that can only discover rules in a single sequence of events, CMRules is able to mine sequential rules in several sequences (e.g. temporal gene expression patterns common to many subjects). Furthermore, CMRules proposes a more relaxed definition of sequential rule with unordered events within each (right/left) part of the rule. This provides the method with a greater ability to recognize that similar rules can describe a same phenomenon; thus avoiding an undesirable loss of information. A detailed description of the employed CMRules algorithm has been described elsewhere [10]. Generally, CMRules starts applying a classic extraction method, such as Apriori, for extracting association rules without taking into account the temporal information and then, the sequential support of the extracted association rules is calculated in order to generate the sequential temporal association rules from them. Two measures for evaluating the interestingness of each mined sequential rule are employed in SRM: the sequential support, defined as seqSup (LHS $\rightarrow$ RHS) = sup(LHS $\blacksquare$ RHS)/|SD|, and the sequential confidence, defined as seqConf(LHS $\rightarrow$ RHS) = sup(LHS $\blacksquare$ RHS)/sup(LHS). The element sup(LHS $\blacksquare$ RHS) refers to the number of sequences from the SD in which all the items of LHS appear before all the items of RHS (note that items within LHS (or RHS) do not need to be in the same transaction nor temporal order for each sequence). The notation sup(LHS) refers to the number of sequences that contains LHS.

The huge number of analyzed genes across the genome brought the curse-of-dimensionality dilemma to our application. As many other machine learning methods working on gene expression data sets, CMRules found difficulties to extract informative gene expression change patterns among all redundant or irrelevant events, background noise, and biased features. The initial number of available probes was thus filtered according to DE genes by time interval and experimental condition. We identified DE genes within experimental condition by assessing the differences in gene expression during each period of the intervention (WL period, comprising end of WL vs baseline; WS period, comprising end of weight stabilization period vs end of WL; or Dietary Intervention (DI) period, comprising the end of weight stabilization vs baseline). Changes were defined as significantly different when the Bonferroni-adjusted P-value was <0.05 and the Log2(FoldChange) was ≥ 1 or ≤ -1 in a paired -test with Bayesian correction. As a result, 431 probes matching 398 unique genes were selected for further analyses.

In order to deal with some other particularities of the study data set (the low number of available temporal records and the specific experimental design with two different diet interventions) a particular modification was implemented in the final step of the algorithm; CMRule was modified to only show those sequential rules which beside fulfilling the condition [seqSup(r) > MinSeqSup & seqConf(r) > MinSeqConf], are further exclusive of each intervention group. That is, sequential rules extracted from the VLCD which are not identified within the LCD intervention, and vice versa.

5 Integration with External Biological Information

Previous works have demonstrated that the use of external biological information during any part of the ARM process is a helpful strategy that enriches the final model and helps biologists to better understand genes and their complex relations [4]. In recent years, a great variety of external databases containing biological knowledge have become available. Among the most robust and reliable resources, it highlights those containing information relative to gene function or molecular interaction (e.g. the gene ontology (GO) project [11] and the Kyoto Encyclopedia of Genes and Genomes (KEGG) [12]). The GO project is an annotation database that provides a structured controlled vocabulary to describe gene and gene product attributes in any organism through the use of three different categories or ontologies (cellular component, molecular function and biological process). On the other hand, KEGG is a bioinformatic resource that integrates current knowledge on molecular interaction networks such as pathways and complexes as well as information about genes and proteins. Both, GO and KEGG resources, have been successfully employed during the ARM process in previous studies on microarray data aiding biological explanation to the extracted association patterns.

Other databases such as TRRUST [13], that includes information relative to transcriptional regulatory relationships, are also of special interest for gene association analysis. TRRUST is a database of human and mouse transcriptional regulatory networks untraveled by the use of sentence-based text mining. The current version of the TRRUST database (v2) contains 8.444 and 6.552 transcription factor (TF)-target regulatory relationships of 800 human TFs and 828 mouse TFs, respectively. Especially for SRM applications to temporal microarray data, the integration of TRRUST information is indispensable if we want to understand the complex gene relations that are illustrated in the form of sequential gene association rules. The main concept motivating this proposal is the fact that the upregulation of a transcription factor (gene A) might induce a later change in the expression levels of a target sequence (gene B) after a given time interval.

In our study, we proposed the incorporation of well-known biological data sources to guide the evaluation of extracted rules in a biological context. Particularly, we proposed a method that integrates biological data from three different resources (metabolic pathway annotation information from KEGG, transcriptional regulatory networks from TRRUST and annotation terms from the three categories of GO) to further compute five new "biological" measures of quality. These new quality measures will take higher values as a significant set of genes within a rule present biological features in common. The integration of this external information allowed us to verify if the final association rules extracted by CMRules were also significant from a biological point of view. On this matter, the associations are intrinsic to data and additional biological verification from other sources reinforces the potential significance of the associations.

6 Software and Tools

All data manipulation and processing steps as well as primary and secondary analyses of expression data were conducted in R environment using the next list of libraries ("*Matrix*", "*lattice*", "*fdrtool*", "*rpart*", "*affy*", "*oligo*", "*affydata*", "*ArrayExpress*", "*limma*", "*Biobase*", "*Biostrings*", "*genefilter*", "*affyQCReport*", "*affyPLM*", "*simpleaffy*", "*ggplot2*", "*dplyr*", "*pd.hugene.1.1.st.v1*", "*FGNet*", "*RGtk2*", "*RDAVIDWebService*", "*topGO*", "*KEGGprofile*", "*GO.db*", "*KEGG.db*", "*reactome.db*", "*org.Hs.e.g.db*", "*arules*", "*arulesViz*"). The implementation of CMRules was carried out through the use the open-source data mining library "*SPMF*" in Java [14].

7 Results and Conclusion

In this study, we aimed to extract sequential association rules from human gene expression data through the application of SRM methods. A specific SRM algorithm named "CMRules" was adapted and applied to longitudinal microarray data derived from obese subjects participating in a dietary intervention [1, 7]. The employed data set consisted on temporal gene expression data collected in three different time points during the course of two dietary interventions (VLCD and LCD). In order to describe which are the WL-induced gene regulatory networks elicited by each diet intervention, CMRules was applied independently to each experimental group (VLCD and LCD). Minimum sequential support and sequential confidence values were set during the rule extraction process by experimental condition (minSeqSup = 0.45 and minSeqConf = 0.4 for the VLCD, and minSeqSup = 0.4 minSeqConf = 0.4 for the LCD group). Standard quality measures usually employed in ARM were computed to estimate the interestingness of each mined rule. Furthermore, new quality measures that integrate external biological knowledge were proposed for the evaluation of extracted rules in a biological framework.

Table 1. Descriptive statistics for mined rules by experimental condition.

	Support	Confidence	Lift	CF	Conviction	BP	CC	MF	TF
	LCD								
Minimum	9	0,58	0.79	−0.2	0.58	0	0.31	0	0
1st quartile	9	0.69	1.18	0.32	1.47	0.42	0.45	0.42	0
Median	9	0.79	1.38	0.46	1.87	0.52	0.54	0.52	0
Mean	9.7	0.78	1.33	0.46	1.7	0.49	0.52	0.5	0.14
3rd quartile	10	0.84	1.5	0.6	2.5	0.54	0.55	0.55	0
Maximum	15	1	1.69	1	*Inf*	1	1	1	1
	VLCD								
Minimum	11	0.65	1.11	0.15	1.18	0	0.25	0.25	0
1st quartile	11	0.78	1.35	0.48	1.95	0.36	0.36	0.36	0
Median	11	0.85	1.47	0.65	2.91	0.42	0.42	0.42	0
Mean	11.07	0.86	1.53	0.68	2.98	0.44	0.45	0.44	0.21
3rd quartile	11	1	1.71	1	*Inf*	0.54	0.54	0.54	0
Maximum	12	1	2	1	*Inf*	1	1	1	2

Fig. 2. Mined ruled emulating plausible transcriptional regulatory relationships. Panel (A) corresponds to the LCD while (B) to the VLCD.

As a result, 50 rules were identified from the LCD group and 350 from the VLCD group. With up to seven times more number of rules mined from the VCLD group than from the LCD group, our results are in accordance with previous findings from Vink [1], which showed a higher impact of the VLCD on the gene expression in AT. Main descriptive statistics for the extracted rules are presented in Table 1. Only rules showing a lift value higher than 1.00 were considered. In general, extracted rules presented good values for the majority of computed quality measures, thus indicating a good performance of the algorithm during the gene association mining process. Top sequential associations by experimental condition were selected based on strict cutoff values for the computed quality measures. Interestingly, top rules involved genes participating in molecular processes previously reported as part of the WL-induced AT response (e.g. mitochondrial function, angiogenesis, immunity and adipogenesis). Of note is that top rules (presenting higher values in the traditional quality measures) also presented good rates in the new proposed biological quality measures.

By the use of a new biological quality measure that integrates information from the database TRRUST, we further selected those mined ruled emulating plausible transcriptional regulatory relationships. As a result, 5 rules were extracted from the LCD group and 15 from the VLCD group (Fig. 2). All of them also presented good rates for standard quality measures (confidence > 0.75, lift > 1.1 and support > 0.4). Among them, it highlights a rule from the LCD group which involves the genes *NOTCH3* and the *EGFL6* (stood out in the panel A of Fig. 2). According to literature, target sites for the *NOTCH3* transcription factor have been reported within the *EGFL6* sequence. Moreover, both genes have been involved in the process of angiogenesis and have been previously associated with obesity [15, 16].

In conclusion, the proposed algorithm has demonstrated a good ability to extract sequential rules from longitudinal microarray data in humans as well as to integrate the mined pattern along with useful biological information. An exhaustive study of all the results is needed otherwise to understand the concrete molecular patterns underlying WL-induced responses in obesity. The application of this method could be successfully extended to other longitudinal microarray data sets derived from human studies.

Acknowledgment. This work was supported by the Mapfre Foundation ("Ayudas a la investigación de Ignacio H. de Larramendi").

References

1. Vink, R.G., et al.: Adipose tissue gene expression is differentially regulated with different rates of weight loss in overweight and obese humans. Int. J. Obes. **41**, 309–316 (2017). https://doi.org/10.1038/ijo.2016.201
2. Brown, T., et al.: Systematic review of long-term lifestyle interventions to prevent weight gain and morbidity in adults. Obes. Rev. **10**, 627–638 (2009). https://doi.org/10.1111/j.1467-789X.2009.00641.x
3. Johansson, L.E., et al.: Differential gene expression in adipose tissue from obese human subjects during weight loss and weight maintenance. Am. J. Clin. Nutr. **96**, 196–207 (2012). https://doi.org/10.3945/ajcn.111.020578

4. Alves, R., Rodriguez-Baena, D.S., Aguilar-Ruiz, J.S.: Gene association analysis: a survey of frequent pattern mining from gene expression data. Briefings Bioinform. **11**, 210–224 (2009). https://doi.org/10.1093/bib/bbp042

5. Wang, L., Meng, J., Xu, P., Peng, K.: Mining temporal association rules with frequent itemsets tree. Appl. Soft Comput. J. **62**, 817–829 (2018). https://doi.org/10.1016/j.asoc.2017.09.013

6. Nam, H., Lee, K.Y., Lee, D.: Identification of temporal association rules from time-series microarray data sets. BMC Bioinformatics **10**, 1–9 (2009). https://doi.org/10.1186/1471-2105-10-S3-S6

7. Vink, R.G., Roumans, N.J.T., Arkenbosch, L.A.J., Mariman, E.C.M., Van Baak, M.A.: The effect of rate of weight loss on long-term weight regain in adults with overweight and obesity. Obesity. **24**, 321–327 (2016). https://doi.org/10.1002/oby.21346

8. Agrawal, R., Imielinski, T., Swami, A.: Mining association in large databases. ACM SIGMOD Record (1993). https://doi.org/10.1145/170036.170072

9. Agrawal, R., Srikant, R.: Mining sequential patterns. In: Proceedings of the Eleventh International Conference on Data Engineering, pp. 3–14. IEEE Computer Society Press (1995)

10. Fournier-viger, P., Faghihi, U., Nkambou, R., Nguifo, E.M.: CMRules: mining sequential rules. Knowl. Based Syst. **25**, 63–76 (2012)

11. Ashburner, M., et al.: Gene ontology: tool for the unification of biology (2000)

12. Kanehisa, M., Goto, S., Sato, Y., Furumichi, M., Tanabe, M.: KEGG for integration and interpretation of large-scale molecular data sets. Nucleic Acids Res. (2012). https://doi.org/10.1093/nar/gkr988

13. Han, H., et al.: TRRUST v2: An expanded reference database of human and mouse transcriptional regulatory interactions. Nucleic Acids Res. (2018). https://doi.org/10.1093/nar/gkx1013

14. Fournier-Viger, P., et al.: The SPMF open-source data mining library version 2. In: Berendt, B., et al. (eds.) ECML PKDD 2016. LNCS (LNAI), vol. 9853, pp. 36–40. Springer, Cham (2016). https://doi.org/10.1007/978-3-319-46131-1_8

15. González-Muniesa, P., Marrades, M., Martínez, J., Moreno-Aliaga, M.: Differential proinflammatory and oxidative stress response and vulnerability to metabolic syndrome in habitual high-fat young male consumers putatively predisposed by their genetic background. Int. J. Mol. Sci. **14**, 17238–17255 (2013). https://doi.org/10.3390/ijms140917238

16. Battle, M., et al.: Obesity induced a leptin-Notch signaling axis in breast cancer. Int. J. Cancer. **134**, 1605–1616 (2014). https://doi.org/10.1002/ijc.28496

Biomedical Engineering

Low Resolution Electroencephalographic-Signals-Driven Semantic Retrieval: Preliminary Results

Miguel Alberto Becerra[1]([⊠]), Edwin Londoño-Delgado[1],
Oscar I. Botero-Henao[1], Diana Marín-Castrillón[2], Cristian Mejia-Arboleda[2],
and Diego Hernán Peluffo-Ordóñez[3]

[1] Institución Universitaria Pascual Bravo, Medellín, Colombia
migb2b@gmail.com
[2] Instituto Tecnológico Metropolitano, Medellín, Colombia
[3] SDAS Research Group, Yachay Tech, Urcuquí, Ecuador
http://sdas-group.com

Abstract. Nowadays, there exist high interest in the brain-computer interface (BCI) systems, and there are multiple approaches to developing them. Lexico-semantic (LS) classification from Electroencephalographic (EEG) signals is one of them, which is an open and few explored research field. The LS depends on the creation of the concepts of each person and its context. Therefore, it has not been demonstrated a universal fingerprint of the LS either the spatial location in the brain, which depends on the variability the brain plasticity and other changes throughout the time. In this study, an analysis of LS from EEG signals was carried out. The Emotiv Epoc+ was used for the EEG acquisition from three participants reading 36 different words. The subjects were characterized throughout two surveys (Becks depression, and emotion test) for establishing their emotional state, depression, and anxiety levels. The signals were processed to demonstrate semantic category and for decoding individual words (4 pairs of words were selected for this study). The methodology was executed as follows: first, the signals were preprocessed, decomposed by sub-bands ($\delta, \theta, \alpha, \beta$, and γ) and standardized. Then, feature extraction was applied using linear and non-linear statistical measures, and the Discrete Wavelet Transform calculated from EEG signals, generating the feature space termed set-1. Also, the principal component analysis was applied to reduce the dimensionality, generating the feature space termed set-2. Finally, both sets were tested independently by multiple classifiers based on the support vector machine and k-nearest neighbor. These were validated using 10-fold cross-validation achieving results upper to 95% of accuracy which demonstrated the capability of the proposed mechanism for decoding LS from a reduced number of EEG signals acquired using a portable system of acquisition.

Keywords: Electroencephalographic signal · Machine learning ·
Semantic retrieval · Semantic category · Signal processing

© Springer Nature Switzerland AG 2019
I. Rojas et al. (Eds.): IWBBIO 2019, LNBI 11466, pp. 333–342, 2019.
https://doi.org/10.1007/978-3-030-17935-9_30

1   Introduction

The semantic retrieval refers to the lexico-semantic (LS) recuperation which is analyzed in this work from electroencephalographic signals acquired using a low-resolution device. This type of systems are essential for human-computer interfaces [9] in multiple applications. Nowadays, the number of investigations in this field can be considered low; therefore, it is regarded as an open research field, which does not have a standard gold for carrying out this type of experiments and their results are still low. Some works in this field are depicted below.

The studies of LS retrieval from EEG signals have been carried out using different approaches. In [3] is discuss a study based on a semantic paradigm which uses the pronunciation of four letters ('a', 'b', 'c', 'd') which were presented to subjects using audio and visual stimuli. These were applied to 11 healthy participants, demonstrating a low correspondence between both stimuli types. In [4] a study on nine subjects based on 64-channel of EEG and EMG signals were accomplished, and support vector machine classifier was used to demonstrate semantic category and individual words. Mean accuracies of 76% (chance = 50%) and 83% were achieved for the decoding semantic categories or individual words respectively. In [5] was conducted an investigation based on invariant semantic representations in the brain, exploiting the capacity of bilinguals to transform between acoustic, phonological, and conceptual neural representations acquiring EEG signals of 16 Dutch subjects with highly proficient in English listened to four monosyllabic and acoustically distinct animal words in both languages and six inanimate object words. Selected from 36 nouns and acquired in 8 trials. Multivariate pattern analysis (MVPA) was applied to identify EEG response patterns for discriminating individual words per language and then it was generalized for both languages. Most interestingly, significant across-language generalization was possible around 550–600 ms, suggesting the activation of common semantic-conceptual representations from the Dutch and English nouns. Besides, low frequency (<12 Hz) of EEG demonstrated its relevance for the neural representation of individual words and concepts. In [7] are decoded spoken words from EEG signals of 32 channels, using frequency-domain features, principal component analysis, and the mean Euclidean for decoding. The performance of the classification systems was around 83.3%. However, some combinations over face motor cortex (FMC), classification accuracy ranged from 53% ('hello' versus 'more') to 100% accuracy ('yes' versus 'no,' 'hot' versus 'hungry,' and others). On the other hand, in [11] a study for identifying object Categories from Event-Related EEG was carried out decoding of conceptual representations in different modalities (spoken, visual representation, and written name). Three monosyllabic semantic categories were used in two relevant categories (animals, tools) and a task category that varied across subjects, either clothing or vegetables. Classifiers based on Bayesian logistic regression with a multivariate Laplace prior were trained to identify semantic categories (animal or tool) and predict semantic category.

Recently, a methodology was proposed to record and classify EEG signals in twelve different semantic categories using a paradigm with images of each one. EEG signals were acquired from 10 subjects using a 14-channel electrode cap. For the experiment, were used six images from each category to present the stimulus, asking the participants to press a button when an image appears with the category that is indicated before starting the test. Before classification with a multiclass SVM, was applied feature selection, obtaining an accuracy ranging between 66.61 to 79.07% from category 2 to 12, and the best performance in category 1 (Animal), with an accuracy of 92.31% when nonlinear features with wavelet coefficients were used [12].

In this study, an exploratory analysis semantic retrieval from EEG signals is conducted on three participants. Thirty-six different expressions were read by the subjects ten times. EEG signals were acquired using Emotiv-Ephoc device, which has been applied in different studies applying similar paradigms for tactile and odor detection and BCI [1, 2, 10]. The signals were processed follow four stages (Quality selection, preprocessing, feature extraction, and classification) for decoding eight semantic categories analyzing four pairs and eight words. Some signals were discard based on quality; then the signals were filtered, decomposed in sub-bands, and standardized. Then, a features extraction was carried out using discrete wavelet transform, and statistical measures. Principal component analysis technique was applied for reducing the dimensionality space. Finally, four Support vector machines and k-NN techniques were tested as classifiers, which were validated using 10-fold cross-validation.

2   Experimental Setup

This study was performed as depicted in Fig. 1. The EEG signals were collected following the paradigm which is explained below in the next section. Then a set of registers were selected and processed. First, a pre-processing stage was carried out eliminating artifacts, and signals of low quality. Besides, a standardization was applied, and a decomposition per sub-bands was performed. The feature extraction stage was carried out calculating linear and non-linear statistical measures, energy of the signals, and discrete wavelet transform (DWT) coefficients, obtaining a features set-1. Also a feature set-2 was obtained applying principal component analysis technique for reducing the dimensionality. Both features sets, were independently tested as inputs of the classifiers (support vector machines - SVM and k-Nearest neighbor - k-NN). Finally, 10 cross-fold validation was accomplished for establishing the performance of the systems.

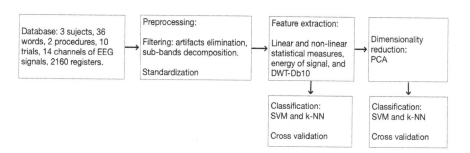

Fig. 1. Experimental procedure

2.1   Signal Acquisition Process

Three mentally and physically healthy volunteers, two men and a woman between 20 and 26 years of age participated in the proposed experiment. The procedure was explained to each subject, and they signed a consent form before the data acquisition. The acquisition of EEG signals was carried out in a controlled environment with low sound and cool temperature. The subjects seated on a comfortable chair and Emotiv Epoc+ device of 14 electrodes was placed on their head for recording the EEG signals. The experiment consisted of the visual presentation individually of 36 written Spanish words or expressions using black letters on a white background. Each word had a semantic meaning for Spanish speakers (see Table 1). These words were chosen considering the usability that could have for people with limited physical capability for communicating. The paradigm is shown in Fig. 2, and it was carried out as follows: each word was presented to the subject for 6 s, where the subject read the word repeatedly. Then, the participant rested for 3 s. Nonetheless, in some cases, this time was upper to maintain the comfort of the participant. The procedure was repeated for the 36 words or expressions. Finally, the participant had a rest higher than 5 min among each trial. This trial was repeated ten times per subject obtaining 100 registers (14 EEG signals per record) from each subject. On the other hand, an heuristic analysis based on k-NN classifier was carried using all words, but due to low results achieved, the team decides to begin this exploratory study using four pairs of words (8 words) which are shown in Table 3. The selection criterion was that the couples were antonyms (Table 2).

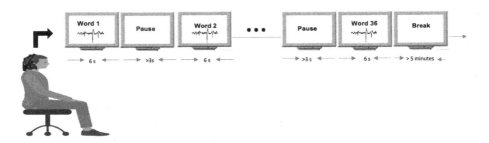

Fig. 2. Paradigm for EEG signals acquisition

Table 1. Applied LS

I want music	Unpleasant	Hot	Sad	Above	Pleasure
Turn off the light	Bad person	Do not	Yes	Happy	Bathroom
I want you to talk to me	Turn on light	Cold	Down	Pain	Please
I want to watch tv	I am sleepy	Zoom in	Help	Boring	Anxiety
I want you to read me	I love you	Ward off	Left	Hello	Hungry
Silence please	Thank you	Goodbye	Fear	Right	Pleasant

Table 2. Selected semantics for classification

Word	w1-Yes	w2-No	w3-Hello	w4-Goodbye	w5-Hot	w6-Cold	w7-Pleasant	w8-Unpleasant

2.2 Preprocessing

The electrodes that showed excessive abnormality in the distribution of their data with respect to the normal distribution were eliminated; the electrodes "T7", and "T8" were eliminated from all the samples due to an abnormality in the acquisition system. The EEG signals were segmented from 1 to 6 s to delimit the study window according to the presentation of the stimuli. Hanning window was applied for segmenting and softing the signal cuts. The offset level was eliminated, and Z-score scaling was applied to standardize the study signals; a FIR filter was applied between (0.5 Hz–70 Hz) for the elimination of low and high frequency artifacts, in the same way a FIR filter was implemented for decomposing the EEG signals into 5 sub-bands δ (0.5 Hz–4 Hz), θ (4 Hz–7.5 Hz), α (8 Hz–13 Hz), β (13 Hz–30 Hz), and γ (>30 Hz) to obtain a comparative study between them.

2.3 Feature Extraction

Feature extraction was carried out using the Linear and non-linear measures shown in Table 3. These measures were calculated directly from the signal, their sub-bands, and DWT coefficients. The DWT [8] was computed using ten levels of decomposition applying the Mother Wavelet Daubechies 10 ($Db10$). The feature vector achieved a size of 110 features per observation combining the characteristics of each wavelet decomposition level which is termed set-1. Besides, principal component analysis (PCA) was applied for reducing the features space with 95% confidence intervals.

Table 3. Measures calculated from EEG, sub-bands, and DWT

Standard deviation	Variance	Kurtosis	Coefficient of variation	Pearson coefficient
Bowley coefficient	Energy	Shannon Entropy	Aperture coefficient	Log energy entropy

2.4 Classification

Support vector machine (SVM) [6] with three different kernels (Linear, Cubic, and Quadratic) and k-NN based on Euclidean distance and k = 3, were tested for classifying the LS stimuli. The selection of these classifiers was made considering the simplicity of k-NN, which allow demonstrating the discriminate capability of the features. On the other hand, the SVM- classifiers was selected taking into account that they have shown high performance in different classification approaches from EEG signals and low dimensional space. The performance of classifiers was established in terms of accuracy and area under the curve, applying 10-fold cross-validation. The classifiers were tested using set-1 and set-2 of each participant and all participant.

3 Results and Discussion

Taking into account the feature estimation and the feature reduction process described in the previous section, in this work were analyzed the results obtained from three different tests. The results presented in Tables 4, 5, and 6 indicate the performance of the SVM and KNN classifiers for some couples of words for subject 1, subject 2, and subject 3 respectively. For this end, features from each sub-band ($\delta, \theta, \alpha, \beta$, and γ) were used separately for the classification before and after feature reduction with PCA. Finally, features calculated from all sub-bands were used together to do the same test, obtaining the best performance to all subjects, compared with the results achieved, using the sub-bands separately.

Table 4. Percentage of accuracy performance - Subject 1 (w1-w2)

Classifier		δ/δ–PCA	θ/θ–PCA	α/α–PCA	β/β–PCA	γ/γ–PCA	All/All–PCA
SVM-L	ACC	77.0/74.5	70.7/71.5	75.7/73.6	66.5/64.4	71.1/72.4	88.33/81.25
	AUC	0.81/0.82	0.77/0.77	0.79/0.78	0.7/0.64	0.71/0.72	0.88/0.81
SVM-Q	ACC	76.2/75.3	78.2/71.5	78.7/78.2	72.8/74.1	70.3/72.8	85.42/86.25
	AUC	0.85/0.84	0.83/0.78	0.83/0.84	0.79/0.83	0.70/0.73	0.85/0.86
SVM-C	ACC	76.6/77.0	78.7/77.0	78.7/88.3	72.8/72.4	69.9/69.5	90.42/96.15
	AUC	0.86/0.84	0.83/0.91	0.84/0.95	0.82/0.8	0.70/0.69	0.90/0.96
k-NN	ACC	81.6/79.9	74.1/77.4	87.4/87.0	72.4/72.8	64.0/67.0	**97.08/95.83**
	AUC	0.82/0.8	0.74/0.77	0.87/0.87	0.72/0.81	0.64/0.67	0.97/0.96

Table 5. Percentage of accuracy performance - Subject 2 (w1-w2)

Classifier		δ/δ–PCA	θ/θ–PCA	α/α–PCA	β/β–PCA	γ/γ–PCA	All/All–PCA
SVM-L	ACC	61.7/61.8	73.3/68.8	75.8/68.8	62.9/58.75	63.8/62.1	78.75/76.67
	AUC	0.62/0.62	0.73/68.8	0.76/0.69	0.63/0.59	0.64/0.62	0.79/0.77
SVM-Q	ACC	67.9/66.3	75.8/75.8	73.3/75.83	66.3/63.8	64.2/63.8	83.33/84.58
	AUC	0.68/0.66	0.76/0.76	0.73/0.76	66.3/0.64	0.64/0.64	0.83/0.85
SVM-C	ACC	62.1/62.9	72.1/71.7	72.1/70.0	55.8/57.1	60.8/62.9	82.5/87.92
	AUC	0.62/0.63	0.72/0.72	0.72/0.7	0.56/0.58	0.61/0.63	0.83/0.88
k-NN	ACC	73.8/72.5	70.0/72.9	78.3/77.9	61.3/62.1	59.6/60.4	**87.5/87.92**
	AUC	73.8/0.73	0.70/0.73	0.78/0.78	0.61/0.62	0.6/0.60	0.87/0.88

Table 6. Percentage of accuracy performance - Subject 3 (w3-w4)

Classifier		δ/δ–PCA	θ/θ–PCA	α/α–PCA	β/β –PCA	γ/γ–PCA	All/All–PCA
SVM-L	ACC	79.6/77.9	73.8/72.9	80.8/79.2	74.2/73.3	74.6/73.8	93.75/92.06
	AUC	0.8/0.78	0.74/0.73	0.81/0.79	0.74/0.73	0.75/0.74	0.94/0.92
SVM-Q	ACC	77.5/77.5	80.4/81.3	80.8/83.3	81.6/82.1	76.25/84.6	93.75/94.58
	AUC	0.78/0.78	0.80/0.82	0.81/0.83	0.82/0.82	0.76/0.85	0.94/0.95
SVM-C	ACC	72.5/73.8	72.5/71.7	77.1/76.3	71.7/72.1	70.0/70.4	93.75/95
	AUC	0.73/0.74	0.73/0.72	0.77/0.76	0.72/0.72	0.70/0.70	0.94/0.95
k-NN	ACC	77.1/79.2	83.3/83.3	86.7/86.7	83.8/83.8	82.92/85.8	**94.58/95**
	AUC	0.77/0.79	0.83/0.83	0.86/0.87	0.84/0.84	0.83/0.86	0.95/0.95

Considering that the best results were obtained using all sub-bands, in Tables 7, 8, and 9 are shown the results for each couple of words, using all sub-bands, applying and without applying PCA, for participant 1, 2, and 3 respectively. The best results were achieved by SVM-Q and k-NN classifiers with accuracies upper to 97% for all couples of subject 1. For subject 2, the best results were achieved by classifiers SVM-Q, SVM-C, and k-NN. However, the variability was low. The results were upper to 80%. For the subject 3, the best results were obtained by SVM-Q and k-NN with accuracies upper to 95%.

Table 7. Subject 1 all bands

Couple	(w3-w4)		(w7-w8)		(w5-w6)		(w1-w2)	
Classifier	–	PCA	–	PCA	–	PCA	–	PCA
SVM-L	95.00	94.17	95.67	86.25	91.25	94.17	88.33	81.25
	0.95	0.94	0.96	0.86	0.91	0.94	0.88	0.81
SVM-C	96.7	96.25	87.08	82.92	94.58	94.18	85.42	86.25
	0.97	0.96	0.87	0.83	0.95	0.94	0.85	0.86
SVM-Q	97.5	**99.17**	91.25	95.42	95.00	**99.17**	90.42	96.15
	0.98	0.99	0.91	0.95	0.95	0.99	0.90	0.96
KNN	98.33	98.75	96.25	**97.5**	98.75	97.5	**97.08**	95.83
	0.98	0.99	0.96	0.98	0.99	0.98	0.97	0.96

Table 8. Subject 2 all bands

Couple	(w3-w4)		(w7-w8)		(w5-w6)		(w1-w2)	
Classifier	–	PCA	–	PCA	–	PCA	–	PCA
SVM-L	84.17	79.17	77.06	78.75	72.92	72.92	78.75	76.67
	0.84	0.79	0.77	0.79	0.73	0.73	0.79	0.77
SVM-C	84.58	84.17	78.75	**80.00**	74.58	75.00	83.33	84.58
	0.85	0.84	0.79	0.80	0.75	0.75	0.83	0.85
SVM-Q	81.67	**88.33**	77.50	74.17	77.5	85.42	82.50	**87.92**
	0.82	0.88	0.78	0.74	0.78	0.85	0.83	0.88
KNN	82.93	84.58	79.17	79.17	83.33	**86.67**	87.50	**87.92**
	0.83	0.85	0.79	0.79	0.83	0.87	0.87	0.88

Table 9. Subject 3 all bands

Couple	(w3-w4)		(w7-w8)		(w5-w6)		(w1-w2)	
Classifier	–	PCA	–	PCA	–	PCA	–	PCA
SVM-L	93.75	92.06	95.67	89.58	95.83	94.58	94.17	90.83
	0.94	0.92	0.96	0.90	0.96	0.95	0.94	0.91
SVM-C	93.75	94.58	90.83	90.83	97.92	97.08	93.33	93.33
	0.94	0.95	0.91	0.91	0.98	0.97	0.93	0.93
SVM-Q	93.75	**95.00**	94.17	96.67	**96.67**	**98.33**	92.92	96.67
	0.94	0.95	0.94	0.97	0.97	0.98	0.93	0.97
KNN	94.58	**95.00**	93.33	93.33	97.5	96.67	97.92	**97.92**
	0.95	0.95	0.93	0.93	0.98	0.97	0.98	0.98

Finally, multiclass classifiers were tested considering all words with and without PCA processing. The results are shown in Tables 10, 11, 12, 13, 14, and 15. In general, some cases achieved high accuracy, but the AUC presented high variability, which demonstrated a low general performance for all subjects. However, the couple w1-w2 (Yes-No) and w3-w4 (Hello-Goodbye) shown the best performance for all subjects and classifiers. K-NN classifiers achieved the best accuracies for all cases.

Table 10. Multiclass - all bands Subject 1 using PCA

	ACC	AUC1	AUC2	AUC3	AUC4	AUC5	AUC6	AUC7	AUC8
SVM-L	71.35	0.87	0.72	0.63	0.55	0.42	0.34	0.29	0.82
SVM-C	69.3	0.87	0.71	0.63	0.56	0.44	0.35	0.29	0.84
SVM-Q	87.00	0.94	0.80	0.68	0.56	0.42	0.30	0.2	0.9
KNN	**91.8**	0.97	0.81	0.68	0.56	0.43	0.3	0.18	0.93

Table 11. Multiclass - all bands Subject 1 without PCA

	ACC	AUC1	AUC2	AUC3	AUC4	AUC5	AUC6	AUC7	AUC8
SVM-L	73.75	0.89	0.75	0.63	0.54	0.42	0.35	0.27	0.85
SVM-C	71.77	0.88	0.72	0.65	0.53	0.43	0.33	0.28	0.83
SVM-Q	74.9	0.91	0.74	0.64	0.56	0.41	0.34	0.27	0.87
KNN	**91.35**	0.98	0.82	0.69	0.56	0.43	0.28	0.19	0.93

Table 12. Multiclass - all bands Subject 2 using PCA

	ACC	AUC1	AUC2	AUC3	AUC4	AUC5	AUC6	AUC7	AUC8
SVM-L	46.94	0.73	0.6	0.59	0.53	0.42	0.41	0.36	0.35
SVM-C	44.9	0.73	0.57	0.55	0.51	0.48	0.41	0.38	0.38
SVM-Q	61.6	0.81	0.68	0.64	0.53	0.46	0.37	0.32	0.8
KNN	**64.6**	0.77	0.71	0.65	0.52	0.45	0.4	0.25	0.75

Table 13. Multiclass - all bands Subject 2 without PCA

	ACC	AUC1	AUC2	AUC3	AUC4	AUC5	AUC6	AUC7	AUC8
SVM-L	48	0.71	0.59	0.62	0.54	0.49	0.39	0.36	0.31
SVM-C	45.83	0.74	0.57	0.56	0.49	0.52	0.38	0.38	0.36
SVM-Q	48.96	0.73	0.63	0.60	0.51	0.52	0.38	0.35	0.29
KNN	**65.31**	0.77	0.72	0.67	0.52	0.47	0.39	0.25	0.79

Table 14. Multiclass - all bands Subject 3 using PCA

	ACC	AUC1	AUC2	AUC3	AUC4	AUC5	AUC6	AUC7	AUC8
SVM-L	76.04	0.83	0.74	0.63	0.56	0.44	0.38	0.23	0.81
SVM-C	74.17	0.83	0.71	0.61	0.57	0.46	0.37	0.25	0.8
SVM-Q	87.08	0.95	0.78	0.67	0.56	0.43	0.32	0.19	0.9
KNN	**89.38**	0.93	0.81	0.69	0.58	0.44	0.31	0.19	0.95

Table 15. Multiclass - all bands Subject 3 without PCA

	ACC	AUC1	AUC2	AUC3	AUC4	AUC5	AUC6	AUC7	AUC8
SVM-L	77.29	0.87	0.73	0.64	0.57	0.44	0.35	0.21	0.81
SVM-C	74.38	0.86	0.72	0.61	0.57	0.43	0.37	0.23	0.79
SVM-Q	78.85	0.91	0.73	0.64	0.55	0.44	0.35	0.22	0.84
KNN	**88.54**	0.92	0.81	0.69	0.57	0.43	0.31	0.2	0.93

4   Conclusions

In this paper, a study of semantic retrieval was presented from EEG signals acquired using a low-resolution system. Support vector machine with the quadratic kernel and k-NN demonstrated a high performance using features extracted of all sub-bands ($\delta, \theta, \alpha, \beta$, and γ) of EEG signals for classifying couples. However, the performance of the multiclass classifiers can be considered low. We demonstrated the direct relationships between semantic retrieval and EEG signals. As future work, we propose to widen the number of subjects and make a spatial analysis for improving the proposed system regarding discernibility and generality. Besides, a study of time traceability of semantic retrieval must be done for establishing if there is a "fingerprint" of these. Moreover, it is necessary to unveil if these EEG responses are constructed based on the experience of the subject and consequently, studies based on adaptive systems must be conducted for semantic retrieval.

References

1. Becerra, M.A., et al.: Odor pleasantness classification from electroencephalographic signals and emotional states. In: Serrano C., J.E., Martínez-Santos, J.C. (eds.) CCC 2018. CCIS, vol. 885, pp. 128–138. Springer, Cham (2018). https://doi.org/10.1007/978-3-319-98998-3_10
2. Becerra, M.A., et al.: Electroencephalographic signals and emotional states for tactile pleasantness classification. In: Hernández Heredia, Y., Milián Núñez, V., Ruiz Shulcloper, J. (eds.) IWAIPR 2018. LNCS, vol. 11047, pp. 309–316. Springer, Cham (2018). https://doi.org/10.1007/978-3-030-01132-1_35
3. Cao, Y., et al.: The effects of semantic congruency: a research of audiovisual P300-speller. BioMed. Eng. OnLine **16**(1), 91 (2017). https://doi.org/10.1186/s12938-017-0381-4

4. Chan, A.M., Halgren, E., Marinkovic, K., Cash, S.S.: Decoding word and category-specific spatiotemporal representations from MEG and EEG. NeuroImage **54**(4), 3028–3039 (2011). https://doi.org/10.1016/j.neuroimage.2010.10.073

5. Correia, J.M., Jansma, B., Hausfeld, L., Kikkert, S., Bonte, M.: EEG decoding of spoken words in bilingual listeners: from words to language invariant semantic-conceptual representations. Front. Psychol. **6**, 71 (2015). https://doi.org/10.3389/fpsyg.2015.00071

6. Cortes, C., Vapnik, V.: Support-vector networks. Mach. Learn. **20**(3), 273–297 (1995). https://doi.org/10.1023/A:1022627411411

7. Kellis, S., Miller, K., Thomson, K., Brown, R., House, P., Greger, B.: Decoding spoken words using local field potentials recorded from the cortical surface. J. Neural Eng. **7**(5), 056007 (2010). https://doi.org/10.1088/1741-2560/7/5/056007

8. Khalid, M.B., Rao, N.I., Rizwan-i Haque, I., Munir, S., Tahir, F.: Towards a brain computer interface using wavelet transform with averaged and time segmented adapted wavelets. In: 2009 2nd International Conference on Computer, Control and Communication, pp. 1–4. IEEE (2009). https://doi.org/10.1109/IC4.2009.4909189

9. Nakamura, T., Tomita, Y., Ito, S., Mitsukura, Y.: A method of obtaining sense of touch by using EEG. In: 2010 IEEE RO-MAN, pp. 276–281. IEEE (2010)

10. Ortega-Adarme, M., Moreno-Revelo, M., Peluffo-Ordoñez, D.H., Marín Castrillon, D., Castro-Ospina, A.E., Becerra, M.A.: Analysis of motor imaginary BCI within multi-environment scenarios using a mixture of classifiers. In: Solano, A., Ordoñez, H. (eds.) CCC 2017. CCIS, vol. 735, pp. 511–523. Springer, Cham (2017). https://doi.org/10.1007/978-3-319-66562-7_37

11. Simanova, I., van Gerven, M., Oostenveld, R., Hagoort, P.: Identifying object categories from event-related EEG: toward decoding of conceptual representations. PLoS One **5**(12), e14465 (2010). https://doi.org/10.1371/journal.pone.0014465

12. Torabi, A., Jahromy, F.Z., Daliri, M.R.: Semantic category-based classification using nonlinear features and wavelet coefficients of brain signals. Cogn. Comput. **9**(5), 702–711 (2017)

A Simpler Command of Motor Using BCI

Yacine Fodil[✉] and Salah Haddab

Signal Processing Laboratory, Department of Electronics,
UMMTO – University Mouloud Mammeri of Tizi-Ouzou, Tizi-Ouzou, Algeria
yacinefodil93@gmail.com, shaddab@yahoo.fr

Abstract. Brain-Computer Interfaces discipline has recently witnessed an unprecedented boom, thanks to technological evolution, which has led to incredible results, such as the control of drones or realistic articulated arms by thought. These outputs are the result of complex algorithms whose conception required years of research. This paper will show that there is a simpler way to achieve a binary BCI in a minimum of time by using the Brain Switch paradigm, Motor Imagery, and Fourier Transform. In the present work we shall demonstrate how it was possible to control a motor by thought, using a digital recorded EEG on an untrained volunteer at the hospital. From this EEG tracing we have extracted on a single electrode a unique feature which is the "Mu" frequency related to motor imaging.

Keywords: BCI · Biomedical engineering · Brain switch · Motor imagery

1 Introduction

Humans have always used their ingenuity to satisfy vital needs, using the available resources surrounding them to create tools necessary for their survival. Over time, their technologies have known various transformations to make daily life easier, notably by reducing or eliminating repetitive tasks for achieving comfort. From this principle of transforming nature, man has created modern technology as we know it, which through studies and discoveries has led to the development of tools for assistance. More recently, the combination of medicine and technology has played an important role in palliating or even replacing failing organs in order to restore their corrupt function, which is how prostheses were born. The addition of on-board electronic systems to the prostheses will lead to the creation of bionic prostheses, which in addition to their support function, are also easy to handle and will offer patients a better autonomy.

The interest shown by scientists and engineers in neurology, and the evolution of functional brain exploration techniques, have made it possible with the combination of signal processing tools to create brain computer interfaces.

2 Brain Computer Interface (BCI)

A BCI offers an alternative way to communicate with the outside world through an artificial communication system that bypasses natural efficient pathways of peripheral nerves and muscles. A BCI provides its user with an alternative method for acting on the world [1]. The history of this tool began almost a century ago with the creation of

I. Rojas et al. (Eds.): IWBBIO 2019, LNBI 11466, pp. 343–350, 2019.
https://doi.org/10.1007/978-3-030-17935-9_31

the EEG in 1929, but the term brain computer interface was introduced by Jacques Vidal only in 1973 to describe any computerized system that provides detailed information on brain functions [2]. The next step was to implant electrodes directly on the motor cortical areas of monkeys, as well as of humans with severe amyotrophic lateral sclerosis [3]. The first functional device came into being a few years later, allowing the movement of a virtual cursor in real time. Studies have continued, thus creating the possibility for silent people to speak again, or for disabled people to regain the use of a defective limb. Unfortunately, these tools are still rarely available to the general public and their use remains restricted for laboratories and test subjects [4].

2.1   Principle of Operation

During brain function, an emission of electrical signals occurs due to the activation of neurons, the process starts with the user's intent, which generates local current flows resulting from the electrical potential differences, caused by the sum of the post-synaptic graduated potentials of pyramidal cells that create electrical dipoles between the soma (neuron's body) and apical dendrites (neuron's branches) [5], this signal ends into the muscle corresponding to the cortical air that controls it.

These signals are sent to the brain computer interface as input through electrodes placed on the patient's scalp (in the non-invasive case), or directly on to the cerebral cortex through surgery (in the invasive case) [6]. The possibility to combine these two methods according to the purpose desired by the operator should be noted.

After the acquisition, conditioning and digitization of these signals, an undulated tracing corresponding to the brain's activity is obtained. The next step is signal processing, which is divided into two parts, the first part is the extraction of characteristics such as spatial filtering, voltage amplitude measurements, or spectral analyses. This analysis extracts the signal features of brain's activity [1]. The second part concerns the development of the classification algorithm, which will make it possible to recognize a typical brain activity, according to the selected characteristics when it appears on the plot, and thus through a program it is possible to generate one or more commands, according to the set goal, and the number of selected characteristics.

From this process can be created as much programmable functions on a BCI as cerebral activities, producing a unique signal corresponding to a specific controllable brain task. As a result, commands can be assigned in the program that will be used to start motors, or make virtual shifts of an object on different axes.

It is necessary to send an informative feedback to the user in order to notify them about the action that is taking place. This feedback is mostly visual, and is represented by the mechanical action performed or a warning light, but it can also be presented under auditory or haptic information.

2.2   Types of BCI

Two types of BCIs can be distinguished:

Asynchronous
During an asynchronous BCI, when an individual decides to interact with the system, they voluntarily modify their brain activity. The BCI detects this modification in the EEG signals and transforms it into commands [7].

Usually, in asynchronous BCIs, the control signals are continuous, which means that they allow a progressive control of the elements present in the interface.

Synchronous

In a synchronous BCI, it is not the spontaneous activity of the brain that is being exploited, but its response to a stimulus. This response is detected in the signals and is transformed into a control system. Due to this innate brain response characteristic, the use of this type of BCI generally requires very limited learning [8].

2.3   Exploited Principles

Brain Switch

In the case of asynchronous BCIs, the user can interact with the interface by generating the appropriate mental task at any time. Generally, BCIs of this type offer the possibility of sending only one binary command (On/Off command), via a single mental task, that why it is called brain switch [9]. The goal is therefore to obtain a simple but powerful control. This type of BCI exploits the increase in power in the Alpha, Beta frequency band following an imagined movement [10], usually the right foot.

Motor Imagery

Motor imagery is a voluntary brain modulation that consists in the mental representation of an action without concrete production of movement. It is a specific activity generation principle widely used in BCIs. The user must generate the specific activity without external stimulation.

The imagination of a limb's movement is used to activate the cortical zone dedicated to it. It is also possible to teach the user to control the activation of a particular area or frequency band, referred to as voluntary brain modulation [11].

2.4   Notions on EEG Rhythms in a Normal Subject

Electroencephalographic rhythms are characterized by their frequency and amplitude, as well as by circumstances of their appearance and disappearance.

The focus will be put only on three rhythms represented by a Greek letter, which are respectively Alpha, Mu and Beta [12].

Alpha

Alpha rhythm has a frequency between 9 and 11 Hz for most adults. This rhythm is found most typically over the posterior portions of the head during wakefulness, but it may also be present in the central or temporal regions. Alpha rhythm is seen best when the patient is resting with the eyes closed [13], without falling asleep. This rhythm is labile, it disappears when the eyes are opened, or during an attention effort or even an emotional reaction.

Mu

This rhythm is characterized by the frequency band 8–12 Hz [1]. It appears during the voluntary movement, or during the imaginary execution of a movement [13] called "Motor Imagery".

Beta
Any frequency activity greater than 13 Hz is considered as Beta rhythm [13]. This
activity appears when the eyes are opened, in the awakened adult, and can be blocked
when a voluntary movement is performed.

3 Experimental Part

The goal of this work was to start an engine, following the exploitation of the "Mu"
wave that occurred during the creation of the motor imagery of biceps contraction,
which was exploited with the brain switch principle, we decided to work on biceps
motor imaging because it is with it that we obtained the best results with the subject.

During the experience, a 23 years old volunteer, in full possession of his mental
abilities was called on. He was a dental surgery student, who was equipped with an
electrode helmet connected to an EEG.

Calibrations on the subject were carried out in order to collect certain information
on its EEG trace such as: the state of relaxation, the state of stimulation (verbal or
luminous), and the state of concentration.

The measurement of "Mu" waves was recorded during the physical movement of
the biceps as shown in Fig. 1.

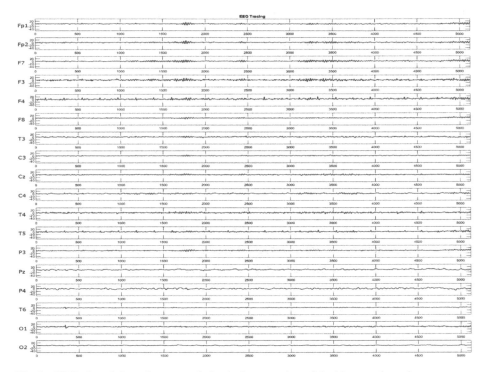

Fig. 1. EEG plot of the voluntary and physical contraction of the biceps, where the appearance
of a "Mu" wave can be noticed.

The brain switch is commonly associated with the imagined movement of the right foot, but we have decided to work on the biceps motor imagery.

A slight delay was observed between the moment of the execution of the movement and its appearance on the EEG tracing, but the waveform can be clearly distinguished when the movement is executed.

The next step was to implement the action of motor imaging, in order to have an EEG trace confirming this brain activity. For this reason, the subject was asked to think about contracting his biceps but without physically executing it. The first few minutes were inconclusive, and there was no manifestation on the EEG plot that could be assimilated to motor imaging, because this exercise was totally unknown to the subject, and required accommodation, so the plot consisted of a series of similar signals with the same shape. However, after a few unsuccessful attempts, an irregular shape was obtained following the request to repeat an umpteenth time the exercise of the imagination of the contraction of his biceps as shown in Fig. 2.

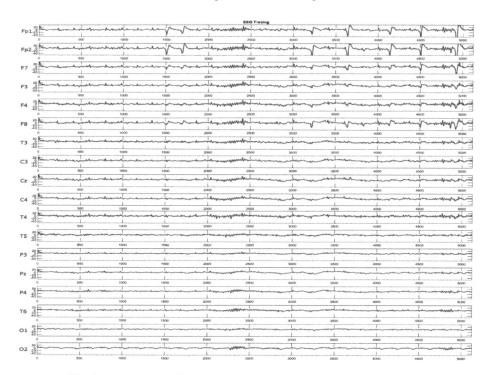

Fig. 2. Appearance of new brain activity during motor imaging exercise

After extracting this trace, and injecting it into signal processing software, in this case MATLAB, the signal of each electrode was analyzed on the frequency domain to confirm that the observed waveform was indeed the "Mu" wave. The next step was to determine which electrode offered the best result. The criterion sought here was the neural synchronization when performing the mental task, i.e. the dominance of the "Mu" frequencies over other brain waves.

The developed procedure was simple, a program was created to analyze the trace of a single electrode second per second, applying a Fourier transform to each sample. The resulting frequency was plotted on a curve and displayed. This process was applied to each of the electrodes of the recording.

From this analysis came the electrode "F7" whose trace is shown in Fig. 3, involving three peaks at very characteristic frequencies of 10 and 11 Hz, as shown in Figs. 4, 5 and 6, so it was opted to the implementation of the BCI.

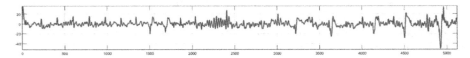

Fig. 3. EEG trace of the "F7" electrode

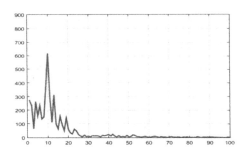

Fig. 4. "Mu" peak frequency at 10 Hz between the 8th and 9th second

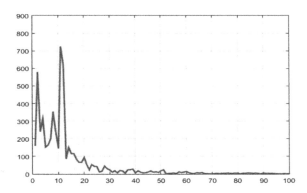

Fig. 5. "Mu" peak frequency at 11 Hz between the 9th and 10th second

The used program was based on the "for" loop, which allows to integrate a variable that recovers a second of the trace (applied on 256 dots) and to apply a Fourier transform to it. This function being embedded in the "for" loop was executed again

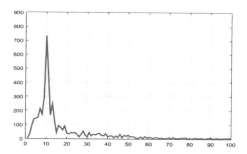

Fig. 6. "Mu" peak frequency at 10 Hz between the 18th and 19th seconds

until the complete analysis of the trace. On each frequency analysis, i.e. each one-second sample, a comparison test was performed between the frequency obtained by the Fourier transform and the "Mu" values. Two possibilities were possible:

- The condition was realized, this implied the identification of the motor imagery. In other words, the mental contraction of the biceps of the subject was concrete.
- The condition was not realized, this implies the non-identification of the motor imagery.

When the condition was met, a high logic level was generated on an "Arduino" card previously connected by USB to the computer. This logic level was then found on a digital output of the card, amplified by a transistor and transmitted to a motor.

4 Results

During the execution of this program on the "F7" electrode, the following results were obtained:

Throughout the plot, the subject was exercising himself with motor imaging by imagining the contraction of his biceps, the "Mu" activity did not appear clearly on the EEG plot, but was hardly manifested during spectral analysis with small amplitudes and only over a very short period of time, scattered over different points on the graph, witnessing the neural desynchronization which is called "Event Related Desynchronization" (ERD).

However, between the 8th and 9th second, i.e. between the [2049–2304] dots (1 s = 256 dots) in Fig. 3, a change started appearing on the plot, and a peak at 10 Hz was detected by the program, indicating neural synchronization which is called "Event Related Synchronization" (ERS), the motor started running briefly, followed by a second detection between the 9th and 10th second (2305–2560 dots), where the motor continued its activity.

On the continuity of the EEG trace no detection was made until the 18th second, when the motor was started one last time before the program stopped, because this EEG trace contained only 20 s.

5 Conclusion

Thanks to this program using a simple "for" loop and the Fourier transform, we were able to create a binary BCI, using motor imagery on the brain switch principle.

This BCI was simple to implement, requiring a minimum of programming time. It allowed to have a first modest look of this technology, which until recently, was part of fiction science.

References

1. Wolpaw, J.R., Birbaumerc, N., McFarlanda, D.J., Pfurtschellere, G., Vaughana, T.M.: Brain-computer interfaces for communication and control. Clin. Neurophysiol. **113**, 767–791 (2002)
2. Vidal, J.J.: Towards direct brain–computer communication. Annu. Rev. Biophys. Bioeng. **2**, 157–180 (1973)
3. Kennedy, P.R., Bakay, R.A.: Restoration of neural output from a paralyzed patient by a direct brain connection. NeuroReport **9**, 1707–1711 (1998)
4. François CA.: Interface Cerveau-Machine principe, fonctionnement application à la palliation du handicap moteur sévère. In: Séminaire LAGIS/LIFL 11 décembre 2012, Equipe Signal & Image Laboratoire d'Automatique, Génie Informatique & Signal. Université de Lille1 (2012)
5. Teplan, M.: Fundamentals of EEG measurement. Meas. Sci. Rev. **2**(2), 1 (2002)
6. He, B.: Neural Engineering, 1st edn. Kluwer Academic Publishers, New York (2005)
7. Mason, S.G., Birch, G.E.: A brain-controlled switch for asynchronous control applications. IEEE Trans. Biomed. Eng. **47**(10), 1297–1307 (2000)
8. Graimann, B., Allison, B.Z., Pfurtscheller, G.: Brain-Computer Interfaces Revolutionizing Human-Computer Interaction. Springer, Heidelberg (2010). https://doi.org/10.1007/978-3-642-02091-9
9. Barachant, A.: Commande robuste d'un effecteur par une interface cerveau machine EEG asychrone. Traitement du signal et de l'image, Université Grenoble Alpes (2012)
10. Pfurtscheller, G., Solis-Escalante, T.: Could the beta rebound in the EEG be suitable to realize a "brain switch"? Clin. Neurophysiol. **120**(1), 24–29 (2009)
11. Pfurtscheller, G., Neuper, C.: Motor imagery and direct brain-computer communication. Proc. IEEE **89**(7), 1123–1134 (2002)
12. Morin, G.: Physiologie du système nerveux central, 6th edn. Masson, Paris (1996)
13. Aminoff, J.: Electrodiagnosis in Clinical Neurology, 5th edn. Elsevier, Atlanta (2005)

A Biomechanical Model Implementation for Upper-Limbs Rehabilitation Monitoring Using IMUs

Sara García de Villa, Ana Jimenéz Martín,
and Juan Jesús García Domínguez[✉]

Department of Electronics, Polytechnics School, University of Alcalá,
Barcelona-Madrid Street, 28805 Alcalá de Henares, Spain
{sara.garciavilla,ana.jimenez,jjesus.garcia}@uah.es

Abstract. Rehabilitation is of great importance in helping patients to recover their autonomy after a stroke. It requires an assessment of the patient's condition based on their movements. Inertial Measurement Units (IMUs) can be used to provide a quantitative measure of human movement for evaluation. In this work, three systems for articular angles determination are proposed, two of them based on IMUs and the last one on a vision system. We have evaluated the accuracy and performance of the proposals by analyzing the human arm movements. Finally, drift correction is assessed in long-term trials. Results show errors of 3.43% in the vision system and 1.7% for the IMU-based methods.

Keywords: Inertial sensor · IMU · Depth sensor ·
Extended Kalman filter · Bio-mechanical model · Rehabilitation ·
Human angle measurement

1 Introduction

Stroke is one of the most common causes of disability. In Europe, prevalence rate ranges from 5% in people aged less than 75 years old to more than 10% in those over 80 [1]. Most of stroke patients lose arm or hand movement skills due to muscle weakness, changes in muscle tone, laxity of joints and other asymmetries associated with the illness [2]. This implies disabilities when reaching, grasping or holding objects, affecting functional independence in the development of activities and, as a result, the quality of life is reduced.

The recovery of upper limb function is of great importance in helping them to increase their autonomy [3]. To regain and maintain body movements, it is important an appropriated physical supervised rehabilitation once the patient has been discharged [4]. Assessment of recovery and of physical exercises performance is a critical aspect of rehabilitation programs [3,4]. Since it is becoming difficult to meet demands for physical therapy, there is a need for home-based intelligent rehabilitation assistants, which should employ monitoring systems [4]. Therefore, the development of motion tracking systems for the quantitative measurement of upper limb movement is necessary.

© Springer Nature Switzerland AG 2019
I. Rojas et al. (Eds.): IWBBIO 2019, LNBI 11466, pp. 351–362, 2019.
https://doi.org/10.1007/978-3-030-17935-9_32

A motion monitoring system for health care applications must be acceptable to patients and clinicians. The system should be transportable, easy to set up, low-priced and have the minimal impact on patients' movements [3]. Existing motion tracking systems are visual or non-visual. Visual systems meet the requirements for upper limb tracking, but they normally are complex, high-priced and require careful setup. The simplest and most economical visual systems have worse characteristics that limit their use. On the other hand, non-visual technologies available are based on inertial, mechanical, acoustic and magnetic sensing strategies. However, mechanical sensing systems are uncomfortable to wear for long periods and acoustic and magnetic sensing are affected by external conditions. As an alternative, Inertial Measurement Units (IMUs) do not present these problems. In addition, IMUs are inexpensive, allow measurements outside the laboratory and have high measurement volume [5]. Therefore, these sensors have become very popular for human movements monitoring.

The use of inertial sensors to estimate trajectories and orientations began in the 70s in space navigation studies [6]. The evolution of microelectromechanical systems (MEMS) has allowed the adaptation of these sensors for measurement applications on human subjects.

The first study addressing the problem of joint kinematic estimation using inertial sensors dates back to 1990 [7]. Since then, IMUs are used to analyze human movements in many published works in the literature [8]. Some of these systems use only accelerometers [9], based on force measurements to estimate the orientation of joints from the gravity vector. These gravity-based measurements are useful for health applications such as postural control. In applications that only use accelerometers it is often assumed that the gravity vector is much greater than that caused by body movement. However, when systems are in motion, it is not valid to ignore the acceleration caused by movement. Also, if accelerometers are used to estimate positions, it is necessary to minimize the drift that occurs by the double integration of linear acceleration [3].

The use of accelerometers for measuring joint angles during movement is very common [8], although the error can be up to $40°$. Other examples shows procedures that estimates joints positions by fusing various accelerometers [3]. However, they use zero velocity update technique for drift compensation. That method assumes that there are intervals of movement with zero angular velocity and it does not always happen.

Combinations of accelerometers and gyroscopes have shown to be reliable enough for joint functionality, tracking and rehabilitation evolution assessing applications [10,11]. These sensors have been fused to measure body parameters such as neck flex-extension angle [12], knee flex-extension angle [13], lower limb exercises assessment [14] and articular arm parameters evaluation [2,3]. Most of them make methods based on the Kalman filter (KF) and derivatives as extended Kalman filter (EKF) to carry out this fusion. However, these algorithms reduce but does not remove drift in their results.

The implementation of biomechanical restrictions on the human body reduces drift and corrects IMU-joint alignment. An arm biomechanical model was used in [15] to estimate elbow and forearm angles. Movements are restricted to the elbow and forearm degrees of freedom (DOF), resulting in root mean square errors (RMSEs) from 2.7° to 3.8° using Euler decomposition. In [16] a leg model with kinematic constraints is proposed. This model has been used for the evaluation of lower extremity exercises [14] getting results with RMSE of 4.3° to 6.5°. However, this model has not been used for upper limbs movements assessment.

Therefore, the primary aim of this study is to develop an elbow and forearm angles estimation system based on IMUs which reduce drift problem. To do this, a biomechanical model with kinematic constraints is implemented. Results from a simple experiment are used for the model validation and to compare the different techniques performance. Thus, we introduce a procedure for elbow and forearm angles estimation that shows limited drift in long-term experiments.

This paper is organized as follows: Sect. 2 describes the proposed algorithms, Sect. 3 details the experimental setups, Sect. 4 shows the current results and their discussion; and finally concluding remarks are presented in Sect. 5.

2 Proposed Systems

We propose three systems to analyze upper-limbs movement. The former is based on a biomechanical model which measures joint angles trough IMUs using an EKF, explained in Sect. 2.1. The second proposal is a modification that adds kinematic constraints which is described in Sect. 2.2. The last one is based on a RGB camera and depth sensor that captures joint motions, see Sect. 2.3.

2.1 Biomechanical Model: *EKF*

The biomechanical model implemented is based on the leg model developed by Lin [16]. The model design is presented in Fig. 1. As shown, a *frame* is associated to each DOF of every joint. In this way, shoulder joint is divided into three *frames* (0-2); and elbow, forearm and wrist joints have only one *frame* apiece (3, 4 and

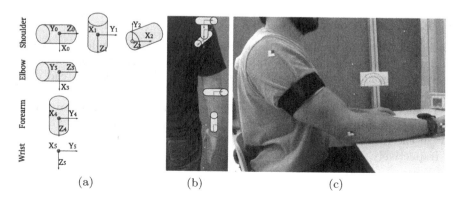

(a)   (b)   (c)

Fig. 1. (a) shows reference coordinates of each i *frame* with the subject in the standing position facing into the page. (b) presents DOF frames over human arm. (c) shows arm instrumented with IMUs and visual labels.

5, respectively). In this model each ith *frame* is related to its previous *frame* $i - 1$ recursively by means of the rotation matrix $R_{i-1,i}$ defined in (1).

$$R_{i-1,i} = \begin{pmatrix} \cos\theta_i & -\cos\alpha_i\sin\theta_i & \sin\alpha_i\sin\theta_i \\ \sin\theta_i & \cos\alpha_i\cos\theta_i & -\sin\alpha_i\cos\theta_i \\ 0 & \sin\alpha_i & \cos\alpha_i \end{pmatrix} \tag{1}$$

According to Denavit-Hartenberg convention [17], θ_i is the rotation angle around the z_{i-1} axis between the x_{i-1} and the x_i axis. α_i is the rotation angle about the x_i axis from z_{i-1} to the z_i axis. Furthermore, $R_{0,i}$ can be calculated as follows:

$$R_{0,i} = R_{0,1}R_{1,2}...R_{i-1,i} \tag{2}$$

The implemented EKF calculates joint angles q_i, angular velocities $\dot{q}_i$ and angular accelerations $\ddot{q}_i$ from each joint using IMUs measurements. Relations between angular velocity $\boldsymbol{\omega}_i$ and linear acceleration $\ddot{\mathbf{x}}_i$ obtained from the IMUs and EKF estimated state vector $[q_i^-;\dot{q}_i^-;\ddot{q}_i^-]$ are expressed in (3) and (4):

$$\boldsymbol{\omega}_i = R_{i-1,i}^T\boldsymbol{\omega}_{i-1} + R_{i-1,i}^T\dot{\mathbf{q}}_i^- \tag{3}$$

$$\ddot{\mathbf{x}}_i = R_{i-1,i}^T\ddot{\mathbf{x}}_{i-1} + \boldsymbol{\alpha}_i \times \mathbf{r}_i + \boldsymbol{\omega}_i \times (\boldsymbol{\omega}_i \times \mathbf{r}_i) + R_{0,i}^T\mathbf{g}, \tag{4}$$

where angular velocity $\boldsymbol{\omega}_i$ depends on the total angular velocity of the preceding *frame* $\boldsymbol{\omega}_{i-1}$ and the joint velocity, vectorized as $\dot{\mathbf{q}}_i^- = [0;0;\dot{q}_i^-]$. A gravity term $\mathbf{g} = [9.8;0;0]$ is also considered in order to model its influence in accelerometer outputs $\ddot{\mathbf{x}}_i$. Similarly, linear acceleration of each i *frame* $\ddot{\mathbf{x}}_i$ is related to the linear acceleration of the previous *frame* $\ddot{\mathbf{x}}_{i-1}$. Additionally, it depends on the angular velocity $\boldsymbol{\omega}_i$ and angular acceleration $\boldsymbol{\alpha}_i$ of the current *frame* i, and the displacement vector $\mathbf{r}_i$ from the origin of the current *frame* i to the origin of the previous *frame* $i - 1$. We differentiate the angular velocity $\boldsymbol{\omega}_i$ from (3) to obtain the angular acceleration $\boldsymbol{\alpha}_i$:

$$\boldsymbol{\alpha}_i = R_{i-1,i}^T\boldsymbol{\alpha}_{i-1} + R_{i-1,i}^T\ddot{\mathbf{q}}_i^- + \boldsymbol{\omega}_i \times R_{i-1,i}^T\dot{\mathbf{q}}_i^- \tag{5}$$

Thus, the flowchart followed for each angle q_i estimation is depicted in Fig. 2.

2.2 Kinematic Constraints: *EKF+KC*

Drift from IMUs measurements can be reduced through an angular acceleration modulation based on anatomic joints ranges of movements (ROMs) of healthy people. This graded parameter is used as an input in the EKF instead of the estimated angular acceleration $\ddot{q}_{k+1}^-$.

The modulation is performed according to a robot controller algorithm [18] implemented in Lin model [16]. We carry out this by calculating the differences between the estimated angles q_{k+1}^-, and upper and lower ROM limits, $\bar{q}_{c,i}$ and $\underline{q}_{c,i}$, respectively, as described in (6) and (7).

$$\bar{\rho}_i = \bar{q}_{c,i} - q_i^- \tag{6}$$

$$\underline{\rho}_i = q_i^- - \underline{q}_{c,i} \tag{7}$$

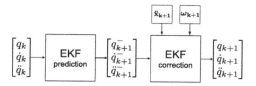

Fig. 2. Flowchart of the EKF. State vector consist of joint angles q_k, angular velocities $\dot{q}_k$ and angular accelerations $\ddot{q}_k$. In the prediction step, an estimated state vector $[q_{k+1}^-; \dot{q}_{k+1}^-; \ddot{q}_{k+1}^-]$ is obtained. This estimation is corrected using IMUs measurements: angular velocity ω_{k+1} and linear acceleration $\ddot{x}_{k+1}$ by means of (3) and (4) to obtain EKF outputs $[q_{k+1}; \dot{q}_{k+1}; \ddot{q}_{k+1}]$.

The ROM limits $\bar{q}_{c,i}$ and $\underline{q}_{c,i}$ are set according to the anatomical parameters found in the literature [19]. The modulation parameter γ used as an EKF input instead of the estimated angular acceleration $\ddot{q}_{k+1}^-$ is calculated trough (8):

$$
\gamma = \begin{cases}
\eta \left(\dfrac{1}{\underline{\rho}_i} - \dfrac{1}{\underline{\rho}_{0i}} \right) \dfrac{1}{\underline{\rho}_i^2} & \text{if } \underline{\rho}_i \leq \underline{\rho}_{0i} \\[2ex]
-\eta \left(\dfrac{1}{\bar{\rho}_i} - \dfrac{1}{\bar{\rho}_{0i}} \right) \dfrac{1}{\bar{\rho}_i^2} & \text{if } \bar{\rho}_i \leq \bar{\rho}_{0i} \\[2ex]
0 & \text{otherwise}
\end{cases}
\tag{8}
$$

where $\bar{\rho}_{0i}$ and $\underline{\rho}_{0i}$ are the minimum differences between measurements and limits for the algorithm to act and η is the constant gain parameter set at value 1. In this case, flowchart followed is shown in Fig. 3.

2.3 RGB and Depht Sensor: *RGB+IR*

We also use an alternative vision system capable of capturing joint movement trough an external vision sensor. Figure 4 illustrates the flowchart followed by the third proposed system. It is based on a RGB camera and a depth sensor that capture user motion in scene and estimates the *elbow-forearm angle* θ_{EF}. Input RGB video is used to track labels located in each user joints. The robust *Lucas-Kanade-Tomasi* (LKT) point label tracker [20] is used to obtain joint motions in the RGB images. The LKT point tracker extracts features about spatial information to search for the position that yields the best match. Depth sensor is calibrated with respect the RGB camera in order to obtain 3D-positions over the RGB image domain. Using the depth information, we compute the 3D-positions of the user joints obtained by the tracker. Specifically, we are interested in obtaining the 3D-position of wrist, elbow and shoulder to compute the *elbow-forearm angle* θ_{EF} as follows:

$$
\theta_{EF} = \arccos \left(\frac{\langle \mathbf{ES}, \mathbf{EW} \rangle}{\|\mathbf{ES}\| \|\mathbf{EW}\|} \right),
\tag{9}
$$

where $\mathbf{ES}$ and $\mathbf{EW}$ are the elbow-shoulder and elbow-wrist vectors, respectively.

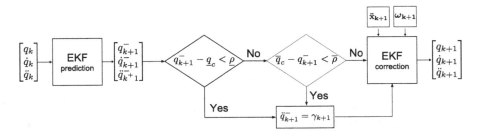

Fig. 3. Flowchart of the EKF taking into account anatomical kinematic constraints. Kinematic constraints impose upper and lower bounds before the correction step of the EKF to reduce IMUs drift.

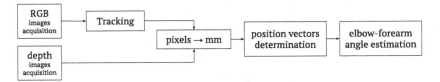

Fig. 4. Stages to determine θ_{EF} from the RGB and depth sensor outputs.

3   Experimental Setup

We carry out three experiments to evaluate the performance of each alternative. We denote the developed system based on IMUs as *EKF*, the kinematic constrained system as *EKF+KC* and the external vision system as *RGB+IR*. In the first experiment, we compare the results of each proposed method versus *ground truth* values (see Sect. 3.2). During the second one, a comparative study between methods based on IMUs and vision system on a simple human arm movement is carried out as shown in Sect. 3.3. Last experiment, shown in Sect. 3.4, evaluates the drift correction of the measurements using *EKF+KC*.

3.1   Implementation Details

EKF and *EKF+KC* systems are implemented by using *MetaMotionR* IMUs from *MbientLab* [21]. We set their sampling rate at 100 Hz and the accelerometer and gyroscope scales are selected at $\pm 8g$ and $\pm 2000°/s$, respectively.

On the other hand, the *RGB+IR* system is carried out by using *Kinect v2* [22], which has a maximum frame rate of 30 fps and a color camera resolution of 1920 × 1080 pixels. Its depth camera has a resolution of 512 × 424 pixels and an operative measure range from 0.5 to 4.5 m .

3.2   Experiment 1: Validation Tests

Experiment 1 consists in making measurements of a hinge structure which is moved in a vertical plane parallel to the camera image plane, as shown in Fig. 5.

Ground truth measurements are obtained with a protractor placed parallel to the cover movement, whose origin is positioned in the hinge structure. Furthermore, the hinge structure is aligned with the origin of RGB camera and perpendicular to camera image plane. The distance between the motion plane and the depth and RGB sensor is set at 0.55 m.

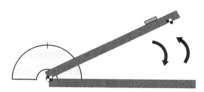

Fig. 5. Diagram of the hinge structure used with the tracker labels (black and white squares) and the IMU (orange) employed for the measurements. Protractor used to obtain reference measurements is drawn behind structure. (Color figure online)

Three repetitions of structure opening and closing between 0° to 127° is carried out. We know a collection of frames when the structure is opened in that range of angles in steps of 5° as well as the exactly frame corresponding to maximum angle. Furthermore, we obtain linear acceleration $\ddot{x}$ and angular velocity ω measurements as well as RGB and depth images during tests executions. We follow the calibration method of [23]. We associate **EW** and **ES** from (9) vectors with hinge-label and horizontal direction vectors, respectively. In this way, hinge angles are obtained from IMUs and external vision sensors.

Synchronization and interpolation are required to compare results achieved by the different techniques employed. We carry out this synchronization by using IMUs ambient light sensor and RGB camera. After signals interpolation, the relative RMSE errors are calculated.

3.3   Experiment 2: Human Arm Movement

We conduct a test on a human arm with the subject seated and his elbow resting on a stable surface. The subject is instrumented with three tracker markers: one on his arm, one on his elbow and one on his forearm, as shown in Fig. 1. In addition, three IMUs are placed on his scapula, arm and forearm.

The subject performs elbow flex-extension with small internal-external forearm rotations. Movements are simple in order to not lose markers with visual system and contrast both techniques. Thus, we assess the performance of *EKF+KC*, *EKF* and *RGB+IR* methods by comparing their results on a human arm.

3.4   Experiment 3: Drift Compensation

Finally, to study drift compensation, we carry out a 160-second test in which the subject performs elbow flex-extension and forearm internal and external rotation movements. The measurements are obtained by IMUs in order to compare *EKF* and *EKF+KC* methods. In this way, the influence of the constraints on long-term tests is evaluated by contrasting outputs of both algorithms.

4   Results and Discussion

4.1   Experiment 1: Validation Tests

Hinge angles determined by different systems are shown in Fig. 6. In this image, reference measurements are drawn in black with star marks. Results from *RGB+IR* method are shown in yellow and *EKF* and *EKF+KC* systems outputs are presented in blue and red, respectively. Furthermore, relative error calculated for each technique is presented in Table 1.

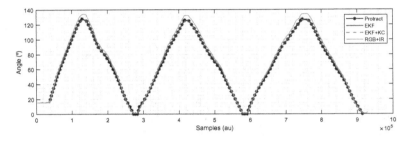

Fig. 6. Angles determined using each employed technique. *Ground truth* is presented in black. Results from *EKF* and *EKF+KC* are shown in blue and dotted red, respectively, as well as *RGB+IR* outputs (yellow line). (Color figure online)

Table 1. Relative RMSE of the evaluated systems versus *ground truth*.

	RGB+IR	EKF	EKF+KC
Relative RMSE (%)	3.43	1.67	1.69

The performance of the different developed systems is evaluated in this experiment. In Fig. 6 it can be seen that we obtain very similar results using all the proposed methods. In addition, these values are very close to the *ground truth*. The results of the *EKF* and *EKF+KC* algorithms are very similar to the *ground truth* and the *RGB+IR* output is the farthest from this. It is especially noticeable at larger aperture angles. Likewise, in Table 1 an error of about 1.7% is associated with systems based on IMUs and 3.43% with *RGB+IR* method. Thus, it is demonstrated that the implemented model works and is consistent with reality. Also, all of them achieve accuracy similar to previous methods in the state of the Art, between 2.7° and 3.8° [15], with a RMSE of 4.4° and 2.1° from *RGB+IR* and IMU-based algorithms, respectively. Therefore, even though an external vision system does not suffer from drift, the accuracy achieved by these systems is lower than using the IMU-based methods.

4.2   Experiment 2: Human Arm Movement

The *elbow-forearm angle* θ_{EF} from the *RGB+IR* method is shown in yellow in Fig. 7. As previously explained, the vision system measures *elbow-forearm angle* θ_{EF} since it does not distinguish between the influence of the two joints involved in this rotation. However, the forearm movement performed is an order of magnitude smaller, so it is assumed that the measurement largely corresponds to the elbow angle. Likewise, the *EKF* and *EKF+KC* results are represented in blue and red, respectively. The elbow and forearm joint angle distinction are executed according to the biomechanical model implemented. These angles are presented in Fig. 7a and b, respectively.

In addition, it is known that the range of flex-extension movement carried out is from 47° to 145° and maximum anatomic amplitude of forearm external rotation is 80°. On the other hand, we obtained elbow rotation angle ranges through the implemented methods from 47.5° to 153.5° with *EKF*, from 47.3° to 146.4° with *EKF+KC* and from 45.0° to 144.7° with *RGB+IR*. With respect to the external rotation angle of the forearm, the maximum angle obtained with *EKF* is 81.6° and with *EKF+KC* 79.3°.

The results on the human arm (Fig. 7) are similar with all developed methods. Although the vision system measures the *elbow-forearm angle* as a whole and not the components of each joint, the result is very close to the elbow joint angle. Therefore, the *RGB+IR* system can be used for movements largely dominated by

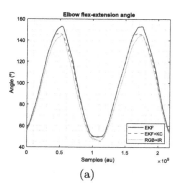
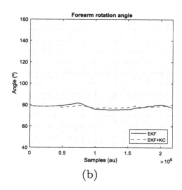

(a)   (b)

Fig. 7. Elbow and forearm rotation angles measure with proposed systems. (a) represents elbow angles in a trajectory computed with *EKF* system (continuous blue line) and adding the kinematic constraints *EKF+KC* (dotted red line). *RGB+IR* output angles are also shown (continuous yellow line). (b) shows forearm rotation angle from both systems based on IMUs, *EKF* and *EKF+KC*. (Color figure online)

one functional DOF. Despite the suitable results obtained by the visual system, it is constrained to planar movements and slow motions. Besides, camera based method needs a previous setup and calibration.

It is also known that the range of flex-extension movement performed is similar to the obtained ranges through the methods implemented. In this way, it is confirmed that the calculated angles by the three systems are similar to the expected values. Although it is noticeable that IMU-based methods require relative measurements from the devices to the rotation or axis center and it demands calibration movements, this systems provides the best outputs. Even so, since the angles obtained by the *EKF+KC* method are the closest to the angles rotated by the subject, the model improvement is proven by the implementation of kinematic constraints.

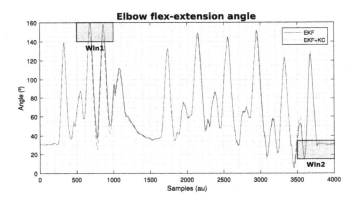

Fig. 8. Elbow joint angle and forearm rotation angle obtained by proposed systems based on IMUs. Elbow angles are shown in a trajectory computed with *EKF* system (continuous blue line) and adding the kinematic constraints *EKF+KC*. (Color figure online)

4.3   Experiment 3: Drift Compensation

Finally, Fig. 8 shows the output angles from *EKF* and *EKF+KC* algorithms from 120 to 160 s of recording. *EKF* results are shown in blue and *EKF+KC* outputs in red. The kinematic constraints performance on more complex motions and longer tests can be seen in the highlighted windows (Win1 and Win2). In Win1 two corrected maximums are highlighted with the maximum anatomical range and in Win2 the correction of drift in the final measurements is observed.

In Win1 we can see that the *EKF+KC* performance is better than the measurements of the *EKF* method since we know that the maximum angle is 145° and this is the maximum value achieved by the former algorithm. Likewise, the effect of the restrictions is observed in Win2 where the measurements are limited to the anatomical minimum angle. Therefore, the improvement achieved through the implementation of these constraints is demonstrated.

5   Conclusions

We propose three systems to determine joint angles, two of them based on IMUs and the last one on a vision system. Outputs from IMU methods have an error of 1.7% whereas vision system entails a higher error of 3.43%. Furthermore, it is confirmed that the limitation of the estimated angles to the anatomical ROMs, we reduce the drift associated to inertial sensors and we improve estimations in long-term experiments. In this way, the results obtained with IMU-based algorithms are very promising. Based on this, as future work it is proposed to assess more complex movements and longer-lasting tests. Also, other sensor fusion techniques are aimed to be evaluated for implementation.

Acknowledgments. This work was supported in part by Junta de Comunidades de Castilla La Mancha (FrailCheck project SBPLY/17/180501/000392) and the Spanish Ministry of Economy and Competitiveness (TARSIUS project, TIN2015-71564-c4-1-R).

References

1. Béjot, Y., Bailly, H., Durier, J., Giroud, M.: Quarterly medical review epidemiology of stroke in Europe and trends for the 21st century. La Presse Medicale **45**(12), e391–e398 (2016)
2. Repnick, E., Puh, U., Goljar, N., Munih, M., Mihelj, M.: Using inertial measurement units and electromyography to qantify movement during action research arm test execution. Sensors **18**(2767), 1–23 (2018)
3. Bai, L., Pepper, M.G., Yan, Y., Spurgeon, S.K.: Quantitative assessment of upper limb motion in neurorehabilitation utilizing inertial sensors. IEEE Trans. Neural Syst. Rehabil. Eng. **23**(2), 232–243 (2015)
4. Huang, B., Giggins, O., Kechadi, T., Caulfield, B.: The limb movement analysis of rehabilitation exercises using wearable inertial sensors, pp. 4686–4689 (2016)
5. de Vries, W., Veeger, H., Cutti, A., Baten, C., van der Helm, F.: Functionally interpretable local coordinate systems for the upper extremity using inertial & magnetic measurement systems. J. Biomech. **43**, 1983–1988 (2010)
6. Bötzel, K., Olivares, A., Cunha, J.P., Górriz Sáez, J.M., Weiss, R., Plate, A.: Quantification of gait parameters with inertial sensors and inverse kinematics. J. Biomech. **72**, 207–214 (2018)
7. Picerno, P.: 25 years of lower limb joint kinematics by using inertial and magnetic sensors : a review of methodological approaches. Gait Posture **51**, 239–246 (2017)
8. Olivares, A., Górriz, J.M., Ramírez, J., Olivares, G.: Using frequency analysis to improve the precision of human body posture algorithms based on Kalman filters. Comput. Biol. Med. **72**, 229–238 (2016)
9. Bennett, T., Jafari, R., Gans, N.: An extended Kalman filter to estimate human gait parameters and walking distance. In: 2013 American Control Conference, pp. 752–757 (2013)
10. Zihajehzadeh, S., Park, E.J.: A novel biomechanical model-aided IMU/UWB fusion for magnetometer-free lower body motion capture. IEEE Trans. Syst. Man Cybern.: Syst. **47**, 1–12 (2016)

11. Leardini, A., Lullini, G., Giannini, S., Berti, L., Ortolani, M., Caravaggi, P.: Validation of the angular measurements of a new inertial-measurement-unit based rehabilitation system: comparison with state-of-the-art gait analysis. J. Neuroeng. Rehabil. **11** (2014). https://doi.org/10.1186/1743-0003-11-136. Article No. 136

12. Morrow, M.M.B., Lowndes, B., Fortune, E., Kaufman, K.R., Hallbeck, M.S.: Validation of inertial measurement units for upper body kinematics. J. Appl. Biomech. **33**(3), 227–232 (2017). https://doi.org/10.1123/jab.2016-0120

13. Seel, T., Schauer, T.: Joint axis and position estimation from inertial measurement data by exploiting kinematic constraints. In: 2012 IEEE International Conference on Control Applications (CCA), pp. 0–4 (2012)

14. Kianifar, R., Lee, A., Raina, S., Kulic, D.: Automated assessment of dynamic knee valgus and risk of knee injury during the single leg squat. IEEE J. Transl. Eng. Health Med. **5**(June), 1–13 (2017)

15. Müller, P., Bégin, M.-A.S.T., Seel, T.: Alignment-free, self-calibrating elbow angles measurement using inertial sensors. IEEE J. Biomed. Health Inform. **21**(2), 312–319 (2017)

16. Lin, J.F., Kulić, D.: Human pose recovery using wireless inertial measurement units. Physiol. Meas. **33**(12), 2099–2115 (2012)

17. Uicker, J., Denavit, J., Hartenberg, R.: An iterative method for the displacement analysis of spatial mechanisms. J. Appl. Mech. **31**(2), 309–314 (1964)

18. Khatib, O.: Real-time obstacle avoidance for manipulators and mobile robots. In: Cox, I.J., Wilfong, G.T. (eds.) Autonomous Robot Vehicles, pp. 396–404. Springer, Heidelberg (1986). https://doi.org/10.1007/978-1-4613-8997-2_29

19. Norkin, C.C., White, D.J.: Measurement of Joint Motion: A Guide to Goniometry. FA Davis, Philadelphia (2016)

20. Jung, B., Sukhatme, G.S.: Detecting moving objects using a single camera on a mobile robot in an outdoor environment. In: International Conference on Intelligent Autonomous Systems, pp. 980–987 (2004)

21. MbientLab: Wearable Bluetooth 9-axis IMUs & environmental Sensors. https://mbientlab.com/

22. Kinect para Windows v2. https://support.xbox.com/

23. Cutti, A.G., Giovanardi, A., Rocchi, L., Davalli, A., Sacchetti, R.: Ambulatory measurement of shoulder and elbow kinematics through inertial and magnetic sensors. Med. Biol. Eng. Comput. **46**(2), 169–178 (2008)

Non-generalized Analysis of the Multimodal Signals for Emotion Recognition: Preliminary Results

Edwin Londoño-Delgado[1], Miguel Alberto Becerra[1(✉)],
Carolina M. Duque-Mejía[1], Juan Camilo Zapata[1], Cristian Mejía-Arboleda[2],
Andrés Eduardo Castro-Ospina[2], and Diego Hernán Peluffo-Ordóñez[3]

[1] Institución Universitaria Pascual Bravo, Medellín, Colombia
migb2b@gmail.com
[2] Instituto Tecnológico Metropolitano, Medellín, Colombia
[3] SDAS Research Group, Yachay Tech, Urcuquí, Ecuador
http://www.sdas-group.com

Abstract. Emotions are mental states associated with some stimuli, and they have a relevant impact on the people living and are correlated with their physical and mental health. Different studies have been carried out focused on emotion identification considering that there is a universal fingerprint of the emotions. However, this is an open field yet, and some authors had refused such proposal which is contrasted with many results which can be considered as no conclusive despite some of them have achieved high results of performances for identifying some emotions. In this work an analysis of identification of emotions per individual based on physiological signals using the known MAHNOB-HCI-TAGGING database is carried out, considering that there is not a universal fingerprint based on the results achieved by a previous meta-analytic investigation of emotion categories. The methodology applied is depicted as follows: first the signals were filtered and normalized and decomposed in five bands (δ, θ, α, β, γ), then a features extraction stage was carried out using multiple statistical measures calculated of results achieved after applied discrete wavelet transform, Cepstral coefficients, among others. A feature space dimensional reduction was applied using the selection algorithm relief F. Finally, the classification was carried out using support vector machine, and k-nearest neighbors and its performance analysis was measured using 10 folds cross-validation achieving high performance uppon to 99%.

Keywords: Emotion recognition · Physiological signals · Signal processing

1 Introduction

Emotions are states that are followed from contact with certain types of stimuli, and which constitutively involve evaluative mental operations [3], and

© Springer Nature Switzerland AG 2019
I. Rojas et al. (Eds.): IWBBIO 2019, LNBI 11466, pp. 363–373, 2019.
https://doi.org/10.1007/978-3-030-17935-9_33

are also associated with some stimuli from external environments, or through personal introspection [18]. They are processes that causes inconsistent perceptions of an object or situation, and are often associated with humor, temperament, personality and disposition. Identifying the emotional states have great interest for psychology and psychiatry and the execution of an adequate diagnosis influences an effective treatment. To determine the emotional state of an individual is of vital importance to find psychological conditions, so several studies have already been done in this field in order to improve the accuracy in the detection of emotions.

In [4] the emotions are classified into 3 categories *(i) motivation:* thirst, hunger, pain, humor *(ii) basic:* Happiness, surprise, disgust, sadness, fear, and anger and *(iii) social:* grief, guilt, pride, and charm. Emotions take an important role in human communication, these can be expressed verbally or through non-verbal signals (facial expressions and gestures [16], intonation of the voice, and body pose) which are relatively easy to collect by means of cameras and microphones. However, human beings can easily suppress these expressions to hide true emotions in certain situations; therefore, to adequately identify emotions is a complex task due to multiple individual variables such as interpersonal, group and cultural relationships. Therefore, it is necessary to use techniques that allow an effective identification of emotion [15,17]. Therefore, some of them are based on physiological measures, which are more difficult to interpret by human beings, for example: sweating due to fear or temperature, heartbeat rate, electroencephalographic signals [2], among others. These modalities offers a great potential for the recognition of emotions [1] and have been widely used for the identification of emotions, and these must have certain conditions for their proper implementation, one of which is the quality, which is a factor that will determine to a large extent the results obtained in the experiments, since many of them are susceptible to acquire signals alien to the original, as happens with the EEG signals, which are very safe signals, but difficult to acquire and susceptible to being contaminated by artifacts [19], such as involuntary muscle movement when performing the acquisition of the signal, the noise produced by the electrical network, the noise produced by the electronic devices and the mismatch or bad contact of the electrodes.

In the literature multiples studies of emotion identification have been reported with variable results that depend on the number of classes and the database used. For example, in [13] 2012 obtained an accuracy of 68.50% in the detection of emotions which was improved in [15] 2014 with an accuracy of 81.45% and in [8] 2016 obtained a precision between 78.8% for neutrality and 100% for anger, surprise and anguish applying the Quadratic Discriminant Analysis (QDA) as a classifier and Human-Computer Interface (HCI) tagging as a benchmark, which demonstrates the advances in this area. However, a method that achieves high reliability to be used as a diagnostic tool has not yet been found. Other approaches have been focused on recognition of the arousal, and valence. There are several studies that focus their efforts on the identification of these factors, in one of the more recent [6] a study was carried out in which were tested 7 algorithms in different public databases in order to identify which had a better accuracy, it was found that none gave the best performance in all data sets, but in two of them the database Ascertain obtained good results with the Support Vector Machine (SVM) algorithm.

Considering that there is not an universal fingerprint based on the results achieved by the meta-analytic investigation of emotion categories presented in [12]. We carried out this study per subject. The MAHNOB-HCI-TAGGING database was applied for this study which was collected from 24 subjects. The methodology of our study is depicted as follows. First, the signals were filtered, and normalized. Then feature extraction was applied for segmented and unsegmented signals. This stage was carried using statistical measure, Cepstral coefficients, and discrete wavelet transform. Besides, the feature selection was performed using Relief F algorithm. Finally, SVM and k-Nearest Neighbors (k-NN) algorithms were applied for classifying 9 different classes of emotions.

2   Experimental Setup

The proposed procedure was applied in four stages, in the first one a preprocessing was carried out for eliminating noise from signals and normalizing them besides, the EEG signals were decomposed in 5 bands. Then, in the second stage, feature extraction allowed obtaining a set of attributes using Discrete Wavelet Transform, Discrete Fourier Transform, Cepstral coefficients and statistical measures of them. In stage third, relevant features were selected using Relief F algorithm. Finally, in stage four, a classification process was carried out using the SVM algorithm using three different kernels (Linear-L, Cubic -C, and Quadratic-Q) and the k-NN algorithm.

2.1   Database

The MAHNOB-HCI-TAGGING database was made with the aim of acquiring knowledge about the natural behavior of 24 healthy adults with ages between 19 and 40 years, interaction with the computer during multimedia observation, designed to elicit powerful reactions. During the experiment, the behavior is recorded using cameras, microphone and gaze tracker. Furthermore, the

Table 1. Label assigned to each emotion.

Label	Emotion name	Label	Emotion name	Label	Emotion name
0	Neutral	5	Sadness	10	Unknown
1	Anger	6	Surpise	11	Amusement
2	Disgust	7	Scream	12	Anxiety
3	Fear	8	Bored		
4	Joy, Happiness	9	Sleppy		

physiological signals recorded were electroencephalogram (EEG), electrocardio-gram (ECG), skin temperature, galvanic skin resistance (GSR) and respiration amplitude. EEG electrodes were placed on cap using the 10–20 system; the ampli-tude was acquired with a respiration belt around the chest, ECG was acquired according to the Einthoven triangle. In Table 1 are shown the emotion labels of the database.

2.2   Pre-processing

The EEG signals were preprocessed, by removing their artefact elicited by elec-tromyography (EMG) and electrooculogram (EOG), In this study of EMG and EOG artifacts in the EEG signals were eliminated, and all the signals were delimited within a range of time to eliminate unnecessary information, or infor-mation that did not present visual or auditory stimuli; a FIR filter of grade 34 was implemented for each brain band. The signals were decomposed in 4 bands theta (Θ), alpha(α), beta (β) and Gamma (γ) sub bands were defined between 4 Hz–7.5 Hz, 8 Hz–13 Hz, 13 Hz–30 Hz and 30 Hz–70 Hz, respectively. In addition, the bandwidth used for filtering the ECG signal was between 0.5 Hz to 100 Hz. Finally, the signals corresponding to galvanic skin, temperature of skin and res-piration were not processed due to low variability between samples. The DC level was eliminated in order to reduce the uncertainty in the extraction of features of ECG and EEG signals. All samples were normalized. The ECG and EEG signals were normalized in a range $[-1,1]$ with the purpose of conserving bipolarity of signals. While, the galvanic skin, skin temperature and respiration was normal-ized in a range [0,1]. In order to reduce calculation of the learning machines, a decimation process was carried out. The signals were resampled from 256 Hz to 128 Hz.

2.3   Feature Extraction

The features of EEG signals were obtained from discrete wavelet transform (DWT), using four levels of decomposition and a daubechies 10 (db10) as mother wavelet. For each level, statistical measures as median, mean, standard deviation, Shannon entropy, kurtosis and variance coefficient were calculated.

The Wavelet Transform. The Wavelet Transform (WT) is a technique for analyzing signals. It was developed as an alternative to the short time Fourier Transform (STFT) to overcome problems related to its frequency and time res-olution properties [14]. The Discrete Wavelet Transform (DWT) has become a powerful technique in biomedical signal processing [7]. It can be written as:

$$W_\psi(s,\tau) = \int_{-\infty}^{+\infty} X(t)\psi_{s,\tau}^*(t)dt \tag{1}$$

Where s and τ are called scale and translation parameters, respectively. W_ψ (s,τ), denotes the wavelet transform coefficients and ψ is the fundamental mother wavelet. DWT uses scale and position values based on powers of two. The values of s and τ are: $s = 2^j$, $\tau = k * 2^j$ and $(j,k)Z^2$ as shown in (2).

$$\psi_{s,t}(t) = \frac{1}{\sqrt{2^j}} \psi \left(\frac{t - k * 2^j}{2^j} \right) \tag{2}$$

2.4 Discrete Fourier Transform

Considering a discrete signal $x[n]$ $\forall n = 1, \ldots, N$, which has been derived from a continuous signal $x(t)$ by sampling at equal time intervals, the discrete Fourier transform is defined as:

$$X[k] = \sum_{n=0}^{N-1} x[n]e^{-j2\pi kn/N} \qquad K = 0, \ldots, N-1 \tag{3}$$

2.5 Classification

In this work two classifiers were used, k-nearest neighbors (k-NN) and Support Vector Machines (SVM). For k-NN the Euclidean distance, City block distance, and distance by cosine were employed to calculate the distance between samples and the number of closest neighbors between 1 and 11 with intervals of 2 neighbors, to evaluate the effect of the number of neighbors in the performance of the classifier. The number of neighbors was always odd in order to avoid draws between classes. While, SVM was employed with strategy one-against-one for multi class approach. Also, it were used three types of kernels, Lineal, quadratic and polynomial of order 3, for evaluating the effect of kernel in SVM performance.

Support Vector Machine. Support vector machine (SVM) used the structural risk minimization to maximize the distance between the margins of two classes by constructing an optimal hyperplane. Initially, SVM was introduced for binary and linear classification [11]. But SVM can be extended to multi class classification using two strategies, One-Against-One (OAO) and One-Against-All (OAA). Also, SVM can be extended to no-linear mode using kernel functions. We used OAA because this strategy is substantially faster to train and seems preferable for problems with very large number of labels [9].

Let $D = \{(x_1, y_1), (x_2, y_2), \ldots, (x_N, y_N)\}$ be a training dataset, where the i-th sample $x_i \in R^n$, such as $y_i \in \{-1, 1\}$. The dual optimization which is the key to extent SVM for no-lineal function, is as follows:

$$min \quad \frac{1}{2} \sum_{i=1}^{N} \sum_{j=1}^{N} y_i y_j K(x_i, x_j) \alpha_i \alpha_j - \sum_{i=1}^{N} \alpha_i \tag{4}$$

$$subject\ to : \begin{cases} \sum_{i=1}^{N} \alpha_i y_i = 0 \\ 0 \leq \alpha_i \leq C, i = 1, \ldots, N \end{cases} \tag{5}$$

Where $K(x_i, x_j)$ is a kernel function that satisfies $K(x_i, x_j) = \phi(x_i)^T = \phi(x_j)$ and $C > 0$ is the cost parameter. The following three types of kernel were used in this study:

$$
\begin{aligned}
Gaussian & \quad K(x_i, x_j) = exp(-\|x_i - x_j\|^2) \\
Linear & \quad K(x_i, x_j) = x_i^T x_j \\
Polynomial & \quad K(x_i, x_j) = (1 + x_i x_j)^q, \, q = 2
\end{aligned}
\tag{6}
$$

Moreover, the classifier decision function can be represented as:

$$
f(x) = sgn \left[\sum_{i=1}^{N} \alpha_i y_i K(x_i, x_j) + b \right]
\tag{7}
$$

The OAA constructs k $(k - 1)/2$ classifiers where each one is trained on data from two classes. For training data from the i and j classes, we solve the following binary classification problem:

$$
min \frac{1}{2} \|w_{ij}\|^2 + C \sum_k \epsilon_k^{ij}
\tag{8}
$$

$$
subject \quad to : \begin{cases} w_{ij}^T \phi(x) + b_{ij} \geq 1 - \epsilon_k^{ij}, \, y_k = i \\ w_{ij}^T \phi(x) + b_{ij} \geq 1 - \epsilon_k^{ij}, \, y_k = j \\ \epsilon_k^{ij} \geq 0 \end{cases}
\tag{9}
$$

k - Nearest Neighbors. The k-Nearest Neighbors (k-NN) calculate the distance between each training sample and test sample in the dataset and returns the k closest samples [5]. This technique assumes that the closer objects in measurement space, it is more likely that they belong to the same class. Besides, k-NN need few parameters to tune, number of neighbors (k) and distance metrics. Euclidean distance (EU) metric is the most used for computing a distance between two samples [10].

3   Results and Discussion

In Table 2 are presented the achieved performances for the SVM model. Three types of kernel were evaluated in order to evaluate the effect of these on the performance of the SVM, the value of box constraint level implemented was 1 for each kernel dimension. The results of sensitivity for SVM with linear kernel are quite poor; classes 1 and 6 obtained the highest results with 92% and 55%, respectively. On the other hand, quadratic SVM shows an acceptable yield and balanced performance in sensitivity and specificity for each class, obtaining a result of accuracy of 80,40%. However, the Cubic SVM show the best performance; it worst recognition per class corresponded to emotion labels 0 and 4, with a sensitivity of 88,6% and 88,7% respectively.

The model based on SVM was used in each EEG sub-band, likewise a fusion of all the sub-bands was also training and validated. For the SVM model trained

Table 2. Performance of SVM with all EEG sub-bands.

SVM kernel	Measure	Label								
		0	1	2	3	4	5	6	11	12
SVM-L	Sensitivity (%)	0.326	0.927	0.331	0.314	0.301	0.363	0.553	0.325	0.358
	Specificity (%)	0.618	0.114	0.056	0.056	0.308	0.198	0.098	0.533	0.050
	Accuracy (%)	32.8 ± 6.25								
SVM-C	Sensitivity (%)	0.751	0.944	0.887	0.893	0.753	0.801	0.929	0.785	0.884
	Specificity (%)	0.849	0.828	0.775	0.724	0.803	0.754	0.813	0.840	0.766
	Accuracy (%)	80.4 ± 3.65								
SVM-Q	Sensitivity (%)	0.886	0.959	0.941	0.932	0.887	0.919	0.938	0.924	0.928
	Specificity (%)	0.937	0.877	0.911	0.884	0.907	0.905	0.892	0.934	0.868
	Accuracy (%)	91.3 ± 1.89								

Table 3. Performance of the k-NN classifier with Euclidean metric distance using all EEG sub-bands

Neighbor	Measure	Class								
		0	1	2	3	4	5	6	11	12
1	Sensitivity (%)	0.939	0.951	0.948	0.944	0.942	0.939	0.938	0.962	0.949
	Specificity (%)	0.944	0.962	0.951	0.944	0.933	0.951	0.94	0.955	0.942
	Accuracy (%)	94.6 ± 1.96								
3	Sensitivity (%)	0.890	0.925	0.925	0.920	0.934	0.947	0.961	0.962	0.957
	Specificity (%)	0.945	0.933	0.939	0.932	0.912	0.926	0.928	0.932	0.924
	Accuracy (%)	93.10 ± 1.99								
5	Sensitivity (%)	0.891	0.954	0.927	0.920	0.914	0.937	0.943	0.954	0.954
	Specificity (%)	0.944	0.926	0.929	0.921	0.914	0.920	0.913	0.927	0.907
	Accuracy (%)	92.60 ± 1.97								
7	Sensitivity (%)	0.876	0.963	0.918	0.920	0.914	0.932	0.937	0.949	0.960
	Specificity (%)	0.945	0.917	0.923	0.913	0.910	0.907	0.903	0.922	0.895
	Accuracy (%)	92.00 ± 2.10								
9	Sensitivity (%)	0.873	0.955	0.914	0.921	0.907	0.921	0.944	0.945	0.960
	Specificity (%)	0.942	0.900	0.914	0.910	0.905	0.905	0.892	0.920	0.893
	Accuracy (%)	92.6 ± 1.99								
11	Sensitivity (%)	0.855	0.956	0.909	0.912	0.892	0.922	0.935	0.936	0.954
	Specificity (%)	0.938	0.882	0.896	0.895	0.893	0.893	0.873	0.914	0.886
	Accuracy (%)	90.50 ± 2.33								

Table 4. Performance of the k-NN classifier with City Block metric distance using all EEG sub-bands

Neighbor	Measure	Class								
		0	1	2	3	4	5	6	11	12
1	Sensitivity (%)	0.997	0.998	0.995	1.000	0.996	0.997	0.997	0.996	0.998
	Specificity (%)	0.995	0.996	0.996	0.998	0.998	0.996	0.997	0.998	0.997
	Accuracy (%)	99.7 ± 0.45								
3	Sensitivity (%)	0.989	0.996	0.993	0.998	0.995	0.997	1.000	0.996	0.999
	Specificity (%)	0.996	0.996	0.992	0.997	0.995	0.994	0.992	0.995	0.994
	Accuracy (%)	99.50 ± 0.49								
5	Sensitivity (%)	0.990	0.998	0.993	1.000	0.997	0.996	1.000	0.997	0.998
	Specificity (%)	0.996	1.000	0.992	0.996	0.995	0.996	0.994	0.995	0.994
	Accuracy (%)	99.50 ± 0.54								
7	Sensitivity (%)	0.991	0.998	0.993	0.998	0.996	0.998	1.000	0.996	0.998
	Specificity (%)	0.996	0.998	0.993	0.996	0.995	0.996	0.994	0.997	0.992
	Accuracy (%)	99.50 ± 0.53								
9	Sensitivity (%)	0.989	0.996	0.994	0.998	0.997	0.999	0.997	0.997	0.998
	Specificity (%)	0.998	1.000	0.993	0.995	0.995	0.995	0.994	0.996	0.990
	Accuracy (%)	99.50 ± 0.60								
11	Sensitivity (%)	0.990	0.991	0.994	0.998	0.997	0.998	0.998	0.995	0.998
	Specificity (%)	0.996	0.996	0.993	0.996	0.995	0.993	0.992	0.997	0.991
	Accuracy (%)	99.4 ± 0.67								

with all sub-bands the best accuracy was 91.30% using a polynomial kernel (order three). The sensitivity and specificity show that the emotion labels with more difficulty of classification are the labels 0, and 4, this is due to the imbalance of classes during the training of the model where some classes present more observations than others. the performance obtained with the use of a polynomial kernel, and quadratic is quite similar. On the other hand, the model based on a linear kernel presents a fairly regular performance showing problems to determine an optimal hyperplane.

The SVM model used for each EEG sub-band showed important training problems, where it was not possible to find an optimal hyperplane, nor to determine any validation measure; the structure and statistical measures of the feature vector was not sufficient to discern the classes in each EEG sub-band individually. In a similar way to the SVM method, the k-NN-based model was implemented in each EEG sub-band, in the same way the fusion of the characteristics of all the EEG sub-bands was validated.

Tables 3, 4, and 5 shows k-NN performance using different amounts of neighbors and different metric distances in order to evaluate the best adequate performance of a particular model; the distance by cosine, by city block, and by

Table 5. Performance of the k-NN classifier with Cosine metric distance using all EEG sub-bands

Neighbor	Measure	Class								
		0	1	2	3	4	5	6	11	12
1	Sensitivity (%)	0.939	0.954	0.947	0.946	0.943	0.940	0.940	0.963	0.947
	Specificity (%)	0.945	0.964	0.952	0.948	0.934	0.953	0.938	0.953	0.941
	Accuracy (%)	94.7 ± 1.27								
3	Sensitivity (%)	0.884	0.932	0.922	0.918	0.934	0.951	0.958	0.963	0.958
	Specificity (%)	0.945	0.942	0.938	0.931	0.912	0.923	0.927	0.931	0.919
	Accuracy (%)	93 ± 1.38								
5	Sensitivity (%)	0.886	0.941	0.927	0.925	0.918	0.939	0.939	0.953	0.961
	Specificity (%)	0.945	0.931	0.930	0.921	0.910	0.918	0.917	0.928	0.906
	Accuracy (%)	92.50 ± 1.50								
7	Sensitivity (%)	0.878	0.958	0.914	0.918	0.913	0.934	0.941	0.946	0.961
	Specificity (%)	0.943	0.913	0.921	0.909	0.909	0.908	0.898	0.924	0.898
	Accuracy (%)	91.90 ± 1.77								
9	Sensitivity (%)	0.870	0.951	0.906	0.915	0.906	0.928	0.945	0.941	0.955
	Specificity (%)	0.942	0.911	0.911	0.901	0.900	0.902	0.889	0.920	0.884
	Accuracy (%)	91.30 ± 1.62								
11	Sensitivity (%)	0.856	0.961	0.909	0.912	0.895	0.922	0.935	0.932	0.955
	Specificity (%)	0.937	0.882	0.902	0.895	0.888	0.897	0.880	0.914	0.883
	Accuracy (%)	90.60 ± 1.99								

Euclidean distance were used. The performance was evaluated using accuracy, sensitivity and specificity. The specificity and sensitivity in all models were reduced as they increased the number of neighbors. For example, for k-NN model by euclidean distance the sensitivity of label 0 (neutral) and label 4 (amusement) show a largest reduction. The label 0 was reduced from 93 to 85, while class 4 changed from 94 to 89 both with 1 neighbor and 11 neighbors, respectively. Besides, the sensibility of labels 1, 3, 6 and 12 were gradually reduced with increasing number of neighbors. This same trend for specificity and sensibility were presented for alpha, beta, theta and gamma bands. However, the performance of k-NN model by cosine distance and City block distance shows a good balance between each class; the sensitivity and specificity for each class are very similar, decreasing analogously as the number of neighbors increases.

In the trained models for each EEG sub-band, very good results were obtained by implementing a smaller number of vectors of characteristics. The performance of the classifier improves as high frequency components are evaluated; for example, the theta band presents the lowest classification performance independently of the metric distance implemented, however it is an acceptable performance reaching an accuracy of 95.8% making use of a distance by City block; On the other hand, the gamma band presents the best results by implementing the Euclidean distance and distance by cosine, reaching an accuracy of 98.6% and 98.7% respectively, surpassing the results shown in the Tables 3, and 5 corresponding to the model trained with all the EEG sub-bands.

4   Conclusions

The results obtained with the proposed classification models establish good bases for the development of methodologies that allow the discrimination of particular emotions. However, it is still necessary to look for alternatives in the method to improve the performance of the classifiers. The analysis of EEG sub-bands shows a classification deficiency in the alpha band; in the other sub-bands, acceptable performance is obtained in both classifiers, obtaining the most relevant results in the beta and gamma sub-bands. Even if, the most promising results are obtained by making use of all the sub-bands applying the methodology proposed to each band individually. It is essential to bear in mind that an acceptable classification result is obtained for emotions subject to a particular mood of a subject; find patterns that can differentiate moods that can be associated with an unpleasant or pleasant set. In the same way, given that the proposed models were trained with signals from different subjects, it can be concluded that in the analyzed physiological signals similar characteristics can be found among the participants of the studied database. The future work will develop an analysis of peripheral and EEG signals for discovering transitions patterns, which could allow proposing new approaches that giving more information to HCI systems and psychology for improving the decision making.

Acknowledgment. The authors acknowledge to the research project "Sistema multimodal multisensorial para detección de emociones y gustos a partir de señales fisiológicas no invasivas como herramienta multipropósito de soporte de decisión usando un dispositivo de registro de bajo costo" supported by Institución Universitaria Pascual Bravo and SDAS Research Group.

References

1. Ackermann, P., Kohlschein, C., Bitsch, J.Á., Wehrle, K., Jeschke, S.: EEG-based automatic emotion recognition: Feature extraction, selection and classification methods. In: 2016 IEEE 18th International Conference on e-Health Networking, Applications and Services, Healthcom 2016 (2016). https://doi.org/10.1109/HealthCom.2016.7749447
2. Aguiñaga, A.R., Ramirez, M.A.L.: Emotional states recognition, implementing a low computational complexity strategy. Health Inform. J. **24**(2), 146–170 (2018). https://doi.org/10.1177/1460458216661862
3. Melamed, A.F.: Las Teorías De Las Emociones Y Su Relación Con La Cognición: Un Análisis Desde La Filosofía De La Mente. Cuadernos de la Facultad de humanidades y Ciencias Sociales- universidad Nacional de Jujuy **49**, 13–38 (2016). http://www.redalyc.org/pdf/185/18551075001.pdf

4. Almejrad, A.S.: Human emotions detection using brain wave signals: a challenging, **44**, 640–659 (2010)
5. Deng, Z., Zhu, X., Cheng, D., Zong, M., Zhang, S.: Efficient KNN classification algorithm for big data. Neurocomputing **195**, 143–148 (2016). https://doi.org/10.1016/j.neucom.2015.08.112
6. Gjoreski, M., Luštrek, M., Gams, M., Mitrevski, B.: An inter-domain study for arousal recognition from physiological signals. Informatica (Slovenia) **42**(1), 61–68 (2018)
7. Haddadi, R., Abdelmounim, E.: https://doi.org/10.1109/ICMCS.2014.6911261
8. Mejia, G., Gomez, A., Quintero, L.: Reconocimiento de Emociones utilizando la Tranformada Wavelet Estacionaria en señales EEG multicanal. In: IFMBEs Proceedings Claib 2016(October), pp. 1–4 (2016)
9. Milgram, J., Sabourin, R., Supérieure, É.D.T.: "One against one" or "one against all": which one is better for handwriting recognition with SVMs? October 2006
10. Saini, I., Singh, D., Khosla, A.: QRS detection using K-Nearest Neighbor algorithm (KNN) and evaluation on standard ECG databases. J. Adv. Res. **4**(4), 331–344 (2013). https://doi.org/10.1016/j.jare.2012.05.007
11. Salimi, A., Ziaii, M., Amiri, A., Hosseinjani, M., Karimpouli, S.: The Egyptian journal of remote sensing and space sciences using a feature subset selection method and support vector machine to address curse of dimensionality and redundancy in hyperion hyperspectral data classification. Egypt. J. Remote. Sens. Space Sci. **21**(1), 27–36 (2018). https://doi.org/10.1016/j.ejrs.2017.02.003
12. Siegel, E.H., et al.: Emotion fingerprints or emotion populations? A meta-analytic investigation of autonomic features of emotion categories. Psychol. Bull. **144**(4), 343–393 (2018). https://doi.org/10.1037/bul0000128
13. Soleymani, M., Lichtenauer, J., Pun, T., Pantic, M.: A multimodal database for affect recognition and implicit tagging. IEEE Trans. Affect. Comput. **3**(1), 42–55 (2012). https://doi.org/10.1109/T-AFFC.2011.25
14. Tzanetakis, G., Essl, G., Cook, P.: 3 The Discrete Wavelet Transform, January 2001
15. Verma, G.K., Tiwary, U.S.: Multimodal fusion framework: a multiresolution approach for emotion classification and recognition from physiological signals. NeuroImage **102**, 162–172 (2014). https://doi.org/10.1016/j.neuroimage.2013.11.007
16. Wang, Z., Yang, X., Cheng, K.T.: Accurate face alignment and adaptive patch selection for heart rate estimation from videos under realistic scenarios. PLoS ONE **13**(5), 12–15 (2018). https://doi.org/10.1371/journal.pone.0197275
17. Wiem, B.M.H., Lacharie, Z.: Emotion classification in arousal valence model using MAHNOB-HCI database. Int. J. Adv. Comput. Sci. Appl. (IJACSA) **8**(3), 318–323 (2017). https://doi.org/10.14569/IJACSA.2017.080344. www.ijacsa.thesai.org
18. Yin, Z., Zhao, M., Wang, Y., Yang, J., Zhang, J.: Recognition of emotions using multimodal physiological signals and an ensemble deep learning model. Comput. Methods Programs Biomed. **140**, 93–110 (2017). https://doi.org/10.1016/j.cmpb.2016.12.005
19. Zapata, J.C., Duque, C.M., Rojas-Idarraga, Y., Gonzalez, M.E., Guzmán, J.A., Becerra Botero, M.A.: Data fusion applied to biometric identification – a review. In: Solano, A., Ordoñez, H. (eds.) CCC 2017. CCIS, vol. 735, pp. 721–733. Springer, Cham (2017). https://doi.org/10.1007/978-3-319-66562-7_51

Relations Between Maximal Half Squat Strength and Bone Variables in a Group of Young Overweight Men

Anthony Khawaja[1,2], Patchina Sabbagh[1,3], Jacques Prioux[2],
Antonio Pinti[4(✉)], Georges El Khoury[1], and Rawad El Hage[1]

[1] Department of Physical Education, Faculty of Arts and Social Sciences,
University of Balamand, Kelhat El-Koura, Lebanon
[2] Movement, Sport, and Health Sciences Laboratory, University of Rennes 2,
Rennes, France
[3] University of Lille, EA 7369 - URePSSS - Unité de Recherche
Pluridisciplinaire Sport Santé Société, Lille, France
[4] I3MTO, EA4708, Université d'Orléans, Orléans, France
antonio.pinti@uphf.fr

Abstract. The purpose of this study was to investigate the relationships between maximal half-squat strength and bone variables in a group of young overweight men. 76 young overweight men (18 to 35 years) voluntarily participated in this study. Weight and height were measured, and body mass index (BMI) was calculated. Body composition, bone mineral content (BMC) and bone mineral density (BMD) and geometric indices of hip bone strength were determined for each individual by Dual-energy X-ray absorptiometry (DXA). Maximal half-squat strength was measured by a classical fitness machine (Smith machine) respecting the instructions of the national association of conditioning and muscular strength (NCSA). Maximal half-squat strength was positively correlated to WB BMC ($r = 0.37$; $p < 0.01$), WB BMD ($r = 0.29$; $p < 0.05$), L1–L4 BMC ($r = 0.43$; $p < 0.001$), L1–L4 BMD ($r = 0.42$; $p < 0.001$), TH BMC ($r = 0.30$; $p < 0.01$), TH BMD ($r = 0.26$; $p < 0.05$), FN BMD ($r = 0.32$; $p < 0.01$), FN cross-sectional area (CSA) ($r = 0.44$; $p < 0.001$), FN cross-sectional moment of inertia (CSMI) ($r = 0.27$; $p < 0.05$), FN section modulus (Z) ($r = 0.37$; $p < 0.001$) and FN strength index (SI) ($r = 0.33$; $p < 0.01$). After adjusting for lean mass, maximal half-squat strength remained significantly correlated to WB BMC ($p = 0.003$), WB BMD ($p = 0.047$), L1–L4 BMC ($p < 0.001$), L1–L4 BMD ($p < 0.001$), TH BMC ($p = 0.046$), FN BMD ($p = 0.016$), FN CSA ($p < 0.001$), FN Z ($p = 0.003$) and FN SI ($p < 0.001$). The current study suggests that maximal half-squat strength is a positive determinant of BMC, BMD and geometric indices of hip bone strength in young overweight men.

Keywords: Maximal strength · DXA · Peak bone mass · Osteoporosis

© Springer Nature Switzerland AG 2019
I. Rojas et al. (Eds.): IWBBIO 2019, LNBI 11466, pp. 374–384, 2019.
https://doi.org/10.1007/978-3-030-17935-9_34

1 Introduction

Worldwide, the prevalence of overweight and obesity is increasing rapidly and appears to be a major risk factor for many chronic diseases [1]. In Lebanon and in the Middle East, the prevalence of obesity and overweight among young men is also increasing [1, 2]. Sibai et al. showed in a study conducted in Lebanon that the prevalence of overweight in adult men was 57.7% [1]. Several studies of bone status in overweight subjects have founded conflicting results [3–8]. Osteoporosis is a systemic skeletal disease characterized by low bone mass and micro-architectural deterioration of bone tissue with a consequent increase in bone fragility and susceptibility to fracture [9]. Historically, osteoporosis has been considered as a "woman's disease"; an appropriate approach to the extent that about 40 to 50% of women suffer from osteoporotic fractures in their life. However, focusing only on women is inappropriate because about 25% of men suffer from a fracture related to osteoporosis during their life [10]. Male osteoporosis, although less common than female osteoporosis, is nevertheless a public health problem, with a prevalence of 13% in men over 50 years of age (compared to 40% in women) and 15% risk lifetime osteoporotic fractures in the same age group. Men differ from women in that their distribution of the fracture by age shows two peaks, one between the ages of 15 and 45 and the other after the age of 70. Men under 45 years of age have a three-fold higher risk of fracture than women of the same age due to increased exposure to injury during sports and work activities. Post-traumatic fractures may be associated with a decrease in bone mineral density (BMD) [11]. Several interconnected factors influence bone mass accumulation during growth. These physiological determinants typically include heredity, vitamin D and bonetropic nutrients (calcium, protein), endocrine factors (sex steroids, IGF-I, 1.25 $(OH)_2D$) and mechanical forces (physical activity and body weight). Thus, quantitatively, the most important determinant seems to be genetically linked [12]. Peak bone mass in the third decade of life is an important determinant of future fracture risk [13]. Thus, increasing the value of peak BMD and maintaining it throughout the aging process is of great importance in the prevention of osteoporosis in adulthood [12, 14]. Mechanical stresses have positive effects on bone mineral content (BMC) and BMD. The practice of impact physical activities is associated with higher BMD values [15]. In addition, the regular practice of physical activities characterized by significant mechanical stress stimulates bone formation and improves BMD in the most solicited sites [16, 17]. In fact, according to the Frost theory known as the mechanostat, the resistance of the bone adapts to the mechanical stresses applied to it [18]. Many studies have shown a significant correlation between BMD and the performances obtained in some physical tests used in current sports practice [19–22]. A previous study suggests that maximal half squat strength (1-RM half-squat) is positively correlated with bone variables in overweight and obese adult women [23]. However, the relationship between 1-RM half-squat and bone variables in young men needs to be elucidated. The purpose of this study was to investigate the relationships between 1-RM half-squat and bone variables (BMC, BMD, hip geometric indices and TBS) in a group of young overweight men. Identification of new determinants of BMC, BMD, hip geometric indices and TBS in young overweight men, would allow screening and early management of future cases of osteopenia and osteoporosis.

2 Materials and Methods

2.1 Subjects and Study Design

Seventy-six overweight and obese (BMI > 25 kg/m2) young men whose ages ranged from 18 to 35 years voluntarily participated in the present study. All participants were nonsmokers and had no history of major orthopedic problems or other disorders known to affect bone metabolism or physical tests of the study. Other inclusion criteria included no diagnosis of comorbidities and no history of fracture. An informed written consent was obtained from the participants. The current study was approved by the University of Balamand Ethics Committee.

2.2 Anthropometrics

Height (in centimeters) was measured in the upright position to the nearest 1 mm with a standard stadiometer. Body weight (in kilograms) was measured on a mechanic scale with a precision of 100 g. The subjects were weighed wearing only underclothes. BMI was calculated as body weight divided by height squared (in kilograms per square meter). Body composition was evaluated by dual-energy X-ray absorptiometry (DXA; GE Healthcare, Madison, WI).

2.3 Bone Variables

BMC (in grams) and BMD (in grams per square centimeter) were determined for each individual by DXA at the whole body (WB), lumbar spine (L1–L4), total hip (TH), and femoral neck (FN; GE Healthcare). FN cross-sectional area (CSA), strength index (SI), buckling ratio (BR), FN section modulus (Z), cross-sectional moment of inertia (CSMI) and L1–L4 TBS were also evaluated by DXA [24–26]. The TBS is derived from the texture of the DXA image and has been shown to be related to bone microarchitecture and fracture risk. The TBS score can assist the healthcare professional in assessing fracture risk [25, 26]. In our laboratory, the coefficients of variation were <1% for BMC and BMD and less than 3% for FN CSA [8, 27–29]. The same certified technician performed all analyses using the same technique for all measurements.

2.4 Maximal Strength

A one-repetition-maximum (1-RM) test, following the protocol established by the National Strength and Conditioning Association, was performed to measure back half-squat maximal strength on a Smith machine [30].

2.5   Statistical Analysis

The means and standard deviations were calculated for all clinical data and for the bone measurements. Associations between 1-RM half-squat and bone data were given as Pearson correlation coefficients. Multiple linear regression analysis models were used to test the relationship of 1-RM half-squat and lean mass (LM) with bone variables, and R^2 values were reported. Statistical analyses were performed using the SigmaStat 3.1 Program (Jandel Corp., San Rafael, CA). A level of significance of $p < 0.05$ was used.

3   Results

3.1   Clinical Characteristics and Bone Data of the Study Population

Age, weight, height, BMI, lean mass, fat mass, fat mass percentage, bone variables and 1-RM half-squat are shown in Table 1.

Table 1. Physical characteristics of the study population

Characteristics	Mean ± SD	Range
Age (yr)	23.8 ± 4.7	18–35
Weight (kg)	100.5 ± 17.1	77–177
Height (m)	1.77 ± 0.06	1.65–2.0
BMI (kg/m^2)	32.0 ± 5.4	25.1–58.4
Lean mass (kg)	64.428 ± 7.168	51.561–86.286
Fat mass (kg)	43.801 ± 16.185	12.200–90.590
Fat mass (%)	31.6 ± 7.6	12.4–50.2
WB BMC (g)	3287 ± 378	2628–4322
WB BMD (g/cm^2)	1.295 ± 0.124	1.073–1.624
L1–L4 BMC (g)	78.4 ± 12.5	52.6–118.4
L1–L4 BMD (g/cm^2)	1.246 ± 0.155	0.956–1.718
L1–L4 TBS	1.358 ± 0.128	0.873–1.621
TH BMC (g)	43.5 ± 6.9	28.8–62.9
TH BMD (g/cm^2)	1.182 ± 0.156	0.831–1.558
FN BMC (g)	6.48 ± 1.08	2.42–9.63
FN BMD (g/cm^2)	1.180 ± 0.147	0.867–1.644
FN CSA (mm^2)	206.5 ± 29.1	153.0–298.0
FN CSMI (mm^2)^2	18914 ± 5288	1669–36157
FN Z (mm^3)	1025 ± 219	596–1755
BR	5.309 ± 2.524	0.000–11.400
FN SI	1.442 ± 0.376	0.700–2.400
1-RM Half-squat (kg)	116 ± 37	80–230

BMI, body mass index; WB, whole body; BMC, bone mineral content; BMD, bone mineral density; TBS, trabecular bone score; TH, total hip; FN, femoral neck; CSA, cross-sectional area; CSMI, cross-sectional moment of inertia; Z, section modulus; BR, buckling ratio; SI, strength index; SD, standard deviation.

3.2 Correlations Between Clinical Characteristics and Bone Variable

1-RM half-squat was positively correlated to WB BMC ($r = 0.37$; $p < 0.01$), WB BMD ($r = 0.29$; $p < 0.05$), L1–L4 BMC ($r = 0.43$; $p < 0.001$), L1–L4 BMD ($r = 0.42$; $p < 0.001$), TH BMC ($r = 0.30$; $p < 0.01$), TH BMD ($r = 0.26$; $p < 0.05$), FN BMD ($r = 0.32$; $p < 0.01$), FN CSA ($r = 0.44$; $p < 0.001$), FN CSMI ($r = 0.27$; $p < 0.05$), FN Z ($r = 0.37$; $p < 0.001$) and FN SI ($r = 0.33$; $p < 0.01$). FM was positively correlated to FN BMC ($r = 0.23$; $p < 0.05$). FM was negatively correlated to FN SI ($r = -0.24$; $p < 0.05$). LM was positively correlated to WB BMC ($r = 0.72$; $p < 0.001$), WB BMD ($r = 0.42$; $p < 0.001$), L1–L4 BMC ($r = 0.50$; $p < 0.001$), L1–L4 BMD ($r = 0.29$; $p < 0.05$), TH BMC ($r = 0.62$; $p < 0.001$), TH BMD ($r = 0.38$; $p < 0.001$), FN BMC ($r = 0.48$; $p < 0.001$), FN BMD ($r = 0.36$; $p < 0.01$), FN CSA ($r = 0.53$; $p < 0.001$), FN CSMI ($r = 0.56$; $p < 0.001$) and FN Z ($r = 0.56$; $p < 0.001$) (Table 2).

BMI, body mass index; FM, fat mass; LM, lean mass; WB, whole body; BMC, bone mineral content; BMD, bone mineral density; TBS, trabecular bone score; TH, total hip; FN, femoral neck; CSA, cross-sectional area; CSMI, cross-sectional moment of inertia; Z, section modulus; BR, buckling ratio; SI, strength index. $*p < 0.05$. $**p < 0.01$. $***p < 0.001$.

3.3 Multiple Linear Regressions

1-RM half-squat remained positively correlated to WB BMC ($p = 0.003$), WB BMD ($p = 0.047$), L1–L4 BMC ($p < 0.001$), L1–L4 BMD ($p < 0.001$), TH BMC ($p = 0.046$), FN BMD ($p = 0.016$), FN CSA ($p < 0.001$), FN Z ($p = 0.003$) and FN SI ($p < 0.001$) after adjusting for LM. 1-RM half-squat was a stronger positive determinant of L1–L4 BMD and SI than LM. LM was a stronger positive determinant of WB BMC, WB BMD, L1–L4 BMC, TH BMC, FN BMD, FN CSA and FN Z than 1-RM half-squat. LM remained positively correlated to WB BMC ($p < 0.001$), WB BMD ($p < 0.001$), L1–L4 BMC ($p < 0.001$), L1–L4 BMD ($p = 0.042$), TH BMC ($p < 0.001$), TH BMD ($p = 0.003$), FN BMC ($p < 0.001$), FN BMD ($p = 0.005$), FN CSA ($p < 0.001$), FN CSMI ($p < 0.001$) and FN Z ($p < 0.001$) after adjusting for 1-RM half-squat (Table 3).

WB, whole body; BMC, bone mineral content; BMD, bone mineral density; TBS, trabecular bone score; TH, total hip; FN, femoral neck; CSA, cross-sectional area; CSMI, cross-sectional moment of inertia; Z, section modulus; BR, buckling ratio; SI, strength index.

Table 2. Correlations between clinical characteristics and bone variables

	WB BMC (g)	WB BMD (g/cm²)	L1–L4 BMC (g)	L1–L4 BMD (g/cm²)	L1–L4 TBS	TH BMC (g)	TH BMD (g/cm²)	FN BMC (g)	FN BMD (g/cm²)	FN CSA (mm²)	FN CSMI (mm²)²	FN Z (mm³)	BR	FN SI
Age (yr)	-0.08	0.02	-0.01	-0.02	-0.37**	-0.15	-0.12	-0.22	-0.20	-0.24*	-0.16	-0.26*	-0.03	-0.10
Weight (kg)	0.42***	0.39***	0.18	0.11	-0.12	0.41***	0.31**	0.35**	0.21	0.22	0.28*	0.24*	0.01	-0.37***
Height (m)	0.45***	0.02	0.48***	0.18	-0.22	0.24*	0.04	0.19	-0.02	0.18	0.40***	0.39***	-0.05	-0.11
BMI (kg/m²)	0.22	0.37**	-0.02	0.03	-0.04	0.30**	0.29*	0.25*	0.21	0.13	0.09	0.05	0.05	-0.34**
FM (kg)	0.04	0.05	-0.04	-0.00	-0.04	0.13	0.05	0.23*	0.00	-0.00	0.10	0.02	0.09	-0.24*
FM %	-0.11	0.09	-0.25*	-0.15	-0.08	-0.04	0.05	0.01	-0.05	-0.23*	-0.18	-0.24*	0.14	-0.56***
LM (Kg)	0.72***	0.42***	0.50***	0.29*	0.12	0.62***	0.38***	0.48***	0.36**	0.53***	0.56***	0.56***	-0.08	-0.04
1-RM Half-squat	0.37**	0.29*	0.43***	0.42***	0.15	0.30**	0.26*	0.14	0.32**	0.44***	0.27*	0.37***	-0.20	0.33**

Table 3. Multiple linear regressions

	Coefficient ± SE	t value	p value
Dependent variable: WB BMC (R^2 = 0.761)			
Constant	710.633 ± 270.303	2.629	0.011
1-RM Half-Squat (kg)	2.436 ± 0.793	3.073	0.003
Lean Mass (kg)	0.0356 ± 0.00421	8.449	<0.001
Dependent variable: WB BMD (R^2 = 0.477)			
Constant	0.786 ± 0.120	6.576	<0.001
1-RM Half-Squat (kg)	0.000708 ± 0.000351	2.019	0.047
Lean Mass (kg)	0.000006 ± 0.00000186	3.548	<0.001
Dependent variable: L1–L4 BMC (R^2 = 0.606)			
Constant	15.151 ± 11.234	1.349	0.182
1-RM Half-Squat (kg)	0.118 ± 0.0329	3.571	<0.001
Lean Mass (kg)	0.000773 ± 0.000175	4.418	<0.001
Dependent variable: L1–L4 BMD (R^2 = 0.474)			
Constant	0.749 ± 0.153	4.897	<0.001
1-RM Half-Squat (kg)	0.00156 ± 0.000449	3.468	<0.001
Lean Mass (kg)	0.00000494 ± 0.000002	2.073	0.042
Dependent variable: L1–L4 TBS (R^2 = 0.186)			
Constant	1.213 ± 0.127	9.553	<0.001
1-RM Half-Squat (kg)	0.000440 ± 0.000370	1.191	0.238
Lean Mass (kg)	0.0000015 ± 0.0000019	0.767	0.446
Dependent variable: TH BMC (R^2 = 0.650)			
Constant	3.026 ± 5.809	0.521	0.604
1-RM Half-Squat (kg)	0.0346 ± 0.017	2.032	0.046
Lean Mass (kg)	0.000567 ± 0.0000905	6.264	<0.001
Dependent variable: TH BMD (R^2 = 0.432)			
Constant	0.593 ± 0.157	3.766	<0.001
1-RM Half-Squat (kg)	0.000826 ± 0.000461	1.791	0.078
Lean Mass (kg)	0.00000766 ± 0.000002	3.125	0.003
Dependent variable: FN BMC (R^2 = 0.485)			
Constant	1.693 ± 1.042	1.624	0.109
1-RM Half-Squat (kg)	0.00180 ± 0.00306	0.590	0.557
Lean Mass (kg)	0.0000709 ± 0.0000162	4.367	<0.001
Dependent variable: FN BMD (R^2 = 0.455)			
Constant	0.634 ± 0.146	4.334	<0.001
1-RM Half-Squat (kg)	0.00106 ± 0.000429	2.481	0.016
Lean Mass (kg)	0.00000657 ± 0.000002	2.885	0.005
Dependent variable: FN CSA (R^2 = 0.642)			
Constant	51.632 ± 24.758	2.085	0.041
1-RM Half-Squat (kg)	0.277 ± 0.0726	3.821	<0.001
Lean Mass (kg)	0.00190 ± 0.000385	4.935	<0.001

(*continued*)

Table 3. (*continued*)

	Coefficient ± SE	*t* value	*p* value
Dependent variable: FN CSMI (R^2 = 0.597)			
Constant	-9130.181 ± 4623.686	−1.975	0.052
1-RM Half-Squat (kg)	26.823 ± 13.559	1.978	0.052
Lean Mass (kg)	0.385 ± 0.0720	5.349	<0.001
Dependent variable: FN Z (R^2 = 0.631)			
Constant	-152.236 ± 185.204	−0.822	0.414
1-RM Half-Squat (kg)	1.692 ± 0.543	3.115	0.003
Lean Mass (kg)	0.0152 ± 0.00288	5.256	<0.001
Dependent variable: BR (R^2 = 0.238)			
Constant	8.193 ± 2.740	2.990	0.004
1-RM Half-Squat (kg)	−0.0152 ± 0.00804	−1.886	0.064
Lean Mass (kg)	−0.0000160 ± 0.000042	−0.375	0.709
Dependent variable: FN SI (R^2 = 0.383)			
Constant	1.394 ± 0.378	3.690	<0.001
1-RM Half-Squat (kg)	0.00379 ± 0.00111	3.418	<0.001
Lean Mass (kg)	−0.0000062 ± 0.000005	−1.062	0.292

4   Discussion

The present study conducted on a group of young overweight and obese men mainly shows that 1-RM half-squat is positively correlated to BMC, BMD and geometric indices of hip bone strength. After adjusting for LM, 1-RM half-squat remained positively correlated to WB BMC, WB BMD, L1–L4 BMC, L1–L4 BMD, TH BMC, FN BMD, FN CSA, FN Z and FN SI. These results highlight the positive influence of maximal strength of the lower limbs on bone variables in young overweight men. Our results are in accordance with those of two previous studies that showed that maximal half-squat strength is a positive determinant of L2–L4 BMD in men [31, 32]. Accordingly, our results highlight the importance of increasing muscle mass in the lower limbs to prevent osteoporosis later in life [31, 32]. Another study conducted on a group of young women confirmed that maximal strength of the lower limbs is a positive determinant of BMD in normal-weighted women [29]. Importantly, 1-RM half-squat was a stronger determinant of L2–L4 BMD and FN SI than lean mass. This result is clinically important since these two parameters (L2–L4 BMD and FN SI) are related to fracture risk in elderly subjects. Powerlifters have higher bone mass compared to sedentary subjects [14]. Tsuzuku et al. [14] conducted a study aimed at studying the effects of high-intensity resistance training on BMD in young male powerlifters and its relationship to muscle strength. They found significant differences between the powerlifters and the control group in lean mass [14]. Powerlifters also showed larger circumferences in the upper body measurements than the control group [14]. The BMD of the whole body, lumbar spine, arm, leg and pelvis was significantly higher in powerlifters than in controls. The lumbar spine BMD was significantly correlated with squat and deadlift performances in powerlifters [14]. They suggest that training with high intensity loads appears to be effective in increasing site-specific BMD in the skeleton [14]. Accordingly, the positive relation between maximal strength and BMD seems to be valid in both athletes and non-athletes. The current study shows that LM is positively correlated to BMC, BMD, FN CSA, FN CSMI and FN Z in young overweight men. LM remained positively correlated to BMC, BMD, FN CSA, FN CSMI and FN Z, after adjusting for 1-RM half-squat. Our results are in accordance with those of many previous studies [31, 33–39]. A study conducted on a group of overweight and obese young men confirms the positive importance of LM on bone mass in this population. Similarly, lean mass appears to be a predictor of BMD and hip geometric indices in overweight/obese men and normal-weighted men [31]. Our study had some limitations. The cross-sectional nature of the present study is a limitation because it cannot evaluate the confounding variables. The second limitation is the small number of subjects in our study group. The third limitation is the 2-dimensional nature of DXA [40, 41]. However, to our knowledge, the present study is one of few studies that aimed to find new determinants of BMC, BMD, and hip geometric indices in overweight men. Some of these determinants are easily calculated when performing simple physical tests.

5 Conclusion

The current study suggests that 1-RM half-squat is a positive determinant of WB BMC, BMD, FN CSA, FN CSMI, FN Z and FN SI in young overweight men. Implementing strategies to increase maximal strength of the lower limbs and LM in young overweight and obese men may be useful for preventing osteoporosis later in life. Therefore, our results may be useful for the prevention and early detection of osteoporosis and osteopenia.

References

1. Sibai, A.M., Hwalla, N., Adra, N., Rahal, B.: Prevalence and covariates of obesity in Lebanon: findings from the first epidemiological study. Obes. Res. **11**(11), 1353–1361 (2003)
2. Nasreddine, L., Naja, F., Chamieh, M.C., Adra, N., Sibai, A.M., Hwalla, N.: Trends in overweight and obesity in Lebanon: evidence from two national cross-sectional surveys (1997 and 2009). BMC Public Health **12**, 798 (2012)
3. Janicka, A., Wren, T.A., Sanchez, M.M., Dorey, F., Kim, P.S., et al.: Fat mass is not beneficial to bone in adolescents and young adults. J. Clin. Endocrinol. Metab. **92**(1), 143–147 (2007)
4. Bredella, M.A., Lin, E., Gerweck, A.V., et al.: Determinants of bone microarchitecture and mechanical properties in obese men. J. Clin. Endocrinol. Metab. **97**(11), 4115–4122 (2012)
5. Rexhepi, S., Bahtiri, E., Rexhepi, M., Sahatciu-Meka, V., Rexhepi, B.: Association of body weight and body mass index with bone mineral density in women and men from Kosovo. Mater. Sociomed. **27**(4), 259–262 (2015)

6. Ornstrup, M.J., Kjær, T.N., Harsløf, T., Stødkilde-Jørgensen, H., Hougaard, D.M., Cohen, A., et al.: Adipose tissue, estradiol levels, and bone health in obese men with metabolic syndrome. Eur. J. Endocrinol. **172**(2), 205–216 (2015)
7. Zhang, P., Peterson, M., Su, G.L., Wang, S.C.: Visceral adiposity is negatively associated with bone density and muscle attenuation. Am. J. Clin. Nutr. **101**(2), 337–343 (2015)
8. El Hage, R., Bachour, F., Sebaaly, A., et al.: The influence of weight status on radial bone mineral density in Lebanese women. Calcif. Tissue Int. **94**(4), 465–467 (2014)
9. Binkley, N.: A perspective on male osteoporosis. Best Pract. Res. Clin. Rheumatol. **23**, 755–768 (2009)
10. Nguyen, T.V., Eisman, J.A., Kelly, P.J., Sambrook, P.N.: Risk factors for osteoporotic fractures in elderly men. Am. J. Epidemiol. **144**, 255–263 (1996)
11. Briot, K., Cortet, B., Trémollières, F., Sutter, B., Thomas, T., Roux, C., et al.: Male osteoporosis: diagnosis and fracture risk evaluation. Joint Bone Spine **76**, 129–133 (2009)
12. Bonjour, J.P., Chevalley, T., Ferrari, S., Rizzoli, R.: The importance and relevance of peak bone mass in the prevalence of osteoporosis. Salud Publica Mexico **51**(suppl 1), S5–S17 (2009)
13. Zakhem, E., El Khoury, G., Feghaly, L., Zunquin, G., El Khoury, C., Pezé, T., et al.: Performance physique et densité minérale osseuse chez de jeunes adultes libanais. J. Med. Liban. **64**(4), 193–199 (2016)
14. Tsuzuku, S., Ikegami, Y., Yabe, K.: Effects of high-intensity resistance training on bone mineral density in young male powerlifters. Calcif. Tissue Int. **63**, 283–286 (1998)
15. El Khoury, G., Zouhal, H., El Khoury, C., Jacob, C., Cabagno, G., Maalouf, G., et al.: Influence du niveau d'activité physique sur les paramètres osseux chez des jeunes hommes en surcharge pondérale. Kinesither Rev. **17**(184), 10–15 (2017)
16. Ainsworth, B.E., Youmans, C.P.: Tools for physical activity counseling in medical practice. Obes. Res. **10**(1), 69S–75S (2002)
17. Nikander, R., Sievänen, H., Heinonen, A., Kannus, P.: Femoral neck structure in adult female athletes subjected to different loading modalities. J. Bone Miner. Res. **20**, 520–528 (2005)
18. Frost, H.M.: Bone's mechanostat: a 2003 update. Anat. Rec. A Discov. Mol. Cell. Evol. Biol. **275**, 1081–1101 (2003)
19. Vicente-Rodriguez, G., Dorado, C., Perez-Gomez, J., Gonzalez-Henriquez, J.J., Calbet, J.A.: Enhanced bone mass and physical fitness in young female handball players. Bone **35**, 1208–1215 (2004)
20. Dixon, W.G., Lunt, M., Pye, S.R., et al.: Low grip strength is associated with bone mineral density and vertebral fracture in women. Rheumatology (Oxford) **44**, 642–646 (2005)
21. Sirola, J., Rikkonen, T., Tuppurainen, M., Jurvelin, J.S., Alhava, E., Kröger, H.: Grip strength may facilitate fracture prediction in perimenopausal women with normal BMD: a 15-year population based study. Calcif. Tissue Int. **83**, 93–100 (2008)
22. Sherk, V.D., Palmer, I.J., Bemben, M.G., Bemben, D.A.: Relationships between body composition, muscular strength, and bone mineral density in estrogen-deficient post-menopausal women. J. Clin. Densitom. **12**, 292–298 (2009)
23. Berro, A,J., Al Rassy, N., Ahmaidi, S., Sabbagh, P., Khawaja, A., Maalouf, G., et al.: Physical performance variables and bone parameters in a group of young overweight and obese women. J. Clin. Densitom. 1–7 (2018, in press). https://doi.org/10.1016/j.jocd.2018.09.008
24. Beck, T.J., Ruff, C.B., Warden, K.E., LeBoff, M.S., Cauley, J.A., Chen, Z.: Predicting femoral neck strength from bone mineral data. A structural approach. Invest. Radiol. **25**(1), 6–18 (1990)

25. Silva, B.C., Broy, S.B., Boutroy, S., et al.: Fracture risk prediction by non-BMD DXA measures: the 2015 ISCD official positions part 2: trabecular bone score. J. Clin. Densitom. 18(3), 309–330 (2015)

26. Harvey, N.C., Glüer, C.C., Binkley, N., et al.: Trabecular bone score (TBS) as a new complementary approach for osteoporosis evaluation in clinical practice. Bone 78, 216–224 (2015)

27. El Hage, R., Zakhem, E., Theunynck, D., et al.: Maximal oxygen consumption and bone mineral density in a group of young Lebanese adults. J. Clin. Densitom. 17, 320–324 (2014)

28. El Hage, R., Bachour, F., Khairallah, W., et al.: The influence of obesity and overweight on hip bone mineral density in Lebanese women. J. Clin. Densitom. 17(1), 216–217 (2014)

29. Zakhem, E., Ayoub, M.I., Zunquin, G., et al.: Physical performance and trabecular bone score in a group of young Lebanese women. J. Clin. Densitom. 18, 271–272 (2015)

30. El Hage, R., Zakhem, E., Moussa, E., et al.: Acute effects of heavy-load squats on consecutive vertical jump performance. Sci. Sports 26, 44–47 (2011)

31. El Khoury, C., Pinti, A., Lespessailles, E., Maalouf, G., Watelain, E., El Khoury, G., et al.: Physical performance variables and bone mineral density in a group of young overweight and obese men. J. Clin. Densitom. 21(1), 41–47 (2018)

32. El Khoury, G., Zouhal, H., Cabagno, G., El Khoury, C., Rizkallah, M., Maalouf, G., et al.: Bone variables in active overweight/obese men and sedentary overweight/obese men. J. Clin. Densitom. 20(2), 239–246 (2017)

33. Petit, M.A., Beck, T.J., Hughes, J.M., Lin, H.M., Bentley, C., Lloyd, T.: Proximal femur mechanical adaptation to weight gain in late adolescence: a six-year longitudinal study. J. Bone Miner. Res. 23, 180–188 (2008)

34. Shea, K.L., Gozansky, W.S., Sherk, V.D., Swibas, T.A., Wolfe, P., Scherzinger, A., et al.: Loss of bone strength in response to exercise-induced weight loss in obese postmenopausal women: results from a pilot study. J. Musculoskelet. Neuronal Interact. 14, 229–238 (2014)

35. MacKelvie, K.J., McKay, H.A., Petit, M.A., Moran, O., Khan, K.M.: Bone mineral response to a 7-month randomized controlled, school-based jumping intervention in 121 prepubertal boys: associations with ethnicity and body mass index. J. Bone Miner. Res. 17(5), 834–844 (2002)

36. Nikander, R., Sievänen, H., Heinonen, A., Kannus, P.: Femoral neck structure in adult female athletes subjected to different loading modalities. J. Bone Miner. Res. 20(3), 520–528 (2005)

37. Lorentzon, M., Mellström, D., Ohlsson, C.: Association of amount of physical activity with cortical bone size and trabecular volumetric BMD in young adult men: the GOOD study. J. Bone Miner. Res. 20(11), 1936–1943 (2005)

38. Bonjour, J.P., Chevalley, T., Rizzoli, R., et al.: Gene environment interactions in the skeletal response to nutrition and exercise during growth. Med. Sport Sci. 51, 64–80 (2007)

39. El Hage, R., Jacob, C., Moussa, E., et al.: Effects of 12 weeks of endurance training on bone mineral content and bone mineral density in obese, overweight and normal weight adolescent girls. Sci. Sports 24(3–4), 210–213 (2009)

40. Beck, T.J.: Measuring the structural strength of bones with dual-energy X-ray absorptiometry: principles, technical limitations, and future possibilities. Osteoporos. Int. 14(Suppl 5), S81–S88 (2003)

41. Beck, T.J.: Extending DXA beyond bone mineral density: understanding hip structure analysis. Curr. Osteoporos. Rep. 5(2), 49–55 (2007)

Brain Hematoma Segmentation Using Active Learning and an Active Contour Model

Heming Yao[1(✉)], Craig Williamson[2], Jonathan Gryak[1],
and Kayvan Najarian[1,3,4]

[1] Department of Computational Medicine and Bioinformatics, University of
Michigan, Ann Arbor, MI, USA
[2] Department of Neurosurgery, University of Michigan, Ann Arbor, MI, USA
[3] Michigan Center for Integrative Research in Critical Care, University of Michigan,
Ann Arbor, MI, USA
[4] Department of Emergency Medicine, University of Michigan, Ann Arbor, MI, USA
{hemingy,craigaw,gryakj,kayvan}@med.umich.edu

Abstract. Traumatic brain injury (TBI) is a massive public health problem worldwide. Accurate and fast automatic brain hematoma segmentation is important for TBI diagnosis, treatment and outcome prediction. In this study, we developed a fully automated system to detect and segment hematoma regions in head Computed Tomography (CT) images of patients with acute TBI. We first over-segmented brain images into superpixels and then extracted statistical and textural features to capture characteristics of superpixels. To overcome the shortage of annotated data, an uncertainty-based active learning strategy was designed to adaptively and iteratively select the most informative unlabeled data to be annotated for training a Support Vector Machine classifier (SVM). Finally, the coarse segmentation from the SVM classifier was incorporated into an active contour model to improve the accuracy of the segmentation. From our experiments, the proposed active learning strategy can achieve a comparable result with 5 times fewer labeled data compared with regular machine learning. Our proposed automatic hematoma segmentation system achieved an average Dice coefficient of 0.60 on our dataset, where patients are from multiple health centers and at multiple levels of injury. Our results show that the proposed method can effectively overcome the challenge of limited and highly varied dataset.

Keywords: Medical image segmentation · Medical image processing ·
Traumatic brain injury · Active learning · Active contour model

1 Introduction

Traumatic brain injury (TBI) is a major cause of death and disability worldwide, especially in children and young adults. Accurate and fast detection and diagnosis of brain damage in the early stage of injury is important for prompt and

© Springer Nature Switzerland AG 2019
I. Rojas et al. (Eds.): IWBBIO 2019, LNBI 11466, pp. 385–396, 2019.
https://doi.org/10.1007/978-3-030-17935-9_35

proper management of TBI patients [1]. To detect the presence and extent of brain hematoma, Computed tomography (CT) is the imaging modality of choice during the first 48 h after injury [1] due to its speed, low-cost, and availability. Previous studies have shown that brain hematoma detection and volume calculation are important for TBI diagnosis [1], mortality and morbidity prediction [2,3], and surgical management [4].

To facilitate TBI patient management, an automatic hematoma segmentation system can provide accurate and quantitative evaluations of brain hematoma. It can decrease medical costs and provide guidance for proper medical treatment [4]. Many segmentation methods have been proposed. A semi-automated method based on a region growing algorithm was proposed for brain hematoma [5] that requires seed points fixed by the user. A level set algorithm was developed [6] where candidate hematoma voxels were identified by an adaptive threshold. An algorithm combining Gaussian Mixture Model (GMM) and Expectation Maximization was proposed to find hematoma component [7]. Most of the previous studies rely on either manual initialization or the distribution of intensity values in brain tissues to segment hematoma regions. However, hematoma intensity varies across imaging protocols and patient conditions, and other anatomical structures such as straight sinus exhibit a similar intensity. Moreover, these techniques ignore textural differences between brain tissues and structures.

In this study, to better segment hematoma, we proposed a fully-automatic hematoma segmentation framework that extracts texture features and integrates a supervised model with an active contour model. One challenge here is although there are plenty of medical images available, annotating these images is very time-consuming. To overcome the shortage of labeled images, an active learning strategy is presented. It started with training an initial support vector machine (SVM) model using one annotated image and then queried the most informative superpixels whose labels may lead to the greatest improvement to the model. After active learning, the superpixel-based classification resulted in a coarse hematoma segmentation. The coarse segmentation was incorporated into an active contour model to generate the final fine segmentation.

Our contributions here are two-fold. First, we proposed an active learning algorithm for image segmentation. Our experiments show that the active learning strategy can effectively select the most informative samples whose labels result in a significantly higher performance improvement compared with random selection. From our results, active learning can help overcome the shortage of annotated data, which is a common problem in medicine. Secondly, we proposed a hematoma segmentation framework and achieved a mean Dice coefficient of 0.60 on a challenging dataset. The dataset consists of CT scans from patients with various health conditions and using different imaging protocol, where hematoma regions are of various shapes, types and occur at different locations.

2   Methodology

As shown in Fig. 1, we first adjusted contrast within the CT images and performed skull removal to extract the soft tissues for further analysis. The brain regions were then over-segmented into superpixels. For each superpixel s_i, statistical and textural features were extracted to generate a feature vector $\boldsymbol{v}_i$. The label of each superpixel is denoted by y_i, with $y_i = 1$ when s_i belongs to a hematoma region (i.e. the majority of pixels in s_i belong to a hematoma region) and $y_i = 0$ otherwise. An SVM classifier was trained to predict superpixel class and generate a coarse segmentation map. Finally, the coarse segmentation was used as a segmentation prior for an active contour model to refine the segmentation boundary.

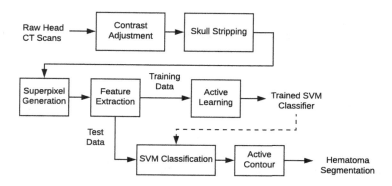

Fig. 1. A diagram of the proposed framework

2.1   Pre-processing

Let us define an image of size $H \times W$ as $I : \Omega \to \mathbb{R}$, where $\Omega = \{1, 2, \ldots, H\} \times \{1, 2, \ldots, W\}$. In the first pre-processing step, a linear transformation

$$I_{HU} = I_{raw} \times slope + intercept \tag{1}$$

was performed to convert the gray values stored in Digital Imaging and Communication in Medicine (DICOM) format to Hounsfield units (HU), where I_{raw} is the image with gray values stored in DICOM format, and I_{HU} is the transformed image in HU. *slope* and *intercept* are parameters retrieved from the DICOM header file.

Next, contrast adjustment was performed by extracting and scaling the predefined range of HU into 8-bit grayscale, as shown in (2).

$$I_{adj}(\boldsymbol{x}) = \begin{cases} 0 & I_{HU}(\boldsymbol{x}) < wc - \frac{ww}{2} \\ (I_{HU}(\boldsymbol{x}) - (wc - \frac{ww}{2})) \times \frac{255}{ww} & wc - \frac{ww}{2} \leq I_{HU}(\boldsymbol{x}) \leq wc + \frac{ww}{2} \\ 255 & I_{HU}(\boldsymbol{x}) > wc + \frac{ww}{2} \end{cases} \tag{2}$$

where $I_{adj}(\boldsymbol{x})$ is the intensity after contrast adjustment at location $\boldsymbol{x} \in \Omega$, and ww and wc are the window width and the window center obtained from the

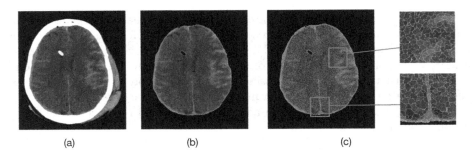

(a)   (b)   (c)

Fig. 2. An illustration of the proposed data preparation (Color figure online). (a) The image after the contrast adjustment. (b) The image after the skull stripping. (c) Superpixel generation.

DICOM header file, respectively. To highlight the appearance and structure of soft tissue while avoiding information loss in pathological tissues, a window width of 80 HU and window center of 80 HU was used in this study.

After the contrast adjustment, a skull stripping method described in [8] was followed to extract brain tissues. For each CT slice, a rectangular contour was initialized around the center of the head. Then, the distance regularized level set evolution algorithm [9] was used to evolve the initialized contour to fit the border of the brain region enclosed by the skull. An example of contrast adjustment and skull stripping is shown in Fig. 2(a)–(b). The image after skull stripping is denoted as I_b.

2.2   Superpixel Generation

After pre-processing, we used the simple linear iterative clustering (SLIC) algorithm [10] to over-segment I_b into superpixels. The SLIC algorithm generates a group of coherent pixel collections based on color and spatial proximity (shown in Fig. 2(c)). There are many advantages of using superpixels. First, instead of processing every image pixel, using superpixels where similar pixels are clustered can reduce computation cost efficiently. Secondly, superpixels divide the entire image into meaningful image patches. Features extracted from superpixels can better characterize regional information. Considering that superpixels adhere to edges within an image (as exhibited in Fig. 2(c)), image segmentation can be performed via superpixel classification. In this work, we performed feature extraction on superpixels and classify those superpixels as belonging to hematoma regions or not. Based on these classification results, a coarse hematoma segmentation can be generated. In this study, I_b was over-segmented into approximately 5000 superpixels (each superpixel includes approximately 30 pixels).

2.3   Feature Extraction

A total of 63 features were extracted to describe the characteristics of superpixels.

Intensity Statistics. The mean, variance, skewness, and kurtosis of intensities in each superpixel were calculated. The mean value measures the average intensity level while the variance measures heterogeneity. The skewness and kurtosis describe the asymmetry and the tailedness, respectively. As different ranges of Hounsfield units correspond to different anatomical structures, these intensity statistics can help to describe superpixels.

Gabor Filters. A Gabor filter is a linear filter used for edge detection and textural analysis. The real component of a Gabor filter (assumed to be centered at zero) can be written as

$$g(x, y; \lambda, \theta, \psi, \sigma, \gamma) = \exp\left(-\frac{x'^2 + \gamma^2 y'^2}{2\sigma^2}\right) \cos\left(2\pi \frac{x'}{\lambda} + \psi\right), \tag{3}$$

$$x' = x \cos\theta + y \sin\theta,$$
$$y' = -x \sin\theta + y \cos\theta,$$

where (x, y) is the location of the Gabor filter. λ and ψ represent the wavelength and phase offset of the sinusoidal wave, respectively. θ and σ represent the orientation and standard deviation of the Gaussian envelope, respectively, while γ is the spatial aspect ratio.

In this study, we used a bank of Gabor filters introduced in [11]. 2-D Gabor filters oriented at 0, 30, 60, 90, 120, and 150° with wavelengths of $2\sqrt{2}, 4\sqrt{2}, 8\sqrt{2}$, and $16\sqrt{2}$ were used to calculate the response map at $\gamma = 0.5$, $\psi = 0$ and $\sigma = 0.5\lambda$. The mean and variance of Gabor responses at each superpixel were calculated as Gabor features as well as the dominant spatial frequency and its orientation.

Saliency. Saliency can be constructed as visual attention. In this study, a low-level approach was employed to determine the saliency of a superpixel by computing the average Euclidean distance of its mean intensity with 50 other superpixels that were randomly selected from the same image. Different from other extracted features, the saliency value contains global information at the slice level. From our observation, CT slices located close to the top of the head have a higher intensity value due to the partial volume effect. The saliency measurement can suppress the effect from this slice-level intensity shift. Also, it can help reduce the variability in intensity for the same tissue across different cases.

Gray-Level Co-Occurrence Matrix. A 16×16 patch around the center of each superpixel was taken to calculate the gray-level co-occurrence matrix (GLCM), which gives the joint probability distribution of gray-level pairs of neighboring pixels. Let $\Omega_p = \{1, 2, \ldots, N_{level}\} \times \{1, 2, \ldots, N_{level}\}$, where N_{level} is the number of levels that gray intensities were quantized into. In this study, $N_{level} = 8$. Second-order statistics of the GLCM were used as features, specifically contrast, energy, and homogeneity, which are calculated as

$$contrast = \int_{\Omega_p} |i - j|^2 p(i, j) di \, dj, \qquad (4)$$

$$energy = \int_{\Omega_p} p(i, j)^2 di \, dj, \qquad (5)$$

$$homogeneity = \int_{\Omega_p} \frac{p(i, j)}{1 + |i - j|} di \, dj, \qquad (6)$$

where $p(i, j)$ is the value of GLCM at location (i, j).

Wavelet Packet Transformation. A two-level discrete Haar wavelet packet transformation [12] was applied to a 16×16 patch around the center of each superpixel. The image patch was decomposed into 8 bands, with each band containing information of different frequencies. The energy of coefficients in each band was computed and the percentages of energy corresponding to the details were used as regional features to characterize each superpixel.

2.4 Active Learning

Active learning is a method [13] to train a supervised classifier with the smallest annotated training dataset possible. As shown in Algorithm 1, the proposed active learning strategy started with training an initial SVM model using the initial training dataset, which consists of superpixels from only one labeled CT scan. After that, the initial model was used to classify superpixels from the pool dataset, which contains CT scans from 49 patients. Based on the predicted possibilities, we calculated the conditional Shannon entropy of each superpixel as

$$H_\Theta(s_i) = - \sum_{\hat{y} \in \{0,1\}} p_\Theta(y_i = \hat{y}|\boldsymbol{v}_i) \log(p_\Theta(y_i = \hat{y}|\boldsymbol{v}_i)), \qquad (7)$$

where Θ denotes the trained SVM model. $\boldsymbol{v}_i$ and y_i are the feature vector and label of s_i, respectively. $p_\Theta(y_i = \hat{y}|v)$ denotes the predicted probability that s_i belongs to the corresponding class.

$H_\Theta(s_i)$ is used as an uncertainty measurement for s_i. A high $H_\Theta(s_i)$ indicates that the trained model is uncertain about which class s_i belongs to. This may occur if s_i is under-represented in the current training dataset. Thus superpixels with high uncertainty values are the most informative samples to update the model. In our work, superpixels were ranked based on their uncertainty measurements in descending order and the top N_{al} superpixels were selected to be annotated and added into the training dataset. Next, an updated SVM model was trained and the uncertainty measurements of the superpixels in the pool dataset were re-calculated. After the final SVM classifier was trained, coarse hematoma segmentation maps were generated by classifying superpixels in brain images.

Algorithm 1. Active Learning Strategy

Input: Labeled training dataset $\mathcal{D}_l$, unlabeled pool dataset $\mathcal{D}_p$, the number of iterations N_{iter}, the number of samples selected for query at each iteration N_{al}
    Output: SVM classifier Θ^*
1: Train an initial SVM model Θ_0 on $\mathcal{D}_t$
2: **for** $k = 1$, k++, while $k <= N_{iter}$ **do**
3:    Use the trained SVM classifier Θ_{k-1} to measure the uncertainty of superpixels in $\mathcal{D}_p$.
4:    Select N_{al} most informative samples $\{v_1^{(k)}, v_2^{(k)}, ..., v_{N_{al}}^{(k)}\}$ from $\mathcal{D}_p$.
5:    Update $\mathcal{D}_p$: $\mathcal{D}_p = \mathcal{D}_p$ - $\{v_1^{(k)}, v_2^{(k)}, ..., v_{N_{al}}^{(k)}\}$
6:    Query the physician for labels $\{l_1^{(k)}, l_2^{(k)}, ..., l_{N_{al}}^{(k)}\}$ of the selected samples
7:    Update $\mathcal{D}_t$: $\mathcal{D}_t = \mathcal{D}_t \bigcup \{(v_1^{(k)}, l_1^{(k)}), (v_2^{(k)}, l_2^{(k)}), ..., (v_{N_{al}}^{(k)}, l_{N_{al}}^{(k)})\}$
8:    Train an SVM model Θ_k on the updated $\mathcal{D}_t$.
9: **end for**
10: Return $\Theta^* = \Theta_{N_{iter}}$

2.5   Active Contour Model

A region-based active contour model [14] was extended by incorporating coarse hematoma segmentation to improve the segmentation performance. In an active contour model, a dynamic contour evolves iteratively by minimizing the energy function. In this work, the energy function is defined as a combination of a region-based term, a prior shape term, and a regularization term. The shape term was added to give a penalty when the evolving contour at the tth iteration deviates from the prior shape.

Let $\phi : \Omega \to \mathbb{R}$, $\Omega = \{1, 2, \ldots, H\} \times \{1, 2, \ldots, W\}$ be the level function of an Euclidean signed distance function. The contour is represented by $\mathcal{C} = \{x \in \Omega | \phi(x) = 0\}$, where points inside the contour have $\phi(x) > 0$ and points outside the contour have $\phi(x) < 0$. Given an image I_b, let $\mathcal{C}_{SVM}$ denote the predicted hematoma contour from the SVM classifier, with ϕ_{SVM} the corresponding level set function. ϕ_{SVM} is used both as an initial contour $\phi^{(0)}$ for curve evolution and a prior shape for shape constraints. The energy function can be written as

$$E(\phi) = E_{region}(\phi) + \alpha_1 E_{shape}(\phi) + \alpha_2 E_{reg}(\phi), \tag{8}$$

$$E_{region}(\phi) = \int_{\Omega} |I_b(x) - c_1|^2 H(\phi(x))dx + \int_{\Omega} |I_b(x) - c_2|^2 (1 - H(\phi(x)))dx, \tag{9}$$

$$E_{shape}(\phi) = \int_{\Omega} (\phi(\boldsymbol{x}) - \phi_{SVM}(\boldsymbol{x}))^2 d\boldsymbol{x}, \tag{10}$$

where $H(\cdot)$ is the Heaviside function. $\nabla(\cdot)$ is the gradient operation. E_{reg} follows the formula proposed in [9], whose role is to maintain the signed distance property $|\nabla\phi(\boldsymbol{x})| = 1$ within the vicinity of the zero level set. c_1 and c_2 in (9) are the average of I_b inside and outside $\mathcal{C}$, respectively. They are calculated as

$$c_1(\phi) = \frac{\int_{\Omega} I_b(\boldsymbol{x})H(\phi(\boldsymbol{x}))d\boldsymbol{x}}{\int_{\Omega} H(\phi(\boldsymbol{x}))d\boldsymbol{x}} \tag{11}$$

and

$$c_2(\phi) = \frac{\int_{\Omega} I_b(\boldsymbol{x})(1 - H(\phi(\boldsymbol{x})))d\boldsymbol{x}}{\int_{\Omega} (1 - H(\phi(\boldsymbol{x})))d\boldsymbol{x}}. \tag{12}$$

The final contour can be obtained by using the gradient descent algorithm to minimize the energy function over ϕ:

$$\phi^{\star} = \operatorname*{argmin}_{\phi} E(\phi). \tag{13}$$

The active contour model proposed in this study was used to smooth the boundary of coarse hematoma segmentation.

2.6   Evaluation Metrics

To evaluate the performance of a trained SVM model on superpixel-based classification, the precision, recall, and accuracy were calculated, as well as the Dice coefficient between the coarse segmentation and manual segmentation. After the active contour algorithm, the final fine segmentation was evaluated by calculating pixel-based precision, recall, accuracy, and Dice coefficient between the final segmentation and manual segmentation. All measures were patient-wise and averaged over patients in the test set.

3   Experimental Results and Discussion

3.1   Dataset

Our dataset consists of 35 head CT scans from the Progesterone for Traumatic brain injury: Experimental Clinical Treatment (ProTECT) study [15] and 27 brain CT scans from University of Michigan Health System. The brain scans are from patients who experienced a moderate to severe head injury and were enrolled in an emergency department within 4 h of their injury. In total, 2433 axial CT images from 62 patients who suffered from acute TBI were used in this study, with image slice thickness ranging from 3.0 to 5.0 mm. To validation our proposed hematoma segmentation framework and active learning strategy, 13 cases were annotated as the test set by an experienced medical expert, who examined 2D cross-sectional slices and then manually drew the boundary around hematoma regions. We used the remaining 49 cases as the training set, wherein each experiment of active learning one slice was randomly selected and annotated as the initial training set while all others were used as the pool set.

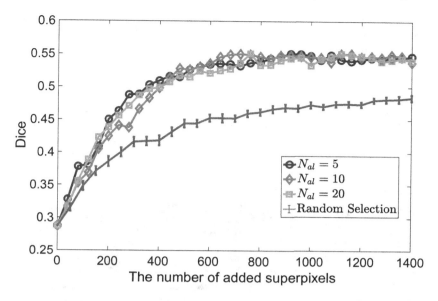

Fig. 3. Comparison of active learning with $N_{al} = 5, 10, 20$ and random selection. The Dice coefficients of random selection with different numbers of added superpixels are averaged over 50 experiments. 95% confidence intervals are also given.

3.2 Active Learning

An initial SVM model was trained on the initial training set using a linear kernel. We then used the active learning strategy to select the most informative superpixels, which were then annotated, gradually improving the performance of the model. Several experiments were performed to explore the performance of classifiers with the same initial training set while varying N_{al}. From Fig. 3, the curves tend to plateau after 1000 newly labeled superpixels are added with active learning, and the effect of N_{al} on the final performance is not significant. In contrast to active learning strategy, an SVM classifier was trained as a baseline on the initial training set and a fixed number of randomly selected superpixels from the pool set (shown as 'Random Selection' in Fig. 3). 50 independent experiments were performed and the results were averaged to represent the performance of using random selection. From Fig. 3, with the same number of added superpixels, the performance of the SVM model trained using the active learning strategy is significantly higher than the baseline. After adding over 1000 additional samples, the model trained with active learning achieved an average Dice coefficient of 0.55 over 13 patients.

To further examine the robustness of active learning algorithm over different initial datasets, we repeated the above active learning algorithm and random selection method for 20 times, respectively. For each time, one slice was randomly selected and annotated as the initial training set while other slices in the training set were used as the pool set. Each SVM classifier using active learning were trained

Table 1. Comparison of the active learning strategy and random selection method using 20 different initial training sets. n is the number of additional superpixels added to the initial training set. The mean and standard derivation (stddev) of evaluation measurements over 20 experiments are given in the format of mean (stddev).

	Dice	Precision	Recall	Accuracy
Active Learning ($n = 1000$)	0.55 (0.01)	0.59 (0.02)	0.60 (0.02)	0.97 (0.01)
Random Selection ($n = 1000$)	0.47 (0.02)	0.45 (0.02)	0.61 (0.03)	0.94 (0.01)
Random Selection ($n = 5000$)	0.54 (0.01)	0.57 (0.02)	0.60 (0.02)	0.96 (0.01)

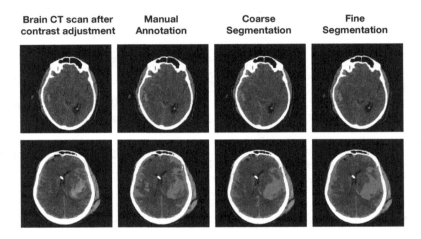

Fig. 4. Segmentation results. Each row gives an example of the segmentation compared with the manual segmentation. The segmented or annotated hematoma regions are shown in red (Color figure online).

on the initial training set and 1000 additional samples selected with $N_{al} = 5$. The comparison of performance metrics between active learning and random selection is shown in Table 1. Our final SVM models from active learning have comparable performance with random selection models trained on the same initial training sets added with five times more annotated superpixels. The standard derivations of measures over experiments are very small.

3.3   Segmentation

After constructing the coarse hematoma segmentation for each slice via a trained SVM classifier, the proposed active contour model was performed to refine the boundary. An additional 10 CT slices in the pool set were annotated to tune the parameters for active contour. Finally, we used $\alpha_1 = 2, \alpha_2 = 0.5$, $\epsilon = 0.05$ and the step size $\mu = 0.2$. From Fig. 4, the coarse segmentation has an accurate hematoma localization while the boundary is rough, which may be due to

Table 2. Segmentation performance comparison. The patient-wise mean and standard derivation (stddev) of evaluation measurements are given in the format of mean (stddev).

	Dice	Precision	Recall	Accuracy
Coarse segmentation	0.55 (0.21)	0.62 (0.23)	0.59 (0.18)	0.97 (0.03)
Fine segmentation	0.60 (0.19)	0.63 (0.23)	0.67 (0.16)	0.98 (0.03)
GMM [7]	0.46 (0.22)	0.45 (0.23)	0.60 (0.18)	0.94 (0.06)

superpixel sampling. The active contour model can help smooth the boundary and improve segmentation accuracy. From Table 2, the proposed active contour model greatly improved the segmentation performance. Additionally, our method significantly outperformed a previous method named GMM [7].

4   Conclusion

Automatic and accurate hematoma segmentation is important for TBI patient management. In this study, a combination of active learning and an active contour model was proposed for hematoma segmentation in acute cases. In supervised SVM training, statistical and textural features were extracted to characterize superpixels over-segmented from brain images. With the proposed active learning strategy, an SVM classifier was trained on a very small amount of annotated data. After that, the coarse segmentation from the SVM classifier was incorporated into the active contour model to generate fine hematoma segmentation. The overall segmentation method achieved a mean dice of 0.60 in a highly heterogeneous dataset. Our experiments show that an active learning strategy can effectively select the most informative data points from a highly imbalanced and varied data pool. From our results, active learning is potential to overcome the shortage of annotated data in medicine.

For future work, we will continue working on the active learning strategy. In our current framework, during the active learning, the selected superpixels will be highlighted in the brain images and presented to clinicians for annotation. Considering annotating 1000 superpixels is still not an easy task, further informativeness measurements will be designed to find the most informative image patches or CT slices to reduce the labor required in the annotation phase of active learning.

Acknowledgment. The work is supported by National Science Foundation under Grant No. 1500124.

References

1. Lee, B., Newberg, A.: Neuroimaging in traumatic brain imaging. NeuroRx **2**(2), 372–383 (2005)

2. Broderick, J.P., Brott, T.G., Duldner, J.E., Tomsick, T., Huster, G.: Volume of intracerebral hemorrhage. a powerful and easy-to-use predictor of 30-day mortality. Stroke **24**(7), 987–993 (1993)
3. Jacobs, B., Beems, T., van der Vliet, T.M., Diaz-Arrastia, R.R., Borm, G.F., Vos, P.E.: Computed tomography and outcome in moderate and severe traumatic brain injury: hematoma volume and midline shift revisited. J. neurotrauma **28**(2), 203–215 (2011)
4. Chesnut, R., Ghajar, J., Gordon, D.: Surgical management of acute epidural hematomas. Neurosurgery **58**(3), S2–7 (2006)
5. Bardera, A., et al.: Semi-automated method for brain hematoma and edema quantification using computed tomography. Comput. Med. Imaging Graph. **33**(4), 304–311 (2009)
6. Liao, C.C., Xiao, F., Wong, J.M., Chiang, I.J.: Computer-aided diagnosis of intracranial hematoma with brain deformation on computed tomography. Comput. Med. Imaging Graph. **34**(7), 563–571 (2010)
7. Soroushmehr, S.M.R., Bafna, A., Schlosser, S., Ward, K., Derksen, H., Najarian, K.: CT image segmentation in traumatic brain injury. In: 37th Annual International Conference of the IEEE Engineering in Medicine and Biology Society (EMBC), pp. 2973–2976. IEEE (2015)
8. Farzaneh, N., et al.: Automated subdural hematoma segmentation for traumatic brain injured (TBI) patients. In: 39th Annual International Conference of the IEEE Engineering in Medicine and Biology Society (EMBC), pp. 3069–3072. IEEE (2017)
9. Li, C., Xu, C., Gui, C., Fox, M.D.: Distance regularized level set evolution and its application to image segmentation. IEEE Trans. Image Process. **19**(12), 3243–3254 (2010)
10. Achanta, R., Shaji, A., Smith, K., Lucchi, A., Fua, P., Süsstrunk, S., et al.: Slic superpixels compared to state-of-the-art superpixel methods. IEEE Trans. Pattern Anal. Mach. Intell. **34**(11), 2274–2282 (2012)
11. Jain, A.K., Farrokhnia, F.: Unsupervised texture segmentation using gabor filters. Pattern Recognit. **24**(12), 1167–1186 (1991)
12. Rioul, O., Vetterli, M.: Wavelets and signal processing. IEEE Signal Process. Mag. **8**(4), 14–38 (1991)
13. Settles, B.: Active learning. Synth. Lect. Artif. Intell. Mach. Learn. **6**(1), 1–114 (2012)
14. Chan, T.F., Vese, L.A.: Active contours without edges. IEEE Trans. Image Process. **10**(2), 266–277 (2001)
15. Wright, D.W., et al.: Protect: a randomized clinical trial of progesterone for acute traumatic brain injury. Ann. Emerg. Med. **49**(4), 391–402 (2007)

DYNLO: Enhancing Non-linear Regularized State Observer Brain Mapping Technique by Parameter Estimation with Extended Kalman Filter

Andrés Felipe Soler Guevara[1]([✉]), Eduardo Giraldo[2], and Marta Molinas[1]

[1] Norwegian University of Science and Technology, Trondheim, Norway
{andres.f.soler.guevara,marta.molinas}@ntnu.no
[2] Universidad Tecnológica de Pereira, Pereira, Colombia
egiraldos@utp.edu.co

Abstract. The underlying activity in the brain can be estimated using methods based on discrete physiological models of the neural activity. These models involve parameters for weighting the estimated source activity of previous samples, however, those parameters are subject- and task-dependent. This paper introduces a dynamical non-linear regularized observer (DYNLO), through the implementation of an Extended Kalman Filter (EKF) for estimating the model parameters of the dynamical source activity over the neural activity reconstruction performed by a non-linear regularized observer (NLO). The proposed methodology has been evaluated on real EEG signals using a realistic head model. The results have been compared with least squares (LS) for model parameter estimation with NLO and the multiple sparse prior (MSP) algorithm for source estimation. The correlation coefficient and relative error between the original EEG and the estimated EEG from the source reconstruction were inspected and the results show an improvement of the solution in terms of the aforementioned measurements and a reduction of the computational time.

Keywords: EEG-based Brain Mapping ·
Extended Kalman Filter EKF ·
Non-Linear Regularized Observer NLO · Dynamic inverse solution

1 Introduction

Electroencephalography (EEG) is a non-invasive technique for recording information of the electrical activity in the brain through the measurement of electrical potentials using electrodes in the scalp. The signals contain information with a high temporal resolution and the analysis of the data has become a useful tool to diagnose different forms of brain disorders like Epilepsy, Parkinson, sleep or memory disorders. In addition, the EEG information can be used to identify the localization of the neural activity in the brain through the use of brain mapping techniques. Nevertheless, the inverse problem technique used for estimating the

© Springer Nature Switzerland AG 2019
I. Rojas et al. (Eds.): IWBBIO 2019, LNBI 11466, pp. 397–406, 2019.
https://doi.org/10.1007/978-3-030-17935-9_36

underlying activity from EEG signals has several challenges due to the ill-posed and the ill-conditioned characteristics of the problem. To help coping with these problems, an infinity configuration of brain activity estimation is proposed to explain the measured EEG signals, and hence, the regularization solutions can be applied to overcome the aforementioned challenges.

Several brain mapping methods have been presented in the last two decades, some of them are based on estimation theory with probabilistic frameworks like the multiple sparse priors MSP [1] and the dynamic multi-model source localization method DYNAMO [2]. Other methods are based on regularized solutions like minimum norm estimation (MNE), weighted minimum norm estimation (WMNE), low resolution tomography (LORETA), iterative regularization algorithm (IRA) and non-linear regularized observer (NLO) [3–7]. The NLO method involves a physiological model that represents the evolution of the activity in the brain. However, this method has a high dependence on the model parameters and these values change between subjects and sessions [7].

The Extended Kalman Filter (EKF) is a widely used estimator in non-linear processes. Due to the non-linear properties of EEG signals, the EKF can be a suitable estimator for identifying parameters in EEG data. Some applications of EKF to EEG and MEG signals have been reported for tracking the dipole source location [8], inverse problem solution and source estimation [2], and for noise reduction and filtering in [9]. This paper considers the implementation of an EKF for the step of model parameter estimation to create a dynamical non-linear regularized observer -DYNLO, to improve the brain mapping solution. The results of the implementation are evaluated comparing the EKF estimated EEG over a dataset of P300 visual evoked potentials VEP using MSP and NLO methods with LS parameter estimation.

2   Materials and Methods

2.1   EEG Forward Model

The equation for relating the EEG signals measured in the scalp with the brain activity is known as the EEG forward problem, and it can be represented as shown in Eq. 1

$$\boldsymbol{y}_k = \boldsymbol{M}\boldsymbol{x}_k + \varepsilon_k \tag{1}$$

Where $\boldsymbol{y}_k \in \mathbb{R}^{d \times N}$ contains the EEG signals of d number of electrodes and N number of samples, $\boldsymbol{x}_k \in \mathbb{R}^{n \times N}$ represents the source activity inside the brain that produces the measured electrical impulses. n is the number of distributed sources considered in the brain. For relating the measured EEG $\boldsymbol{y}_k$ and the neural activity $\boldsymbol{x}_k$, the lead field matrix $\boldsymbol{M} \in \mathbb{R}^{d \times n}$ is introduced, this matrix is obtained from magnetic resonance images (MRI) and numerical methods e.g. Finite Element Method (FEM) or Boundary Element Method (BEM), which usually involves the skin, skull, cerebrospinal fluids (CSF) and brain matter to establish the position of sources and their relationship with electrodes. In addition, a white noise with zero mean and C_ε covariance is considered. The subscript k represents the time instant of the sampled data.

2.2 Non-Linear Brain Activity Model

The activity in the brain exhibits a highly non-linear behaviour and can be viewed as a non-linear dynamical system as stated in [10] where the evolution of the activity considers previous samples as expressed in the following equation

$$x_k = a_1 x_{k-1} + a_2 x_{k-2} + a_3 x_{k-\tau} + a_4 x_{k-2}^{\circ 2} + a_5 x_{k-2}^{\circ 3} + \eta_k \tag{2}$$

The terms $x_k^{\circ 2}$ and $x_k^{\circ 3}$ represent the Hadamard power of x_k and a_i terms are the model parameter of the non-linear representation. Equation 2 shows that the actual activity depends of $i - th$ model parameters and on the previously sampled data x_{k-1}, x_{k-2} and $x_{k-\tau}$, where the $\tau - th$ sample is a feedback due to the activity of nearby neurons and depends on the sampling frequency of the EEG data. η_k is the noise in the activity, which is considered to follow a normal distribution with zero mean and covariance C_η. To simplify the notation, the Eq. 2 can be represented as a multiplication of the matrix G_k with ω_k, as shown below

$$x_k = G_k \omega_k + \eta_k \tag{3}$$

Where the parameters a_i are represented in a vector ω_k denominated model parameter vector as in Eq. 4 and the temporal activity matrix G_k can be formed by the concatenation of the activity in previously sampled data as in Eq. 5.

$$\omega_k = \begin{bmatrix} a_1 & a_2 & a_3 & a_4 & a_5 \end{bmatrix}^T \tag{4}$$

$$G_k = \begin{bmatrix} \hat{x}_{k-1} & \hat{x}_{k-2} & \hat{x}_{k-\tau} & \hat{x}_{k-2}^{\circ 2} & \hat{x}_{k-2}^{\circ 3} \end{bmatrix} \tag{5}$$

2.3 Extended Kalman Filter for Model Parameter Estimation

The non-linear model parameter vector ω_k that represents the dynamical behavior of the brain activity can be estimated using the EKF, where two standard Kalman filter steps are developed: the prediction and the correction steps. Initially in the prediction step, the *a priori* information is calculated in each instance k using the following equations

$$\hat{\omega}_{\bar{k}} = \hat{\omega}_{k-1} \tag{6}$$

$$P_{\bar{\omega}_k} = P_{\omega_{k-1}} + R_{k-1}^r \tag{7}$$

Where the term $P_{\omega_k}^-$ is the *a priori* predicted covariance of model parameters, $\omega_{\bar{k}}$ is the *a priori* predicted model parameters, $P_{\omega_{k-1}}$ is the predicted covariance in the previous sample, and R_{k-1}^r is a diagonal matrix called the innovation covariance and is defined as shown below.

$$R_{k-1}^r = (\gamma^{-1} - 1) P_{\omega_{k-1}} \tag{8}$$

Being $\gamma^{-1} \in \mathbb{R}^{[0,1]}$ a forgetting factor that represents the dependence of previous information, e.g. when it value tends to zero, the estimation has a strong dependence of previous estimations. The EKF gain matrix $\boldsymbol{K}_{\omega_k}$ can be computed as it is shown in Eq. 9 below

$$\boldsymbol{K}_{\omega_k} = \boldsymbol{P}_{\omega_k}^- \boldsymbol{G}_k^T \boldsymbol{M}^T (\boldsymbol{G}_k \boldsymbol{M} \boldsymbol{P}_{\omega_k}^- \boldsymbol{G}_k^T \boldsymbol{M}^T + \boldsymbol{R}^e)^{-1} \tag{9}$$

The correction step allows to estimate the set of parameters $\hat{\omega}_k$ and update the covariance of the parameters $\boldsymbol{P}_{\omega_k}$ using following Eqs. 10 and 11.

$$\hat{\omega}_k = \omega_{\bar{k}} + \boldsymbol{K}_{\omega_k}(\boldsymbol{y}_k - \boldsymbol{M} \boldsymbol{G}_k \hat{\omega}_{k-1}) \tag{10}$$

$$\boldsymbol{P}_{\omega_k} = (\boldsymbol{I} - \boldsymbol{K}_{\omega_k} \boldsymbol{M} \boldsymbol{G}_k) \boldsymbol{P}_{\omega_k}^- \tag{11}$$

2.4   Non-Linear Regularized Observer

The non-linear regularized observer is defined by Eq. 12, where the inverse problem can be addressed as an optimization problem with constrains based on l_2 norm. The cost function is formed by three terms, where the first term represents the basic inverse solution with least squares; adding the second term, the solution involves a spatial constraint like in brain mapping methods MNE and LORETA presented in [5] and [6] respectively. The third term considers the time evolution of the activity and is treated as a temporal constraint as presented in [3] and [7]. The minimization problem involves the non-linear model, taking into account the activity in previously sampled data as shown in Eqs. 2 to 5.

$$J = ||\boldsymbol{M} \boldsymbol{x}_k - \boldsymbol{y}_k||_2^2 + \rho_k^2 ||\boldsymbol{x}_k||_2^2 + \lambda_k^2 ||\boldsymbol{x}_k - \boldsymbol{G}_k \omega_k||_2^2 \tag{12}$$

The variable ρ_k is the spatial regularization parameter and λ_k is the temporal regularization parameter. The estimation of the activity can be performed by Eq. 13, where the adaptive solution depends on the model parameter estimated with the EKF.

$$\hat{\boldsymbol{x}}_{k(\rho_k, \lambda_k, \hat{\omega}_k)} = (\boldsymbol{M}^T \boldsymbol{M} + \rho_k^2 \boldsymbol{I} + \lambda_k^2 \boldsymbol{I})^{-1} (\boldsymbol{M}^T \boldsymbol{y}_k + \lambda_k^2 \boldsymbol{G}_k \hat{\omega}_k) \tag{13}$$

The lead field matrix M can be decomposed by singular values decomposition (SVD), where M can be represented by $M = USV^T$. By applying SVD of M on Eq. 13 it is possible to reduce the computational cost to estimate the activity, especially when the inverse of $(\boldsymbol{M}^T \boldsymbol{M} + \rho_k^2 \boldsymbol{I} + \lambda_k^2 \boldsymbol{I})$ in Eq. 13 is computed. The estimated neural activity $\hat{\boldsymbol{x}}$ involving the M SVD is shown in Eq. 14

$$\hat{\boldsymbol{x}}_{k(\rho_k, \lambda_k, \hat{\omega}_k)} = \boldsymbol{V}(\boldsymbol{S}^2 + \rho_k^2 \boldsymbol{I} + \lambda_k^2 \boldsymbol{I})^{-1} \boldsymbol{V}^T (\boldsymbol{M}^T \boldsymbol{y}_k + \lambda_k^2 \boldsymbol{G}_k \omega_k) \tag{14}$$

3   Experimental Framework

For evaluating the proposed method, a dataset of P300 visual evoked potentials described in [11] was used. The protocol of the records consisted on six images displayed in a screen, which were flashed in random sequences with a duration of 100 ms with a resting time of 300 ms in between images. The subjects of this experiment were requested to count the times that a specific image appeared. The EEG signals from 8 subjects (four of them with neurological deficit called dysarthria or hypophonia) were recorded from 32 channels localized according to the 10–20 international system with a sample rate of 2048 Hz.

A head model is required for solving the inverse problem, therefore, we use a realistic brain model with $n = 8196$ distributed sources in the cortical surface. This model was computed with 70 electrodes on the scalp using the 10-10 system layout. The used model corresponds to the first subject of the dataset presented in [12]. The head model has 30 common electrodes with the EEG, hence, the distributed model was reduced to 30 according to the 10–20 system used in the EEG dataset. In addition, the EEG signals were organized to coincide with the channels' positions of the brain model.

Figure 1 shows the 30 electrodes and their distribution according to the 10–20 layout used in the EEG recordings. It additionally shows the 8196 distributed sources and how the electrodes are located in the scalp around the brain. The procedure followed for processing the data of each subject is explained by the next steps:

– *Pre-processing:* The average signal from the two mastoid electrodes was used for referencing each one of the channels. In addition, the EEG channels were organized according to the head model order for electrodes, where the two electrodes Fp1 and Fp2 from dataset were discarded, because the used head model does not consider them in the forward model.

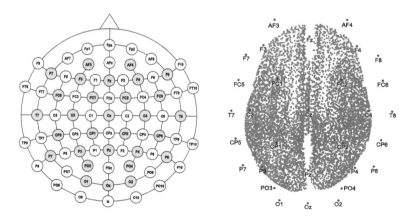

Fig. 1. Left: Head model electrodes layout according to 10–10 system and 30 electrodes (yellow) used under 10–20 layout. Right: Brain model with $n = 8196$ distributed sources (blue) and the 30 electrodes position (red) (Color figure online)

- *Inverse Solution:* The EKF is iteratively used to estimate the model parameters $\hat{\omega}_k$ for the NLO method (DYNLO). NLO with LS and MSP methods
  were used for estimating the neural activity $\hat{x}_k$, where the activity for each
  one of the 8196 distributed sources were found.
- *Forward Problem:* The EEG signals were estimated using the following equation, similar to Eq. 1.

$$\hat{y}_k = M\hat{x}_k \tag{15}$$

- *Evaluation:* The estimated EEG signals using DYNLO, NLO-LS and MSP,
  were computed and compared with the original EEG signals used for the
  source estimation procedure. Two performance measurements were assessed:
  the relative error and the correlation coefficient shown in Eqs. 16 and 17
  respectively. These measurements were calculated for the 192 available runs.

$$\varepsilon_r = \frac{||\hat{y}_k - y_k||_2^2}{||y_k||_2^2} \tag{16}$$

In addition, the correlation coefficient was calculated for evaluating the similarity between the estimated signal and the original referenced EEG. To compute
the variable, the following equation is used:

$$C_c = \frac{1}{N-1} \sum_{k=1}^{N} \left(\frac{y_k - \mu_{y_k}}{\sigma_{y_k}} \right) \left(\frac{\hat{y}_k - \mu_{\hat{y}_k}}{\sigma_{\hat{y}_k}} \right) \tag{17}$$

Where the term μ and σ represent the mean and standard deviation respectively, being N the number of samples of y_k.

4   Results and Discussion

Figure 2 depicts the mean and the standard deviation of the relative error for
the 192 runs with each one of the three methods. The mean of the relative errors

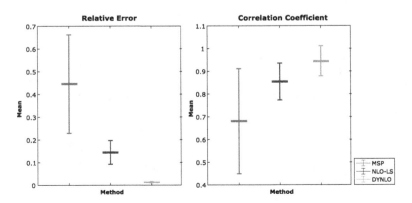

Fig. 2. Performance measurements over 192 runs of EEG data. Left: Mean and standard deviation of Relative error. Right: Mean and standard deviation of Correlation
Coefficient

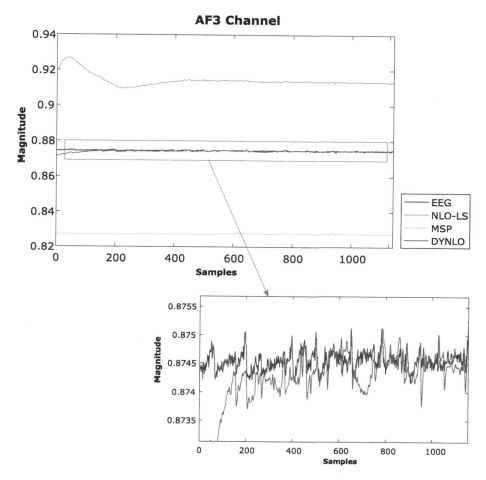

Fig. 3. Top: AF3 channel in measured EEG (blue), estimated with MSP (yellow), NLO-LS (red) and DYNLO (purple). Bottom: zoom of estimated AF3 channel with NLO EKF and original EEG (Color figure online)

were $\varepsilon_{r(MSP)} = 0,4459$, $\varepsilon_{r(NLO-LS)} = 0,1446$ and $\varepsilon_{r(DYNLO)} = 0,01271$. The lowest value of relative error was obtained when using DYNLO. In addition, the standard deviation was the lowest with DYNLO.

Additionally, Fig. 2 also shows the mean and the standard deviation of the correlation coefficient for the 192 runs with each one of the three methods. The mean of the correlation coefficients were $C_{c(MSP)} = 0,6803$, $C_{c(LS)} = 0,8544$ and $C_{c(EKF)} = 0,9459$. Considering that in this case a higher value of correlation is desired, the DYNLO has obtained the best performance correlation and standard deviation.

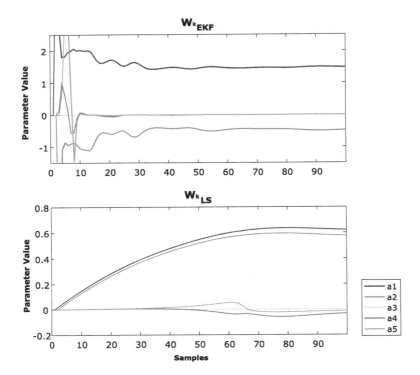

Fig. 4. Evolution of model parameter estimation by EKF in DYNLO (Top) and LS in NLO-LS (Bottom)

The estimated EEG signal for a single channel with each one of the methods and the original EEG signal are shown in Fig. 3. This figure shows that the DYNLO has the best fitting with the original EEG signal compared to NLO-LS and MSP. It is also evident in the lower plot that as time passes, the fitting is improving. In addition, it is noticeable that DYNLO has a good estimation of the signal amplitude, where NLO-LS and MSP present a higher difference.

Figure 4 shows the time evolution of the estimated model parameters using the EKF and LS for one of the runs, where the model values describe an asymptotic behavior. However, the EKF stabilizes the parameters faster than LS,

Table 1. 192 runs mean parameters estimated with EKF and LS and static NLO model parameter values

Method/Parameters	a_1	a_2	a_3	a_4	a_5
Static parameters	1.9023	−0.9100	0.0067	1.1921e-04	−2.3842e-05
EKF	1.6477	−0,6472	−0.0004717	−1.2045e-06	1.4701-07
LS	0.5994	0.5637	−0.003596	−0.004754	0.002082

situation that translates in a better performance in the correlation coefficient and the relative error. Furthermore, as stated in [7], the model parameters depend of the sampling frequency and the type of activity to be analyzed. i.e normal activity or seizure. In the case of the EEG dataset used in this study, the activity is considered normal and the sampling frequency is 2048 Hz. Therefore, the parameters can be calculated according to [7], where their values of the dataset frequency are presented in Table 1. The table also includes the mean parameters of the estimations with EKF and LS methods.

When evaluating and comparing the computational costs of the two the iterative methods, DYNLO and NLO-LS, the computational time for estimating the model parameter vector $\hat{\omega}_k$ using EKF in DYNLO is 0.835 ms per sample, meanwhile, with LS in NLO, the required time per sample is 1.494 ms. From these results it is seen a reduction of computational time by 45% in the calculation of $\hat{\omega}_k$. The time measurements where taken in a computer with the following characteristics: RAM 16GB , processor Core i7-4790, OS windows 10, 64-bit and executing the algorithms in Matlab® 2016b.

5   Conclusions and Future Work

The estimation of EEG signal parameters using EKF improves the brain activity estimation with NLO. A much better performance in the correlation between signals and a lower relative error were obtained with the DYNLO. In addition to these advantages in performance, the computational time was lower when using the EKF, which is a desirable feature to enable faster results in clinical analysis and essential for the use of brain mapping techniques in brain computer interfaces (BCI).

According to the results presented in this paper, the DYNLO approach provides faster brain mapping solutions which can be useful for real-time applications to study brain disorders/diseases, emotions, and memory processing evoked responses.

Generally, brain mapping methods require the information of high number of channels, which will result in high computational times. Nevertheless, the computational time and the response time towards real-time applications, could be improved by reducing the number of electrodes and using a lower sampling frequency. New experiments are currently being performed with the aforementioned reduction of channels and sampling frequency, and will be reported in the near future.

Acknowledgment. This work was carried out under the funding of Norwegian University of Science and Technology NTNU project "David and Goliath: single-channel EEG unravels its power through adaptive signal analysis" and the Departamento Administrativo Nacional de Ciencia, Tecnología e Innovación (Colciencias). Research project: 111077757982 "Sistema de identificación de fuentes epileptogénicas basado en medidas de conectividad funcional usando registros electroencefalográficos e imágenes de resonancia magnética en pacientes con epilepsia refractaria: apoyo a la cirugía resectiva".

References

1. Friston, K., et al.: Multiple sparse priors for the M/EEG inverse problem. NeuroImage **39**(3), 1104–1120 (2008)
2. Antelis, J.M., Minguez, J.: Dynamo: concurrent dynamic multi-model source localization method for EEG and/or MEG. J. Neurosci. Methods **212**(1), 28–42 (2013)
3. Giraldo-Suarez, E., Martinez-Vargas, J.D., Castellanos-Dominguez, G.: Reconstruction of neural activity from EEG data using dynamic spatiotemporal constraints. Int. J. Neural Syst. **26**(07), 1650026 (2016)
4. Grech, R.: Review on solving the inverse problem in EEG source analysis. J. NeuroEngineering Rehabil. **5**, 25 (2008)
5. Hauk, O.: Keep it simple: a case for using classical minimum norm estimation in the analysis of EEG and MEG data. NeuroImage **21**(4), 1612–1621 (2004)
6. Pascual-Marqui, R.D.: Review of methods for solving the EEG inverse problem. Intl. J. Biomagn. **1**(1), 75–86 (1999)
7. Soler, A.F., Muñoz-Gutiérrez, P.A., Giraldo, E.: Regularized state observers for source activity estimation. In: Wang, S., et al. (eds.) BI 2018. LNCS (LNAI), vol. 11309, pp. 195–204. Springer, Cham (2018). https://doi.org/10.1007/978-3-030-05587-5_19
8. Yao, Y., Swindlehurst, A.L.: Tracking single dynamic meg dipole sources using the projected extended kalman filter. In: 2011 Annual International Conference of the IEEE Engineering in Medicine and Biology Society, pp. 4365–4368, August 2011
9. Walters-Williams, J., Li, Y.: Comparison of extended and unscented Kalman filters applied to EEG signals. In: IEEE/ICME International Conference on Complex Medical Engineering, pp. 45–51, July 2010
10. Soler, A.F., Giraldo, E.: An adaptive nonlinear regularized observer for neural activity reconstruction. In: 2018 IEEE ANDESCON, pp. 1–5, August 2018
11. Hoffmann, U., Vesin, J.-M., Ebrahimi, T., Diserens, K.: An efficient P300-based brain-computer interface for disabled subjects. J. Neurosci. Methods **167**(1), 115–125 (2008). Brain-Computer Interfaces (BCIs)
12. Henson, R.N., Wakeman, D.G., Litvak, V., Friston, K.J.: A parametric empirical bayesian framework for the EEG/MEG inverse problem: generative models for multi-subject and multi-modal integration. Front. Human Neurosci. **5**(August), 1–16 (2011)

Analysis of the Behavior of the Red Blood Cell Model in a Tapered Microchannel

Mariana Ondrusova[(⊠)] and Ivan Cimrak

Faculty of Management Science and Informatics,
Cell-in-Fluid Biomedical Modeling and Computations Group,
University of Zilina, Zilina, Slovakia
mariana.ondrusova@fri.uniza.sk

Abstract. Red blood cells are flexible during their movement in microchannels, they adapt easily to their immediate environment and eventually return to their relaxed shape as soon as the environment exerts no forces on the cell. This behaviour is determined by elastic properties of red cell's membrane which must be carefully taken in consideration when creating the computational model of red blood cell.

In our work we use previously developed model of red blood cell that employs five basic elastic moduli, each representing a different part of the overall elastic properties. The aim of this work is to assess the validity of such model. To this end we analyse behaviour of cells in tapered channels. By adapting to the flow conditions, cell begins to perform a certain type of repetitive motion. We focus on tank-treading and stretching. In this article we show that the model of red blood cell is capable of reconstruction these motions and we show quantitative and qualitative measures comparing the experimental and simulation data. We provide the analysis and description of these movements, and we study the effect of the cell's initial position on the type of motion to be performed.

Keywords: Computational modeling · Red blood cell · Simulation

1 Introduction

Biological experiments have their limitations. For example, when working with red blood cells, it is their viability. Examples of such biological experiments include for example the stretching of red blood cell by optical tweezers [1], or shear-induces release of specific proteins from red blood cells [2]. One has to ensure the possibility of repeating the experiment, its reproducibility. With biological experiments, this has been a major issue. This is why people have been looking for other options, for example using computer simulations. Simulations of biological processes stem from biological experiments, but they do not suffer from reproducibility issues due to the problems with cells' viability. Simulations also offer us wider possibilities. Variation of results in computational models is larger than it would be using the experiments. Finally, their great advantage is that they do not need blood samples or other biological samples.

Computer models we use are based on complex solutions of mathematical equations. As well, elastic properties of membranes are based on calculations involving

© Springer Nature Switzerland AG 2019
I. Rojas et al. (Eds.): IWBBIO 2019, LNBI 11466, pp. 407–417, 2019.
https://doi.org/10.1007/978-3-030-17935-9_37

description of mechanical phenomena. We call them elastic moduli. Each module is characterized by different property of the red blood cell membrane. An example is a local modulus of elasticity that preserves the elasticity of the membrane at the site of its stretching and a gradual return of the cell to its original shape. Another example would be a volume conservation modulus that retains the total cell volume at any shape change. More about elastic modules can be found in [3–5].

The model of a red blood cell can be embedded in a different simulation environments. Under the notion simulation environment, we understand the physical dimensions and geometry of micro channels, the fluidic conditions like velocity, or pressure gradient. The channels may vary by larger or smaller dimensions, they can include narrowing channels, or channels with various obstacles. Inside the channel, a suspension of cells is modelled together with fluid solver mimicking the real cell flow in the microfluidic channels. This way, the behavior of simulation models is directly comparable with biological experiments.

By adapting to the simulation environment and conditions, the cell begins to show a certain type of repetitive motion in dynamic processes. For example tumbling, tank-treading, or stretching. We focus on the description of these movements, on the percentage of occurrence under certain conditions in the simulation channel, and on the effect of the cell's initial position on the type of motion to be performed.

In Sect. 2 we will describe the basics of our simulation model with description of the elastic properties. In Sect. 3, we focus on the simulation environment, including channel walls, initial conditions, fluid properties and various cell settings. Section 4 describes the results that compare the cell stretching in the biological experiment and in the simulation, subject to flow in different channel types. Finally in Sect. 5 we draw conclusions.

2   Model of Red Blood Cell

During simulation, different interactions occur at each time step:

- between the objects and the fluid,
- between the fluid and the walls of the channel,
- between the objects and the walls of the channel

In simulations with multiple objects, we also have interactions:

- between objects themselves

Objects move by means of fluid. Liquid in the channel is described by the lattice-Boltzmann method. The space in which the fluid is located consists of points in the cubic grid. The grid remains fixed throughout the simulation. The fluid is formed by fictitious points that interfere with each other and thus transfer information about themselves - the direction of movement and speed. You can find more information in [6].

Objects in our case represent red blood cells. The basis of each red blood cell model is the flexible triangular network located on the surface of the cell - Fig. 1. We define different forces between the elements of the network (edges, triangles, angles between triangles) to model different elastic moduli that are representatives of elastic properties. We have five elastic moduli:

- stretching – preserves the lengths of individual edges and thus contributes to the preservation of the overall shape of the cell when stretched
- bending – while deforming, it maintains the angles between the neighbouring triangles
- local area conservation – preserves the area of the triangles
- global area conservation – preserves the total surface area of the cell
- volume conservation – preserves the total cell volume

Analysis of individual moduli can be found in [7–11]. Concrete set up for the red blood cell model is provided in Sect. 3.4 in Table 1.

For relation object - liquid it is important to mention the friction coefficient. Note that the fluid consists of points in the fixed lattice grid. However, the position of the points of the triangular network of cells does not always match the positions of the grid points. The friction coefficient equalizes the velocity of the fluid points and the velocity of the cell points [11].

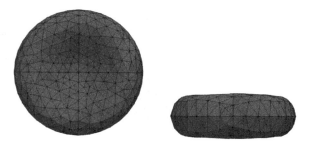

Fig. 1. Simulation model of cell with triangular network.

3 Simulation Environment

With different setting of the simulation environment, we can easily affect cell behavior. This accounts for inner geometry of the channels such as wall heights, widths, position and size of obstacles, etc.

3.1 Channel Parameters

Inspiration for our simulation environment was the biological research presented in [2] that uses a channel, which gradually narrows. [2] describes two different simulation environments with two different lengths of the narrow part. We tested the flow of cells in both, but for detailed information, we only worked with one type. Both types are described in Fig. 2.

For the first type of the simulation environment, they used the following dimensions: initial channel width is $w = 100$ μm, in the place of constriction this width reduces to $w_c = 20$ μm and the length of the constriction is $l_c = 100$ μm. In the second type, some dimensions remain unchanged such as w – initial channel width and w_c – constriction's width, but some are changed: l_c – length of the constriction from 100 to 800 μm. Overall height was preserved over the entire height of the channel being 38 μm.

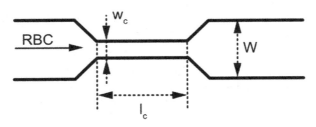

Fig. 2. Schematic view of the channel from the top. Parameters of the channel are chosen as in [2]: initial width $w = 100$ μm, width of the narrow part $w_c = 20$ μm, height of channel was 38 μm, l_c was variable due to current experiments from 100 to 800 μm.

The simulation environment can be seen in Figs. 3, 4 and 5. Figure 3 represents the total closed channel for $l_c = 100$ μm. For a better idea, the top and bottom walls are drawn

Fig. 3. Channel with length in constriction $l_c = 100$ μm without fluid, the top and bottom walls are transparent for better visualization.

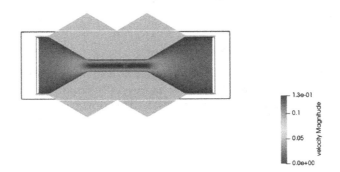

Fig. 4. Channel with constriction length $l_c = 100$ μm, without top and bottom walls, with fluid.

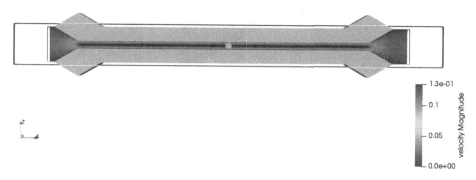

Fig. 5. Channel with length in constriction l_c = 800 μm, without top and bottom walls, with fluid.

transparently. Figure 4 is a view of the inside of the channel for l_c = 100 μm with simulation fluid, Fig. 5 shows simulation channel for l_c = 800 μm with fluid. The magnitude of fluid velocity in the channel is color-coded with the scale depicted on the right.

3.2   Fluid in the Channel

In our simulations, the fluid is moved either by setting a constant inflow over the channel boundaries, or by setting an external fluid force density over the entire channel. We use the latter approach.

We want to obtain the same fluid velocity in our simulations as in [2]. There, the cells were injected into the channel under a specific pressure achieving a steady flow rate 3 μl/min. To reach the same flow rate in our simulations, we used different external fluid force densities until we determined a value that gave us desired flow rate that corresponds to the biological experiment. More details about the individual values for the simulation environment are presented in Table 1.

3.3   Positions and Rotation of Cells in the Channel

To provide variety of initial cell seedings similar to the biological experiment, we placed the cells in twelve different locations in the channel and we used twenty positions with different rotations of the cells. The following section shows and explains all options. First, position of cells (Fig. 6) were chosen to cover the substantial channel width.

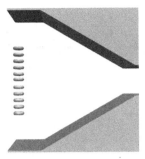

Fig. 6. Eleven positions of cells in their initial position.

Then we chose twenty different rotated positions Fig. 7.

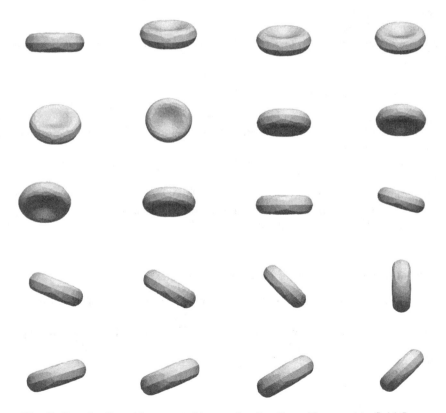

Fig. 7. Set of cell positions rotated in x and z-direction with respect to fluid flow.

3.4   Other Settings

In the previous section, we have already described the channel parameters, the full settings for our simulations are shown in Table 1.

Table 1. Simulation setup

Timestep	10^{-7} s
Stretching modulus	5.56×10^{-6} N/m
Bending modulus	3.4×10^{-18} N.m
Local area conservation modulus	2.54×10^{-5} N/m
Global area conservation modulus	7×10^{-4} N/m
Volume conservation modulus	9×10^2 N/m^2
Radius of RBC	4×10^{-6} m
Friction coefficient	3.39×10^{-9} N.s/m
Fluid density	9.9×10^2 kg/m^3
Fluid viscosity	8.9×10^{-7} m^2/s
External fluid force for $l_c = 100$ μm	1.139×10^{-12} N
External fluid force for $l_c = 800$ μm	2.086×10^{-12} N

4   Results

While moving in dynamic systems, the shape of cell must change in order to adapt to the simulation environment, obstacles, fluid pressure, etc. Simulation results showed that the cells after entering the narrow part perform two basic movements: the stretching and the tank-treading. These two movements were the same for all cells regardless of the initial position or rotation. The cells retained their approximate position relative to the width of the channel after entering the narrowed part: Cells moving from the upper half of the channel remained in the upper part of the narrowed section, while the cells entering the constriction from the lower half remained below. Cells in the central part stayed all the way in the center of the channel.

4.1   Quantitative Comparison with Experiments

The cell has its relaxed state and also the state in which it is maximally stretched. The relative stretching of a cell can be quantified by the ratio of cell stretch size from relaxed to maximal position:

$$L = \frac{L_{max}}{L_0} \tag{1}$$

where L_{max} is the diameter of maximally stretched cell and L_0 is length of cell in relaxed state.

In our case, maximum stretching of the cell was reached at the entrance to the narrowed part of the channel with a width $w_c = 20$ μm. The cell moving from any position and any rotation reached the maximum extension upon entering the channel. Deformation is due to the fact that the cell at the front acts at a higher fluid velocity than the cell at the back. When exiting from the narrowed section, the cell again changes its shape under the influence of fluid velocity. The front part will be pushed back into the cell, while the back part of the cell still retains the original velocity. The cell velocity and the resulting shape are based on a preset flow rate. For example, we chose a random cell with a random position and rotation. More about changing cell shapes in Fig. 8.

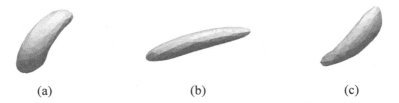

(a)   (b)   (c)

Fig. 8. Snapshots of a cell initially seeded in one third of the channel width. The change in the shape of one of the cells in the transition from the initialization position (a) to the inside of the narrowed portion (b). The cell begins to spin under the influence of the fluid before the narrow part (a), and further, at the time of entering the narrowed part (b). The shape of the cell at the exit from the narrowed part and the re-entry into the wider part (c).

Our cell dimensions are $L_0 = 8.00$ μm and $L_{max} = 12.70$ μm. The lengths of the cell in a relaxed state and after the maximum extension are consistent with [2]. Comparison of biological experiment and simulation model is on Fig. 9 and on Fig. 10. The cell velocity in the narrowed part gradually decreases. The lengths of the cell at the exit from the narrowed to wide channel are 10.09 μm. This confirms that the maximum stretching of the cell occurs at the entrance to the narrowed part of the channel.

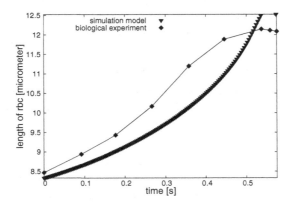

Fig. 9. Gradual stretching of the simulation model of a cell compared to a cell from a biological experiment during time. Shows an increase from lowest to highest value in channel with constriction length $l_c = 100$ μm.

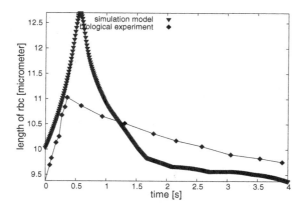

Fig. 10. Stretching of the simulation model of a cell compared to a cell from a biological experiment during time. Figure shows value for stretching of the cells from the beginning to the half width of the narrowed section in channel with constriction length $l_c = 800$ μm.

4.2   Qualitative Comparison with Experiments

Our aim is to verify whether the modelled cell expresses the same type of behaviour as was observed in the experiments. In the previous section we already described one type of the behaviour, the stretching.

The second behaviour is tank-treading. This movement is documented in several research studies [12]. Tank-treading can be described as rotation of the cell membrane around its interior. The cell is also stretched in this case.

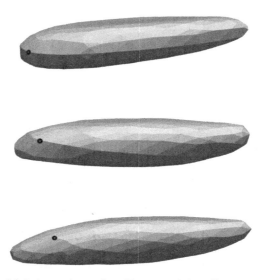

Fig. 11. Our cell model during tank-treading. The core of the cell appears static, we see motion on the membrane.

Tank-treading is caused by the fluid velocity difference at the location on top and bottom of the cell. Such difference occurs nearly in any non-uniform flow and can be quantified by the fluid shear rate. In cases of very low shear rate, tank-treading is replaced by tumbling when the cell almost preserves its biconcave shape while rotating. We performed simulations with $11 \times 20 = 220$ different starting positions of the cell, resulted from combinations of 11 different locations and 20 different rotations of cell. We analyzed data from all these experiments and we categorized each movement whether it is single stretching, or tank treading. After entering the narrowed section, tank-treading motion in simulations was observed for the cells with off-center initial location. This is in accordance with experimental observations from [13] and is linked to the shear rate. In central section of the channel, the shear rate is almost negligible while towards the channel walls, shear rate increases to its maximal value. An example of tank-treading cell is depicted in Fig. 11.

5   Conclusion

Our simulation results show interesting findings. Apparently, the initial cell orientation does not influence whether the cell only stretches or it also tank-treads. The initial location however does plays role and it in fact determines, whether the cell tank-treads or not. We showed that only cells travelling exactly along the central axis of the channel do not undergo tank-treading and that they solely stretch.

We also confirmed that the cell undergoes the largest deformation upon the entrance to the narrow part. This is clearly visible in Figs. 9 and 10, where a drop of the maximal prolongation occurs immediately after the cell enters the narrowed portion of the channel. Although the curves in Figs. 8 and 9 do not fit experimental data perfectly, the trend is clearly visible. The misfit can be caused by variable elastic properties of the red cells due to its size or age [14].

Acknowledgements. This work was supported by the Slovak Research and Development Agency (contract number APVV-15-0751) and by the Ministry of Education, Science, Research and Sport of the Slovak Republic under the contract No. VEGA 1/0643/17.

References

1. Dao, M., Lim, C.T., Suresh, S.: Nonlinear elastic and viscoelastic deformation of the human red blood cell with optical tweezers. Mol. Cell. Biomech. **1**(3), 169–180 (2004)
2. Wan, J., Ristenpart, W.D., Stone, H.A.: Dynamics of shear-induced ATP release from red blood cells. PNAS **105**(43) 16432–16437(2008)
3. Ondrušová, M., Cimrák, I.: Dynamical properties of red blood cell model in shear flow. In: IDT, pp. 288–292 (2017)
4. Kovalčíková, K., Bohiniková, A., Slavík, M., Mazza Guimaraes, I., Cimrák, I.: Red blood cell model validation in dynamic regime. In: Rojas, I., Ortuño, F. (eds.) IWBBIO 2018. LNCS, vol. 10813, pp. 259–269. Springer, Cham (2018). https://doi.org/10.1007/978-3-319-78723-7_22

5. Cimrák, I., Gusenbauer, M., Jančigová, I.: An ESPResSo implementation of elastic objects immersed in a fluid. Comput. Phys. Commun. **185**(3), 900–907 (2014)
6. Dünweg, B., Ladd, A.J.C.: Lattice Boltzmann simulations of soft matter systems. In: Holm, C., Kremer, K. (eds.) Advanced Computer Simulation Approaches for Soft Matter Sciences III. Advances in Polymer Science, vol. 221, pp. 89–166. Springer, Heidelberg (2009). https://doi.org/10.1007/978-3-540-87706-6_2
7. Jančigová, I., Tóthová, R.: Scalability of forces in mesh-based models of elastic objects. In: 2014 ELEKTRO, Rajecke Teplice, pp. 562–566 (2014)
8. Slavík, M., Bachratá, K., Bachratý, H., Kovalčíková, K.: The sensitivity of the statistical characteristics to the selected parameters of the simulation model in the red blood cell flow simulations. In: 2017 International Conference on Information and Digital Technologies (IDT), Zilina, pp. 344–349 (2017)
9. Bachratý, H., Bachratá, K., Chovanec, M., Kajánek, F., Smiešková, M., Slavík, M.: Simulation of blood flow in microfluidic devices for analysing of video from real experiments. In: Rojas, I., Ortuño, F. (eds.) IWBBIO 2018. LNCS, vol. 10813, pp. 279–289. Springer, Cham (2018). https://doi.org/10.1007/978-3-319-78723-7_24
10. Tóthová, R., Jančigová, I., Bušík, M.: Calibration of elastic coefficients for spring-network model of red blood cell. In: 2015 International Conference on Information and Digital Technologies, Zilina, pp. 376–380 (2015)
11. Bušík, M.: Development and optimization model for flow cells in the fluid. Dissertation thesis, p. 114 (2017)
12. Krüger, T., Gross, M., Raabe, D., Varnik, F.: Crossover from tumbling to tank-treading-like motion in dense simulated suspensions. Soft Matter **9**, 9008–90015 (2013)
13. Shi, L., Yu, Y., Glowinski, R.: Oscillating motions of neutrally buoyant particle and red blood cell in Poiseuille flow in a narrow channel. Phys. Fluids **26**, 041904 (2014)
14. Ward, K.A., Baker, C., Roebuck, L., et al.: Red blood cell deformability: effect of age and smoking. AGE **14**, 73 (1991)

Radiofrequency Ablation for Treating Chronic Pain of Bones: Effects of Nerve Locations

Sundeep Singh[1] and Roderick Melnik[1,2(✉)]

[1] MS2Discovery Interdisciplinary Research Institute, Wilfrid Laurier University, 75 University Avenue West, Waterloo, ON N2L 3C5, Canada
rmelnik@wlu.ca
[2] BCAM - Basque Center for Applied Mathematics, Alameda de Mazarredo 14, 48009 Bilbao, Spain

Abstract. The present study aims at evaluating the effects of target nerve location from the bone tissue during continuous radiofrequency ablation (RFA) for chronic pain relief. A generalized three-dimensional heterogeneous computational model comprising of muscle, bone and target nerve has been considered. The continuous RFA has been performed through the monopolar needle electrode placed parallel to the target nerve. Finite-element-based coupled thermo-electric analysis has been conducted to predict the electric field and temperature distributions as well as the lesion volume attained during continuous RFA application. The quasi-static approximation of the Maxwell's equations has been used to compute the electric field distribution and the Pennes bioheat equation has been used to model the heat transfer phenomenon during RFA of the target nerve. The electrical and thermo-physical properties considered in the present numerical study have been acquired from the well-characterized values available in the literature. The protocol of the RFA procedure has been adopted from the United States Food and Drug Administration (FDA) approved commercial devices available in the market and reported in the previous clinical studies. Temperature-dependent electrical conductivity along with the piecewise model of blood perfusion have been considered to correlate with the *in-vivo* scenarios. The numerical simulation results, presented in this work, reveal a strong dependence of lesion volume on the target nerve location from the considered bone. It is expected that the findings of this study would assist in providing *a priori* critical information to the clinical practitioners for enhancing the success rate of continuous RFA technique in addressing the chronic pain problems of bones.

Keywords: Bone · Chronic pain · Continuous radiofrequency ablation · Nerve ablation · Computational modeling · Pennes bioheat equation

1 Introduction

Chronic pain is one of the most common problems of advancing age. Although, conservative management (physical therapy and analgesics such as nonsteroidal anti-inflammatory drugs) is effective in chronic pain treatments, it only confers short-term benefits, as it is expensive, and may have significant adverse effects. The primary goal

© Springer Nature Switzerland AG 2019
I. Rojas et al. (Eds.): IWBBIO 2019, LNBI 11466, pp. 418–429, 2019.
https://doi.org/10.1007/978-3-030-17935-9_38

of the physicians responsible for the management of chronic pain is to target long-term solutions rather than short-lived interventions [1]. Nonsurgical minimally invasive options for treating the chronic pain have also surged during the past decades that are focused on targeting nerves transmitting pain signals. Importantly, for ablation of nervous system elements during chronic pain management, there are several main nonsurgical treatment modalities, viz., cryoablation (use of extreme cold), various laser therapies, including recent "point-and-shoot" techniques, high temperature radiofrequency, and chemical neurolysis, such as alcohol or phenol. Among the available nonsurgical modalities to alleviate chronic pain, radiofrequency ablation (RFA) offers the advantage of being precise, reproducible, cheap and effective to a great extent [1].

The application of RFA is well pronounced for treating various tumors in liver, lung, kidney, bones, prostate, and breast [2]. However, with regard to the chronic pain management, the first study of RFA application was reported in 1960s [3] and subsequently several noticeable studies have been reported for the treatment of chronic low back pain [4], chronic hip pain [5], chronic knee pain [6], and chronic head ache [7]. Importantly, during RFA, a high frequency alternating current in the frequency range of 500 kHz is applied in the vicinity of a nerve via an electrode, leading to neurodestructive thermocoagulation, thereby degrading its ability to conduct pain signals. The lesion produced due to the resistive heating during the RFA procedure may give pain relief for 12–18 months or longer, with minimal side effects and associated complications [8]. Further, since the largest area of thermal lesion is produced along the axis of electrode during continuous RFA, the radiofrequency electrode is precisely placed parallel to the target nerve so as to maximize the damage to the adjacent nerve fibers. Several patient specific clinical studies of RFA modality in treating chronic pain of bones have already been performed and reported in the literature, though questions regarding anatomical targets, selection criteria, and evidence for effectiveness, are still prevalent [9].

Numerical modeling and simulations provide a powerful tool to predict such important characteristics as the temperature distribution and lesion volume during RFA. They give a quick, convenient and inexpensive *a priori* information during the treatment planning stage of the modality to the medical practitioners. The present study is one of the initial efforts, focusing on the mathematical modeling of RFA in treating chronic pain within the bones. The main motivation and important novelty of this study is to quantify the effect of target nerve distance from the bone on the efficacy of the continuous RFA procedure.

2   Mathematical Modeling of the RFA Procedure

A three-dimensional simplified model comprising of muscle, bone and nerve tissue has been considered in the present numerical study as depicted in Fig. 1. The RFA procedures have been performed utilizing a 22-gauge needle monopolar electrode inserted parallel to the periphery of target nerve. The active tip length of the RF electrode has been considered to be 5 mm [10]. In what follows, the effect of spacing between the target nerve and the bone tissue on the efficacy of the continuous RFA will be quantified. Importantly, these studies have been performed for different values of spacing between the nerve and the bone, viz., 0 mm (no spacing), 3 mm and 5 mm. The material properties considered in the present study are given in Table 1 [11–13]. The initial voltage and initial temperature of the entire computational domain have been considered to be 0 V and 37 °C, respectively. A constant voltage source has been applied at the active tip length of the RF electrode. The dispersive ground electrode has been modelled by utilizing a zero voltage electric potential on the outer boundaries of the analyzed domain. At each interface of our computational domain, electrical and thermal continuity boundary conditions have been imposed.

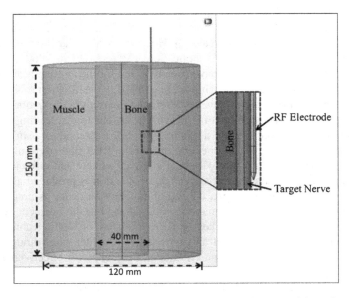

Fig. 1. Three-dimensional heterogeneous computational domain comprising of muscle, bone, nerve and RF electrode. (Color figure online)

Table 1. Electric and thermo-physical properties of different materials applied in this study (500 kHz) [11–13].

Material (Tissue/Electrode)	Electrical conductivity σ [S/m]	Specific heat capacity c [J/(kg · K)]	Thermal conductivity k [W/(m · K)]	Density ρ [kg/m³]	Blood perfusion ω_b [s⁻¹]
Muscle	0.446	3421	0.49	1090	6.35×10^{-4}
Bone	0.0222	1313	0.32	1908	4.67×10^{-4}
Nerve	0.111	3613	0.49	1075	3.38×10^{-3}
Plastic	10^{-5}	1045	0.026	70	–
Electrode	7.4×10^6	480	15	8000	–
Blood	–	3617	–	1050	–

A simplified version of Maxwell's equations, known as quasi-static approximation, has been used to compute the electric field distribution within the computational domain. It is given by

$$\nabla \cdot (\sigma(T)\nabla V) = 0, \tag{1}$$

where σ is the temperature-dependent electrical conductivity (S/m) and V is the electric potential (V), which is related to the electric field "E" (V/m) by the standard potential field relationship:

$$\mathbf{E} = -\nabla V. \tag{2}$$

Further, the current density "J" (A/m^2) can be obtained from the electrical conductivity and field as follows:

$$\mathbf{J} = \sigma \mathbf{E}. \tag{3}$$

The volumetric heat generation rate Q_p (W/m^3) due to RFA is evaluated by

$$Q_p = \mathbf{J} \cdot \mathbf{E}. \tag{4}$$

The heat transfer phenomenon during RFA of target nerve has been analyzed based on the application of Pennes bioheat equation

$$\rho c \frac{\partial T}{\partial t} = \nabla \cdot (k\nabla T) - \rho_b c_b \omega_b (T - T_b) + Q_m + Q_p, \tag{5}$$

where ρ is the density (kg/m^3), c is the specific heat (J/(kg $\cdot$ K)), k is the thermal conductivity (W/(m $\cdot$ K)), ρ_b is the density of blood (1050 kg/m^3), c_b is the specific heat capacity of blood (3617 J/(kg $\cdot$ K)), ω_b is the blood perfusion rate (1/s), Q_p is the heat source generated by RF power (W/m^3) and is computed by using Eq. 4 above, Q_m is the metabolic heat generation (W/m^3) that has been neglected in present study, see, e.g., [12] for relevant motivations, T_b is the blood temperature (37 °C), T is the tissue temperature computed from Eq. 5 and t is the time (s).

We note that the model for the heat transfer used here allows several straightforward generalizations. Firstly, a fully coupled thermoelastic model, accounting for possible nonlinear effects, would be a natural way to improve on the present consideration of the Pennes bioheat equation. Such models, along with efficient numerical procedures for their implementations, were developed before in the context ultrashort-pulsed lasers (e.g., [13–15]), with an increasing range of medical applications currently in place. A number of them are connected with nanosecond pulsed laser heating techniques [16], as well as with thermoplasmonics applications in medicine [17], where the goal is to create highly specific therapies by inactivating dysfunctional protein molecules (the field known as molecular hyperthermia). Another avenue for an extension of the current model is to account for the finite speed of heat propagation through thermal relaxation models (e.g., [18, 19]). Such extensions have not been developed in the context of RFA procedures and they warrant a separate investigation.

In the present computational study, the tissue blood perfusion rate has been modelled using a piecewise model [20], whereby the blood perfusion rate remains at a constant predefined value below the tissue temperature of 50 °C and beyond that temperature it ceases due to the collapse of microvasculature, and is given by

$$\omega_b(T) = \begin{cases} \omega_{b,0} & \text{for } T < 50\,°C \\ 0 & \text{for } T \geq 50\,°C \end{cases}, \tag{6}$$

where $\omega_{b,0}$ is the constant blood perfusion rate of the different tissues given in Table 1 and T is the unknown temperature computed from Eq. 5.

It is noteworthy to mention that, the lethal temperature range during RFA of nerve ablation is considered to be at or above 45–50 °C [10]. Henceforth, in the present numerical study, the ablation volume (V) has been quantified by the isotherm of 50 °C (i.e. the volume within the computational domain having temperature ≥ 50 °C for the post-RFA procedure). It is given by [21]

$$\dot{V} = \iiint_\Omega dV(\text{mm}^3) \quad (\text{where } \Omega \geq 50\,°C). \tag{7}$$

A finite element method based on the COMSOL Multiphysics 5.2 software [22] has been used to solve the coupled thermo-electric problem of the nerve ablation for treating chronic pain of bones. The computational domain has been discretized using a heterogeneous tetrahedral mesh elements using COMSOL's built-in mesh generator. A further refinement closer to the active tip of the electrode has been applied so as to accurately capture the electrical and thermal gradients. A convergence analysis has been carried out with the mesh refinement study. The final mesh comprises of 174486 elements and 476384 degrees of freedom. Further, the tissue thermal conductivity (k) has been modeled as constant [12], while the tissue electrical conductivity has been modeled by a linearly increasing (+2% per °C) function of temperature [2]. The coupled thermo-electric problem has been solved using the "multifrontal massively parallel sparse direct solver" (MUMPS) for estimating the electric field and the iterative conjugate gradient method for solving the temperature field. The maximum absolute tolerance has been set to be 10^{-3} for the time-dependent solver and the method used was based on the backward differentiation formula (BDF).

3   Results and Discussion

During the RFA procedure, ionic agitation is induced as the high-frequency alternate current interacts with the biological tissue, which is transformed into heat and propagate into the more peripheral areas by virtue of thermal conduction. The generated ionic (frictional) heating within the biological tissue leads to thermal coagulation within a few seconds above the temperature of 50 °C. Moreover, the temperatures beyond 100 °C during the RFA procedure would result in tissue boiling, vaporization and charring that usually leads to a drastic decline in electrical and thermal conductivities of the biological tissues [23, 24]. The charring phenomenon results in the abrupt rise in the electrical impedance of the tissue, thereby limiting any further conduction of the thermal energy to more peripheral areas from the RF electrode [2]. Thus, the charred tissue around the RF electrode acts as a barrier that limits the energy deposition and reduces the lesion size generated during the RFA procedure. To address this charring problem of RFA, in our computational study we carry out a sensitivity analysis to estimate the applied voltage at the active length of the RF electrode. The goal of this sensitivity analysis has been to find an appropriate value of the applied voltage whereby the maximum temperature does not exceeds 100 °C (i.e. leading to charring), at least till the first 50 s of the RFA procedure. The applied voltage values during the sensitivity study have been varied from 10 to 25 V, that basically lies within the limit of United States Food and Drug Administration (FDA) approved commercial RFA devices available in the market and reported in the previous clinical studies [25]. The time at which charring temperatures occurred has been found to be 50 s, 10 s and 4 s for the applied voltages of 15 V, 20 V and 25 V, respectively. Further, for the applied voltage of 10 V, although no charring has taken place till the 3 min duration of RFA, the maximum temperature for this voltage was limited to only 72.67 °C. Based on the conducted sensitivity study, applied voltage of 15 V has been found to be the most appropriate level of applied energy for the computational study that can restrict the occurrence of charring at least close to the first minute of the RFA procedure.

As mentioned earlier, one of our main motivations for the present numerical study is to evaluate the effect of target nerve location from the considered bone on the efficacy of the RFA procedure. Primarily, three locations have been selected, viz., (a) no gap between the target nerve and the bone (0 mm gap), (b) 3 mm distance between the target nerve and the bone, and (c) 5 mm distance between the target nerve and the bone. The choice of selection of such distances between the target nerve and the bone during the RFA application for chronic pain relief has been motivated by [26]. Figure 2 depicts the variation of the temperature at the tip of RF electrode with ablation time for the three cases mentioned above. As evident from Fig. 2, the temperature profile is on the lower side when there is no gap between the nerve and the bone. For the cases based on 3 mm and 5 mm distances between the target nerve and the bone, the time at which charring occurred has been found to be 35 s and 31 s, respectively, as compared to the 50 s for the case of 0 mm gap.

Further, the propagation of the damage to the target nerve (corresponding to the isotherm of 50 °C) with treatment time has been presented in Fig. 3. As follows from the analysis of this figure, the ablation volume after 30 s of the RFA procedure is on a lower side for the case having no gap between the target nerve and the bone. This can be attributed to the fact that the bones have lower thermal and electrical conductivities as compared to those of nerve and muscle (see Table 1). The lower electrical and thermal conductivities of the bones in the heterogeneous computational model do not allow an efficient conduction of the heat on one side of the electrode. This would results in either, more requirements on the input energy with the same treatment time or in an increase in the treatment time with the same input energy, for attaining an identical volume of nerve ablation. Thus, the distance between the bone and the target nerves significantly effects the efficacy of the RFA procedure during nerve ablation, by either resulting in the increment of the treatment time or requirements of more energy input.

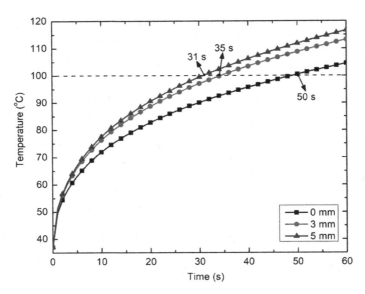

Fig. 2. Temperature distribution as a function of time monitored at the tip of the RF electrode for three different cases, viz., 0 mm, 3 mm and 5 mm distance between the target nerve and the bone. (Color figure online)

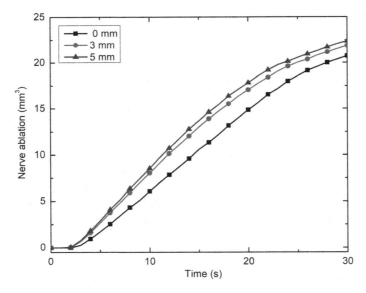

Fig. 3. Variation of nerve damage as a function of treatment time for different cases considered in the present study. (Color figure online)

Figure 4 presents the temperature distribution at a point that lies within the centre of the nerve and along a perpendicular direction from the tip of the RF electrode. As evident from Fig. 4, the time required for attaining the lethal temperature of 50 °C at the centre of the target nerve during the RFA procedure varies significantly. The time required for attaining the lethal temperature at the centre of nerve slightly decreases as the distance between the target nerve and the bone increases during the RFA procedure.

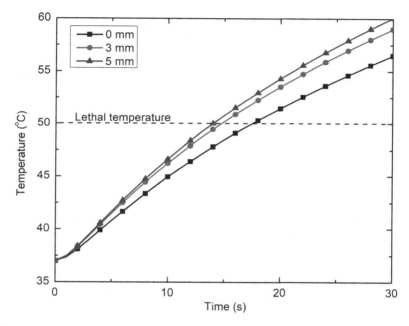

Fig. 4. Temperature distribution as a function of time monitored at the centre of nerve along perpendicular axis from the tip of the RF electrode for different cases considered in the present study. (Color figure online)

Figure 5 shows the variation of temperature along a line drawn perpendicular to the electrode axis from the tip of the RF electrode after 30 s of the RFA procedure. It can be seen from Fig. 5 that there prevails an asymmetry in the temperature profile from the electrode axis on the nerve side. Further, the asymmetric variation is more pronounced for the cases where the nerve is positioned closer to the bone tissues. This asymmetric nature could be partly attributed to the lower electrical and thermal conductivities of the bone surrounding the nerve tissue and partly due to the higher blood perfusion rate of the nerve in comparison to the muscle tissue, as seen from the analysis of Table 1. The consequence of this heterogeneous variation in the electrical and thermo-physical properties of considered tissues is that the ionic (frictional) heat generated during RFA would be easily propagated to the muscle side as compared to the nerve side. Thus, one of the severe drawbacks of such a technique could be excessive damage to the muscle tissue or some critical structures surrounding this area. This justifies an increasing demand in patient-specific modelling and simulations for performing safe and reliable RFA of the target nerve by optimising the thermal dosage. It is expected that future studies will be addressing this issue by developing more realistic patient-specific models in collaboration with hospitals and clinicians.

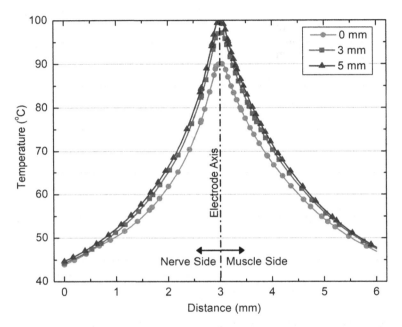

Fig. 5. Temperature distribution obtained after 30 s of the RFA procedure as a function of distance from the RF electrode axis (measured perpendicular to electrode tip). (Color figure online)

Finally, we note that Fig. 6 represents the propagation of the lesion volume attained with different time steps during the RFA procedure for the case having no gap between the target nerve and the bone. Again, the asymmetric variation can be clearly seen from the analysis of this figure, which is more pronounced for lower time steps and diminishes with the passage of time. One of the limitations of the present computational study is the lack of experimental validation of the proposed model. However, to the best of authors' knowledge, no experimental data is available in the literature for addressing this issue, specifically while considering heterogeneous surroundings. Although several experimental studies previously reported in the literature are pertinent to the present study, they have been performed utilizing only homogeneous agar-based phantom [12], egg white phantom [27], and ex-vivo liver [10].

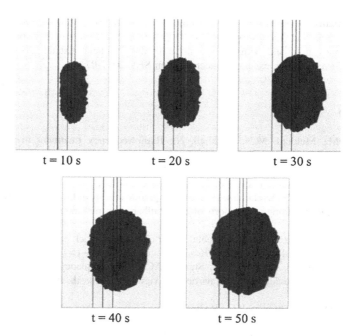

t = 10 s   t = 20 s   t = 30 s

t = 40 s   t = 50 s

Fig. 6. Propagation of the lesion volume (corresponding to the 50 °C isotherm) obtained for different time steps during the RFA procedure. (Color figure online)

4 Conclusion

In this work, a numerical study has been carried out to evaluate appropriate values of the applied voltage suitable for performing the continuous RFA of the target nerve mitigating the occurrence of charring, an undesirable phenomenon, observed at temperatures greater than 100 °C. Further, the effect of the distance between the target nerve and the analyzed bone has been evaluated in the context of the efficacy of the continuous RFA. Based on the results obtained from this study, it has been found that there prevails a strong dependence of the distance between the target nerve and the bone on the outcomes of the RFA. The nerve ablation volume decreases for the target nerve located closer to the bone and vice versa. It is expected that the results presented in this study would assist the researchers to better recognize the variation in the outcomes of RFA associated with the target nerve location. Subsequently, as our study demonstrates, it is critical to focus on evaluating the effect of heterogeneous surrounding in the analysis of nerve ablation for treating chronic pains, rather than focusing only on oversimplified studies based on the homogeneous surrounding assumption. We also expect that future studies in this field will increasingly connect to patient-specific data, assisting the clinicians in better optimizing the thermal dosages for enabling safe and reliable RFA applications in pain mitigation.

Acknowledgements. Authors are grateful to the NSERC and the CRC Program for their support. RM is also acknowledging support of the BERC 2018–2021 program and Spanish Ministry of Science, Innovation and Universities through the Agencia Estatal de Investigacion (AEI) BCAM Severo Ochoa excellence accreditation SEV-2017-0718.

References

1. Soloman, M., Mekhail, M.N., Mekhail, N.: Radiofrequency treatment in chronic pain. Expert Rev. Neurother. **10**(3), 469–474 (2010)
2. Singh, S., Repaka, R.: Temperature-controlled radiofrequency ablation of different tissues using two-compartment models. Int. J. Hyperthermia. **33**(2), 122–134 (2017)
3. Sweet, W.H., Mark, V.H., Hamlin, H.: Radiofrequency lesions in the central nervous system of man and cat: including case reports of eight bulbar pain-tract interruptions. J. Neurosurg. **17**, 213–225 (1960)
4. Leggett, L.E., et al.: Radiofrequency ablation for chronic low back pain: a systematic review of randomized controlled trials. Pain Res. Manage. **19**(5), 146–154 (2014)
5. Bhatia, A., Yasmine, H., Philip, P., Steven, P.C.: Radiofrequency procedures to relieve chronic hip pain: an evidence-based narrative review. Reg. Anesth. Pain Med. **43**(1), 72–83 (2018)
6. Bhatia, A., Philip, P., Steven, P.C.: Radiofrequency procedures to relieve chronic knee pain: an evidence-based narrative review. Reg. Anesth. Pain Med. **41**(4), 501–510 (2016)
7. Abd-Elsayed, A., Kreuger, L., Wheeler, S., Robillard, J., Seeger, S., Dulli, D.: Radiofrequency ablation of pericranial nerves for treating headache conditions: a promising option for patients. Ochsner J. **18**(1), 59–62 (2018)
8. Collighan, N., Richardson, J.: Radiofrequency lesioning techniques in the management of chronic pain. Anaesth. Intensive Care Med. **9**(2), 61–64 (2008)
9. Jamison, D.E., Cohen, S.P.: Radiofrequency techniques to treat chronic knee pain: a comprehensive review of anatomy, effectiveness, treatment parameters, and patient selection. J. Pain Res. **11**, 1879 (2018)
10. Cosman Jr., E.R., Cosman Sr., E.R.: Electric and thermal field effects in tissue around radiofrequency electrodes. Pain Med. **6**(6), 405–424 (2005)
11. Hasgall, P.A., Gennaro, F.D., Baumgartner, C., et al.: IT'IS database for thermal and electromagnetic parameters of biological tissues. Version 4.0, 15 May 2018. https://doi.org/10.13099/vip21000-04-0.itis.swiss/database
12. Ewertowska, E., Mercadal, B., Muñoz, V., Ivorra, A., Trujillo, M., Berjano, E.: Effect of applied voltage, duration and repetition frequency of RF pulses for pain relief on temperature spikes and electrical field: a computer modelling study. Int. J. Hyperthermia. **34**(1), 112–121 (2018)
13. Zhang, S., Dai, W., Wang, H., Melnik, R.V.N.: A finite difference method for studying thermal deformation in a 3D thin film exposed to ultrashort-pulsed lasers. Int. J. Heat Mass Transf. **51**(7–8), 1979–1995 (2008)
14. Wang, H.J., Dai, W.Z., Melnik, R.V.N.: A finite difference method for studying thermal deformation in a double-layered thin film exposed to ultrashort pulsed lasers. Int. J. Therm. Sci. **45**(12), 1179–1196 (2006)
15. Wang, H., Dai, W., Nassar, R., Melnik, R.V.N.: A finite difference method for studying thermal deformation in a thin film exposed to ultrashort-pulsed lasers. Int. J. Heat Mass Transf. **49**(15–16), 2712–2723 (2006)

16. Liu, J., et al.: Behavior of human periodontal ligament cells on dentin surfaces ablated with an ultra-short pulsed laser. Scientific Reports **7**(1) (2017). Article number 12738
17. Kang, P., et al.: Molecular hyperthermia: spatiotemporal protein unfolding and inactivation by nanosecond plasmonic heating. Small **13**(36), (2017). Article number 1700841
18. Melnik, R.V.N., Strunin, D.V., Roberts, A.J.: Nonlinear analysis of rubber-based polymeric materials with thermal relaxation. Numer. Heat Transf. Appl. **47**(6), 549–569 (2005)
19. Strunin, D.V., Melnik, R.V.N., Roberts, A.J.: Coupled thermomechanical waves in hyperbolic thermoelasticity. J. Therm. Stresses **24**(2), 121–140 (2001)
20. Scott, S.J., Salgaonkar, V., Prakash, P., Burdette, E.C., Diederich, C.J.: Interstitial ultrasound ablation of vertebral and paraspinal tumours: parametric and patient-specific simulations. Int. J. Hyperthermia. **30**(4), 228–244 (2014)
21. Singh, S., Repaka, R., Al-Jumaily, A.: Sensitivity analysis of critical parameters affecting the efficacy of microwave ablation using Taguchi method. Int. J. RF Microw. Comput. Aided Eng. e2158 (2018). https://doi.org/10.1002/mmce.21581
22. COMSOL Multiphysics® v. 5.2. www.comsol.com. COMSOL AB, Stockholm
23. Makimoto, H., Metzner, A., Tilz, R.R., et al.: Higher contact force, energy setting, and impedance rise during radiofrequency ablation predicts charring: new insights from contact force-guided in vivo ablation. J. Cardiovasc. Electrophysiol. **29**(2), 227–235 (2018)
24. Zhang, B., Moser, M.A.J., Zhang, E.M., Luo, Y., Liu, C., Zhang, W.: A review of radiofrequency ablation: large target tissue necrosis and mathematical modelling. Physica Med. **32**(8), 961–971 (2016)
25. Calodney, A., Rosenthal, R., Gordon, A., Wright, R.E.: Targeted radiofrequency techniques. In: Racz, G.B., Noe, C.E. (eds.) Techniques of neurolysis, pp. 33–73. Springer, Cham (2016). https://doi.org/10.1007/978-3-319-27607-6_3
26. Franco, C.D., Buvanendran, A., Petersohn, J.D., Menzies, R.D., Menzies, L.P.: Innervation of the anterior capsule of the human knee: implications for radiofrequency ablation. Reg. Anesth. Pain Med. **40**(4), 363–368 (2015)
27. Heavner, J.E., Boswell, M.V., Racz, G.B.: A comparison of pulsed radiofrequency and continuous radiofrequency on thermocoagulation of egg white in vitro. Pain Physician **9**(2), 135–137 (2006)

Biomedical Image Analysis

Visualization and Cognitive Graphics in Medical Scientific Research

Olga G. Berestneva[1,2], Olga V. Marukhina[1,2(✉)],
Sergey V. Romanchukov[1], and Elena V. Berestneva[1]

[1] Tomsk Polytechnic University, Lenina Avenue 30, 634050 Tomsk, Russia
Marukhina@tpu.ru
[2] Tomsk State University, Lenina Avenue 36, 634050 Tomsk, Russia

Abstract. The aim of the research is to apply methods of structural analysis of multidimensional data using different approaches to visualizing the results of experimental studies. To solve applied problems, the authors used NovoSparkVisualizer (demo) system, as well as the R scripting language. The leading approach to the study of this problem is the mapping of multidimensional experimental data in the form of generalized graphic images on the basis of original methods and approaches developed by the authors. The paper presents the results of solving two applied problems illustrating the effectiveness of the method for visualization of multidimensional experimental data:

(1) analysis of the dynamics of the physiological state of pregnant women;
(2) study of breathing parameters in patients with bronchial asthma.

In the first case, the use of cognitive graphics tools made it possible to propose an effective way of displaying the dynamics of the state of the bio-object (for example, comparing the patient's condition before and after treatment). In the second case, it helped to reveal some previously unknown patterns of physiological response of the bronchial-pulmonary system to psycho-physiological exposure. The results of the research allow the authors to state that the methods and approaches presented in the paper can be regarded as promising directions in the analysis and presentation of multidimensional experimental data.

Keywords: Visualization methods · Cognitive graphics · Cluster analysis · Spectral representations

1 Introduction

Data visualization is the presentation of data in a form that guarantees their most effective study. [1] Data visualization is widely used in scientific and statistical research (in particular, in forecasting, intellectual analysis of data, business analysis). Data visualization is related to visualization of information and scientific data, exploratory data analysis and statistical graphics.

In the first half of the XIX century, there was a significant increase in the number of fields, which used graphic data display. By the middle of the century, all the main types of data representation were invented: columnar and circular diagrams, bar charts, linear graphs, time series graphs, contour diagrams, etc. [2].

© Springer Nature Switzerland AG 2019
I. Rojas et al. (Eds.): IWBBIO 2019, LNBI 11466, pp. 433–444, 2019.
https://doi.org/10.1007/978-3-030-17935-9_39

The growth trend began to decline in the early XX century, giving way to mathematics. Nevertheless, it was during this period that textbooks and courses on graphical methods of data presentation began to appear, and researchers started to use graphs not only for presenting results, but also for researching information and putting forward hypotheses in astronomy, physics, biology, and other sciences [3].

A new stage of visualization started in the third quarter of the 20th century. This development was caused by three events [4]: the appearance of John Tuke's work [5] devoted to exploratory data analysis; the appearance of Jacques Bertin's "Graphic semiology" (Fr. Sémiologie graphique) [6].

At this stage there appeared an ability to visualize data using computers: the emergence of effective output tools (pen plotters, graphic terminals), as well as ergonomic means of data entry (encoding tablet, mouse). The impact of interactive computer graphics (ICG) led to the emergence of a new issue in the problem of artificial intelligence, called cognitive (i.e. aimed at accumulating knowledge) computer graphics. This term was introduced in the Russian science in the 1990's by Zenkin. Cognitive graphics is a set of tools and methods of visual representation of the conditions of the problem, which allows either to immediately see the solution, or to get a hint for finding it. Cognitive graphics allow the user to see a new way or approach to solving a problem that was not visible when using traditional data visualization tools. Pospelov formulated three main tasks of cognitive computer graphics [7]:

- the creation of models of information representation in which it would be possible to represent with same means both objects that are characteristic of logical thinking and images/pictures with which figurative thinking operates,
- visualization of the human knowledge for which it is not yet possible to find textual descriptions,
- the search for ways of transition from the image-pictures to the formulation of a hypothesis concerning those mechanisms and processes that lie behind the dynamics of the observed patterns.

In his work [8], he points out that "the cognitive function of images was used in science before the advent of computers. Visual representations associated with the concepts of graph, tree, net, etc. helped to prove many new theorems, the Euler circles allowed us to visualize the abstract relation of Aristotle's syllogism, the Venn diagrams made visual analysis of the functions of the algebra of logic."

The technology of cognitive modeling, designed for analysis and decision-making in unclear situations, was proposed by the American researcher Axelrod. The main direction of his activity was the evolution of cooperation in living systems [9]. In the course of research of this issue he came to the concept of cognitive perception [10]. These studies gave an impulse to the development of cognitive modeling. As a result, there emerged a modern viewpoint on the use of cognitive modeling with the help of cognitive maps [11].

Later it became clear that graphic images can activate the associative logic of the subconscious processes of thinking in the human brain, which allows us to quickly find original and often unexpected solutions with the help of cognitive graphics [12].

Using cognitive graphics enables the user to draw conclusions without analyzing a large amount of information. Information can be presented in a cognitive way: by a sector, a bar chart, a cross, a circle, etc., parts of which have different colors each having a certain meaning [13]. A separate field of cognitive graphics are poorly structured problem areas, such as socio-psychological and medical field. Visualization of the current state and characteristics allows to ensure continuous control over the condition of groups of individuals or an individual.

The purpose of the research was to visualize the images represented by numerical data and to show the possibility of revealing the relationships between them on the basis of observations of the set of images obtained.

2   Materials and Methods

The most complete description of the approaches used here, accompanied by detailed references to key papers, is presented in "Computer data analysis" by Berestneva et al. [14]. "Information technologies in biomedical research" by Duke and Emanuel [15] gives a classification of the main methods for analyzing the structure of multidimensional data:

1. Data visualization: linear methods of diminishing dimensionality, non-linear mappings, multidimensional scaling, space-filling curves.
2. Automatic grouping: factor and cluster analysis of objects and attributes, hierarchical grouping, definition of "condensation points".

The basis of the classification above is a criterion reflecting the degree of participation of the experimenter in distinguishing the features of the relationship between the objects and the characteristics under investigation. The application of data visualization methods aims at finding the most expressive images of all the objects under investigation for the subsequent maximum utilization of the potential of the experimenter's visual analyzer.

Computer data processing involves mathematical transformation of data using certain software tools. For this, it is necessary to have an idea of both the mathematical methods of data processing and the corresponding software tools [16].

The methods of visualization allow the researcher to visually detect features, reveal regularities and anomalies in large amounts of information. The main task of data visualization is the task of obtaining a visual image that totally corresponds to the data set [17]. The basis of the visualization approach [18, 19] is the linear transformation of the values of the multidimensional observation A into a two-dimensional curve $f_A(t)$, i.e. $A \leftrightarrow f_A(t)$. It is guaranteed that visually close images/curves $f_A(t)$ and $f_B(t)$ will correspond to observations A and B that are close in values, while for images that differ greatly in the values of their observations, the curves will be markedly different.

In the case under consideration, the most general form of data representation is the vector of the finite-dimensional space R_n:

$$A = (a_0, a_1, a_2, \ldots a_{n-1}) \in R_n. \tag{1}$$

To move from this vector to the visual image, the authors use a basis of orthonormal functions $\{\varphi_i(\tau)\}_{i=0}^{\infty}$. As a basis, known functions can be used, in particular, orthonormal Legendre polynomials on the interval $[0, 1]$, the set of which is denoted by $\{l_i(\tau)\}_{i=0}^{\infty}$.

Legendre polynomials $l_n(x)$ are orthogonal polynomials on the interval $[-1, 1]$, which are found by the formula:

$$l_n(x) = \frac{1}{2^n \cdot n!} \cdot \frac{d^n}{dx^n} (x^2 - 1)^n. \tag{2}$$

In this case, a point with coordinates $A = (a_0, a_1, a_2, \ldots a_{n-1})$ can be associated with a function:

$$F_{(A)}(\tau) = \sum_{i=0}^{n-1} a_i l_i(\tau). \tag{3}$$

The formation of vector A is connected with the transformation of the data. The values of a multidimensional object's coordinates play an important role in its characterization. In most cases, each indicator has its own unit of measure, and its value will affect the function $F_A(\tau)$. In order to exclude the effect of a difference in the indexes on the form of the function $F_A(\tau)$, it is necessary to move to dimensionless units using one of the known ways.

It should be noted that the order of inclusion of indicators in vector A, will also affect the form of the function $F_A(\tau)$. The difference between formulas (1) and (3) is that for vector A from formula (1) only an analytical representation is possible, while for the function $F_A(\tau)$ a representation in the form of a graph is possible. Between formulas (1) and (3) there is a one-to-one relationship in both directions, that is, a one-to-one mutual relationship. If we introduce the second vector $B = (b_0, b_1, b_2, \ldots b_{n-1}) \in R_n$ into consideration, it will be associated with the function $F_{(B)}(\tau) = \sum_{i=0}^{n-1} b_i l_i(\tau)$.

By way of illustration, the present paper uses several examples from "Visual interpretation of quantitative characteristics of biosystems" by Volovodenko et al. [18]. Suppose there are two 10-dimensional observations named A and B with the following values:

A: {53.78, 1, 17.56, 2.54, 6.36, 0.16, 4.63, 8.1, 3.28, 1.9},
B: {50.53, 1.4, 19.05, 2.34, 5.95, 1.53, 3.63, 7.82, 2.98, 2.48}.

The following curves are visual representations of observations A and B and the combination of these two images (Fig. 1). This means that the initial observations are also very close to each other.

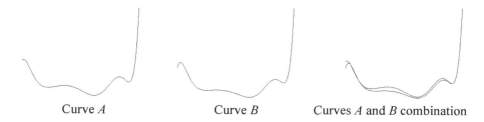

Curve *A*   Curve *B*   Curves *A* and *B* combination

Fig. 1. Images/curves of observations *A* and *B* and combination.

The more the curves are indistinguishable from each other, the more identical the observations they represent, i.e. the method establishes a one-to-one correspondence between the rows in the data set and their curves.

Many interesting properties can be revealed if the observation curves are displayed in three-dimensional space using a third dimension ("Z-dimension") as a distance in a multidimensional space or the time interval between two observations. The Z axis is the axis of the image movement. In the simplest case, the values of Z coincide with the image number, but this coordinate can be given the value of the distance in the feature space from the origin to the image (object) (Fig. 2).

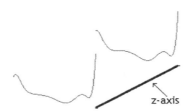

Fig. 2. Curves A and B in three-dimensional space.

NovoSpark Visualizer package implements the presented approach [19]. On its basis, the authors have successfully solved a number of applied problems of analysis and interpretation of multidimensional data in the social sphere [20] and medicine [21].

The results of solving two applied problems that illustrate the effectiveness of the presented method of visualization of multidimensional experimental data are presented below.

3   Results

3.1   Analysis of the Dynamics of the Physiological State of Pregnant Women

As initial information of the first problem, we have selective information about pregnant women and the corresponding results of laboratory and clinical studies (a detailed description of the experimental data is presented in [21]). The experiment includes three groups of pregnant women (30 people) aged 18 to 44 years.

The first (control) group consisted of 10 virtually healthy pregnant women aged 20 to 42 years who had not received any treatment. The second (comparison group) consisted of 10 pregnant women aged 18 to 40 years who had somatic diseases had not received any treatment. The third (the main group) consisted of 10 pregnant women aged 19 to 43 years who had somatic diseases and had received a complex of treatment.

For each patient, we introduce the designation – W_i, in order to hide confidential information and simplify the notation of the corresponding observation [21].

$$W_i = \{ttg, t_3, t_4, kor, ins, mda, e, gem, tr, bel, fil\} \tag{4}$$

where *ttg* is a thyroid stimulating hormone;

t_3 – triiodothyronine;
t_4 – thyroxin;
kor   – cortisol;
ins   – insulin;
mda   – malonic dialdehyde;
e   – vitamin E;
gem   – hemoglobin;
tr   – platelets;
bel   – protein;
fil   – fibrinogen

The authors assessed the change in their condition in the first and third trimesters of pregnancy according to a set of biomedical indicators. As a reference set, a control group was selected. Below are observations from each group: control group (1), comparison group (2), main group (3).

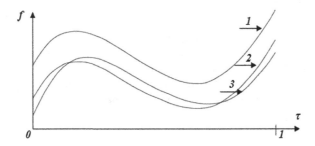

Fig. 3. Observations in the I trimester

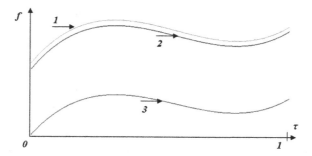

Fig. 4. Observations in the III trimester

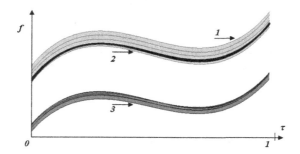

Fig. 5. Reference group observations

Figure 3 shows that the observations of the main group and the comparison group are very similar and significantly differ from the control group's observation, which indicates that the initial set of pregnancy condition variables is "suitable" for assessing the dynamics of their state during a complex of treatment. Figure 4 shows the curves of observations in the III trimester.

Figures 4 and 5 show that the observation of the main group strongly approached the control group, and the observation of the comparison group remained practically unchanged.

This means that by the third trimester, when receiving a set of health-improving measures, the indicators of pregnant women with somatic diseases are much closer to those of virtually healthy pregnant women.

3.2 Investigation of Respiratory Parameters in Patients with Bronchial Asthma

During the last dozens of years, the fields that study the complex mechanisms of bronchial asthma have been rapidly developing. In recent years, many different both medical and psychological works have been written and published. Basically, they concern the study of the role of psychosocial and emotional factors affecting the course of the disease and the patient's condition. Although at the moment, psychosomatic disorders have not been sufficiently studied, it can be reasonably assumed that the psychological factor participates both in the pathogenesis of the disease and changes the entire social situation in general.

There is a great variety of classifications of the disease: depending on the origin, severity of the disease, etc. Currently, many causes of bronchial asthma have been identified, but the question of rational and generally accepted classification is still open [22, 23].

In the course of a long-term follow-up of patients, analysis of life events, an anamnesis of life, Yazykov and Nemerov [23] identified a group of patients with high sensitivity to psychotraumatic life situations. In 2009–2012, they obtained and published results confirming this hypothesis. They hypothesized that among people suffering from bronchial asthma, there are patients with special psychobiological reactivity. Psychosocial stress effects trigger the development of the disease and further aggravation of its symptoms [23]. Scientists noted that it is possible to single out a special group of patients, with whom emotional factors play a great role in the pathogenesis. These observations allowed Yazikov and Nemerov to propose a new classification of bronchial asthma [23]:

PIBA   – psychogenically-induced bronchial asthma;
SPBA   – somato-psychogenic bronchial asthma;
NPBA   – non-psychogenic bronchial asthma;
PD   –psychogenic dyspnea

Classification is one of the fundamental processes in science. Application of multidimensional data analysis in solving problems associated with the classification of bronchial asthma will enhance the effectiveness of medical care and create more comfortable conditions for diagnosis and treatment of patients. The introduction of a new classification of bronchial asthma can lead to a more complete, accurate, in-depth and timely understanding of the problems associated with the treatment of patients.

The implementation of the approach proposed in "Visualization and analysis of multidimensional data using the NovoSparkVisualizer package" by Volovodenko, Eidzon [19] allows to display multidimensional objects in the form of "spectra". "Spectral representations" in this method emphasize the distinctive features of each curve and help to examine in more detail their visual properties. The color palette accentuates the levels of change in the values of the curves. Locating the curves along the axis, and looking at the result of this operation from above, one can obtain colored strips representing the spectrum of each observation [19]. The color scheme is defined in the map, where the reddish tones correspond to the highest values of the function, and the coldest, blue shades correspond to the smallest values.

The screenshots below represent a comparison of the sample data for four forms of bronchial asthma according to the visual proximity of the observation spectra (Fig. 6).

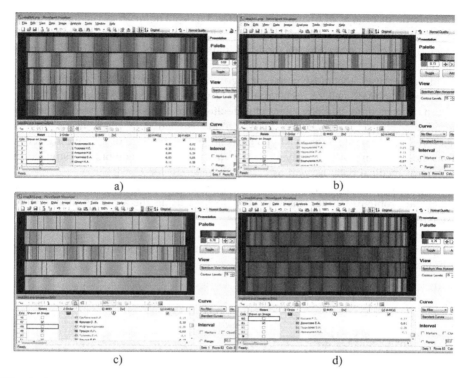

Fig. 6. Spectral representations of data for patients with different diagnoses (a) PIBA; (b) SPBA; (c) NPBA; (d) PD

Each color strip in the spectral form corresponds to the indicators of one patient. At Fig. 6, each screenshot presents five color strips, corresponding to five patients with different forms of bronchial asthma. The color "spectra" of patients diagnosed with PIBA and PD are similar. The same concerns patients diagnosed with SPBA and NPBA. According to Fig. 6, patients with SPBA and PD have the closest indicators. The most pronounced differences can be observed in patients with PIBA.

Thus, the use of cognitive graphics enabled us to identify some of the previously unknown patterns of physiological reactions of the broncho-pulmonary system to psychophysiological effects.

3.3 Classification of Forms of Bronchial Asthma

Cluster analysis reflects the features of multivariate analysis in the classification most clearly. A cluster is the union of several homogeneous elements, which can be considered as an independent unit with certain properties. The main purpose of cluster analysis is the partitioning of the set of objects and attributes under investigation into homogeneous groups or clusters. This means that the problem of classifying data and identifying their structure is being solved. Cluster analysis methods can be used in a variety of cases, even when it is a simple grouping, in which everything comes down to the formation of groups of quantitative similarity [14]. The main advantage of cluster analysis is that it allows you to split objects not by one parameter, but by a whole set of characteristics. Using various methods of cluster analysis, we can obtain various solutions for the same data.

This paper uses the most common algorithm of k-means. The algorithm splits objects into k groups, where each group represents one cluster. The main difference between this method and the hierarchical cluster analysis is the need to initially determine the number of clusters to which the studied data set is to be divided. Therefore, it is desirable to have a hypothesis about the structure of the investigated set before the beginning of the analysis. Using this method, the authors investigated the structure of medical data, as well as cases of clustering for $k = 4$.

As a result of the comparative analysis, R scripting language was chosen for solving the problem. Cluster analysis is performed using the k-means method for 4 clusters using the k-means function:

> clus<-k-means (data, 4)

Figure 7 shows the result of clustering according to the physiological parameters of the bronchopulmonary system.

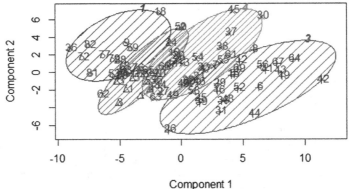

Component 1
These two components explain 53.38 % of the point variability.

Fig. 7. Results of cluster analysis according to the initial ("background") physiological indicators

The first cluster includes, mainly, patients diagnosed with "psychogenic-induced bronchial asthma", the second includes patients with nonpsychogenic bronchial asthma and psychogenic dyspnea, in the third group are patients with psychogenically induced and somato-psychogenic bronchial asthma. And, finally, in the fourth cluster there is practically the same number of representatives of each of the diseases. At the same time, the results show that clustering based only on the parameters of the bronchopulmonary system failed to single out clusters that uniquely match the diagnoses of PIBA, SPBA, NPBA and PD. Apparently, more research using socio-psychological indicators is necessary.

4 Discussions

The representation of a multidimensional observation in the form of a two-dimensional image (curve) ensures that visually close images/curves correspond to objects close in their characteristics. For strongly differing objects, their images/curves will noticeably differ. It is now becoming possible to automatically classify observations, identify the most important variables in a model, cluster data, visually compare individual observations and whole sets of data, and perform many other tasks in working with multidimensional data.

The spectral representation of the visual image is a more "subtle" tool, emphasizing the differences or similarities of images, than the traditional methods characterizing these properties at the level of numerical parameters. This circumstance allows the researcher-operator to be more careful about the differences and to identify "similarities" in a wider range of properties. The operator can ignore common color differences by choosing a monochromatic view.

5 Conclusion

The results of the research allow the authors to confidently state that the methods and approaches presented in the article are promising fields of study in the analysis and visual representation of multidimensional experimental data.

Methods, algorithms and software developed in this research, are used in solving applied research problems by the authors of the article together with researchers in different subject areas (social sphere, medicine).

Funding. The study was carried out with the partial financial support of the Russian Foundation for Basic Research (RFBR) within scientific projects No. 18-07-00543, 18-37-00344.

References

1. Paklin, N.B., Oreshkov, V.I.: Vizualizatsiya dannykh [Visualization of data]. In: Business Analytics. From data to Knowledge, 2nd edn, Peter, St. Petersburg, pp. 173–210 (2013)
2. Friendly, M.: Milestones in the history of thematic cartography, statistical graphics, and data visualization, 24 August 2009. Accessed 21 July 2017. http://www.math.yorku.ca/SCS/Gallery/milestone/milestone.pdf
3. Krum, R.: Cool Infographics: Effective Communication with Data Visualization and Design, 348 p. Wiley, Indianapolis (2014)
4. Iliinsky, N., Steele, J.: Designing Data Visualizations. O'Reilly, Sebastopol (2011)
5. Tukey, J.W.: Exploratory Data Analysis, p. 688. Pearson, Reading (1977)
6. Bertin, J., Barbut, M.C.: Sémiologie Graphique. Les diagrammes, les réseaux, les cartes, 431 p. Gauthier-Villars, Paris (1967)
7. Zenkin, A.A., Pospelov, D.A.: Kognitivnaja kompjuternaja grafika. [Cognitive computer graphics], 192 s. Nauka, Moscow (1991)

8. Pospelov, D.A.: Desyat «goryachikh tochek» v issledovaniyakh po iskusstvennomu intellektu. [Ten "hot spots" in research on artificial intelligence]. Intellect. Syst. (MSU). 1(1–4), 47–56 (1996)
9. Axelrod, R., Hamilton, W.D.: The evolution of cooperation. Science 211(4489), 1390–1396 (1981)
10. Axelrod, R.: Schema theory: an information processing model of perception and cognition. Am. Polit. Sci. Rev. 67(04), 1248–1266 (1973)
11. Axelrod, R. (ed.): Structure of Decision: The Cognitive Maps of Political Elites. Princeton University Press, Princeton (2015)
12. Massel, L.V., Massel, A.G., Ivanov, R.A.: Kognitivnaya grafika I semanticheskoye modelirovaniye dlya geoprostranstvennykh resheniy v energetike. [Cognitive graphics and semantic modeling for geospatial solutions in the energy sector]. In: Materials of the International Conference InterCarto/InterGIS 2015, vol. 1, no. 21, pp. 496–503 (2015). https://doi.org/10.24057/2414-9179-2015-1-21-496-503
13. Berestneva, O.G., Osadchaya, I.A., Nemerov, E.V.: Metody issledovaniya struktury meditsinskikh dannykh. [Methods for studying the structure of medical data]. Bull. Siberian Sci. 1(2), 333–338 (2012)
14. Berestneva, O.G., Muratova, E.A., Urazaev, A.M.: Kompyuternyy analiz dannykh. [Computer data analysis], 204 p. Publishing House TPU, Tomsk (2003)
15. Duke, V., Emanuel, V.: Informatsionnye tekhnologii v mediko-biologicheskikh issledovaniyakh. [Information technologies in biomedical research], 528 p. Peter, St. Petersburg (2003)
16. Sharopin, K.A., Berestneva, O.G., Shkatova, G.I.: Vizualizatsiya rezultatov eksperimentalnykh issledovaniy. [Visualization of the results of experimental research]. Izvestiya Tomsk Polytech. Univ. 316(5), 172–176 (2010)
17. Berestneva, O.G., Pekker, Y.S.: Simulation and evaluation of biological systems, adaptive capabilities. In: Mechanical Engineering, Automation and Control Systems, 16–18 October 2014, Tomsk. TPU Publishing House, Tomsk (2014)
18. Volovodenko, V.A., Berestneva, O.G., Sharopin, K.A.: The Visual Conception of Quantitative Characteristics of Biosystems. Fundamental medicine: from scalpel towards genome, proteome and lipidome. In: Proceedings of the 1st International Conference, 25–29 April 2011, Kazan (2011)
19. Volovodenko, V.A., Eidzon, D.V.: Vizualizatsiya I analiz mnogomernykh dannykh s ispolzovaniyem paketa «NovoSparkVisualizer». [Visualization and analysis of multidimensional data using the "NovoSparkVisualizer" package] (2008). Accessed 25 Nov 2011. www.novospark.com, http://www.tsu.ru/storage/iro/k020410/s4/s4.doc
20. Berestneva, O.G., Volovodenko, V.A., Gerget, O.M., Sharopin, K.A., Osadchaya, I.A.: Multidimensional medical data visualization methods based on generalized graphic images. World Appl. Sci. J. 24, 18–23 (2013)
21. Gerget, O.M., Marukhina, O.V., Cherkashina, Y.A.: System for visualizing and analyzing multivariate data from medico-social research. In: Key Engineering Materials, vol. 685, pp. 957–961 (2016)
22. Ovcharenko, S.I.: Bronkhialnaya astma: diagnostika I lecheniye. [Bronchial asthma: diagnosis and treatment]. RMJ. 10(17), 766–767 (2002)
23. Nemerov, E.V., Kazakov, G.G.: K voprosu izucheniya lichnostnykh svoystv v psikhofiziologicheskoy reaktivnosti bolnykh bronkhialnoy astmoy na audiovizualnuyu stimulyatsiyu. [On the study of personal properties in the psychophysiological reactivity of patients with bronchial asthma on audiovisual stimulation]. Bull. Tomsk State Pedagogical Univ. 6(108), 134–137 (2011)

Detection of Static Objects in an Image Based on Texture Analysis

Frantisek Jabloncik[(✉)], Libor Hargas, Jozef Volak, and Dusan Koniar

Department of Mechatronics and Electronics, Faculty of Electrical Engineering,
University of Zilina, Zilina, Slovakia
frantisek.jabloncik@fel.uniza.sk

Abstract. The article deals with the design of a method for the automatic detection of static objects in the image captured by an optical microscope. The search algorithm for static objects in the image - non-moving cilia is based on texture description methods. The texture of the image is described by statistical values, where it can be noticed that background texture, cells and cilia have different mathematical statistical parameters. Just based on the different statistical parameters of the textures, the classification for each texture parameter was done separately. As a result, the resulting classification considers the most predominant group to which the pixel has been assigned. In the end, the obtained mask was adjusted by morphological operations to obtain the boundary of the area, where the algorithm automatically evaluated that one was about Cilia. This work is supported by medical specialists from Jessenius Faculty of Medicine in Martin (Slovakia) and proposed tools would fill the gap in the diagnostics in the field of respirology in Slovakia.

Keywords: Cilia · Image segmentation · K-NN classification · Texture · Harlick features

1 Introduction

The point of observation is focused to observe individual frames of video sequences. These sequences were captured in an optical microscope with a high-speed camera. Video sequences came from the medical environment, from Clinic of Children and Adolescents by Jessenius Faculty of Medicine, Comenius University in Martin, Slovakia. The subject of interest is presented by respiratory epithelium samples. The image from the optical microscope contains a coating of the respiratory epithelium. The respiratory epithelium contains on its surface a small microscopic body called cilia. Cilium has a length of about 6–10 μm and a thickness of fewer than 1 μm. The task of the cilia is to move the mucus, dust particles, bacteria, viruses and other foreign bodies from the bottom of the breathing system outwards through the oral cavity or nose. Cilia are moving at a frequency of approximately 18 to 30 Hz.

To capture moving cilia, there are a number of techniques derived from either the optical flow or, for example, methods based on changes in the brightness of a given pixel or neighborhood over time. However, it is much harder to describe cilia that are not moving in the image, whether due to a pathological condition or for other reasons.

© Springer Nature Switzerland AG 2019
I. Rojas et al. (Eds.): IWBBIO 2019, LNBI 11466, pp. 445–457, 2019.
https://doi.org/10.1007/978-3-030-17935-9_40

For static detection, for example, an LBP (local binary pattern) method, which compares the pixel with its surroundings, can help. Each texture is described by various parameters, such as texture energy, dissimilarity, contrast value, entropy value, and more. Based on these parameters, it is possible to construct a classifier system that will classify individual parts of the image into classes.

2   Statistical Description of the Texture

The statistical method of the description describes the textures in a form suitable for further statistical processing and other suitable purposes. Each structure is described by the property vector function. The key values of the structure are most often derived based on the gray level of the particular image. The central k-moment of function p (l) is defined by the relation:

$$\mathbf{m_k} = \sum\nolimits_{l=0}^{L-1} (l - \mu_f)^k \mathbf{p(l)} \tag{1}$$

When l = 0, 1, 2… L − 1 and represents the gray levels in the image f and μ_f the equation describes:

$$\mu_f = \sum\nolimits_{l=0}^{L-1} l p(l) \tag{2}$$

The second central torque can help in determining inhomogeneities and at the same time is a deviation of the gray level. Is given by the relationship:

$$\sigma_f^2 = m_2 = \sum\nolimits_{l=0}^{L-1} \left(l - \mu_f\right)^2 p(l) \tag{3}$$

The third and fourth normalized moments, skewness (4) and kurtosis (5) are defined as:

$$skewness = \frac{m_3}{m_2^{3/2}} \tag{4}$$

$$kurtosis = \frac{m_4}{m_2^2} \tag{5}$$

Both describe asymmetry and uniformity. High-resolution moments serve only to describe the gray levels in the image.

Determination of Texture by Haralick

Based on generalized co-occurrence matrices (GCM) Robert Haralick suggested several relationships that describe the properties of structures. Because of their definitions, it is also necessary to know the normalized GCM and these are our case given by the relationship:

- L - the number of gray levels 256
- l - takes values from 0 to 255... 0 to L − 1
- P (l, l) - the normalized frequency of occurrence of the pair
- P (l, l) - the probability with which it acquires the determined value
- d - distance
- θ - angle

Haralick has defined several basic algorithms to calculate image properties. Specifically:

1. **Energy** F1 is also defined as a homogeneity measure. If the image has few inputs along diagonals (diagonals) with large values, it is a homogeneous image. Conversely, the non-homogeneous image will have small values distributed over the entire matrix and this will result in a low energy value.

$$F_1 = \sum_{l_1=0}^{L-1} \sum_{l_2=0}^{L-1} p^2(l_1, l_2) \tag{6}$$

2. **Contrast** F2

$$F_2 = \sum_{k=0}^{L1} k^2 \sum_{l_1=0}^{L-1} \sum_{l_2=0}^{L-1} p^2(l_1, l_2), |l_1 - l_2| = k \tag{7}$$

3. **Correlation** F3 represents the linear grain dependence, where μ_x a μ_y are mean values, σ_x a σ_y are standard deviations from px a py.

$$F_3 = \frac{1}{\sigma_x \sigma_y} \left[\sum_{l_1=0}^{L-1} \sum_{l_2=0}^{L-1} l_1 l_2 p(l_1, l_2) - \mu_x \mu_y, \right] \tag{8}$$

4. **Variation** F4, whichever applies, μ_f is the average gray value in the image.

$$F_4 = \sum_{l_1=0}^{L-1} \sum_{l_2=0}^{L-1} \left(l_1 - \mu_f \right)^2 p(l_1, l_2) \tag{9}$$

5. **The inverse moment** of the F5 difference describes the local homogeneity value.

$$F_5 = \sum_{l_1=0}^{L-1} \sum_{l_2=0}^{L-1} \frac{1}{1 + (l_1 - l_2)^2} p(l_1, l_2) \tag{10}$$

6. **The average sum** F6

$$F_6 = \sum_{k=0}^{2(L-1)} k p_{x+y}(k) \tag{11}$$

7. **The sum variance** F7

$$F_7 = \sum_{k=0}^{2(L-1)} (k - F_6)^2 p_{x+y}(k), \tag{12}$$

8. **The entropic sum** is the degree of non-uniformity in the image, i.e. it describes the complexity of the texture and is defined:

$$F_8 = - \sum_{k=0}^{2(L-1)} p_{x+y}(k) \log_2 \left[p_{x+y}(k) \right] \tag{13}$$

3   Proposed Segmentation Technique

The original image was captured by an optical microscope and has a size of 405×371 px. The sample under the microscope contained various artifacts, such as dust particles, but these were not investigated, nor was the image modified in any way with a focus on image properties. The original image is shown on Fig. 1.

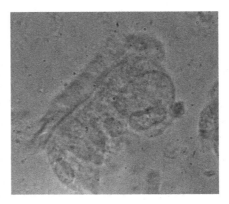

Fig. 1. Original microscopic image of respiratory epithelium with cilia. Size of the original image was 405×371 px

In the first step, the image was by doctor manually bound to three groups, the eye of the observed cilia being indicated in red, the cell of the respiratory epithelium marked with a blue borderline. And the last part of the image is everything else, so in a simplified form, it can be called a background. This situation is shown in Fig. 2. Splitting into three areas makes sense to classify correctly to determine the classification set.

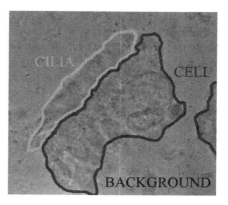

Fig. 2. Original image with manually overline regions for needs of determination of classification set of image

The basis of the algorithm is wavelet image transformation using Haar transformation. For further steps, the LL Haar transformation layer was selected (Fig. 3).

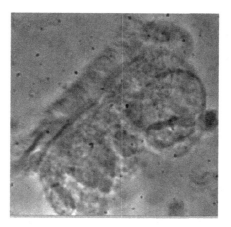

Fig. 3. Image after wavelet transform using Haar mather wave

Based on the LL image matrix, the statistical parameters for the texture were calculated. From whole Haralicks statistical parameters were selected these, where in which the individual groups are sufficiently different to be able to classify the data at a glance of the selected parameters. These numerical parameters were arranged in a two-dimensional matrix corresponding to 14×13 pixels. This is because the statistical parameters were calculated for windows with a size of 30×30 px and when the original dimensions of the image were divided by the size of the window, one parameter was calculated for each window and the total statistical parameters filled with a 14×13 px matrix.

Thus, the following statistical properties of the texture were calculated: Dissimilarity, Entropy, Contrast, Homogeneity, Correlation. For each property of the texture, exactly 15 representative values were selected, which entered the classification as a classification set for the particular image group - background, cell, and cilia. Number 15 was selected because the exact 15 statistic values were for the cilia when the numerical value matrix was bounded by the boundaries of the Fig. 1. The representative values for the background and the cell were randomly selected over the entire range of the image belonging to that group.

For the classification, a simple K-NN classifier of the 5th grade was chosen, where the Euclidean distances were calculated between the classification set and the specific value for the pixel. Classification for each parameter separately for Dissimilarity, Entropy, Contrast, Homogeneity, and Correlation has been applied. The whole image was divided into three groups, where group #1 was represented by Background, group #2 represented Cell, and group #3 determined Cilia. Therefore, more statistical parameters were chosen to achieve the highest accuracy with the desired result. Thus, the classifier output was 5 classification, where e.g. for the pixel at position [1, 9], according to Dissimilarity, it was in group #1, according to Entropy in a group #2, according to Contrast in a group no. 2, Homogeneity in a group no. 2 and according to Correlation in group no.1. In this case, it is clear that 3 out of 5 classifications included this pixel in group #2 (Cell). In case the same number of classifications occurred in groups, the result was a group with a smaller sum of Euclidean distances. The overall classification result is shown in Fig. 4.

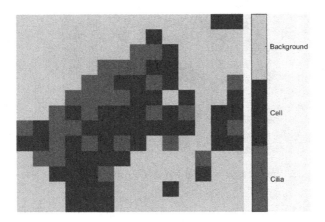

Fig. 4. Final results of classification based on 5 statistical parameters of image texture

From the previous figure, only those values that belonged to group#3 (Cilia) were exceeded. The mask thus created also contained unwanted components caused by noise or still not an ideal classification set (Fig. 5).

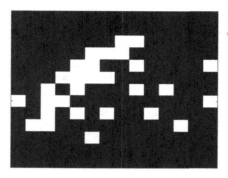

Fig. 5. Created mask only from pixels belonging to group #3 meaning the cilia

It is precisely for this reason that all the pixels in the mask whose area was less than 2 were neglected. This step can be seen in Fig. 6.

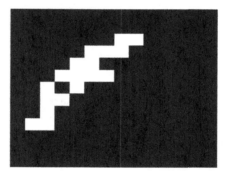

Fig. 6. Edited mask after removing small objects with their area less than 2 pixels

The last step that was needed to convert the mask was to enlarge it to the original image size and thus 405×371 px. There are several ways to scale the image, for our needs, bilinear transformation scaling has been chosen as it decreases the stairway effect. The resulting mask is shown in Fig. 7.

Fig. 7. Scaled mask to dimensions of the original image. In scaling it was used bilinear interpolation with smoothing of edges

As the last step, the transformation of the mask into the border region and the overlay of the original image with the boundary thus created was performed. Subjectively, it is possible to conclude that algorithm has been successful, as has been attempted on other similar images, retaining the original classification set from the first image. Original image with overlayed cilia boundaries is shown in Fig. 8.

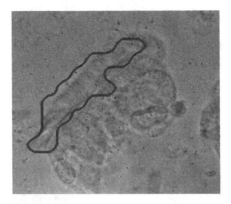

Fig. 8. Original image (image 1) of respiratory epithelium with overlayed automatically determined cilia area

As the algorithm was applied to the original image from which the classification kits were also created, the algorithm was also applied to other images from other, similar sequences of capturing cells and cilia. Using the same classification set and proposed algorithm (all steps as already mentioned). Achieved resulting cilia marked areas are shown in Fig. 9.

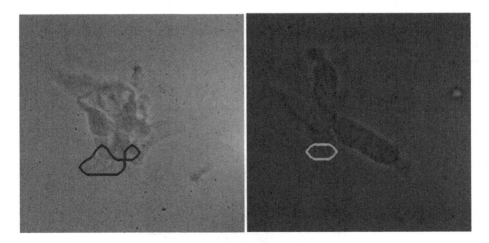

Fig. 9. Original image (left - image #2, right – Image #3) of respiratory epithelium with overlayed automatically determined cilia area. Size of images was 450 × 450 px

4   Evaluation of the Proposed Algorithm

In order to evaluate the success of the algorithm, it is necessary to know how the ciliary boundaries match the proposed algorithm and boundaries by doctors.

We simply chose the contours of cilia based on the proposed algorithm and based on a doctor's observation and overlay them into one image. As seen in Fig. 11, four overlapping areas were created. The largest part of the image, colored in white, corresponds to the True Negative part of the confusion matrix. This section tells us that there are no real searched structures, cilia, and the algorithm correctly identified the area as negative for ciliate finding. The True Positive area, filled in light green, says there are real cilia in the area, and the algorithm has confirmed that they are here. with the light blue color is filled False Negative area, where the cilia are found by the doctor, but the algorithm did not find them. The last area, orange, is called False Positive. Here, on the contrary, there are no cilia, but the algorithm incorrectly evaluated this area as positive (Fig. 10).

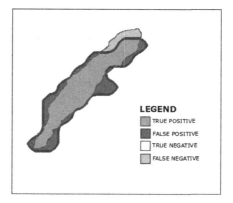

Fig. 10. Four areas created by overlay ciliary boundaries by a doctor and ciliary boundaries by the proposed algorithm

As with the reference image, we also proceeded with additional images and overlapped the ciliary area with the doctors and the ciliary area according to the proposed algorithm.

On the left side are the resulting four areas of the algorithm for Image #2. Image #3 and its algorithm evaluation is in the right.

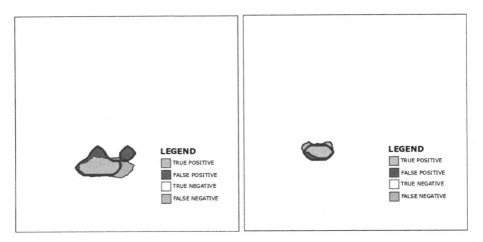

Fig. 11. Four areas created by overlay ciliary boundaries by a doctor and ciliary boundaries by the proposed algorithm. Left- image #2, right- image #3

It is not just graphically representation for accurate, percentage assessment of the success of the algorithm. Therefore, it is necessary to calculate the areas of the individual segments and, based on different statistical approaches, to evaluate the proposed algorithm. Pixel areas are shown in Table 1. For each image separately in the next row. Shortcut TP means True Positive, TN means True Negative, as FP is written False Positive and False Negative is shortened as FN.

Table 1. Areas of four parts of the confusion matrix for each image separately

	TP	TN	FP	FN
Image #1	10 011 px	136 442 px	2 805 px	997 px
Image #2	2 198 px	198 762 px	842 px	698 px
Image #3	910 px	201 340 px	88 px	162 px

The most commonly used parameter is Accuracy. It expresses the proportion of all correctly detected areas (both True Positive and True Negative) to the total area of the image. The disadvantage of this parameter is that it does not take into account the fact that the algorithm has found the right cilia, but in addition, the large area where they are located is incorrectly evaluated.

$$Accuracy = \frac{TP + TN}{TP + TN + FP + FN} \tag{14}$$

For the calculation of the success of the algorithm, we specify the Precision parameter. This parameter only counts for positive areas where it gives proportionally correctly detected positive areas and the sum of all real areas containing cilia. Since we are interested in cilia, we may consider this parameter to be credible.

$$Precision = \frac{TP}{TP + FP} \tag{15}$$

Parameter Recall (also referred to as sensitivity) is the ratio of the number of cilia found to the total number of cilia, which was marked by a doctor. Recall can be labeled as the probability with which the cilia will be selected.

$$Recall = \frac{TP}{TP + FN} \tag{16}$$

There is an indirect proportionality between Recall and Precision, the greater the Recall, the lower the Precision, and vice versa. Because of the indirect proportionality between Recall and Precision, it is common to use a composite parameter F1 that describes a compromise between Precision and Recall.

$$F1\ score = \frac{2TP}{2TP + FP + FN} \tag{17}$$

For the overall evaluation of the success of the algorithm, one of the above-mentioned evaluation parameters can be used. As shown in Table 2, the highest values reached Accuracy, on average for the three images up to 0.988, which is almost 99%. However, the Precision and Recall parameters are much higher, with average values ranging from 80% to 84%. However, the most significant parameter for us is parameter F1, which takes into account the previous 2 parameters. We evaluate the resulting success of the proposed algorithm at 82.1%.

Table 2. Computed selected parameters of proposed algorithm evaluation

	Accuracy	Precision	Recall	F1 score
Image #1	0.975	0.781	0.909	0.840
Image #2	0.992	0.723	0.758	0.741
Image #3	0.998	0.911	0.848	0.879
Average	**0.988**	**0.805**	**0.839**	**0.821**

In the future, it is planned to improve the training set of data and to use a larger test set of data. In addition to changing data, other texture features will also be tried.

5 Conclusion

Detection of static objects in the image, based on the description of the texture properties, brings with it many challenges. The basis is a good selection of given parameters, as there are several, and one texture structure can be better demineralized by the Dissimilarity parameter, otherwise, it may not pay at all. That is why we have chosen such an approach that we have made a classification for 5 selected texture properties and we have subsequently evaluated them. Also changes the grade of the K-NN classifier, causing a change in the overall classification, as the system will otherwise be maintained by counting the 5 shortest Euclidean distances and otherwise by counting the 9 shortest Euclidean distances.

We consider the subjective impairment of the success of the algorithm so designed to be a successful one with the potential to improve the method in the future. Proper Classification of Dynamic or Static Cilia is of major importance for medical evaluation whether it is one of the diseases or not (for example, Primary Ciliary Dicinase).

In addition to the K-NN classifier, there are other types of classifications we plan to work with them in the future, such as the K-mean method or the methods using neural networks. Necessary task for further work either based on statistical description of texture or based on other approach is creation of suitable dataset of images containing cilia. This is important, because nowadays it doesn't exist any similar dataset of images.

Acknowledgment. Authors of this paper wish to kindly thank all supporting bodies, especially to grant APVV-15-0462: Research on sophisticated methods for analyzing the dynamic properties of respiratory epithelium's microscopic elements.

References

1. Russ, J.C.: The Image Processing Handbook, 6th edn. CRC Press, Boca Raton (2011). ISBN 978-1-4398-4063-4
2. Javorka, K.: A Kol. Lekárska fyziológia. Osveta, Martin (2001). ISBN 80-8063023-2
3. Nečas, E., Šulc, K, Vokurka, M.: Patologická Physiologie orgánových systémů. Nakladatelství Karolinum, Praha (2006). ISBN 80-246-0675-5
4. Koniar, D.: Vyšetrovanie kinematiky mikroskopických objektov vysokorýchlostným zobrazovaním. EDIS, Zilina (2013)
5. Yu, H., Kim, S.: SVM Tutorial—classification, regression, and ranking. In: Rozenberg, G., Bäck, T., Kok, J.N. (eds.) Handbook of Natural Computing, pp. 479–506. Springer, Heidelberg (2012). https://doi.org/10.1007/978-3-540-92910-9_15. ISBN 978-3-540-92909-3
6. Kataria, A., Singh, M.D.: A review of data classification using K-nearest neighbour algorithm. Int. J. Emerg. Technol. Adv. Eng. 3(6), 354–360 (2013). ISSN 2250-2459
7. Haralick, R.M., Shanmugam, K., Dinstein, I.: Textural features for image classification. IEEE Trans. Syst. Man Cybern. 3, 610–621 (1973)
8. Haralick, R.M.: Statistical and structural approaches to texture. Proc. IEEE 67, 786–804 (1979)

9. Gonzalez, R.C., Woods, R.E.: Digital Image Processing, 3rd edn. Pearson, London (2008). ISBN 978-0131687288
10. Umbaugh, S.E.: Computer Imaging: Digital Image Analysis and Processing. CRC Press, Boca Raton (2000). ISBN 0-8493-2919-1
11. Mikulova, Z., Duchon, F., Dekan, M., Babinec, A.: Localization of mobile robot using visual system. Int. J. Adv. Robot. Syst. **14**(5). Article Number: 1729881417736085, ISSN 1729-8814
12. Martins, A.C.G., Rangayyan, R.M., Ruschioni, R.A.: Audification and sonification of texture in images. J. Electron. Imaging **10**(3), 690–705 (2001)
13. Rosenfeld, A.: Digital Picture Analysis. Topics in Applied Physics, vol. 11. Springer, Heidelberg (1976). https://doi.org/10.1007/3-540-07579-8
14. Sonka, M., Hlavac, V., Boyle, R.: Image Processing, Analysis, and Machine Vision. Thomson, Iowa (2008)

Artefacts Recognition and Elimination in Video Sequences with Ciliary Respiratory Epithelium

Libor Hargas[1]([⊠]), Zuzana Loncova[2], Dusan Koniar[1],
Frantisek Jabloncik[1], and Jozef Volak[1]

[1] Department of Mechatronics and Electronics,
Faculty of Electrical Engineering, University of Zilina, Zilina, Slovakia
libor.hargas@fel.uniza.sk
[2] Division of Bioinformatics, Biocenter, Medical University of Innsbruck,
Innsbruck, Austria

Abstract. The ciliary respiratory epithelium analysis is performed to detect of possible malfunction of cilia. The moving cilia is investigated and their movement is automatically evaluated. The areas with moving cilia is marked in video sequences. When the moving cilia is searched in some cases the false detection can be occur. It means that area with no cilia is marked as ciliated epithelium. These errors are caused by artefacts. The most frequent artefacts are erythrocytes and air bubbles. Article deals with techniques that helps to find artefacts which are responsible for false detection of movement. The used techniques for artefacts detection are based on pattern matching and geometrical matching. The results of designed algorithms are compared in the conclusion of this article. The main idea of this work is to create complex diagnostic tool for evaluation of ciliated epithelium in airways. This work is supported by medical specialists from Jessenius Faculty of Medicine in Martin (Slovakia) and proposed tools would fill the gap in the diagnostics in the field of respirology in Slovakia.

Keywords: Cilia · Image segmentation · Pattern matching ·
Geometrical matching

1 Introduction

Artefacts analysis algorithms in video sequences with ciliary epithelium represents the support parts of the overall algorithm for detection of both dynamic and static cilia (Fig. 1). Moving (beating) cilia have in one's airways an irreplaceable function: these structures assign the movement of mucus and so clean the whole respiratory apparatus and remove undesirable foreign particles out of airways (this is called a mucociliary transport) [1]. Without this mechanism or just when cilia do not work properly or stopped working, a serious health problem arises.

Recording of the respiratory epithelium is performed by using special hardware elements. The individual parts of the hardware are assembled of microscope with a powerful lighting module, a high-speed camera and a computer that can handle large data streams.

© Springer Nature Switzerland AG 2019
I. Rojas et al. (Eds.): IWBBIO 2019, LNBI 11466, pp. 458–468, 2019.
https://doi.org/10.1007/978-3-030-17935-9_41

The microscope must include a powerful illumination module because it needs sufficient light output to capture quick changes in observed sample. For this purpose, an LED light module has been developed in the past. With this module it is possible to control the light of the observed sample continuously.

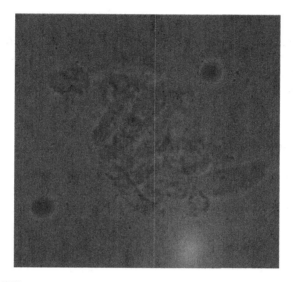

Fig. 1. Ciliary structures and artefacts (erythrocytes) in the video sequence

A high speed camera is used to record the data from a microscope. Parameters of the camera must ensure that the analysis process can be performed by image analysis methods. It means that the suitable frame rate and light conditions are selected. Because the changes in the sample are relatively fast, it is necessary to ensure a high number of frames per second (fps). The high speed camera works in different recording modes, but most often with 256 fps. These settings are suitable for further analysis of the captured video.

Recording computer is equipped with software and hardware capabilities to record and process a large amounts of data. The computer and camera connection must be created using appropriate hardware means. The CameraLink or USB 3.0 Vision connection is generally used for this purpose. The large amount of data is received from the high speed camera and stored in RAM memory. Acquisition software must ensure that acquired data is not lost. The data is then stored on the hard disk. Data archiving as well as the possibility of further analysis of the record is ensured in this way. When viewing the sample on the microscope online, it is possible to determine the speed of movement of individual objects using the acquisition software.

Investigated video sequences are in fact high-speed camera records which captures the scene within microscopic section. The section consists of a sample taken from human respiratory epithelium, usually from nasal or tracheal mucosa, which in a short time after removal from organism (approx. 10 min) contains oscillating hair-like structures, called cilia of respiratory epithelium. Its function lies in cleaning of airways, i.e. removal of foreign bodies such as dust particles, viruses, bacteria or allergens. Malfunction or dysfunction of these structures leads to often, chronical inflammations of respiratory system and to other related serious or even life-threatening diseases. The main objective of monitoring the proper function of cilia using the high-speed camera records is to properly diagnose the patients suspicious to suffer from ciliary pathologies.

Cilium is a microscopic structure with length of approx. 6–10 μm and width less than 1 μm. Each cell of respiratory epithelium carries on its surface about 200–300 cilia.

Cilia oscillate synchronously with frequency of 18–30 Hz, while the speed of their beating is referred as CBF (Ciliary Beat Frequency) and represents a crucial factor when determining ciliary pathologies [2, 3].

2   Artefacts and Their Origin

Videos with microscopic samples of respiratory epithelium's cilia are often affected by various artefacts. They sometimes move across the video scene and so causes wrong detections of motile cilia and errors of diagnose determining. Searching for artefacts and their elimination will increase the precision of diagnostics using automatic evaluation of ciliary cinematics. Artefacts which occur in microscopic video records of respiratory epithelium can be according to their origin divided into two classes. First class consists of records, where the presence of artefacts is caused by incorrect setting of a microscope and acquisition system, the other class includes artefacts that arise during obtaining the microscopic section.

2.1   Artefacts Caused by Instrumentation

The acquisition instrumentation has a huge effect to an output image quality, especially settings of the microscope. Among artefacts caused by instrumentation belong:

- non-homogenous illumination of the scene,
- vibrations.

Non-homogenous illumination of the scene (Fig. 2) is usually caused by incorrect setting of the microscope's field of vision. There is also an active cooler as a part of the microscope and its operation causes vibrations which influences the analysis of ciliary beating. This artefact can be suppressed using some software tools.

2.2   Artefacts Which Occur During Obtaining the Microscopic Section

Obtaining the microscopic sample and preparation of microscopic section can also significantly contribute to artefacts occurrence. Often artefacts which occur during making the microscopic sections are:

- presence of erythrocytes – red blood cells (Fig. 3),
- presence of air bubbles (Fig. 4).

Red blood cells occur within the section as a result of bleeding during taking the samples. When sampling the epithelium of respiratory tract, a slight damage to its mucosa may occur and this causes small bleeding. Making a more precise and gentle brushing can limit the number of present erythrocytes in the section, however, it can never be completely avoided. Presence of air bubbles cannot be influenced; they arise when putting the tissue into microscopic glass.

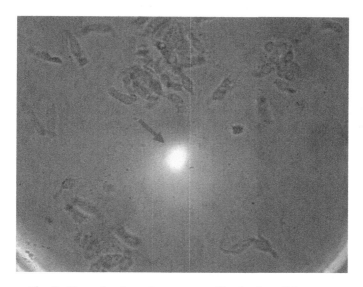

Fig. 2. Example of non-homogenous illumination of the scene

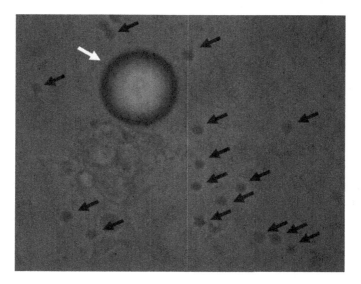

Fig. 3. Example of image artefacts: erythrocytes (red arrows) and a big air bubble (yellow arrow). (Color figure online)

3   Segmentation Methods for Identification of Image Artefacts

Selected algorithms described in this chapter serve for searching of image artefacts which cause distortion of resulting segmentation and false-positive detections of objects of interest [4, 5]. Such objects are usually of a circular shape (or very close to it) and their size is in various video sequences approximately the same and these facts use also the below mentioned techniques. These techniques were selected based on shape of artefacts. A lot of artefacts have circular shape. So the authors supposed that these methods will be effective to detect artefacts.

3.1   Pattern Matching

The pattern matching method is able to quickly find the areas which match with the predefined template within a grayscale image. The defined template represents the objects that is wanted to be find by the user. Principle of this method lies in normalized cross-correlation between the template and image [6].

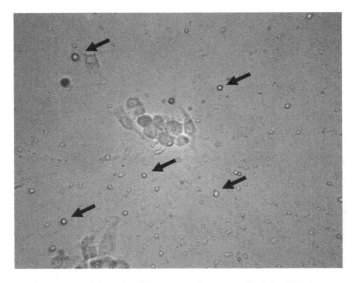

Fig. 4. Example of image artefacts: small air bubbles

In general, it can be assumed that correlation expresses a dependency between two (or more) particular features or properties of two (or more) images. Let assume an image $f(x, y)$ sized M x N and a template $w(x, y)$ sized K x L, while $K \leq M$ and

L $\leq$ N; then a cross-correlation between the template $w(x, y)$ and image $f(x, y)$ in certain point defined by coordinates (i, j) can be written as follows:

$$C(i,j) = \sum_{x=0}^{L-1} \sum_{y=0}^{K-1} w(x,y)f(x+i,y+j) \tag{1}$$

where $i = 0, 1, \ldots, M{-}1$; $j = 0, 1, \ldots, N{-}1$.

Basic correlation is very sensitive to amplitude changes in the image, such as intensity, and in the template. For example, if the intensity of the image f is doubled, so are the values of C. One can overcome sensitivity by computing the normalized correlation coefficient, which is defined as:

$$R(i,j) = \frac{\sum_{x=0}^{L-1} \sum_{y=0}^{K-1} ((w(x,y) - \bar{w})(f(x+i,y+j) - \bar{f}(i,j)))}{\left[\sum_{x=0}^{L-1} \sum_{y=0}^{K-1} (w(x,y) - \bar{w})^2\right]^{1/2} \left[\sum_{x=0}^{L-1} \sum_{y=0}^{K-1} (f(x+i,y+j) - \bar{f}(i,j))^2\right]^{1/2}} \tag{2}$$

3.2 Geometrical Matching

Geometric matching is specialized to locate templates that are characterized by distinct geometric or shape information. Such specific geometric and shape information can be for example simple features such as edges or curves or more complex features, e.g. whole geometric shapes (circles, rectangles,). Principle of geometric matching method consists of two phases: a learning phase and a matching phase.

The learning stage consists of the following three main steps: curve extraction, feature extraction, and representation of the spatial relationships between the features. A curve is a set of edge points that are connected to form a continuous contour. Curves typically represent the boundary of the part in the image. The curve extraction process is quite complicated and one needs to know lots of information [7].

4 Results for Artefacts Recognition and Elimination

When designing the algorithm for detection of artefacts present in images of respiratory epithelium, the main focus is put in detection of artefacts of circular shape, the red blood cells. The properties of such structures (also known as erythrocytes) which are very useful for image processing techniques are: always regular (almost circular) shape, always the same size using the same zoom of a microscope (experimental measurements of red blood cells show that their average radius in investigated images is approximately 20 pixels) and as they do not contain core, they are of homogenous color. These facts enable the usage of two selected objects' detection methods pattern matching and geometric matching.

Table 1. Table of possible results of testing an arbitrary condition

	Elements satisfying the condition	Elements dissatisfying the condition
Positive results of a test	true positives (t_p)	false positives (f_p)
Negative results of a test	false negatives (f_n)	true negatives (t_n)

In order to measure how successful are particular segmentation methods and compare obtained results from different types of images, it is necessary to define some evaluation parameters (Table 1). Among the main measures of effectiveness of searching for elements in a defined set of results belong the parameters called recall and precision [8–10].

Parameter recall (known also as sensitivity) is defined as a fraction of found relevant objects to the number of all relevant objects in a certain class:

$$R = \frac{t_p}{(t_p + f_n)} \tag{3}$$

In fact, recall R means a probability of choosing the relevant object.

Parameter precision is defined as a fraction of found relevant objects to the number of all found objects in a certain class:

$$R = \frac{t_p}{(t_p + f_p)} \tag{4}$$

Precision P means a probability that the chosen object will be a relevant one. There is an inverse proportion between parameters R and P, i.e. the higher is the value of recall, the smaller is precision and vice versa. Another parameter that describes the fraction of correctly found objects is accuracy. Accuracy A can be mathematically defined as:

$$A = \frac{t_p + t_n}{(t_p + f_p + t_n + f_n)} \tag{5}$$

Accuracy can be understood as a measure of errors.

Examples of usage pattern and geometric matching and obtained results are shown in following (Fig. 5) and (Tables 2 and 3).

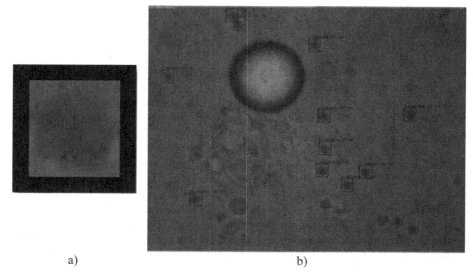

a)   b)

Fig. 5. Pattern matching method applied to image containing the respiratory epithelium, (a) defined template, (b) obtained result (erythrocytes are marked with red squares). (Color figure online)

Table 2. Obtained results of pattern matching method applied to selected image of respiratory epithelium

Matching score	Number of matches	t_p	f_p	f_n	Accuracy [%]
100%	1	1	0	18	5,26
90%	4	4	0	15	21,05
80%	15	15	0	4	78,95
70%	18	16	2	3	76,19
60%	27	17	10	2	58,62
50%	41	17	24	2	39,53

Table 3. Obtained results of geometric matching method applied to selected image of respiratory epithelium

Matching score	Number of matches	t_p	f_p	f_n	Accuracy [%]
100%	1	1	0	18	5,26
90%	5	5	0	14	26,32
80%	12	10	2	9	47,62
70%	22	11	11	8	36,66
60%	24	11	13	8	34,38
50%	24	11	13	8	34,38

Figure 5 is used as tested image where the pattern and geometrical matching is tested for its accuracy for artefacts detection. The various matching score was set for each method and results are shown in Tables 2 and 3.

The Table 2 shows the results for the pattern matching detection method for various settings of matching score. It can be seen that the results with best accuracy (78,95%) are obtained with score set to 80%.

Results obtained from geometric matching (Table 3) shows that this method is not the best suitable one for such an application as its accuracy (47,62%) is quite low with score set to 80%.

The proposed segmentation method for artefacts has been tested in 7 selected types groups of images of respiratory epithelium, representing the samples of various types (e.g. including artefacts, with overlapping epithelial cells, ones with small ciliary areas, ones with large areas, ones with different proportion of background and epithelium, etc.). Assessment criteria such as precision, recall or accuracy were also evaluated. These methods were used with another training set of ciliary images. The algorithm of artefacts detections was created as combination of these mentioned methods. After testing the several matching scores and comparing the obtained result, the score of 70% match was selected as an optimal one. The other ten templates were created and the efficiency of algorithm was incremented.

The next figure (Fig. 6) and Table 4 shows efficiency of proposed algorithm as a combination of pattern and geometrical matching.

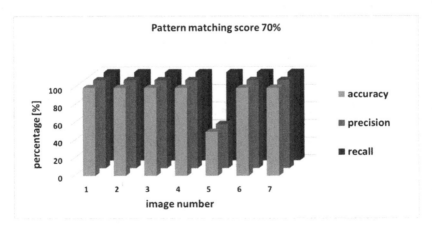

Fig. 6. Graphical representation of results obtained in proposed method for detection of artefacts in images of respiratory epithelium for matching score set to 70%

Table 4. Obtained results of proposed algorithm applied to investigated images, matching score 70%

Images with artefacts	Number of matches	t_p	f_p	f_n	Accuracy
# 1	2	2	0	0	2 of 2
# 2	1	1	0	0	1 of 1
# 3	3	3	0	0	3 of 3
# 4	2	2	0	0	2 of 2
# 5	1	1	1	0	1 of 1 and 1 extra
# 6	1	1	0	0	1 of 1
# 7	1	1	0	0	1 of 1

5 Conclusion

A reasonable selection of an appropriate algorithm for detection of artefacts is provided, considering the possible future real-time implementation of segmentation of ciliary areas together with artefacts elimination. Proposed method operates quick enough, even when comparing the investigated images with ten templates. When looking at the results, it is obvious that the proposed method with not such a strict score (70%) works reliably enough. Sometimes it might happen that when the matching score is too benevolent, then lots of false positive detection occurred, however, this is not the case of the proposed method – only 1 f_p occurred in one image. Well, but to be objective it should be mentioned here that images tested for the purposes of this article contains rather small number of artefacts (ranging 1–3) and this might distort the obtained accuracy. However, the selected set of tested images was chosen as a representative sample and is purposely the same for each group of proposed algorithms.

Acknowledgement. Authors of this paper wish to kindly thank to all supporting bodies, especially to grant APVV-15-0462: Research on sophisticated methods for analysing the dynamic properties of respiratory epithelium's microscopic elements.

References

1. Trojan, S., et al.: Lékařská fyziologie, 4th edn. Grada Publishing, Praha (2003). ISBN 80-247-0512-5
2. Nečas, E., Šulc, K., Vokurka, M.: Patologická fyziologie orgánových systémů. Nakladatelství Karolinum, Praha (2006). ISBN 80-246-0675-5
3. Silbernagl, S., Despopoulos, A.: Atlas fyzilogie člověka, 8th edn. Grada Publishing, Praha (2016). ISBN 978-80-247-4271-7
4. Babinec, A., Jurisica, L., Hubinsky, P., Duchon, F.: Visual localization of mobile robot using artificial markers, modelling of mechanical and mechatronic systems. Procedia Eng. **96**, 1–9 (2014). https://doi.org/10.1016/j.proeng.2014.12.091
5. Yum, Y.J., Hwang, H., Kelemen, M., Maxim, V., Frankovsky, P.: In-pipe micromachine locomotion via the inertial stepping principle. J. Mech. Sci. Technol. **28**(8), 3237–3247 (2014). https://doi.org/10.1007/s12206-014-0734-x

6. NI Vision Concepts Manual. National Instruments (2007)
7. Bow, S.T.: Pattern Recognition and Image Preprocessing, 2nd edn. Marcel Dekker, New York City (2002). ISBN 0-8247-0659-5
8. Zhang, X., Feng, X., Xiao, P., He, G., Zhu, L.: Segmentation quality evaluation using region-based precision and recall measures for remote sensing images. ISPRS J. Photogram. Remote Sens. **102**, 73–84 (2015). ISSN 0924-2716
9. Morstatter, F., Wu, L., Nazer, T.H., Carley, K.M., Liu, H.: A new approach to bot detection: striking the balance between precision and recall. In: Advances in Social Networks Analysis and Mining (2016). ISSN: 1869-5469
10. Yingchareonthawornchai, S., Nguyen, D.N., Valapil, V.T., Kulkarni, S.S., Demirbas, M.: Precision, recall, and sensitivity of monitoring partially synchronous distributed systems. In: Falcone, Y., Sánchez, C. (eds.) RV 2016. LNCS, vol. 10012, pp. 420–435. Springer, Cham (2016). https://doi.org/10.1007/978-3-319-46982-9_26

Cross Modality Microscopy Segmentation via Adversarial Adaptation

Yue Guo[1(✉)], Qian Wang[2], Oleh Krupa[3], Jason Stein[4], Guorong Wu[2], Kira Bradford[1], and Ashok Krishnamurthy[1]

[1] Renaissance Computing Institute,
Chapel Hill, NC, USA
{yueguo,ashok,kcbradford}@renci.org
[2] Department of Psychiatry,
University of North Carolina at Chapel Hill, Chapel Hill, NC, USA
{qian_wang,guorong_wu}@med.unc.edu
[3] Department of Biomedical Engineering,
University of North Carolina at Chapel Hill, Chapel Hill, NC, USA
ok37@email.unc.edu
[4] Department of Genetics,
University of North Carolina at Chapel Hill, Chapel Hill, NC, USA
jason_stein@med.unc.edu

Abstract. Deep learning techniques have been successfully applied to automatically segment and quantify cell-types in images acquired from both confocal and light sheet fluorescence microscopy. However, the training of deep learning networks requires a massive amount of manually-labeled training data, which is a very time-consuming operation. In this paper, we demonstrate an adversarial adaptation method to transfer deep network knowledge for microscopy segmentation from one imaging modality (e.g., confocal) to a new imaging modality (e.g., light sheet) for which no or very limited labeled training data is available. Promising segmentation results show that the proposed transfer learning approach is an effective way to rapidly develop segmentation solutions for new imaging methods.

Keywords: Transfer learning · Generative adversarial networks · Microscopy segmentation

1 Introduction

In the last decade, various cell segmentation methods have been developed for images from electron microscopy [18], confocal [9], and light sheet imaging [12]. Recently, there have also been successful applications of cell segmentation using deep learning techniques [16], such as U-Net [14].

It is well known that the success of deep learning is dependent upon having a substantial number of labeled training samples [3]. Manual labeling of

© Springer Nature Switzerland AG 2019
I. Rojas et al. (Eds.): IWBBIO 2019, LNBI 11466, pp. 469–478, 2019.
https://doi.org/10.1007/978-3-030-17935-9_42

training samples, however, is prohibitively expensive in terms of time and labor. Therefore, having a sufficient number of training samples presents a major challenge in developing an automatic cell segmentation algorithm for a new imaging modality.

Various transfer learning methods [4,6,17] were proposed to tackle the lack of training data challenge. Notably, R-CNN [4] first proposed fine-tuning deep features to transfer deep representations from a labeled dataset to a dataset where the labels are limited. A similar approach was adopted for neonatal video analysis in [6]. Recent transfer learning methods rely on Generative Adversarial Network (GAN) [5], which consists of a discriminative model and a generative model, and they are trained in an adversarial fashion: the generative model generates data to confuse the discriminative model while the goal of the latter is to distinguish the generated data from the real data. This paradigm has been applied to the state-of-the-art transfer learning method by adversarial adaptation [17], i.e., minimizing the disparity between the two models so that they cannot differentiate between the source dataset (i.e., the existing labeled dataset) and target dataset (i.e., the unlabeled dataset). One drawback of this method is that the underlying deep neural network is LeNet which is designed for classification and is not suitable for image segmentation.

There have been several applications to leverage GAN-based approaches in the microscopy segmentation field [2,15,19]. In [15], a multi-scale GAN was proposed with post processing for bright-field microscopy image segmentation, while [2] developed a GAN architecture with multiple deep neural network blocks in the discriminator. However, both methods were trained in a supervised fashion and still required manually labeled data. [19] utilized GAN to model the gap between the labeled and unlabeled data, yet their iterative training process required the labeled and unlabeled data to come from the same modality, and this model cannot be extended to a cross-modality situation.

Our contribution is two-fold. First, we present an unsupervised solution for cross-modality microscopy segmentation, inspired by the adversarial adaptation method [17]. Specifically, we trained a segmentation model for light sheet images by utilizing confocal images. To further validate this model, we manually labeled a limited number of light sheet images for evaluation purposes. Secondly, we extended the adversarial adaptation method [17] by (a) replacing the underlying deep model with an optimized U-Net [14] to improve segmentation performance; and (b) incorporating batch normalization [8] to expedite the training process.

2   Related Work

Recent transfer learning work has witnessed much progress thanks to the success of deep learning [4–7,17]. R-CNN [4] fine-tuned pre-trained deep features to transfer knowledge from an image classification dataset to an object detection dataset. [6] also utilized pre-trained deep features with a hidden Markov model for neonatal video analysis. [7] further extended the idea by averaging fine-tuned deep features from the K nearest categories to approximate the adaptations.

Generative Adversarial Network (GAN) [5] related work has been extensively explored for transfer learning. Prior work focused on generative tasks, e.g., DCGAN [13] used pre-trained discriminators in GAN to generate human face images or CGAN [11] introduced additional conditioning on both generators and discriminators in GAN to generate image tags. Later, CoGAN [10] applied GAN for transfer learning by proposing two independent GANs for source and target images to learn a joint distribution from different datasets. [17] simplified CoGAN by removing the generative model of CoGAN and designed a discriminative model. However, these methods are tested on classification tasks and not suitable for microscopy segmentation.

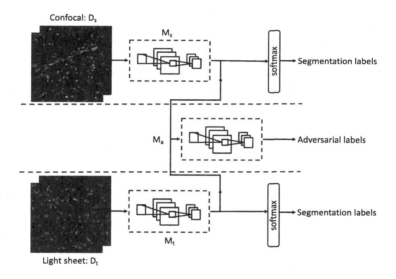

Fig. 1. The overview of the proposed framework. Note that M_s and M_t share the same deep network structure (number of layers, number of units, connectivity), while M_a can have a different structure

Several methods have used GAN for microscopy segmentation [2,15,19]. [2] applied GAN by proposing a multiple input architecture in the discriminators of GAN, which takes as input both microscopy images and corresponding annotated segmentation. [15] extended GAN by replacing the generative model with a multi-scale segmentation network. One limitation of these method, however, is that they still require manually annotated images during training. [19] followed the dual-network structure in [15] and expanded it with an iterative training process, which could gradually update the segmentation network to generate correct results for images without manual annotations, but could not handle the cross-modality scenario.

Inspired by [17], our work applied GAN with an optimized U-Net [14] and batch normalization [8] for cross-modality microscopy segmentation.

3   Methods

Our work integrates an optimized U-Net [14] with adversarial adaptation [17]. In addition, batch normalization [8] is adopted to facilitate the training process. Each component of our proposed model is described in more detail in the following sections.

3.1   Adversarial Adaptation

Our unsupervised adversarial adaptation framework consists of three parts: a source model M_s, a target model M_t and an adversarial model M_a, as shown in Fig. 1. We assume that there is an existing source data set D_s with labels L_s that has been used to train a model M_s using supervised learning to an acceptable performance. In our case, D_s are confocal images and the labels L_s are binary assignments (nucleus/background) for each pixel in each of the images in D_s. The labels L_s are manually generated. Also given is the target dataset D_t, which is unlabeled. The goal of the adversarial adaptation framework is to create a model M_t that can assign labels (nucleus/background) to each pixel of each image in D_t. In our application, the images in D_t are light-sheet images. The adversarial network approach is thus used to transfer the knowledge from the model M_s to the model M_t. While we focus on transfer learning from confocal image segmentation to light sheet image segmentation in this case, this approach is generalizable to other cases. For example, D_s could be images with a particular setting of microscope acquisition parameters, while D_t could be the same imaging modality, but with a different setting of microscope acquisition parameters.

The source model is first trained using D_s and L_s. The loss function used in training is

$$
\begin{aligned}
\mathcal{L}_s(D_s, L_s, M_s) = \\
- \sum_{D_s} 1_{L=L_s} \log S(M_s(D_s)) \\
- \sum_{D_s} (1 - 1_{L=L_s}) \log(1 - S(M_s(D_s))),
\end{aligned}
\tag{1}
$$

which is a standard cross entropy function for a binary segmentation task. S represents the softmax layer, as illustrated in Fig. 1. The target and source model share the same architecture but are trained separately. The source model is trained using the labeled source data D_s, while the target model training is through adversarial adaptation.

The trained source model M_s is used to train the target model via adversarial adaptation. The objective of adversarial adaptation is to minimize differences of these two models so that the target model can learn a discriminative mapping from the source domain to the target domain. We use the GAN-based loss [5] for this goal

$$\mathcal{L}_a(D_s, D_t, M_t) =$$
$$-\sum_{D_s} \log M_a(M_s(D_s))$$
$$-\sum_{D_t} \log(1 - M_a(M_t(D_t))). \tag{2}$$

This is also a standard binary cross entropy loss function, where the label 1 and 0 are assigned for the source dataset and the target dataset, respectively.

In the training process, the parameters of the source model M_s is fixed to avoid oscillation [17] and M_s is also used for the initialization of the target model M_t. In addition, inverted labels [5], i.e., assigning opposite labels for source images and target images, were adopted using the loss function.

$$\mathcal{L'}_a(D_s, D_t, M_a) =$$
$$-\sum_{D_t} \log M_a(M_t(D_t))$$
$$-\sum_{D_s} \log(1 - M_a(M_s(D_s))). \tag{3}$$

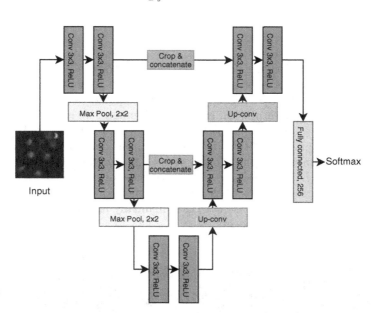

Fig. 2. The structure of our optimized U-Net

In summary, we have two independent loss functions in the adversarial learning process. Equations 2 and 3 optimize the target model M_t , and the adversarial model M_a, respectively. The dual-loss-function setting proves to be an efficient training technique; using a single loss function can cause vanishing gradients [17].

3.2   U-Net

U-Net [14] consists of two networks, a contracting network and an upsampling network. This symmetric structure is capable of capturing precise localization information of the images, and can achieve state-of-the-art results on several datasets. One drawback of the network, however, is that it is unable to segment the boundary of images due to the pooling process. Therefore, we propose to use a sliding window approach to extensively scan the whole image and only segment the center pixel of the window. The filter size and layer number are adjusted based on the input image size. Figure 2 shows our optimized U-Net. We use this model as our implementation choice of M_s and M_t in Sect. 2. The TensorFlow model in [1] is used for implementation.

3.3   Batch Normalization

To mitigate the problem of covariate shift, our work also incorporates batch normalization [8], which uses the mean and variance within each batch to normalize the activation values of the non-linear layers so that the activation values can achieve a standard Gaussian distribution during training. A noticeable performance improvement was observed when applying this technique in the adversarial network M_a.

4   Dataset

We are continuing acquiring new data for research and we have obtained nine confocal images and four light sheet images so far. The confocal images (D_s) with a resolution of 800×600 $(0.4613 \times 0.4613\,\mu m^2/\text{pixel})$ were used to train the source model M_s. For each pixel in the confocal images, we extracted a 31×31 sub-image centered on that pixel, and then assigned the binary label of that pixel, i.e., nucleus or background, as the label for this sub-image. The same image batch generating paradigm was applied to the 620×520 $(0.4853 \times 0.4853\,\mu m^2/\text{pixel})$ light sheet images (D_t). To evaluate the effectiveness of the transfer learning, we used a limited set of manually-labeled data for the light sheet image. Example images are shown in Fig. 3. Please note that manually labeled light sheet images are *not* used in the training process.

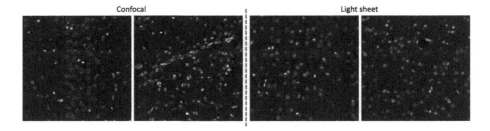

Fig. 3. Example confocal (left) and light sheet (right) images

5   Experiments

We conducted three sets of experiments to prove the effectiveness of our proposed model. We first performed an ablation study of our model with the state-of-the-art adversarial adaptation [17] as the baseline. Second, we compared our unsupervised method with the supervised state-of-the-art segmentation method, U-Net. Finally, we explored a bi-directional adaptation, i.e., from confocal images segmentation to light sheet image segmentation and vice versa. Sørensen-Dice similarity coefficient (DICE), $2TP/(2TP + FP + FN)$, was used to evaluate the segmentation performance. The adversarial model has 3 fully connected layers with Leaky ReLU activation function. All of the layers have 1024 hidden units with Batch Normalization. All of the models were trained over 10000 iterations with a batch size of 128 and no further improvement was observed.

Table 1. Experimental results of unsupervised adaption from confocal images to light sheet images. Light sheet image labels were NOT used in either experiment

Experimental settings	DICE
Adversarial adaptation [17]	0.593
Adversarial adaptation + U-Net	0.672
Adversarial adaptation + U-Net + Batch Normalization	0.709

Table 2. Experimental results of supervised methods and our unsupervised method. For the supervised methods, cross validation is applied and averaged results are shown

Model	Experimental settings	DICE
U-Net	Trained on 1 light sheet images, tested on 3 light sheet images	0.577
U-Net	Trained on 3 light sheet images, tested on 1 light sheet image	0.676
Ours	Tested on 4 light sheet images	0.709

5.1   Ablation Study

Adversarial adaptation [17] was employed as a baseline to validate the performance of the adversarial learning. [17] uses LeNet with ReLu activation function as the source and target model. We replaced the LeNet with our optimized U-Net and added batch normalization in the model. The results are provided in Table 1, which reveal the incremental improvement of our proposed approach.

5.2   Comparison with U-Net

For this experiment, we compared the performance of the proposed method with U-Net, which trained on light sheet images at a variety of settings in terms of the number of the available training samples. We had to reduce the size of the network by removing the last fully connected layer to avoid severe over-fitting. Table 2 presents the results of the supervised method. The results show that the performance of U-Net suffers when training data is limited while our proposed method can take advantage of data from other domains to tackle this problem.

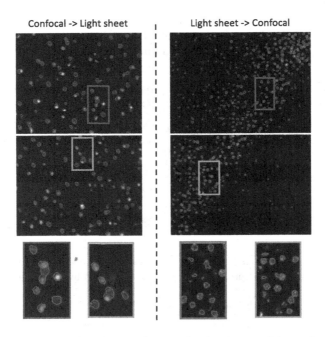

Fig. 4. Segmentation results of our adversarial adaption model: confocal domain to light sheet domain (left) and light sheet domain to confocal domain (right). The red contour obtained using MATLAB `imcontour` function on the binary segmentation predictions. The bottom row shows the zoomed results . (Color figure online)

5.3   Bi-Directional Adaptation

In addition to transfer the segmentation knowledge from confocal domain to light sheet domain, we conducted another experiment for the opposite domain, i.e. light sheet to confocal to further demonstrate the capabilities of our method. The results of this bi-directional experiment are shown in Table 3. Example visual results are also illustrated in Fig. 4.

Table 3. Experimental results of Bi-directional adaptation

Source domain	Target domain	DICE
Confocal	Light sheet	0.709
Light sheet	Confocal	0.680

6 Conclusion

This paper demonstrates the capabilities of an unsupervised adversarial adaptation framework for cross-modality transfer learning. Our model leverages the dataset from confocal images to train a segmentation model for light sheet images and achieves strong results in this challenging task. We also investigate the generality of our model by exploring bi-directional adaptation, indicating potential for other cross-modality applications in imaging research. Finally, our work shows promise in applying transfer learning to image segmentation problems in the neuroscience domain.

Acknowledgements. This work was supported in part by the National Science Foundation under grants OCI-1153775 and OAC-1649916.

References

1. Akeret, J., Chang, C., Lucchi, A., Refregier, A.: Radio frequency interference mitigation using deep convolutional neural networks. Astron. Comput. **18**, 35–39 (2017)
2. Arbelle, A., Raviv, T.R.: Microscopy cell segmentation via adversarial neural networks. In: ISBI 2018, pp. 645–648. IEEE (2018)
3. Deng, J., Dong, W., Socher, R., Li, L.J., Li, K., Fei-Fei, L.: ImageNet: a large-scale hierarchical image database. In: CVPR, pp. 248–255. IEEE (2009)
4. Girshick, R., Donahue, J., Darrell, T., Malik, J.: Rich feature hierarchies for accurate object detection and semantic segmentation. In: CVPR, pp. 580–587 (2014)
5. Goodfellow, I., et al.: Generative adversarial nets. In: NIPS, pp. 2672–2680 (2014)
6. Guo, Y., Wrammert, J., Singh, K., Ashish, K., Bradford, K., Krishnamurthy, A.: Automatic analysis of neonatal video data to evaluate resuscitation performance. In: ICCABS, pp. 1–6. IEEE (2016)
7. Hoffman, J., et al.: LSDA: large scale detection through adaptation. In: Advances in Neural Information Processing Systems, pp. 3536–3544 (2014)
8. Ioffe, S., Szegedy, C.: Batch normalization: accelerating deep network training by reducing internal covariate shift. In: ICML, vol. 37, pp. 448–456. PMLR (2015). http://proceedings.mlr.press/v37/ioffe15.html
9. Liu, M., et al.: Adaptive cell segmentation and tracking for volumetric confocal microscopy images of a developing plant meristem. Mol. Plant **4**(5), 922–931 (2011)
10. Liu, M.Y., Tuzel, O.: Coupled generative adversarial networks. In: NIPS, pp. 469–477 (2016)
11. Mirza, M., Osindero, S.: Conditional generative adversarial nets. arXiv preprint arXiv:1411.1784 (2014)

12. Packard, R.R.S., et al.: Automated segmentation of light-sheet fluorescent imaging to characterize experimental doxorubicin-induced cardiac injury and repair. Sci. Rep. **7**(1), 8603 (2017)
13. Radford, A., Metz, L., Chintala, S.: Unsupervised representation learning with deep convolutional generative adversarial networks. arXiv preprint arXiv:1511.06434 (2015)
14. Ronneberger, O., Fischer, P., Brox, T.: U-Net: convolutional networks for biomedical image segmentation. In: Navab, N., Hornegger, J., Wells, W.M., Frangi, A.F. (eds.) MICCAI 2015. LNCS, vol. 9351, pp. 234–241. Springer, Cham (2015). https://doi.org/10.1007/978-3-319-24574-4_28
15. Sadanandan, S.K., Karlsson, J., Wählby, C.: Spheroid segmentation using multiscale deep adversarial networks. In: ICCVW, pp. 36–41. IEEE (2017)
16. Shen, D., Wu, G., Suk, H.I.: Deep learning in medical image analysis. Ann. Rev. Biomed. Eng. **19**(1), 221–248 (2017). pMID: 28301734
17. Tzeng, E., Hoffman, J., Saenko, K., Darrell, T.: Adversarial discriminative domain adaptation. In: CVPR, vol. 1, p. 4 (2017)
18. Yang, H.F., Choe, Y.: Cell tracking and segmentation in electron microscopy images using graph cuts. In: ISBI, pp. 306–309. IEEE (2009)
19. Zhang, Y., Yang, L., Chen, J., Fredericksen, M., Hughes, D.P., Chen, D.Z.: Deep adversarial networks for biomedical image segmentation utilizing unannotated images. In: Descoteaux, M., Maier-Hein, L., Franz, A., Jannin, P., Collins, D.L., Duchesne, S. (eds.) MICCAI 2017. LNCS, vol. 10435, pp. 408–416. Springer, Cham (2017). https://doi.org/10.1007/978-3-319-66179-7_47

Assessment of Pain-Induced Changes in Cerebral Microcirculation by Imaging Photoplethysmography

Alexei A. Kamshilin[1]([envelope]), Olga A. Lyubashina[2,3],
Maxim A. Volynsky[1], Valeriy V. Zaytsev[1], and Oleg V. Mamontov[1,4]

[1] ITMO University, St. Petersburg 197101, Russia
alexei.kamshilin@yandex.ru,
maxim.volynsky@gmail.com, zaytsevphoto@gmail.com,
mamontoffoleg@gmail.com
[2] Pavlov Institute of Physiology of the Russian Academy of Sciences,
St. Petersburg 199034, Russia
laglo2009@yandex.ru
[3] Valdman Institute of Pharmacology, Pavlov First Saint Petersburg State
Medical University, St. Petersburg 197022, Russia
[4] Almazov National Medical Research Centre, St. Petersburg 197341, Russia

Abstract. Evaluation of the dynamics of cerebral blood flow during surgical treatment and during experiments with laboratory animals is an important component of assessment of physiological reactions of a body. None of the modern methods of hemodynamic assessment allows this to be done in the continuous registration mode. Here we propose to use green-light imaging photoplethysmography (IPPG) for assessment of cortical hemodynamics. The technique is capable for contactless assessment of dynamic parameters synchronized with the heart rate that reflect the state of cortical blood flow. To demonstrate feasibility of the technique, we carried out joint analysis of the dynamics of systemic arterial pressure and IPPG synchronized with electrocardiogram during visceral or somatic stimulation in an anesthetized rat. IPPG parameters were estimated from video recordings of the open rat brain without dissecting the dura mater. We found that both visceral and somatic painful stimulation results in short-term hypotension with simultaneous increase in the amplitude of blood pulsations (BPA) in the cerebral cortex. BPA changes were bigger in the primary somatosensory cortical area but they correlated with other areas of the cortex. Apparently, change of BPA is in the reciprocal relationship with variations of mean systemic and pulse pressure in femoral artery that is probably a consequence of cerebrovascular reflex regulating the cerebral blood flow. Therefore, BPA reflects dynamic properties of cortex microcirculation that are synchronized with systemic arterial pressure and depend from cortical area specialization.

1 Introduction

Developing a non-invasive tool capable of quantifying the physiological response of the brain to pain would greatly benefit clinical practice and research. Cortical activity is usually monitored by using optical imaging of intrinsic signals, which are slow varying

© Springer Nature Switzerland AG 2019
I. Rojas et al. (Eds.): IWBBIO 2019, LNBI 11466, pp. 479–489, 2019.
https://doi.org/10.1007/978-3-030-17935-9_43

changes in the light intensity reflected from the cortex [1, 2]. This simple technique uses a camera to observe the exposed cortex under visible light and it is also referred to as wide-field optical mapping (WFOM) [2]. Recently, another optical technique referred to as imaging photoplethysmography (IPPG) became very popular in applications for evaluation of cardiovascular parameters in a very convenient way [3]. From the hardware point of view, both techniques are quite similar consisting of a digital camera and illumination system. The main difference is in processing of the video data. In contrast to WFOM, the IPPG system exploits variations in light modulation at the heartbeat frequency, which requires video recordings at higher frame rate. There is only few works devoted to study vascular activity of the exposed cortex by means of IPPG system [4]. In this paper, we report on preliminary experiments to quantify the changes in the microcirculation of the rat cerebral cortex in response to various painful stimuli compared to innoxious ones.

2 Method

2.1 Animal Preparation

The experiments were carried out with the open cortex of a rat's brain. Twelve weeks old male Wistar rat was used in this study. During surgical preparation, the rat was anesthetized with a mixture of urethane (800 mg/kg; ICN Biomedicals, Aurora, OH, USA) and α-chloralose (60 mg/kg; MP Biomedicals, Solon, OH, USA). The left femoral artery and left femoral vein were cannulated for continuous monitoring of blood pressure and drug administration, respectively. The trachea was exposed and a tracheal cannula was inserted for measurements of respiratory airflow and end-tidal carbon dioxide. Then anesthetized rat was mounted into a stereotactic frame for craniectomy and remained there during video recording after the surgery. Body temperature was maintained at approximately 38 °C with a heating pad. Dorsal parts of the left frontal and parietal skull bones were removed so that maximum care was taken to preserve integrity of meninges and prevent bleeding. Innoxious (physiological) visceral stimulation was produced by inflating a rubber balloon in the lumen of the rat's colorectum to a pressure of 40 mmHg. To induce visceral pain, the colorectal balloon was inflated rapidly to a pressure of 90 mmHg. Innocuous (tactile) somatic stimulation was delivered to the rat's tail with forceps. Somatic pain was caused by firmly squeezing the base of the tail using forceps with a fixed lock mechanism. All animal procedures were approved by the Institutional Animal Care and Use Committee of the Pavlov Institute of Physiology of the Russian Academy of Sciences. The experiments were conducted in accordance with the guidelines set by the European Community Council Directives 86/609/EEC, with the Ethical Guidelines of the International Association for the Study of Pain, and EU Directive 2010/63/EU for animal experiments. After the end of experiment, the rat was sacrificed by intravenous injection of a lethal dose of urethane (>3 g/kg).

2.2   Experimental Arrangement

A custom-made IPPG system was used to continuously record the focused image of the open rat's cortex. The system consisted of a digital monochrome CMOS camera (8-bit model GigE uEye UI-5220SE of the Imaging Development System GmbH) and eight light-emitting diodes (LED) operating at 530 nm (green light) for illuminating the cortex. All LEDs were assembled around the camera lens (25 mm focal length) providing uniform illumination of the open brain area (see Fig. 1). Both the camera lens and LEDs were covered with crossed polarization films thus forming the polarization filter, which reduces the skin specular reflections and motion artefact influence on the detected signals [5]. All videos were recorded at 100 frames per second with resolution of 752 × 480 pixels and saved frame-by-frame in the hard disk of a personal computer. Electrocardiogram (ECG) was recorded by a digital electrocardiograph at the data acquisition frequency of 1 kHz. To synchronize ECG and video recordings, synch pulses of the each frame were recorded in one of the ECG channel providing the synchronization accuracy of 1 ms. Arterial blood pressure in the femoral artery was continuously monitored with a pressure transducer (MLT844, ADInstruments Inc., Colorado Springs, USA) and recorded in the personal computer using Spike2 software (Cambridge Electronic Design, Cambridge, UK).

Fig. 1. Photograph of the custom-made IPPG system. (Color figure online)

2.3   Procedure Protocol

After the surgery, the rat was kept in rest during 40 min to minimize the effect of postsurgical reaction. The total duration of the experiment was 30 min. It consisted of four successive series in which the following stimulations were applied: (1) innoxious visceral, (2) tactile somatic, (3) painful visceral, (4) painful somatic. The minimum time interval between series was 5 min. In each series, the data containing the cortex images, ECG, and arterial blood pressure were recorded continuously while the rat was consistently in the following physiological states: baseline, stimulation for 1 min, and relaxation.

2.4   Data Processing

All recorded video frames from the cameras were processed off-line by using custom software implemented in the Matlab® platform. First, the images were digitally stabilized to compensate involuntary rat's brain motions. To this end, the algorithm described in our recent paper [6] was applied. Second, we manually selected the area of the cortex in the recorded image. This area was completely covered by small regions of interest (ROI) sizing 3×3 pixels, which corresponds to the area of 40×40 μm^2 at the rat's cortex. Each ROI was chosen to have a common border with adjacent ROIs without overlapping. Third, we calculated PPG waveform as frame-by-frame evolution of average pixel value in every chosen ROI. Typically, it consists of alternating component (AC), which follows to the heartbeats, and slowly DC varying component. Both components are proportional to the incident light intensity [6]. To compensate unevenness of illumination, we calculated AC/DC ratio, deduced the unity from the calculated ratio, and inverted the sign. These transformations are typical in photoplethysmography providing the waveform to correlate positively with variations of arterial blood pressure [7]. All PPG waveforms were filtered to remove noise and DC components by means of a band-pass filter (0.12–20 Hz), which was implemented using the *filtfilt* function in Matlab® to perform zero-phase digital filtering of the waveforms. An example of the filtered PPG waveform from one of the small ROIs along with the respective ECG signal is shown in Fig. 2a. As one can see, each oscillation of the waveform follows the R-peak of ECG signal. These oscillations are plotted together in Fig. 2b by thin colored lines so that every respective R-peak is at the beginning of the time scale. Apparently, all twenty oscillations measured in this ROI are synchronized with R peaks. Thick red line in Fig. 2b shows a mean waveform obtained by averaging these oscillations.

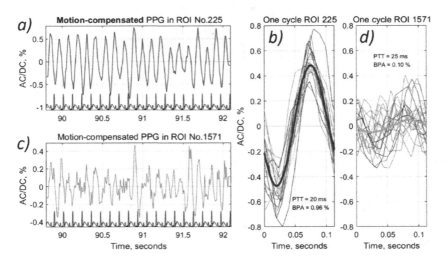

Fig. 2. PPG waveforms in two different ROIs with size of 3×3 pixels. (Color figure online)

However, not all ROIs exhibit such synchronization of individual parts of the waveform. An example of irregular PPG waveform measured in another small ROI is shown in Fig. 2c whereas individual oscillations of this waveform are shown in Fig. 2d. As seen, despite pronounced modulation at the heart rate, individual oscillations have a random phase with respect to the R-peaks of ECG. Consequently, a mean one-cardiac-cycle waveform (shown by thick blue line in Fig. 2d) has significantly smaller amplitude than individual oscillations. The one-cycle PPG waveforms averaged over 20 cardiac cycles (such as those shown by thick lines in Fig. 2b, d) allows us to estimate two parameters characterizing the vascular activity, blood pulsation amplitude (BPA) and pulse transit time (PTT). BPA was estimated as the difference between the maximum and minimum values of the mean one-cardiac-cycle PPG waveform. PTT was calculated as the time delay between the R-peak and the minimum of the mean waveform because the latter corresponds to the beginning of the anacrotic wave with fast blood-pressure increase.

Parameters PTT and BPA were calculated for each small ROI thus allowing us their mapping on the image of the rat's cortex. Since the heart rate of the rat was about 520 beats per minute, 20 cardiac cycles lasted 2.3 s. Typical BPA map calculated using PPG waveforms averaged for 20 cardiac cycles between 90-th and 92-th s of the experiment is shown in Fig. 3. Taking into account possibility of asynchronous oscillations, we adopted the threshold of 0.1% below which the PPG of the ROI is considered as unreliable. BPA in Fig. 3 is coded by pseudo colors with the scale shown on the right whereas unreliable ROIs are left empty. Similar maps were calculated for the whole experiment with the step of 2.5 s.

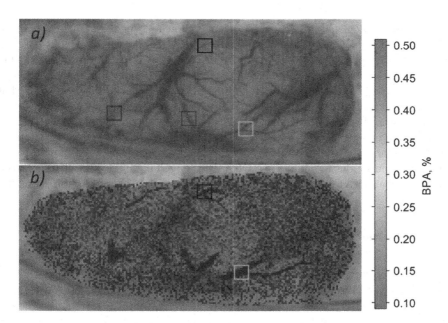

Fig. 3. (*a*) One of the video frames showing the open rat's brain. (*b*) The map of BPA parameter overlaid with the image shown in (*a*). Colored squares show location of big ROIs. The color scale on the right shows BPA in percent. (Color figure online)

As one can see in Fig. 3b, the amplitude of blood pulsations is unevenly distributed over the cortex. BPA in the areas colored red is several times higher than in blue areas. Note that areas of the increased BPA are usually located in the vicinity of the arteries. We have found that the spatial distribution of BPA does not remain constant but it is varying in the time course. To monitor the BPA dynamics, four big ROIs were selected in areas with increased pulsations. While blue, red, and magenta ROIs were located in various zones of the primary somatosensory cortical area, the black ROI was in the primary motor area of the cortex. Each big ROI contains 9×9 small ROIs or 27×27 pixels that corresponds to the region of 1.1 mm^2 at the rat's cortex. These big ROIs are shown by the colored squares in Fig. 3. BPA maps similar to that shown in Fig. 3b were calculated each 2.5 s keeping the location of the big ROIs. The mean BPA of each big colored ROI was calculated by averaging BPA of reliable small ROIs only.

3 Results

During each session of rat's stimulation, video frames of the open brain were recorded continuously and simultaneously with ECG and arterial blood pressure measured by a beat-to-beat invasive method. In the following graphs, we show dynamics of BPA calculated from the recorded video using ECG as the reference together with dynamics of arterial blood pressure.

3.1 Innocuous Visceral Stimulation

BPA dynamics calculated in selected big ROIs is shown in Fig. 4a so that the color of each thin line coincides with the color of big ROI in Fig. 3. Innocuous visceral stimulation was applied between 59 and 130 s. Thick brown line in Fig. 4a shows the dynamics of the mean blood pressure measured in the left femoral artery. Due to a technical problem, monitoring of arterial blood pressure in this session was stopped at 119-th s. No statistically significant change of vascular parameters between the baseline and impact was observed. Variations of BPA in big ROIs shown by thin colored curves correlate each other ($0.66 < r < 0.84$; $P < 0.001$). We hypothesize that these variations reflect vascular activity of the brain. However, no correlation between arterial blood pressure and pulsations amplitude in ROIs was observed in this session of stimulation, which is probably caused by the problem of the pressure monitoring in this particular session.

3.2 Tactile Somatic Stimulation

Dynamics of the vascular parameters in the session of somatic stimulation is shown in Fig. 4b. Like in the previous case, there is no statistically significant changes of both blood pressure and BPA between the baseline and impact. Dynamics of BPA calculated in the same big ROIs correlate each other, as well: $0.51 < r < 0.83$, $P < 0.001$. However, one can see in Fig. 4b that fluctuations of the arterial blood pressure are opposite to the dynamics of BPA that means their correlation with the negative sign: $-0.55 < r < -0.47$, $P < 0.001$ for all big ROIs.

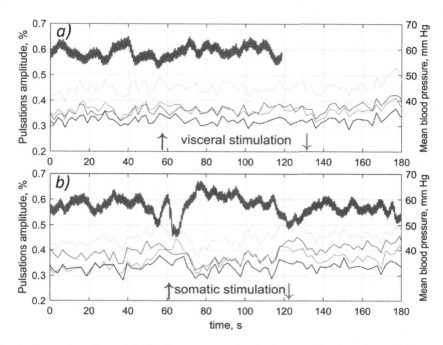

Fig. 4. Dynamics of arterial blood pressure and pulsations amplitude estimated from PPG waveforms in the cortex during sessions of innocuous visceral (*a*) and tactile somatic (*b*) stimulations. Thin lines of the same color as squares in Fig. 3 show BPA dynamics in selected big ROIs (Y-labels on the left) whereas brown lines show variations of arterial blood pressure (Y-label on the right). The red and blue arrows indicate the beginning and end of each stimulation, respectively. (Color figure online)

3.3   Painful Visceral Stimulation

Variations of the vascular parameters during two sessions of painful stimulation are shown in Fig. 5. As one can see in Fig. 5*a*, visceral pain leads to significant decrease of the arterial blood pressure and respective increase of pulsations amplitude (BPA parameter) that results in significant negative correlation: $-0.94 < r < -0.53$, $P < 0.001$. Like in previous cases, correlation among BPA dynamics in big ROIs remains at the high level: $0.68 < r < 0.95$, $P < 0.001$. It is worse noting that the relative increase of BPA is not the same in the selected ROIs. While the mean BPA in the blue ROI, which is located in the primary somatosensory cortical area, during the impact is 1.45 times higher than that in the baseline, this ratio is just 1.11 in the black ROI located in the primary motor area of the cortex.

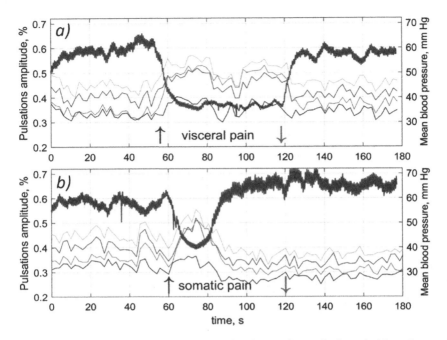

Fig. 5. Dynamics of the vascular parameters in the sessions of visceral (*a*) and somatic (*b*) painful stimulation. The red and blue arrows indicate the beginning and end of each stimulation, respectively. Like in Fig. 4, thin colored lines show fluctuations of BPA with Y-labels on the left, whereas brown curve is the dynamics of arterial blood pressure with Y-labels on the right. (Color figure online)

3.4   Painful Somatic Stimulation

As can be seen from Fig. 5*b*, the hemodynamic response to somatic pain simulation qualitatively differs from the visceral one shown in Fig. 5*a*. While the visceral pain leads to permanent increase BPA throughout impact and returned to the baseline level after the stimulation was released, somatic pain causes relatively short-term BPA increase (about 20 s), after which the amplitude of blood pulsations is reduced to significantly below the baseline, and remaining at this level until the end of the recording. It is clearly seen that during somatic stimulation (Fig. 5*b*) variations of arterial blood pressure also correlate negatively with BPA changes: $-0.88 < r < -0.77$, $P < 0.0001$ in all ROIs. Like in the case of the visceral pain, here the largest change in BPA in respect to the baseline is observed in the blue ROI primary somatosensory cortical area), whereas the smallest changes are in the black ROI located in the primary motor cortex (1.33 versus 1.07). In the session of painful somatic stimulation, BPA in all ROIs during relaxation period is significantly smaller than in the baseline, e.g. in the blue ROI it is 0.31 ± 0.01 and 0.36 ± 0.02, $P < 0.001$, respectively.

It should be underlined that during painful stimulations the mean pulse pressure measured in the femoral artery is significantly smaller than in the respective baseline. The pulse pressure is 5.39 ± 0.73 versus 6.04 ± 0.60 mmHg, $P < 0.0001$, and

5.78 ± 0.75 versus 6.91 ± 1.14 mmHg, $P < 0.0001$ in the case of visceral and somatic painful stimulations, respectively. Despite of diminished pulse pressure, the parameter BPA is significantly increased during the impact as it is seen in Fig. 5.

A reliable response of both the BPA parameter and blood pressure was registered during painful visceral and somatic stimulations (Fig. 5) whereas no significant change was observed during innocuous visceral and tactile somatic stimulation (Fig. 4). Both visceral and somatic pain irritation resulted in significant changes in blood pressure with simultaneous opposite alterations in BPA. Note that the hypotension reaction on the visceral pain was three times longer than similar response to somatic pain irritation.

4 Discussion

A wide variety of optical methods is used to evaluate noninvasively cerebral blood flow in real time. Among them the methods based on video recordings of the exposed brain under different illumination are very popular due to their low cost, simplicity of implementation, and no use of any dies [2, 8]. These methods are referenced to as "optical intrinsic signals" (OIS) imaging [1, 9] or wide-field optical mapping, WFOM [2]. These techniques reveals slowly varying changes in the recorded video images of the open brain, which are assumed to be proportional to changes in blood volume [10, 11]. Temporal modulation of OIS at the heartbeat frequency is typically considered as an artifact which have to be filtered out to increase the signal-to-noise ratio [12, 13]. In contrast, heartbeat related modulation is the primary source of information in IPPG systems [7, 14]. These systems are widely used for assessment of cardiovascular parameters from video images of the human skin [3, 15] but not enough attention was paid to use IPPG for studying brain hemodynamics.

Notwithstanding rapidly growing use of IPPG systems, the reasons of light modulation at the heart rate (especially for the green light) remains the subject of continuing debates [16–18]. Anatomical-physiological structure of blood vessels is significantly different from that of the skin [19]. Arteries and arterioles are situated at the upper surface of the cortex [19] whereas in the skin they are located below the layer of capillaries, which prevent the interaction of green light with pulsating vessels [16] due to a small depth of green light penetration. The conventional model assumes that the modulation mainly originates from the relative changes of blood volume in arteries and arterioles, which modulate the light absorption [7]. This model well explains the observation of increased BPA in arteries and arterioles (see red areas in Fig. 3b). However, heart related pulsations are also observed far from the arteries (see green areas in Fig. 3b). These pulsations could be explained in the alternative model suggesting origin of light modulation being mechanical compression/decompression of the capillary bed due to pulsatile transmural arterial pressure [16]. Both theoretical models assume pulsations of arteries to be the primary source of the light modulation.

In this study, we carried out concurrent measurements of cortical blood flow using contactless, noninvasive IPPG system and arterial blood pressure in femoral artery using standard invasive method during four different peripheral stimulations. It was shown for the first time to the best of our knowledge that the amplitude of heart-related oscillations of the optical signal (BPA) can be used as a measure of cortical hemodynamics. As seen in Figs. 4 and 5, mean arterial blood pressure and BPA are in reciprocal relationships. We suggest that the observed negative correlation of the blood pressure and BPA is due to cerebrovascular reflex manifested by a decrease in cerebral vascular tone. Such a decrease is response to systemic hypotension aimed at maintaining constant blood flow in the cerebral cortex. It can be assumed that PPG reflects the state of cerebral blood flow, and BPA is a measure of cerebral vascular tone, which is significantly reflexively reduced during systemic hypotension caused by painful stimulation.

Furthermore, the present study has disclosed for the first time that hemodynamics in the primary somatosensory cortex is more responsive to both visceral and somatic painful stimuli than that in the primary motor cortical area. However, the effect of visceral painful stimulation on microcirculation in the somatosensory area qualitatively differs from the somatic pain.

5   Conclusions

In this paper, we have demonstrated that contactless, noninvasive, and cost efficient IPPG method allows us to monitor variations of cerebral vascular tone and its response on various peripheral stimulations. It was shown that the amplitude of the heart-related pulsations of optical signal, BPA, is a measure of cortical hemodynamics. Apparently, change of BPA is in the reciprocal relationship with variations of mean arterial pressure and mean pulse pressure. The reaction of BPA in the same cortical area on visceral and somatic painful simulation is clearly different. It is also found that different areas of the brain react differently on the same peripheral stimulation.

Acknowledgements. Financial support provided by the Russian Science Foundation (grant 15-15-20012) is acknowledged.

References

1. Grinvald, A., Lieke, E., Frostig, R.D., Gilbert, C.D., Wiesel, T.N.: Functional architecture of cortex revealed by optical imaging of intrinsic signals. Nature **324**, 361–364 (1986)
2. Ma, Y., Shaik, M.A., Kim, S.H., Kozberg, M.G., Thibodeaux, D.N., Zhao, H.T., Yu, H., Hillman, E.M.C.: Wide-field optical mapping of neural activity and brain haemodynamics: considerations and novel approaches. Philos. Trans. R. Soc. B **371**, 20150360 (2016)
3. Zaunseder, S., Trumpp, A., Wedekind, D., Malberg, H.: Cardiovascular assessment by imaging photoplethysmography - a review. Biomed. Eng./Biomed. Tech. **63**, 529–535 (2018)
4. Teplov, V., Shatillo, A., Nippolainen, E., Gröhn, O., Giniatullin, R., Kamshilin, A.A.: Fast vascular component of cortical spreading depression revealed in rats by blood pulsation imaging. J. Biomed. Opt. **19**, 046011 (2014)
5. Sidorov, I.S., Volynsky, M.A., Kamshilin, A.A.: Influence of polarization filtration on the information readout from pulsating blood vessels. Biomed. Opt. Express. **7**, 2469–2474 (2016)

6. Kamshilin, A.A., Krasnikova, T.V., Volynsky, M.A., Miridonov, S.V., Mamontov, O.V.: Alterations of blood pulsations parameters in carotid basin due to body position change. Sci. Rep. **8**, 13663 (2018)
7. Allen, J.: Photoplethysmography and its application in clinical physiological measurement. Physiol. Meas. **28**, R1–R39 (2007)
8. Bouchard, M.B., Chen, B.R., Burgess, S.A., Hillman, E.M.C.: Ultra-fast multispectral optical imaging of cortical oxygenation, blood flow, and intracellular calcium dynamics. Opt. Express **17**, 15670–15678 (2009)
9. Schiessl, I., Wang, W., McLoughlin, N.: Independent components of the haemodynamic response in intrinsic optical imaging. Neuroimage **39**, 634–646 (2008)
10. Malonek, D., Grinvald, A.: Interactions between electrical activity and cortical microcirculation revealed by imaging spectroscopy: implications for functional brain mapping. Science **272**, 551–554 (1996)
11. Frostig, R.D., Lieke, E., Ts'o, D.Y., Grinvald, A.: Cortical functional architecture and local coupling between neuronal activity and the microcirculation revealed by in vivo high-resolution optical imaging of intrinsic signals. Proc. Natl. Acad. Sci. USA **87**, 6082–6086 (1990)
12. Ma, H., Zhao, M., Schwartz, T.H.: Dynamic neurovascular coupling and uncoupling during ictal onset, propagation, and termination revealed by simultaneous in vivo optical imaging of neural activity and local blood volume. Cereb. Cortex **23**, 885–899 (2013)
13. Stewart, R.S., Huang, C., Arnett, M.T., Celikel, T.: Spontaneous oscillations in intrinsic signals reveal the structure of cerebral vasculature. J. Neurophysiol. **109**, 3094–3104 (2013)
14. Kamshilin, A.A., Miridonov, S., Teplov, V., Saarenheimo, R., Nippolainen, E.: Photoplethysmographic imaging of high spatial resolution. Biomed. Opt. Express **2**, 996–1006 (2011)
15. Sun, Y., Thakor, N.: Photoplethysmography revisited: from contact to noncontact, from point to imaging. IEEE Trans. Biomed. Eng. **63**, 463–477 (2016)
16. Kamshilin, A.A., et al.: A new look at the essence of the imaging photoplethysmography. Sci. Rep. **5**, 10494 (2015)
17. Volkov, M.V., et al.: Video capillaroscopy clarifies mechanism of the photoplethysmographic waveform appearance. Sci. Rep. **7**, 13298 (2017)
18. Moço, A.V., Stuijk, S., de Haan, G.: New insights into the origin of remote PPG signals in visible light and infrared. Sci. Rep. **8**, 8501 (2018)
19. Nonaka, H., Akima, M., Nagayama, T., Hatori, T., Zhang, Z., Ihara, F.: Microvasculature of the human cerebral meninges. Neuropathology **23**, 129–135 (2003)

Detection of Subclinical Keratoconus Using Biometric Parameters

Jose Sebastián Velázquez-Blázquez[1], Francisco Cavas-Martínez[1(✉)],
Jorge Alió del Barrio[2,3], Daniel G. Fernández-Pacheco[1],
Francisco J. F. Cañavate[1], Dolores Parras-Burgos[1], and Jorge Alió[2,3]

[1] Department of Graphical Expression, Technical University of Cartagena,
30202 Cartagena, Spain
francisco.cavas@upct.es

[2] Keratoconus Unit of Vissum Corporation Alicante, 03016 Alicante, Spain

[3] Department of Ophthalmology, Miguel Hernández University of Elche,
03202 Alicante, Spain

Abstract. The validation of innovative methodologies for diagnosing kerato-conus in its earliest stages is of major interest in ophthalmology. So far, sub-clinical keratoconus diagnosis has been made by combining several clinical criteria that allowed the definition of indices and decision trees, which proved to be valuable diagnostic tools. However, further improvements need to be made in order to reduce the risk of ectasia in patients who undergo corneal refractive surgery. The purpose of this work is to report a new subclinical keratoconus detection method based in the analysis of certain biometric parameters extracted from a custom 3D corneal model.

This retrospective study includes two groups: the first composed of 67 patients with healthy eyes and normal vision, and the second composed of 24 patients with subclinical keratoconus and normal vision as well. The proposed detection method generates a 3D custom corneal model using computer-aided graphic design (CAGD) tools and corneal surfaces' data provided by a corneal tomographer. Defined bio-geometric parameters are then derived from the model, and statistically analysed to detect any minimal corneal deformation.

The metric which showed the highest area under the receiver-operator curve (ROC) was the posterior apex deviation.

This new method detected differences between healthy and sub-clinical keratoconus corneas by using abnormal corneal topography and normal spectacle corrected vision, enabling an integrated tool that facilitates an easier diagnosis and follow-up of keratoconus.

Keywords: Ophthalmology · Early keratoconus · Computational modelling · Scheimpflug technology

1 Introduction

Debilitating primary corneal ectasia shows changes in the corneal structure, including stromal thinning and progressive change of shape related to breakage of Bowman's membrane [1]. These morphological changes seriously affect the rigidity and elasticity

© Springer Nature Switzerland AG 2019
I. Rojas et al. (Eds.): IWBBIO 2019, LNBI 11466, pp. 490–501, 2019.
https://doi.org/10.1007/978-3-030-17935-9_44

of the tissues that form part of the corneal surface and behave very sensitively to variations in intraocular pressure [2]. Corneal shape shows changes such as conical focal corneal deformation located in the central or paracentral region [3]. This is followed by a progressive myopization, astigmatic shift with irregular astigmatism and progressive visual loss [4]. This disease usually begins in puberty and progresses to the third or fourth decade of life [5].

Several systems have been described in the literature for diagnosing and classifying keratoconus (KC) severity. Most of these grading systems have been developed by considering the patient's optical and geometric parameters [2, 31]. These systems have proven essential as a therapeutic approach to manage keratoconus [6].

A sample of this disease with a located steepening pattern exists, but neither clinical signs of the disease, nor other causes that could explain an altered topographical pattern, are manifested, and patients present the sharpness of normal visual correction. Subclinical keratoconus was a term introduced by Amsler [7] and is a condition in which the clinician suspects the potential clinical development of keratoconus. Identifying keratoconus in early stages is crucial to monitor disease progression [6, 8, 9], to perform genetic studies with the patient [10], to indicate therapeutic approaches to avoid its evolution [11], even most importantly, when making the preoperative screening of candidates for refractive surgery. The last reason has been identified as one of the main risk factors for ectasia after LASIK, after PRK [6–10], and even after low myopic ablations [12], because refractive surgery can weaken corneal tissues, and failure can be detected in cornea's biomechanics in totally asymptomatic individuals [13]. For these reasons, it is a fundamental challenge for the ophthalmic community to improve the strategies, tools and techniques employed to detect those individuals at potential risk of developing this pathology.

So far, subclinical keratoconus diagnosis has been made by combining several clinical criteria: age, family history, genetic predisposition [7] and corneal topography [14]. The topographic data and optical parameters obtained in this way have defined univariate and multivariate indices for detection, and for the decision trees or neural networks based on these indices, now valuable tools for detecting subclinical keratoconus [9, 15, 32].

However, further improvements should be made to detect subclinical keratoconus in order to enhance the strategies that diagnose this disease and to avoid the risk of ectasia in patients who undergo corneal refractive surgery. Progression from subclinical to clinical keratoconus is more likely to occur in patients aged 10–20 years, and is less likely to occur in patients over the age of 30 [7]. These cases are difficult to diagnose because symptoms are lacking in early disease stages. When patients come to consultations, the disease may have already advanced to the clinical keratoconus stage. Many studies are attempting to find an objective way to detect these cases and to treat them before their visual function is affected [11].

Accordingly, a subclinical keratoconus detection technique could be based on analysing the biometric morphometry of discrete landmarks [16] in the region where the focused curvature was initially manifested on corneal surfaces, because no other clinical signs are observed in this stage and patients present normal corrected visual acuity. In the biology and modern medicine fields, a geometric morphometric analysis works on a multidimensional image of a discrete data set obtained from a three-dimensional reconstruction of biological structures [17]. These studies are character-ized by both high sensitivity and specificity for structure characterization or pattern recognition, regardless of the technology used to generate the virtual model. This procedure could prove to be a methodological analysis process because the virtual environment provides a large number of hypotheses, which can avoid using complex analytical methods and can considerably reduce in vivo experimentation costs [17].

We have previously demonstrated that morphogeometric analysis of a custom virtual model of the cornea could be useful for the study of keratoconus characteri-zation, and has been previously validated for the evaluation of disease progression across its different clinical stages [18, 19]. It has also been recently used in ophthal-mology, specifically in the cornea biomechanics field, where some authors propose a customized model of the cornea obtained from interpolating the topographic data, which is further used as a basis for biomechanical analyses [20–22].

The aim of this study is to provide new rational and objective indices that accu-rately quantify the morphogeometric changes associated to the clinical evidence of corneal ectatic disease, enabling the description of early local ectasia in a preclinical stage, and to allow healthy corneas to be differentiated from corneas with the so-called subclinical keratoconus in order to avoid idiopathic progression of the disease.

2 Materials and Methods

2.1 Participants

This observational case series study evaluated 91 subjects (only one cornea per patient, randomly selected to avoid interference) divided into two groups: the first group (healthy corneas) presented no ocular pathology and included 67 patients (36.51 ± 14.99 years). In the second group were included 24 patients diagnosed with subclinical keratoconus (33.99 ± 10.97 years). Participants with any kind of ocular pathology were excluded from both groups.

The classification protocol for normal or subclinical keratoconus cases was run according to reported state of the art of clinical and topography evaluations [23].

These evaluations were made in Vissum Hospital (Alicante, Spain). The study follows ethical standards of the Declaration of Helsinki and was approved by the local Clinical Research Ethics Committee with informed consent.

2.2 Examination Protocol

Examination of all the selected patients was performed following a previous validated protocol created by our research group and using Sirius System® (CSO, Florence, Italy) [18, 19]. In this protocol, only the data from the first stage of the tomographic data acquisition procedure, so called raw data, is registered. Using the tomographer's vision algorithm in this first stage, we obtained a finite set of discrete, real and rep-resentative spatial data from the corneal surfaces [18, 19]. For obtaining the average values that were used for later analyses, a set of three successive measures were taken, always by the same expert optometrists.

2.3   Detection Procedure

The procedure proposed in this article consisted in two stages: first, using raw data from the corneal tomographer, a virtual 3D model was reconstructed throughout computational geometry techniques; second, geometric parameters were determined from this model and analysed in order to characterize the cornea morphology.

First Stage: Virtual 3D Modelling

The reconstruction of the cornea was performed by following the steps below (Fig. 1):

(i) *Acquisition of data from the corneal tomographer.*

The Syrius device provides two 3D point clouds that make up both the anterior and the posterior corneal surface, respectively (see Fig. 1). With this procedure we obtained useful data avoiding the interpolation that manufacturers use to fill or substitute wrongly scanned data [24].

We have taken into account that major irregularity levels in the corneal morphology of keratoconic eyes for both the corneal surfaces were presented between radii of 0–4 mm, which encompasses 97% of all keratoconus cases [25]. Any case in which the data provided by the Sirius tomographer had some erroneous point in the study area (r = 0–4 mm) was discarded.

Spatial points for both surfaces are obtained in polar format. These points are distributed in regular circles (256 points for each circle) in the XY plane whose radii are incremented in intervals of 0.2 mm. [18, 19]. To proceed with the reconstruction process, points' elevation data were then converted to Cartesian coordinates (X, Y, Z) using the following equations:

$$X = 0.2i \cdot \cos\left(j\frac{360°}{256}\right); \ Y = 0.2i \cdot \sin\left(j\frac{360°}{0256}\right); \ Z = \text{value in the exported table}$$

$$(1)$$

For this study, an algorithm programmed in Matlab V R2015 (Mathworks, Natick, USA) was implemented to perform this task.

(ii) *Geometric Surface Reconstruction and Solid Modelling.*

In a second stage, after the conversion of the spatial points into a Cartesian format, data were imported to the surface reconstruction CAD software Rhinoceros® V 5.0 (MCNeel & Associates, Seattle, USA). Non-uniform rational B-splines were used to generate surfaces applying point grid function, which allows a fitting of the reconstructed surface with regard to the point cloud of about 4.91 × 10–16 ± 5.19 × 10–16 mm.

The generated surface was then imported into the software SolidWorks V 2017 (Dassault Systèmes, Vélizy-Villacoublay, France), which allowed the generation of an in vivo solid model representing the custom biometrics of each cornea.

Second Stage: Biometric Parameters Analysis

The final 3D virtual model of the cornea was then used to run an analysis of the determined biometric parameters. These geometric parameters studied herein, along

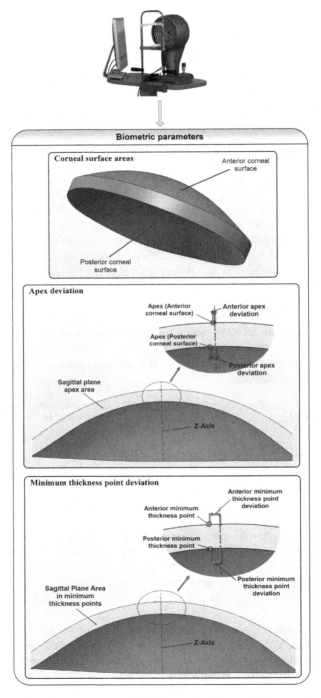

Fig. 1. Corneal surface areas, apex deviation and minimum thickness point deviation

with their characteristics have been previously described [19], and are shown in Table 1, being used in this case for the first time to detect subclinical keratoconus.

2.4 Statistical Analysis

A Kolmogorov-Smirnov test was run to assess the data engagement scores. According to this test and thereafter, a Student's t-test or U-Mann Whitney Wilcoxon test was employed, as and when appropriate. ROC curves were established to determine what parameters could be used to classify the diseased corneas by calculating optimal cut-offs, sensitivity and specificity [26, 27]. All the analyses were performed by the Graphpad Prism V 6 (GraphPad Software, La Jolla, USA) and SPSS V 17.0 software (SPSS, Chicago, USA).

3 Results

Most of the modelled parameters showed statistically significant differences when comparing healthy and subclinical corneas, as shown in Table 1 below.

Table 1. Descriptive values and differences in the modelled biometric parameters among the normal and subclinic KC groups (SD: standard deviation, P: statistical test, Z: z-score).

Biometric parameters	Normal group (n = 67)				Subclinical KC group (n = 24)				Z	P
	Mean	SD	Min	Max	Mean	SD	Min	Max		
Total corneal volume (mm^3)	25.80	1.57	21.29	29.39	24.09	1.69	19.79	28.21	−8.39	0.00
Anterior corneal surface area (mm^2)	43.10	0.16	42.69	43.40	43.18	0.16	43.01	43.48	−6.88	0.00
Posterior corneal surface area (mm^2)	44.19	0.31	43.38	44.91	44.31	0.34	43.57	44.79	−5.80	0.00
Total corneal surface area (mm^2)	103.94	1.25	100.80	106.09	103.50	1.20	101.31	105.49	−3.17	0.00
Sagittal plane apex area (mm^2)	4.29	0.25	3.60	4.99	4.02	0.31	3.24	4.86	−9.02	0.00
Sagittal plane area in minimum thickness point (mm^2)	4.30	0.49	0.00	5.01	3.99	0.30	3.22	4.81	−9.01	0.00

(*continued*)

Table 1. (*continued*)

Biometric parameters	Normal group (n = 67)				Subclinical KC group (n = 24)				Z	P
	Mean	SD	Min	Max	Mean	SD	Min	Max		
Anterior apex deviation (mm)	0.00	0.00	0.00	0.01	0.00	0.01	0.00	0.21	−6.01	0.00
Posterior apex deviation (mm)	0.07	0.02	0.02	0.15	0.14	0.07	0.05	0.40	−10.42	0.00
Anterior minimum thickness point deviation (mm)	0.84	0.3	0.42	2.21	1.2	0.29	0.61	1.79	−4.72	0.00
Posterior minimum thickness point deviation (mm)	0.79	0.26	0.46	2.01	1.1	0.29	0.49	1.69	−4.89	0.00
Net deviation from centre of mass XY (mm)	0.05	0.03	0.01	0.09	0.08	0.04	0.00	0.12	−4.20	0.00
Centre of mass X (mm)	0.05	0.03	0.01	0.10	0.04	0.03	0.00	0.08	−3.46	0.28
Centre of mass Y (mm)	0.04	0.03	0.00	0.11	0.01	0.02	0.00	0.10	−7.70	0.14
Centre of mass Z (mm)	0.81	0.02	0.72	0.83	0.78	0.03	0.70	0.81	−0.82	0.08
Volume of corneal cylinder (mm^3) with Radius 0.5 mm	0.47	0.29	0.34	3.21	0.4	0.03	0.29	0.48	−1.02	0.00
Volume of corneal cylinder (mm^3) with Radius 1 mm	1.69	0.13	1.39	1.99	1.49	0.15	1.19	1.9	−1.58	0.00
Volume of corneal cylinder (mm^3) with Radius 1.5 mm	3.89	0.44	3.31	7.29	3.6	0.3	2.9	4.3	−1.9	0.00
Volume of corneal cylinder (mm^3) with Radius 2 mm	7.11	0.45	5.94	8.3	6.55	0.5	5.25	7.8	−2.3	0.00

3.1   Roc Analysis

The predictive value of the modelled parameters was established by a ROC analysis. Six biometric parameters were identified with an area under the ROC (AUROC) above 0.7 (see Fig. 2 and Table 2):

Table 2. Parameters with an AUROC above 0.7

Biometric parameters	AUROC	Sensitivity	Specificity	95% Confidence interval	
				Lower limit	Upper limit
Anterior corneal surface area (mm^2)	0.719	91.5	19.0	0.560	0.779
Posterior corneal surface area (mm^2)	0.701	90.0	20.4	0.521	0.749
Anterior apex deviation (mm)	0.767	67.0	99.9	0.639	0.869
Posterior apex deviation (mm)	0.891	91.6	36.0	0.799	0.959
Anterior minimum thickness point deviation (mm)	0.751	91.4	23.0	0.641	0.849
Posterior minimum thickness point deviation (mm)	0.761	88.1	19.1	0.649	0.858

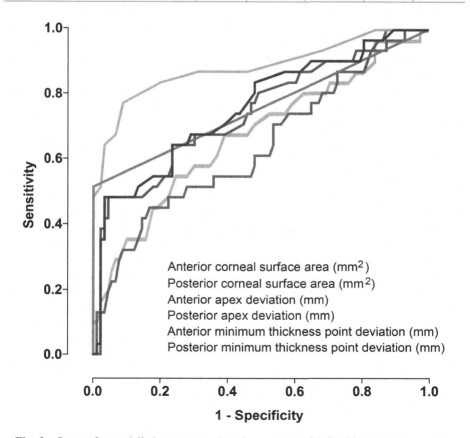

Fig. 2. Curves for modelled parameters detecting subclinical KC with AUROC over 0.7.

4 Discussion

This study obtained good accuracy reconstruction of the intrinsic biometric morphology of the human cornea as a biological structure, making aberrometric analysis unnecessary, and creating a new concise global understanding of early corneal pathology in keratoconic eyes.

The deterioration process in keratoconus is characterized by a significantly reduced total corneal volume compared with healthy eyes, which is triggered by an alteration in corneal collagen fibres that causes stromal thinning and breaks in Bowman's membrane from disease subclinical stages [1]. The studied volumetric parameters and the volume of the analysed corneal cylinders (0.5–1–2 mm) showed a statistically significant reduced total corneal volume in the subclinical group compared with healthy eyes. A similar volumetric reduction of corneas has been described in several studies as a characteristic parameter for the differentiation of subclinical eyes [28–30]. Therefore, the 3D corneal model defined in this work for subclinical keratoconus, allows an accurate characterisation of small changes in its architecture, when the degree of alteration in corneal morphology is low.

Presence of corneal irregularities due to local steepening by a reduced curvature radius leads to increased corneal surface [1]. Eyes with subclinical keratoconus show significant differences for both the anterior and posterior surfaces compared with healthy corneas. In the disease group, these surfaces were larger given their local structural weakening, as fewer collagen fibres are present in each lamella [2]. However, the total cornea area, which included the area of the peripheral region, was larger in healthy corneas, because of their constant thickness compared with the thinning noted in the pathological group due to the influence of intraocular pressure on their weakened structure. This structural weakening is produced by the reduction in the number of stromal lamellae and a reduction in the interconnecting layer's area.

The average distances from the Optical axis to the apex of the anterior and posterior corneal surfaces differed between groups, with the largest deviations found in the group of eyes with subclinical keratoconus. This has been previously described when the disease has been clearly established [1].

The aforementioned presence of an irregular corneal surface, which created a protrusion in the keratoconic eye, also led to incremented corneal curvature [33] and, therefore, to an increase in the deviation distances of corneal apex (maximum curvature) and in the deviation distances of the minimum thickness points, each increase produced in both anterior and posterior surfaces (Table 1). The existence of structural instability in subclinical keratoconic corneas explained why the best disease discrimination results were obtained for the posterior apex deviation variable (ROC area: 0.891, p < 0.000, std. error: 0.039, 95% CI: 0.799–0.959). The posterior corneal surface is more susceptible to variations given the forces exerted on tissue [34]. This is why the posterior apex deviation is one of the variables that most reliably represents early changes in patients in first disease stages. Several studies have concluded the importance of, and interest in, the posterior corneal surface [35]. Upon disease onset, structural changes occur on the posterior surface of the cornea, so analysing this surface can positively identify early subclinical keratoconus [29].

Due to all the mentioned above, this new technique could help improve the widely used and well-accepted assessment corneal irregularity methods, in line with the main conclusions drawn by the group of world experts in keratoconus who gathered for the Project "Global Consensus on Keratoconus and Ectatic Diseases" [36]. The existing limitations of current topographic methods (i.e., data interpolation, unknown internal algorithms, etc.) difficult study comparisons and data sharing, and could also lead to relevant information losses. The method proposed in this study is based in a non-finite biometric parameters analysis of the human cornea, and it also does not depend on any restricted commercial algorithm, so could be implemented in all corneal topographers, allowing data comparison between different devices. This technique allows a good and accurate characterization of the corneal health status that will enable new detection paths and to follow-up corneal pathologies.

Funding. This publication has been carried out in the framework of the Thematic Network for Co-Operative Research in Health (RETICS) reference number RD16/0008/0012 financed by the Carlos III Health Institute-General Subdirection of Networks and Cooperative Investigation Centers (R&D&I National Plan 2013–2016) and the European Regional Development Fund (FEDER).

Additional Information
Conflict of Interest: The authors declare no conflict of interest.

Financial Disclosure: Neither author has a financial or proprietary interest in any material or method mentioned.

References

1. Rabinowitz, Y.S.: Keratoconus. Surv. Ophthalmol. **42**, 297–319 (1998)
2. Piñero, D.P., Alió, J.L., Barraquer, R.I., Michael, R., Jimenez, R.: Corneal biomechanics, refraction, and corneal aberrometry in keratoconus: an integrated study. Invest. Ophthalmol. Vis. Sci. **51**, 1948–1955 (2010). https://doi.org/10.1167/iovs.09-4177
3. Egorova, G.B., Rogova, A.: Keratoconus. diagnostic and monitoring methods. Vestn. oftalmol. **129**, 61–66 (2013)
4. Krachmer, J.H., Feder, R.S., Belin, M.W.: Keratoconus and related noninflammatory corneal thinning disorders. Surv. Ophthalmol. **28**, 293–322 (1984)
5. Kennedy, R.H., Bourne, W.M., Dyer, J.A.: A 48-year clinical and epidemiologic study of keratoconus. Am. J. Ophthalmol. **101**, 267–273 (1986)
6. Belin, M.W., Duncan, J.K.: Keratoconus: the ABCD grading system. Klin. Monbl. Augenheilkd. **233**(06), 701–707 (2016). https://doi.org/10.1055/s-0042-100626
7. Amsler, M.: Kératocône classique et kératocône fruste; arguments unitaires. Ophthalmologica **111**, 96–101 (1946). https://doi.org/10.1159/000300309
8. Belin, M.W., Duncan, J., Ambrósio Jr., R., Gomes, J.A.P.: A new tomographic method of staging/classifying keratoconus: the ABCD grading system. Int. J. Kerat. Ect. Cor. Dis. **4**, 55–63 (2015)
9. Cavas-Martinez, F., De la Cruz Sanchez, E., Nieto Martinez, J., Fernandez Canavate, F.J., Fernandez-Pacheco, D.G.: Corneal topography in keratoconus: state of the art. Eye Vis. (Lond) **3**, 5 (2016). https://doi.org/10.1186/s40662-016-0036-8

10. McGhee, C.N., Kim, B.Z., Wilson, P.J.: Contemporary treatment paradigms in keratoconus. Cornea **34**(Suppl 10), S16–S23 (2015). https://doi.org/10.1097/ico.0000000000000504
11. Muftuoglu, O., Ayar, O., Hurmeric, V., Orucoglu, F., Kilic, I.: Comparison of multimetric D index with keratometric, pachymetric, and posterior elevation parameters in diagnosing subclinical keratoconus in fellow eyes of asymmetric keratoconus patients. J. Cataract Refract. Surg. **41**, 557–565 (2015). https://doi.org/10.1016/j.jcrs.2014.05.052
12. Malecaze, F., Coullet, J., Calvas, P., Fournie, P., Arne, J.L., Brodaty, C.: Corneal ectasia after photorefractive keratectomy for low myopia. Ophthalmology **113**, 742–746 (2006). https://doi.org/10.1016/j.ophtha.2005.11.023
13. Ambrosio Jr., R., Dawson, D.G., Belin, M.W.: Association between the percent tissue altered and post-laser in situ keratomileusis ectasia in eyes with normal preoperative topography. Am. J. Ophthalmol. **158**, 1358–1359 (2014). https://doi.org/10.1016/j.ajo.2014.09.016
14. Sonmez, B., Doan, M.P., Hamilton, D.R.: Identification of scanning slit-beam topographic parameters important in distinguishing normal from keratoconic corneal morphologic features. Am. J. Ophthalmol. **143**, 401–408 (2007). https://doi.org/10.1016/j.ajo.2006.11.044
15. Parker, J.S., van Dijk, K., Melles, G.R.: Treatment options for advanced keratoconus: a review. Surv. Ophthalmol. **60**, 459–480 (2015). https://doi.org/10.1016/j.survophthal.2015.02.004
16. Wilson, L.A.B., Humphrey, L.T.: A virtual geometric morphometric approach to the quantification of long bone bilateral asymmetry and cross-sectional shape. Am. J. Phy. Anthropol. **158**, 541–556 (2015). https://doi.org/10.1002/ajpa.22809
17. Bartocci, E., Lio, P.: Computational modeling, formal analysis, and tools for systems biology. PLoS Comput. Biol. **12**, e1004591 (2016). https://doi.org/10.1371/journal.pcbi.1004591
18. Cavas-Martínez, F., Fernández-Pacheco, D.G., De La Cruz-Sánchez, E., Martínez, J.N., Cañavate, F.J.F., Alio, J.L.: Virtual biomodelling of a biological structure: the human cornea. Dyna (Spain) **90**, 647–651 (2015). https://doi.org/10.6036/7689
19. Cavas-Martínez, F., Bataille, L., Fernández-Pacheco, D.G., Cañavate, F.J.F., Alió, J.L.: A new approach to keratoconus detection based on corneal morphogeometric analysis. PLoS ONE **12**(9), e0184569 (2017). https://doi.org/10.1371/journal.pone.0184569
20. Ariza-Gracia, M.A., Zurita, J.F., Piñero, D.P., Rodriguez-Matas, J.F., Calvo, B.: Coupled biomechanical response of the cornea assessed by non-contact tonometry. a simulation study. PLoS ONE **10**, e0121486 (2015). https://doi.org/10.1371/journal.pone.0121486
21. Roy, A.S., Dupps Jr., W.J.: Patient-specific modeling of corneal refractive surgery outcomes and inverse estimation of elastic property changes. J. Biomech. Eng. **133**, 011002 (2011). https://doi.org/10.1115/1.4002934
22. Simonini, I., Pandolfi, A.: Customized finite element modelling of the human cornea. PLoS ONE **10**, e0130426 (2015). https://doi.org/10.1371/journal.pone.0130426
23. Li, X., Yang, H., Rabinowitz, Y.S.: Keratoconus: classification scheme based on videokeratography and clinical signs. J. Cataract Refract. Surg. **35**, 1597–1603 (2009). https://doi.org/10.1016/j.jcrs.2009.03.050
24. Ramos-Lopez, D., et al.: Screening subclinical keratoconus with placido-based corneal indices. Optom. Vis. Sci. **90**, 335–343 (2013). https://doi.org/10.1097/OPX.0b013e3182843f2a
25. Wilson, S.E., Klyce, S.D.: Quantitative descriptors of corneal topography. a clinical study. Arch. Ophthalmol. **109**, 349–353 (1991)

26. Lasko, T.A., Bhagwat, J.G., Zou, K.H., Ohno-Machado, L.: The use of receiver operating characteristic curves in biomedical informatics. J. Biomed. Inform. **38**, 404–415 (2005). https://doi.org/10.1016/j.jbi.2005.02.008
27. Pepe, M.S.: The Statistical Evaluation of Medical Tests for Classification and Prediction. Oxford, New York (2004)
28. Piñero, D.P., Alió, J.L., Aleson, A., Escaf Vergara, M., Miranda, M.: Corneal volume, pachymetry, and correlation of anterior and posterior corneal shape in subclinical and different stages of clinical keratoconus. J. Cataract Refract. Surg. **36**, 814–825 (2010). https://doi.org/10.1016/j.jcrs.2009.11.012
29. Saad, A., Gatinel, D.: Topographic and tomographic properties of forme fruste keratoconus corneas. Invest. Ophthalmol. Vis. Sci. **51**, 5546–5555 (2010). https://doi.org/10.1167/iovs. 10-5369
30. Saad, A., Lteif, Y., Azan, E., Gatinel, D.: Biomechanical properties of keratoconus suspect eyes. Invest. Ophthalmol. Vis. Sci. **51**, 2912–2916 (2010). https://doi.org/10.1167/iovs.09-4304
31. Cavas-Martínez, F., Fernández-Pacheco, D.G., Cañavate, F.J.F., Velázquez-Blázquez, J.S., Bolarín, J.M., Alió, J.L.: Study of morpho-geometric variables to improve the diagnosis in Keratoconus with mild visual limitation. Symmetry **10**(8) (2018). https://doi.org/10.3390/sym10080306
32. Cavas Martinez, F., et al.: Detección De Queratocono Temprano Mediante Modelado 3D Personalizado Y Análisis De Sus Parámetros Geométricos. Dyna Ingenieria E Industria **94**(1), 175–181 (2019). https://doi.org/10.6036/8895
33. de Rojas Silva, V.: Clasificación del queratocono. In: Albertazzi, R. (ed.) Queratocono: pautas para su diagnostico y tratamiento. Buenos Aires, Argentina: Ediciones Científicas Argentinas, pp. 33–97 (2010)
34. Sloan Jr., S.R., Khalifa, Y.M., Buckley, M.R.: The location- and depth-dependent mechanical response of the human cornea under shear loading. Invest. Ophthalmol. Vis. Sci. **55**, 7919–7924 (2014). https://doi.org/10.1167/iovs.14-14997
35. Fukuda, S., et al.: Comparison of three-dimensional optical coherence tomography and combining a rotating Scheimpflug camera with a Placido topography system for forme fruste keratoconus diagnosis. Br. J. Ophthalmol. **97**, 1554–1559 (2013). https://doi.org/10.1136/bjophthalmol-2013-303477
36. Gomes, J.A., et al.: Global consensus on keratoconus and ectatic diseases. Cornea **34**, 359–369 (2015). https://doi.org/10.1097/ico.0000000000000408

A Computer Based Blastomere Identification and Evaluation Method for Day 2 Embryos During IVF/ICSI Treatments

Charalambos Strouthopoulos and Athanasios Nikolaidis[(✉)]

Department of Informatics Engineering,
Technological Educational Institute of Central Macedonia, Serres, Greece
{strch,nikolaid}@teicm.gr

Abstract. Human embryos evaluation is one of the most important challenges in vitro fertilization (IVF) programs. The morphokinetic and the morphology parameters of the early cleaving embryo are of critical clinical importance. This stage spans the first 48 h post-fertilization, in which the embryo is dividing in smaller blastomeres at specific time-points. The morphology, in combination with the symmetry of the blastomeres, seem to be powerful features with strong prognostic value for embryo evaluation. To date, the identification of these features is based on human inspection in timed intervals, at best using camera systems that simply work as surveillance systems without any precise alerting and decision support mechanisms. This paper aims to develop a computer vision technique to automatically detect and identify the most suitable cleaving embryos (preferably at Day 2 post-fertilization) for embryo transfer (ET) during treatments. To this end, texture and geometrical features were used to localize and analyze the whole cleaving embryo in 2D grayscale images captured during in-vitro embryo formation. Because of the ellipsoidal nature of blastomeres, the contour of each blastomere is modeled with an optimal fitting ellipse and the mean eccentricity of all ellipses is computed. The mean eccentricity in combination with the number of blastomeres form the feature space on which the final criterion for the embryo evaluation is based. Experimental results with low quality 2D grayscale images demonstrated the effectiveness of the proposed technique and provided evidence of a novel automated approach for predicting embryo quality.

Keywords: IVF · Day 2 embryo · Blastomere

1 Introduction

Epidemiological data indicate that infertility is a problem of global proportions that trends to increase, affecting one-fifth (20%) of couples trying to conceive worldwide. Over the past 30 years, reliable methodologies have been developed to help infertile couples. Assisted Reproductive Technology (ART) and associated techniques have become the predominant treatment of infertility over the past years in most developed countries. Nevertheless, pregnancy and delivery rates are approximately 28.5% and 21% respectively, irrespectively of the number of embryos transferred and the day of

© Springer Nature Switzerland AG 2019
I. Rojas et al. (Eds.): IWBBIO 2019, LNBI 11466, pp. 502–513, 2019.
https://doi.org/10.1007/978-3-030-17935-9_45

embryo transfer [1]. According to the method and the current practice the main goals include: (a) to perform a single-embryo transfer to prevent multiple pregnancies and (b) to achieve higher overall pregnancy rates. However, the ability to identify the most viable embryo in a cohort remains a challenge despite the numerous scoring systems currently in use [2]. Embryologists, as a first-line approach for embryo evaluation, still depend on morphological assessment in combination with cleavage rate just using light microscopy, while research in this field involves mainly non-invasive methodologies for grading embryos and scoring them according to their ability to implant. The last five years, attention is being focused on time-lapse evaluation. This specific method is based on monitoring embryo development and helps pick the best embryo based on morphokinetic changes [3–6]. This imaging technique allows early selection of cleavage embryos with high implantation potential within shorter periods of incubation [7]. Besides, prolonged embryo culture has been associated with significant epigenetic changes [8, 9] and increased risk of preterm delivery when compared to embryos transferred on day 3 or day 5 [10]. [11] that examined a large set of zygotes coming from in vitro fertilization, it was found that success in progression to the blastocyst stage can be predicted with >93% sensitivity and specificity by evaluating three dynamic, noninvasive imaging parameters before embryonic genome activation (EGA). Nevertheless, the latest Cochrane review shows that the time-lapse monitoring systems did not improve the live birth rate or the clinical pregnancy rate (OR = 1.11, 95% CI = 0.45–2.73 and OR = 1.23, 93% CI = 0.96–1.59, respectively) [12].

The use of a semi-automatic computer controlled system (Ferti-Morph system, morphological analyses software) proved to be very useful for the analysis of embryo morphology [13]. Although it has not been evaluated extensively with the use of the Hoffman Modulation Contrast (HMC) microscopy a segmentation method based on an image model for the zona pellucida (ZP) was also proposed [14]. In that method, the segmentation of ZP was divided in two parts: (a) the segmentation of the embryo from the background, in order to find the ZP external boundary, (b) the segmentation of the ZP internal boundary from the rest of the embryo. In another study [15], using appropriate software, the morphological characteristics of zygotes was measured. The proposed method used threshold and ellipse fitting and provided measurements of 24 automatically, semi-automatically or manually defined features, useful for the assessment of embryo viability. Linear discriminant analysis was performed to predict embryo implantation in low and high implantation probability embryos. Images from three different clinics were used. The average classification error of the embryos and the average classification error of the method was 39% and 36%, respectively. An algorithm for automatic ZP thickness identification was proposed [16] to take advantage of the fact that the variation of the ZP thickness is directly related to the embryo implantation rate. Their algorithm was based on an active contour technique (snake), after the application of image enhancement and Canny edge detection. The dataset used in this research consisted of 76 images taken on day 2 and day 3, at fixed magnification and brightness levels. The method yielded 91.65% accuracy in localizing the boundaries (compared to manual segmentation), but it was just an approach to detect the ZP and no other features. In another study [17] that was based on a different type of Bayesian classification that considered variables based on morphology (thickness of ZP, number of cells, level of fragmentation) in combination with the clinical data of the patients yielded an accuracy ranging between 63.49% and 71.43% for predicting the outcome of a cohort of embryos. Giusti and colleagues [18] proposed an edge-based technique to segment the surrounding of the zygote from the rest of the image, although this did not detect any of the cells inside the zygote. The segmentation process was divided into two steps: during the first step the approximate location of the cell center is determined; during the second step the cell location is converted to polar coordinates and the shortest-path formulation is used to recover the actual zygote contour. As this approach filled the inside of the contour, it was not able to detect the blastomeres, and so was limited to images of day 1 embryos. Giusti and colleagues [19] focused on 4-blastomere stage embryos using image sets taken in different focus planes that were converted into polar coordinate images with an energy value associated to each pixel. A graph was then constructed leading to the estimation of the blastomeres' contours by following a minimal energy path of the graph. A total of 53 embryos analyzed in the 4-blastomere stage, it was found that a 71% had all 4 blastomeres. The goal of the method is the segmentation and 3D morphology measurements of early embryos with the use of HMC contrast image stacks. That method is not applicable in 2D digital images because the focus is impossible. None of the above publications focused on the automatic identification of day-2 blastomeres in 2D digital images. Recently El-Shenawy [20] focused on different techniques to automatically detect and classify day 2 embryos based purely on image processing. El-Shenawy applied a Hough-based technique to detect potential cells and attained a success percentage between 60.8% and 65.9%. Template matching-based techniques were also employed using two approaches that yielded a detection percentage between 46.8% and 47.8% respectively. Further modifications improved the percentage of the algorithm in the range between 75.5% and 76.5%. A number of empirical restrictions are used in images of low noise. In order to automate the inspection process and to improve the early detection accuracy we tried to evaluate embryos during the second developmental day with a new bioinformatics image-based approach, modeling embryo morphology with a more accurate manner. Because of the similarity of blastomere contours with circles or ellipses, the basic idea is to model the blastomere shapes with ellipses and to compute the mean eccentricity. In our proposal the mean blastomere eccentricity is estimated in order to evaluate the quality of day-2 embryos. In addition, the defined criteria in the proposed paper can be used as additional features in embryo classification systems. The novelty in the proposed method is in a "smart" combination of mature image processing techniques to deal with the specific highly challenging problem.

2   Image-Based Evaluation of Day 2 Embryos

In this section the proposed technique is described step by step using a representative example of a day 2 embryo low quality 2D grayscale digital image (Fig. 1). Such an image is complex and noisy due to the translucent nature of the observed object and the video capture conditions. It should be clear that the 2D images represent projections of a complete 3D nearly spherical object that consists of ellipsoidal blastomeres, so overlapping, interference and occlusion among blastomeres is expected to occur. The method and algorithm that has been developed to provide identification of blastomeres and evaluation of the day 2 embryos is based on mature image processing techniques and consist of three basic stages, namely: (a) localization of embryos, (b) identification and localization of blastomeres and (c) modeling of blastomeres' contours. Figure 2 presents a simplified overview of the proposed Day 2 embryo localization, blastomere modeling and evaluation method. In the following paragraphs the algorithm is unfolded in detail and is presented using a typical real world sample.

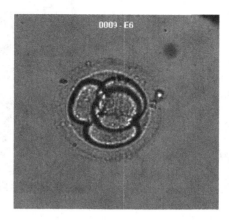

Fig. 1. Typical low resolution Day 2 embryo 2D grayscale image captured in incubators

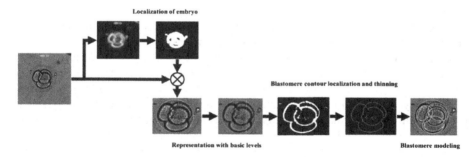

Fig. 2. Simplified overview of the proposed Day 2 embryo localization and evaluation method

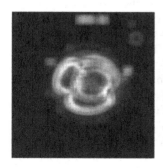

Fig. 3. Representation of the local gray-level variance

Fig. 4. Identification of objects in the image

Fig. 5. Localization mask of the embryo in the image

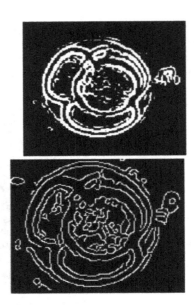

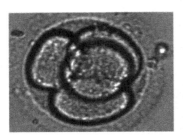

Fig. 6. Localized embryo grayscale sub-image

Fig. 7. Edge detection results (a) using Sobel and (b) Canny edge detectors

At the first stage, localization of the embryo is performed to separate the object of interest from the image background. The benefit of this segmentation lies in the limitation of blastomeres search region and the speeding up of subsequent steps.

The approach makes two basic assumptions (heuristics) regarding the characteristics of the embryo image region: (a) there are significant more variations in brightness in the embryo region rather than in the background region, which in most cases is almost uniform; (b) the embryo has a circular shape and a significant area relative the overall image size. The first assumption is mathematically quantified by the estimation of a statistical spatial texture feature, which can simply be expressed by computing the standard deviation for each m × m non-overlapping region in the original image. The regions in our example were selected to have m = 15 pixels, which corresponds to the 20% of an image with 50 pixels/μm resolution and size 300 × 300 pixels. This percentage is suitable to retain the shape details of the image contents. The computation results in the matrix that is depicted as a grayscale image (Fig. 3). Dark values in the image correspond to regions of low (approximately zero values) standard deviation. Apparently, after this computation, the extraction of the embryo region can be achieved using a gray level thresholding method. The application of Otsu [21] thresholding results in the binary image shown in Fig. 4, in which white pixels correspond to the object and black to the background. This image consists of connected components which are groups of object pixels having at least one path leading to other pixels in the group. The connected components are determined with the use of a fast contour following algorithm without any morphological restrictions [22]. After this step a number of connected components may be identified. It is clear that the components that do not correspond to the embryo have to be identified and masked. To this end the circularity of the embryo component is taken into consideration. The circularity C of a connected component is given by

$$C = \frac{4\pi A}{P^2} \tag{1}$$

where A is the area of the component and P is the perimeter. The maximum value for C corresponds to a circle. A simple classification scheme based on the area and the circularity determines the embryo region and a bounding box is created to denote the correct image region, as shown in Fig. 5. The region of the initial image that corresponds to this bounded subimage R becomes the new image that will be processed in subsequent algorithm stages (Fig. 6).

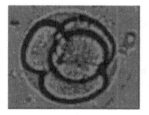  

Fig. 8. Reduced gray-level representation of the image

Fig. 9. Rough blastomeres' boundaries identification

Fig. 10. Fine localization of blastomeres' boundaries

Now the image is prepared for the second stage of processing, which targets the identification of blastomeres' contours. Applying classical edge detection techniques on the embryo grayscale image region R, a number of double or complicated and intersected edges emerge, which describe undesirable structural details. Figure 7a and b show the detected edges by the Sobel and Canny edge detectors respectively. Observing the contour shape of a blastomere, we noted that it was a wide zone of dark pixels. In order to isolate these pixels a gray level image multi-thresholding method were applied on the embryo region [23]. This method reduces the image levels to the four dominant gray levels. Specifically, the initial gray values are clustered, in a predefined number of clusters, using the Kohonen SOFM neural network [24]. Figure 8 shows the image of the embryo region using four dominant gray level values. The smallest value corresponds to the dark pixels that define the blastomere contour. Masking the image to keep only these pixels results in the binary image depicted in Fig. 9. Apparently, the contours are thick and a thinning binary morphological operation [25] has to be applied to better localize the blastomeres (Fig. 10) and simplify the computational complexity in the subsequent ellipse detection stage.

The final stage of the algorithm targets the modeling of the contours using optimally fitted ellipses. In this stage, the thin blastomere contours are modeled as ellipses by assuming that the projection of a 3D blastomere on the camera plane represents an ellipse. In order to identify the number of the ellipses and their parameters that optimally model the blastomeres, an ellipse detection method [26] is applied on the image of Fig. 10. Figure 11 shows the obtained ellipses for this example. If N is the number of ellipses, the eccentricity of the nth ellipse ($n = 1...N,$) is given by the relation

$$e(n) = \sqrt{1 - \left(\frac{b(n)}{a(n)}\right)^2}, \quad e(n) \in [0\ 1) \tag{2}$$

where $a(n)$ is the length of the semi-major axis and $b(n)$ is the length of the semi-minor axis. A zero value for the eccentricity corresponds to a circle. The eccentricity increases as the ellipse elongates and tends to the unit. N takes the values 4, 3 or 2 according to the number of blastomeres.

Finally, the mean eccentricity $\bar{e}$ of the N ellipses is computed. For the embryo of the example the number of ellipses is 4 and the mean eccentricity $\bar{e} = 0.29$. Apparently, a mean eccentricity of $\bar{e} = 0$ represents a perfect circle, which in turn corresponds to excellent, shaped blastomeres and thus an excellent embryo. The number of identified blastomeres N and their mean eccentricity $\bar{e}$ formed the criteria, on which based the assessment of the quality and viability of the embryos according to the described method. The first one is a common morphological criterion for every embryo and the second one is a similar characteristic to the regularity/irregularity of the embryos. The above criteria can be combined in a single score value s according the following relation

$$s = N - \bar{e} \tag{3}$$

For the embryo of the example s = 3.71

An embryo with four blastomeres is better of an embryo with three blastomeres and an embryo with three blastomeres is better of an embryo with two blastomeres. For two embryos with the same number of blastomeres the best is that having the smaller mean eccentricity. Since N takes the values 4, 3 or 2 and $\bar{e} \in [0\ 1)$ it is obvious that $s \in [4\ 1)$. The embryo quality is analog to the value of s.

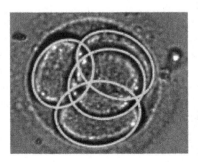

Fig. 11. Final modeling of the identified blastomeres in a Day 2 embryo grayscale image

3 Results

The described method was tested against a number of different cases to evaluate its functionality and usefulness. Several Day 2 images of developing embryos were isolated from time-lapse videos and went through the described identification and grading system. The proposed technique was applied to a number of fifty, day-2 embryo digital images. The obtained results were evaluated in terms of the correct number of cells detected and their corresponding position and shape. Correct results were obtained in 42 cases for all of the above-mentioned terms, i.e. 82% of the total. Errors occurred in the images of high noise or images where the bounds of overlapped cells were not distinguishable. Among the relevant publications, El-Shenawy [20] can be used for comparison reasons because it aims to detect the number and the position of cells in the 2D digital image of day-2 embryo. As the author claims from a test set of 39 images, only in 12 was it possible to get correct counts of cells. Some indicative experimental results are shown and discussed in the following.

Specifically, in the example of Fig. 12 is presented a morphologically excellent embryo, which is graded by the proposed system with a value of 3.76 (it is reminded here that the score of an excellent embryo is zero).

In Fig. 13, which shows apparently a morphologically excellent embryo, the proposed system gives a score of 3.73, because the blastomeres are not so equally positioned as in the previous case and therefore the mean eccentricity of the blastomeres is a little bit different.

In Fig. 14 the score of the embryo is 3.71 and morphologically it is slightly different from the previous case because the blastomeres in the new case display a lesser regularity and this is successfully captured by the proposed method.

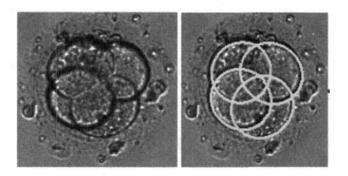

Fig. 12. A morphologically excellent Day 2 embryo, graded 3,76

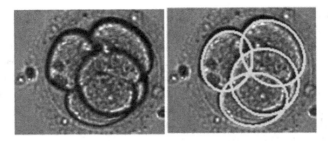

Fig. 13. A morphologically excellent Day 2 embryo, graded 3.73

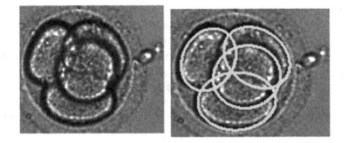

Fig. 14. A morphologically almost excellent Day 2 embryo, graded 3.71 (one blastomere is not totally regular and another is smaller than the rest regular blastomeres)

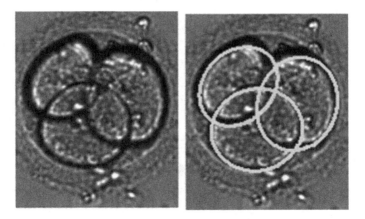

Fig. 15. A three-blastomere Day 2 embryo, grade 2.68 (one blastomere has not been divided yet and one blastomere is not so regular).

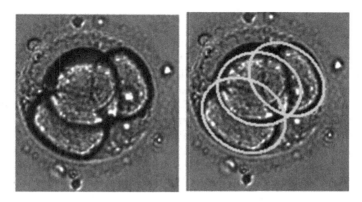

Fig. 16. A three-blastomere Day 2 embryo, graded 2.66 (one blastomere has not been divided yet and two blastomeres are not so regular).

In the subsequent Figures (Figs. 15 and 16) another aspect of the embryo development that the proposed method captures is demonstrated. In both cases, the embryos include only three blastomeres, although they are running their second developmental day and most embryos are expected to have completed their second mitosis and therefore they should include four blastomeres. Apparently in this case the quality of these embryos is inferior of the embryos demonstrated in the previous cases. Specifically, for Fig. 15, although the regularity of the blastomeres is high, a score of $2.68 = 3-0.32$ is obtained due to the fact that there are only three blastomeres. This is in line with the knowledge that embryos with three regular blastomeres during day 2 are characterized as embryos with asynchrony division and high incidence of aneuploidy. Last, Fig. 16 demonstrates an embryo with three blastomeres with a small deviation regarding their regularity and takes a score of $2.66 = 3-0.34$.

4 Conclusions

In the present study, it was attempted to evaluate embryo morphology with the use of bioinformatics aiming to improve the assessment of embryo quality at the early developmental stages.

The new approach employed two criteria for grading the developmental embryos. The first one was the number of blastomeres, which is a common morphological criterion for every embryo and the second one was the mean eccentricity of each blastomere, which is a similar characteristic to the regularity/irregularity of the embryos. This characteristic was modified according to a bioinformatics approach to score an embryo. Bearing in mind that the irregularity of the embryos is due to the incorrect division as well as due to the incidence of high percent of fragmentation (and therefore the shape of the remaining blastomeres will appear with increased irregularity), then the mean eccentricity used in the proposed method is expected to give a useful score for day 2 embryos. Apparently, this bioinformatics scoring system improves the assessment of embryo quality at the very early developmental days. Moreover, it is a non-invasive embryo scoring system based on morphological characteristics and provides accurate embryo scoring, better than those used only with time-lapse or even with snapshots. Last but not least, the proposed approach provides a quantitative score that can be used in statistics and significantly deviates from simplified qualitative assessments like "good" or "poor", which, of course are ambiguous.

In conclusion, the new described bioinformatics approach seems to be a promising method for evaluating embryos coming from IVF/ICSI treatments. It is a non-invasive method and very reliable since it covers all the morphological characteristics of the developing embryos. The use of the mean eccentricity of the blastomeres in combination with the number of blastomeres formed the algorithm that based our bioinformatics approach. Nevertheless, a large-scale study will provide more evidence about the new bioinformatics approach.

Acknowledgments. The authors want to thank the Technological Educational Institute (TEI) of Central Macedonia Greece for the kind contribution and help to complete this work, as well as George Anifandis for addressing the issue in concern and providing the dataset for the experiments.

References

1. Kupka, M.S., D'Hooghe, T., Ferraretti, A.P., et al.: Assisted reproductive technology in Europe, 2011: results generated from European registers by ESHRE. European IVF-monitoring consortium (EIM), European society of human reproduction and embryology (ESHRE). Hum. Reprod. **31**, 233–248 (2016)
2. Magli, M.C., Jones, G.M., Lundin, K., van den Abbeel, E.: Atlas of human embryology: from oocytes to preimplantation embryos. Hum. Reprod. **27**(Suppl 1), i1 (2012)
3. Meseguer, M., Herrero, J., Tejera, A., et al.: The use of morphokinetics as a predictor of embryo implantation. Hum. Reprod. **26**(10), 2658–2671 (2011)
4. Kirkegaard, K., Agerholm, I.E., Ingerslev, H.J.: Time-lapse monitoring as a tool for clinical embryo assessment. Hum. Reprod. **27**(5), 1277–1285 (2012)
5. Kirkegaard, K., Kesmodel, U.S., Hindkjær, J.J., Ingerslev, H.J.: Time-lapse parameters as predictors of blastocyst development and pregnancy outcome in embryos from good prognosis patients: a prospective cohort study. Hum. Reprod. **28**(10), 2643–2651 (2013)
6. Basile, N., Vime, P., Florensa, M., et al.: The use of morphokinetics as a predictor of implantation: a multicentric study to define and validate an algorithm for embryo selection. Hum. Reprod. **30**(2), 276–283 (2015)
7. Milewski, R., Kuć, P., Kuczyńska, A.: A predictive model for blastocyst formation based on morphokinetic parameters in time-lapse monitoring of embryo development, J. Assist. Reprod. Genet. **32**(4), 571–579 (2015)
8. Calle, A., Fernandez-Gonzalez, R., Ramos-Ibeas, P., et al.: Long-term and trans-generational effects of in vitro culture on mouse embryos. Theriogenology **77**(4), 785–793 (2012)
9. Anifandis, G., Messini, C.I., Dafopoulos, K., Messinis, I.E.: Genes and conditions controlling mammalian pre- and post-implantation embryo development. Curr. Genomics **16**(1), 32–46 (2015)

10. Dar, S., Librach, C.L., Gunby, J., Bissonnette, F., Cowan, L.: IVF directors group of canadian fertility and andrology society. Increased risk of preterm birth in singleton pregnancies after blastocyst versus Day 3 embryo transfer: Canadian ART Register (CARTR) analysis. Hum. Reprod. **28**(4), 924–928 (2013)
11. Wong, C.C., Loewke, K.E., Bossert, N.L., et al.: Non-invasive imaging of human embryos before embryonic genome activation predicts development to the blastocyst stage. Nat. Biotechnol. **28**(10), 1115–1121 (2010)
12. Armstrong, S., Arroll, N., Cree, L.M., Jordan, V., Farquhar, C.: Time-lapse systems for embryo incubation and assessment in assisted reproduction, Cochrane. Database. Syst. Rev. **27**(2), CD011320 (2015)
13. Hnida, C., Engenheiro, E., Ziebe, S.: Computer-controlled, multilevel, morphometric analysis of blastomere size as biomarker of fragmentation and multinuclearity in human embryos. Hum. Reprod. **19**(2), 288–293 (2004)
14. Karlsson, A., Overgaard, N.C., Heyden, A.: Automatic segmentation of zona pellucida in HMC images of human embryos. In: Proceedings of the 17th International Conference on IEEE (ICPR 2004), vol. 3, pp. 518–521 (2004)
15. Beuchat, A., Thévenaz, P., Unser, M., et al.: Quantitative morphometrical characterization of human pronuclear zygotes. Hum. Reprod. **23**(9), 1983–1992 (2008)
16. Morales, D.A., Bengoetxea, E., Larranaga, P.: Automatic segmentation of zona pellucida in human embryo images applying an active contour model. In: Proceedings of the 12th Annual Conference on Medical Image Understanding and Analysis, pp. 209–213 (2008)
17. Morales, D.A., Bengoetxea, E., Larrañaga, P., et al.: Bayesian classification for the selection of in vitro human embryos using morphological and clinical data. Comput. Methods Programs Biomed. **90**(2), 104–116 (2009)
18. Giusti, A., Corani, G., Gambardella, L.M., Magli, C., Gianaroli, L.: Segmentation of human zygotes in hoffman modulation contrast images. In: Procceedings of MIUA (2009)
19. Giusti, A., Corani, G., Gambardella, L., Magli, C., Gianaroli, L.: Blastomere segmentation and 3D morphology measurements of early embryos from hoffman modulation contrast image stacks. In: 2010 IEEE International Symposium on Biomedical Imaging: From Nano to Macro, pp. 1261–1264. IEEE (2010)
20. El-Shenawy, M.A.: An automatic detection and identification of cells in digital images of day 2 IVF embryos. University of Salford (2013)
21. Otsu, N.: A threshold selection method from gray-level histograms. IEEE Trans. Syst. Man Cybern. **9**(1), 62–69 (1979)
22. Pratt, W.K.: Digital Image Processing, p. 583 (2001). ISBN0-471-37407-5
23. Strouthopoulos, C., Papamarkos, N.: Multithresholding of mixed type documents. Eng. Appl. Artif. Intell. **3**, 323–343 (2000)
24. Kohonen, T.: Self-organized formation of topologically correct feature maps. Biol. Cybern. **43**(1), 59–69 (1982)
25. Lam, L., Seong-Whan, L., Ching, Y.S.: Thinning methodologies-a comprehensive survey. IEEE Trans. Pattern Anal. Mach. Intell. **14**(9), 879 (1992)
26. Goneid, A., El-Gindi, S., Sewisy, A.: A method for the hough transform detection of circles and ellipses using a 1-dimensional array. In: 1997 IEEE International Conference on Systems, Man, and Cybernetics. Computational Cybernetics and Simulation (1997)

Detection of Breast Cancer Using Infrared Thermography and Deep Neural Networks

Francisco Javier Fernández-Ovies[1],
Edwin Santiago Alférez-Baquero[2(✉)],
Enrique Juan de Andrés-Galiana[1,3], Ana Cernea[1],
Zulima Fernández-Muñiz[1], and Juan Luis Fernández-Martínez[1(✉)]

[1] Group of Inverse Problems, Optimization and Machine Learning,
Department of Mathematics, University of Oviedo,
C/Federico García-Lorca, 18, 33007 Oviedo, Spain
jlfm@uniovi.es
[2] Department of Mathematics, Technical University of Catalonia,
Barcelona, Spain
santiago.alferez@upc.edu
[3] Department of Informatics and Computer Science, University of Oviedo,
Oviedo, Spain

Abstract. We present a preliminary analysis about the use of convolutional neural networks (CNNs) for the early detection of breast cancer via infrared thermography. The two main challenges of using CNNs are having at disposal a large set of images and the required processing time. The thermographies were obtained from Vision Lab and the calculations were implemented using Fast.ai and Pytorch libraries, which offer excellent results in image classification. Different architectures of convolutional neural networks were compared and the best results were obtained with resnet34 and resnet50, reaching a predictive accuracy of 100% in blind validation. Other arquitectures also provided high classification accuracies. Deep neural networks provide excellent results in the early detection of breast cancer via infrared thermographies, with technical and computational resources that can be easily implemented in medical practice. Further research is needed to asses the probabilistic localization of the tumor regions using larger sets of annotated images and assessing the uncertainty of these techniques in the diagnosis.

1 Introduction

Breast cancer is one of the main causes of female mortality in developed countries. Early detection is of utmost importance in order to increase the survival rates of the patients and reducing the cost of the health systems [1]. Medical imaging technologies (X-rays, ultrasounds and magnetic resonance) has a fundamental role in the early detection of breast cancer. Infrared thermography is an inexpensive and non-invasive tool for the early detection of breast cancer. Nevertheless, this technique is known for introducing a high number of false positives and false negatives (around 10%) and the affected area is not located accurately [2, 3]. The detection of regions with high

© Springer Nature Switzerland AG 2019
I. Rojas et al. (Eds.): IWBBIO 2019, LNBI 11466, pp. 514–523, 2019.
https://doi.org/10.1007/978-3-030-17935-9_46

temperature gradients, the automatic detection of regions of interest in each breast [2] and the analysis of asymmetries [4, 5] were important breakthroughs in the use of this technique. Automation of this analysis via AI methodologies has been performed by Silva et al. [6] and Acharya et al. [7] using texture attributes and support vector machines (SVM). Other authors used k-NN classifiers [2] or SVM-RFB [8, 9]. The implementation of open- access databases for infrared breast thermography has allowed the comparison of different AI techniques [10], which are based in detecting the thermal effects of tumors in breasts [11].

Convolutional neural networks (CNN) have been widely used in image processing with excellent results. Up to our knowledge the application of this technique to infrared thermography, and more specifically in the early detection of breast carcinoma, is novel. The aim of this work is to evaluate the use of CNNs in the early detection of breast cancer via thermography, and particularly in differentiating patients with breast cancer from healthy controls, showing that the confusion matrix can become almost diagonal (false positives and negatives minimization) when the optimum CNN architecture is adopted. For that purpose, we have implemented the CNNs using the Fast.ai library, achieving a blind validation accuracy of 100%. Besides, these accuracies appear to be very stable. CNNs improved the results obtained with other machine learning algorithms, such as SVM and k-NN.

Breast cancer thermography detection has important clinical advantages, such as, to be suitable for very early detection of breast cancer, and being and innocuous, non-invasive and very economic technique with am easy technical implementation and high-speed of execution. Besides, hormonal changes do not affect the result that has been obtained. Its main drawbacks are the high false-positive rate, which can result in performing the standard mammogram anyway, and also a high false-negative rate, which can lead to avoidance of the standard mammogram. These facts cause that thermographies are rarely covered by medical insurances, hampering their development. We expect that the use of larger sets of annotated images will serve to locate the tumors regions probabilistically with high accuracy, that is, to perform the probabilistic segmentation of the thermographies (see for instance [12]).

2 Materials and Methods

This research was carried out on thermographies obtained from Vision Lab [6] using the Fast.ai library executed on paperspace servers (https://www.paperspace.com). The images were organized in three groups (training, testing and validation) with the objective of comparing the predictive accuracy of different CNN architectures to detect breast cancer in early stages.

The classification problem is binary and consists in predicting the class of the patients (Healthy/Sick) based on their respective thermographies:

$$\mathbf{F}(t): t \in R^{mxn} \rightarrow C\{Healthy, Sick\}.$$

where **t** are color images (480 × 640 × 3) which identify the temperature maps of the thermographies. The location of the tumor regions was not addressed since these images were not annotated by medical doctors.

It is necessary to construct the model **F** that facilitates this classification. To achieve this objective we have used convolutional neuronal networks, and we have evaluated the predictive accuracy of different CNN architectures.

The methodology can be summarized is as follows:

First, we started by constructing an ordered set of labeled thermographies $\mathbf{T} = \mathbf{H} \cup \mathbf{S}$, being **H** the ordered set of thermographs corresponding to the control group (Healthy women) and **S** those of the patients labeled as Sick. We try to build balanced sets, that is, the number of thermographies in both groups are similar, since CNNs are sensitive to unbalanced groups. For the CNNs evaluation a cross validation will be used, starting from a division into two groups (training-test and blind validation):

$$H = H_t \cup H_b, \ S = S_t \cup S_b.$$

The training-test group was divided into 5 groups with the same number of thermographies:

$$H_t = H_1 \cup H_2 \cup H_3 \cup H_4 \cup H_5,$$

$$S_t = S_1 \cup S_2 \cup S_3 \cup S_4 \cup S_5.$$

These groups (H_i and S_i) are used to perform cross validation with a set of CNNs architectures (A_k):

$$N(A_k, T_m, \cup_{j \neq i} H_j, \cup_{j \neq i} S_j) \rightarrow F_{i,k}, \tag{1}$$

$A_k \in$ CNN architectures set,
$T_m \subset$ Set of CNN techniques applied.

This approach provide different CNN models for each of the n architectures that were tested:

$$\mathbf{M} = \{\mathbf{F}_{1,1}, \mathbf{F}_{1,2}, \ldots \mathbf{F}_{1,n}, \mathbf{F}_{2,1}, \mathbf{F}_{2,2} \ldots \mathbf{F}_{5,n}\}. \tag{2}$$

1. The next step consists in applying each of these CNN models to the test set:

$$F_{i,k}(t) : t \ \epsilon (H_i \cup S_i) \rightarrow R_{i,k} \in C\{Healthy, Sick\}. \tag{3}$$

2. Finally we average the accuracy of each architecture, selecting the model with the highest accuracy and testing its accuracy blindly in the validation group, training previously with the set of thermographs of the group training-test:

$$N(A_k, T_m, H_t, S_t) \rightarrow F_{t,k}, \tag{4}$$

$$F_{t,k}^*(t) : t \in (H_b \cup S_b) \rightarrow R^* \in C\{Healthy, Sick\}. \tag{5}$$

Given $F_{t,k}^*$, we also obtain the confusion matrix that can be used to analyze the groups of false positives and false negatives.

In the case of 2D images remember the convolution of an image X and a filter W (called kernel) is defined as follows:

$$Y(i,j) = (X * W)(i,j) = \sum_m \sum_n X(i+m, j+n) W(m,n). \tag{6}$$

The CNN architectures are designed based on convolution operations. Figure 1 shows the conceptual scheme valid for all the CNN architectures.

Fig. 1. Conceptual scheme of CNN

The convolution region of the CNNs includes the convolution, the activation and the pooling layers. The convolutional layer takes into account the spatial structure of the input. The activation layer introduces the necessary connections among neurons, while pooling reduces the dimension of the features, the number of parameters and the cost of computing. Finally we have the output layer.

The learning or training process consists in finding optimum values for the parameters for the cost function f of the CNN architecture. This is typically achieved by gradient descent:

$$W_{t+1} = W_t - \alpha \nabla f(W_t, X). \tag{7}$$

where W_t are the weights in iteration t, and α is the learning rate, which indicates the size of the step in the negative gradient direction of the cost function $\nabla f(W_t, X)$. It should be noted that there are different improvements to the gradient descent method, which try to optimize the calculation of the learning rate, thus reducing computing costs. Optimization is performed via the "mini batch Stochastic Gradient Descent" (mini batch SGD), which essentially calculates the gradient for subset of the parameters. This algorithm needs moresteps to reach convergence but reduces considerably the computing cost.

2.1 Thermographies

Breast cancer is basically of two types [2]: ductal (present in the ducts that carry milk from the breast to the nipple) and lobular (present in the lobules that produce milk), being more infrequent in other areas Fig. 2.

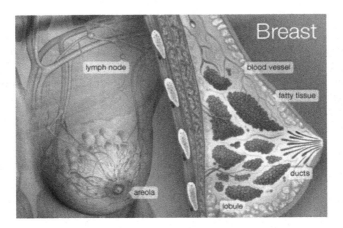

Fig. 2. Anatomy and location of breast cancer (from Pérez et al. [2]).

In tumors and specifically in breast carcinoma, metabolic changes occur that cause greater activity in cancer cells that induce changes in the temperature. These variations can be identified by infrared thermography, allowing for an early detection of breast cancer with accuracies higher than 90% [2]. This technique serves to anticipate the results of the anatomical analysis provided by other techniques such as mammograms. It is a non-invasive method that measures the temperatures in frontal, oblique and lateral views of the entire breasts including the armpits, where the ganglia are located. Previous works tried to identify the regions of interest studying possible asymmetries in these thermographies. The objective of this work is to perform the classification of the patients (sick or healthy) via CNN by using the attributes extracted from the analysis of the thermographies. This is one of its main advantages: the automatic extraction of the discriminatory attributes via the different filters of the CNN architecture.

This technique also allows the comparison of the thermographies obtained in each breasts (differential thermography), looking for differences to reveal symptoms of the disease. In this paper, we used a set of 5.604 thermographic data files corresponding to 216 women (downloaded from the public server Vision Lab). Figure 3 shows one of these thermographies. In this dataset, only 15% of the women (41) were diagnosed with breast cancer. In order to increase the number of images we have used different photograms of the same breasts. Although the dataset is limited in size, this preliminary research aims at analyzing the potential of CNNs to automatically interpret thermographies and detecting breast cancer in early stages. In this work we only appraise the binary classification problem: sick or healthy. It is expected that the use of larger sets of annotated images will serve to locate the tumor regions with high accuracy.

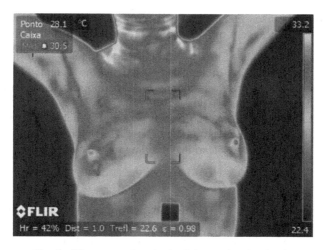

Fig. 3. Thermography obtained from Visual Lab [6].

2.2   Fast.ai- CNNs

To perform the classification with CNN, we took advantage of the resources freely available at Fast.ai, an initiative of Rachel Thomas and Jeremy Howard to promote the use of Artificial Intelligence to solve a big variety of technological and scientific problems. This platforms also allow working with GPUs (Graphics Processing Units) to make these calculations faster. The processing was done using the Paperspace cloud servers.

Training CNNs from scratch requires a large number of cataloged images and equipment with high processing capacity. The community collaborates by exchanging the weights of architectures previously trained to take advantage of more specific investigations. The weights of the first layers have more general features, acquiring more abstract knowledge as the analysis of the image progresses through the different layers of the CNN architecture. In this work, we analyzed the predictive accuracy of different architectures classifying thermographies of patients previously evaluated, in some cases only by mammography.

2.3   Data Processing

The first step was to download the thermographies of 216 patients, corresponding to 175 healthy women and 41 women with breast cancer, identified with a BI-RADS of 4 or higher. A total of 5.604 thermographs were obtained, corresponding to frontal shots, left 45°, right 45°, left 90° and right 90°. Subsequently, all these thermographies were converted to 480 × 640 images to facilitate the application of the Fast.ai functions. A previous filtering was done to select the ones that were suitable for the study. Only thermographies correspond to frontal views were selected, being limited to those showing the two breasts and eliminating those of the patients who had used deodorant since it interferes in the predictions of the diagnosis. Thermographies with reduced sharpness were also eliminated due to acquisition problems. Therefore, a total of 2.411 images of healthy patients and 534 images of diseased patients were used.

A balanced randomized selection was performed on 500 healthy patients and 500 sick patients. The images were organized into three groups as follows:

- 80% (400 Healthy and 400 Sick) were allocated to training and testing, that is, the processes of extracting the attributes and optimizing the weights of the filters. This set was divided into 2 groups: training (320 thermographies in each class) and testing (80 thermographies in reach class).
- 20% (100 thermographies in each class) were reserved for blind validation and establishing the final predictive accuracy of these architectures using a database of images that was not used for learning (training and testing).

Different techniques were used to improve the accuracy:

- Learning rate tuning, which controls the update of the CNNs weights. This parameter greatly impacts the result and performance of the CNN model [13]. Figure 4 shows the learning rate graph in log scale for one of the architectures used. The neural networks are updated via the stochastic gradient.

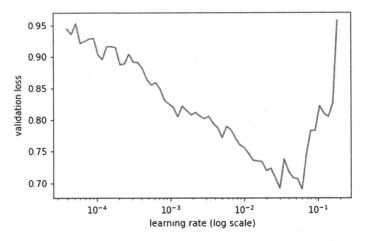

Fig. 4. Learning rate estimation

- Data Augmentation, that consists in the realization of different random operations of rotation, translation and zoom on the images to avoid overfitting and improve generalization [14]. Random modifications (zoom, rotation, displacements) of the images that might eliminate the regions affected by the carcinoma should be avoided.
- Fine-Tuning [15], that is based on defrosting the weights of the convolutional layers, allowing the network to train in its integrity. It is applied using a "differential learning rate", that is, introducing three different and successively higher values of the learning rate, thus taking into account the differential knowledge of the layers. The last layers are the ones that require the least modifications because they have a more generalized knowledge.

3 Results

The procedure was applied to different CNN architectures: resnet18, resnet34, resnet50, resnet152 [16], vgg16 and vgg19 [17]. Table 1 shows the results of a 5-fold test cross-validation in testing and also the accuracy in blind validation. We also provide the median and interquartile range of the these architectures.

Table 1. Accuracy of different architectures in testing and validation (median and IQR).

	Resnet18	Resnet34	Resnet50	Resnet152	Vgg16	Vgg19
K = 1	97.50	98.13	98.75	95.63	93.75	90.00
K = 2	98.75	98.75	98.75	98.75	87.50	89.38
K = 3	96.88	98.13	97.50	95.63	88.75	86.25
K = 4	99.38	98.13	100.0	98.75	94.38	86.88
K = 5	96.88	96.25	98.13	93.75	89.38	90.00
Testing						
Median	97.50	98.13	98.75	96.63	89.38	89.38
IQR	2.08	0.63	1.09	3.59	5.47	3.28
Validation						
Accuracy	99.50	100.0	100.0	97.50	93.00	94.50
Median	98.50	98.25	97.50	97.00	93.50	91.50
IQR	1.00	1.00	1.00	1.00	2.50	1.0

Resnet architectures showed higher accuracies in testing than Vgg architectures. Resnet50 and Resnet34 provided the highest median accuracies in the 5 testing experiments (98.75% and 98.13% respectively) with interquartiles ranges of 0.63% and 1.09%. Therefore, although Resnet50 provided the highest accuracy is less stable than Resnet 34 (1.09% vs 0.63%). These two architectures show the highest accuracies in blind validation (100%). Also, Resnet18 provides an almost perfect blind validation accuracy (99.50%). Table 2 provides the corresponding confusion matrices in validation. It can be observed that the amount of false positives and negatives depend on the architecture that it is used (see for instance Vgg16 and Vgg19). Resnet34 and Resnet50 achieve a diagonal confusion matrix (perfect classification).

To analyze the stability of theses signatures whose accuracy was provided in blind validation we have tested these architectures, applied them to ten different random hold-outs composed of 200 thermographies (100 sick/100 healthy) selected on the whole database. Although some of these images have been used in testing, this procedure provides a good idea of the accuracy variability that we will get in real practice. Interesting, resnet18 is the architecture that shows the highest stability (median accuracy of 98.50%), outperforming Resnet34 and Resnet50, although it shows ·a lower median predictive (97.50%) and higher IQR (2.08%) in testing. This is an illustration of the

Table 2. Confusing matrices of different CNN architectures in blind validation.

Resnet18	Healthy	Sick
Healthy	100	0
Sick	1	99
Accuracy:	0.995	

Resnet34	Healthy	Sick
Healthy	100	0
Sick	0	100
Accuracy:	1	

Resnet50	Healthy	Sick
Healthy	100	0
Sick	0	100
Accuracy:	1	

Resnet152	Healthy	Sick
Healthy	98	2
Sick	3	97
Accuracy:	0.975	

Vgg16	Healthy	Sick
Healthy	89	11
Sick	3	97
Accuracy:	0.93	

Vgg19	Healthy	Sick
Healthy	98	2
Sick	9	81
Accuracy:	0.945	

principle of parsimony: the simplest the better. These results show that CNN outperform other machine learning algorithms that were previously used to classify thermographies, such as k-NN or SVM-RFB [2, 8, 9], providing very good and stable results in validation. The uncertainty analysis of these classifiers has to be performed [18].

4   Conclusions

This paper analyzed the predictive accuracy of different CNN architectures to classify patients with breast cancer vs healthy controls using thermographies. We have shown that Resnet residual architectures show excellent results in the classification of carcinoma from thermographies, especially Resnet18, Resnet34 and Resnet50, being Resnet18 the most stable architecture among these three. Increasing the number of layers does not improve the validation accuracy, when compared to the results obtained for Resnet152. Resnet architectures show a superior performance to Vgg architectures. The highest predictive accuracy provided by CNNs is related to the way in which the attributes of the thermographies are extracted, compared to other machine learning algorithms that need to learn which are the discriminatory attributes or fixed them a priori. Future research will be performed to simplify these architectures and understanding which are the attributes that make this diagnosis successful, tryin to provide transparency to these black-box classifiers. Therefore, CNNs applied to thermographies is a very efficient method for breast cancer screening. Future research will be devoted to locate the affected regions accurately, in very early stages. It is expected that the use of larger sets of annotated images for training and validation will increase the accuracy of the CNNs in locating the tumor regions, since CNNs provide optimum results when the number of training examples increase. The design of reliable and economical non-invasive methods of diagnosis and prevention is very important to increase the survival rates of women with breast cancer and lowering the costs of treatment in public healthcare systems. Based on the results of this research, we believe that thermographies analyzed via CNN can help to achieve this goal.

References

1. Anderson, B.O., et al.: Breast J. **12**(Suppl 1:S3-15) (2005). PMID: 16430397
2. Pérez, M.G., Conci, A., Aguilar, A., Sánchez, A., Andaluz, V.H.: Detección temprana del cáncer de mama mediante la termografía en Ecuador (2014)
3. Araújo, M.C., Lima, R.C.F., de Souza, R.M.C.R.: Interval symbolic feature extraction for thermography breast cancer detection (2014)
4. Gogoi, U.R., Majumdar, G., Bhowmik, M.K., Ghosh, A.K., Bhattacharjee, D.: Breast abnormality detection through statistical feature analysis using infrared thermograms (2015)
5. Mejía, T.M., Pérez, M.G., Andaluz, V.H., Conci, A.: Automatic segmentation and analysis of thermograms using texture descriptors for breast cancer detection (2015)
6. Silva, L.F., et al.: A new database for breast research with infrared image. Banco de imágenes Visual Lab (2014). http://visual.ic.uff.br/dmi
7. Acharya, U.R., Ng, E.Y.K., Tan, J.-H., Sree, S.V.: Thermography based breast cancer detection using texture features and support vector machine (2012)
8. Ali, M.A.S., Hassanien, A.E., Gaber, T., Silva, L.: Detection of breast abnormalities of thermograms based on a new segmentation method (2015)
9. Sathish, D., Kamath, S., Prasad, K., Kadavigere, R., Martis, R.J.: Asymmetry analysis of breast thermograms using automated segmentation and texture features (2016)
10. Guerrero, S.R., Loaiza, H., Retrepo, A.D.: Automatic segmentation of thermal images to support breast cáncer diagnosis (2014)
11. Kandlikar, S.G., et al.: Infrared imaging technology for breast cancer detection – Current status, protocols and new directions (2017)
12. Fernández-Martínez, J.L., Xu, S., Sirieix, C., Fernández-Muniz, Z., Riss, J.: Uncertainty analysis and probabilistic segmentation of electrical resistivity images: the 2D inverse problem. Geophys. Prospect. **65**, 112–130 (2017)
13. Smith, L.N.: Cyclical learning rates for training neuronal networks (2014)
14. Takahashi, R., Matsubara, T., Uehara, K.: Data Augmentation using Random Image Cropping and Patching for Deep CNNs (2018)
15. Montone, G., O'Regan, J.K., Terekhov, A.V.: Gradual Tuning: a better way of Fine Tuning the parameters of a Deep Neural Network (2017)
16. He, K., Zhang, X., Ren, S., Sun, J.: Deep residual learning for image recognition (2015)
17. Simonyan, K., Zisserman, A.: Very deep convolutional networks for large-scale image recognition (2014)
18. Fernández-Muñiz, Z., Khaniani, H., Fernández-Martínez, J.L.: Data kit inversion and uncertainty analysis. J. Appl. Geophys. **161**, 228–238 (2019)

Development of an ECG Smart Jersey Based on Next Generation Computing for Automated Detection of Heart Defects Among Athletes

Emmanuel Adetiba[1,2(✉)], Ekpoki N. Onosenema[1], Victor Akande[1],
Joy N. Adetiba[3], Jules R. Kala[4], and Folarin Olaloye[1]

[1] Department of Electrical and Information Engineering, Covenant University,
Canaan-Land, P.M.B 1023, Ota, Nigeria
emmanuel.adetiba@covenantuniversity.edu.ng
[2] HRA, Institute for Systems Science, Durban University of Technology,
P.O. Box 1334, Durban, South Africa
[3] Department of Nursing, Durban University of Technology, P.O. Box 1334,
Durban, South Africa
[4] Department of Information Technology, Durban University of Technology,
P.O. Box 1334, Durban, South Africa

Abstract. Heart defects have remained one of the top causes of death in the world. As much as sporting activities and exercises are considered beneficial for the health of an individual, there are also a few but significant demerits as huge populations of athletes suffer from heart defects. This is especially pronounced among those who participate in strenuous activities for long periods of time. One of the oldest and most useful diagnostic tools for heart disease is the electrocardiograph. However, they remain bulky, heavy and not wearable. They also often require the help of medical professionals to interpret the electrocardiogram. As a result, the preliminary results of the implementation of a prototype ECG Smart Jersey using Next Generation Computing (which include IoT, Machine Learning and Android App) for the automated detection of heart defects among athletes is presented in this paper. The prototype is a proof of concept, which could be further enhanced to ensure that a wearable and automated ECG monitoring, analysis and interpretation is accomplished for athletes in order to reduce the burden of sudden deaths among them.

Keywords: ECG · IoT · Machine Learning · Next Generation Computing · Smart Jersey · Sudden Cardiac Deaths

1 Introduction

All around the globe, heart defects are known to be the major causes of Sudden Cardiac Deaths (SCD). The first recorded incidence of SCD occurred in 490 B.C. during the Persian army's attack against Greece in the battle of Marathon. However, just towards the end of the twentieth century there was a drastic increase in the death rate of high profiled athletes due to SCD. This drew more research attention among physicians, sport officials and other professionals on the causes of SCD [1].

© Springer Nature Switzerland AG 2019
I. Rojas et al. (Eds.): IWBBIO 2019, LNBI 11466, pp. 524–533, 2019.
https://doi.org/10.1007/978-3-030-17935-9_47

Maron, in a 2003 study compiled several heart defects, which were found to be the causes of SCD in about 387 American Athletes aged 12 to 35. He noted that one of the main triggers for SCD was the participation in sport in the presence of cardiovascular diseases [1]. Furthermore, the higher risk of SCD in athletes than in non-athletes have being linked to certain effects associated with sporting activities that lead to the existence of basic cardiovascular irregularities, such as congenital coronary anomalies, cardiomyopathies and ion-channelopathies. Blunt chest trauma may also cause ventricular fibrillation in a structurally normal heart (i.e., commotio cordis) [1, 15]. However, the most common cause of SCD has been discovered to be Hypertrophic Cardiomyopathy (HCM), a situation, where the myocardium, or heart muscle thickens and as a result, taking more space in the heart and consequently pumping less blood [17].

SCD accounts for about 20% of all deaths in the western world. Approximately 450,000 humans die from SCD in the USA only, with a yearly occurrence of 0.1%–0.2% [11]. This is related to cases in other western developed countries where Coronary Artery Disease (CAD) is common. Thus, over 7 million lives are usually lost to SCD per year. Furthermore, the improper eating habits of people in developing nations contribute a great deal to the increased occurrence of SCD [11]. Adetiba et al. [13] indicated that the epidemiology of SCD in Africa is estimated to be about 300,000 cases each year. In codicil, a research was carried out at the Obafemi Awolowo University Teaching Hospitals Complex in Ile-Ife Nigeria to review all cases of medico-legal autopsies over a ten-year period. 2,529 records were reviewed and 79 of them were found to be cases of SCD [4]. In addition to several reported cases of sudden death among Nigeria Athletes, Ugochukwu Ehiogu, a Nigerian- born Athlete who played in the premier league on several occasions and was known to be a star in sports died on April 20, 2017 due to a heart defect at the training ground of Tottenham Hotspur [13].

Electrocardiography (ECG) is the first stage for detecting many heart defects. ECG is the graphic representation of the electrical activity of the heart with time displayed on the x-axis and voltage displayed on the y-axis. ECG records the alteration of potentials between standardized sites on the body surface, routinely termed the Einthoven-, Goldberger-, and Wilson-lead systems. It provides a noninvasive procedure for monitoring and recording of electrical changes in the heart. It is the main measurement used for tests such as exercise tolerance and to evaluate the symptoms and causes of heart defects that lead to SCD [20]. Usually, the professional knowledge of a cardiologist is needed to interpret electrocardiograms. However, a real-time electrocardiography analysis and monitoring system embedded in a sport man's Jersey (herein referred to as Smart Jersey), could detect these defects in athletes as they occur, even in the absence of a professional cardiologist and further help to monitor and reduce SCD.

A Smart Jersey is a stretchy clothing in which ECG is weaved to alert patients of their heart activities and abnormalities on their mobile devices after it has been monitored by the clothing containing the ECG [6]. Smart jersey was also regarded in [7] as an ECG T-shirt, which can be worn for sensing ubiquitous vital signs. It was further described by Boehm et al. [8] as an ECG T-shirt using a portable recorder for long-term and unobtrusive multichannel ECG monitoring with electrodes.

This paper presents the preliminary results of the development of an ECG Smart Jersey for the automated detection of heart defects using Next Generation Computing (NGC) technologies such as IoT, Machine Learning, Open Microcontroller and Android app. Section 2 contains the materials and methods, Sect. 3 presents the results and discussion while conclusion is drawn in Sect. 4.

2 Materials and Methods

2.1 System Architecture

A system architecture shown in Fig. 1 was designed in order to illustrate the interconnection of the different layers of the NGC based Smart Jersey in this study. As shown in Fig. 1, the first layer is the *ECG Smart Jersey*, which contains the ECG

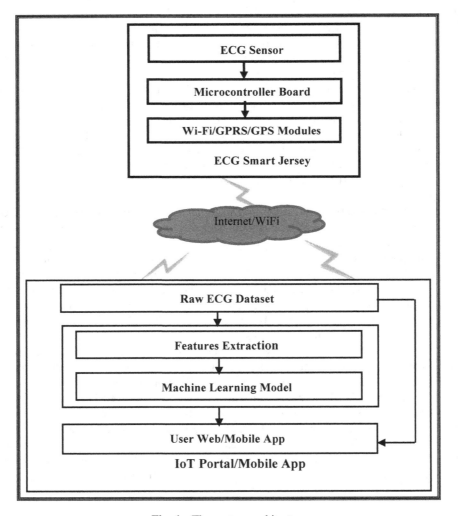

Fig. 1. The system architecture

sensor with probes for contact to relevant body parts (e.g. ECG AD8232), microcontroller board (e.g. Arduino UNO or Raspberry Pi) and Wi-Fi/GPRS/GPS modules for ECG raw data as well as position information transmission to the second layer.

The second layer is the *IoT Portal/Mobile App*, with components such as raw ECG dataset storage, features extraction algorithm and machine learning model. These three components are the toolkits within the system architecture for automated detection of the heart condition that is inherent in the ECG dataset. The second layer also contains the User Web/Mobile App module, which provides web/mobile based interface with information on the heart condition of the Smart Jersey wearer.

2.2   ECG Smart Jersey Circuit Design

The circuit diagram that underlies the electronics of the ECG Smart Jersey is presented in Fig. 2. As shown, the selected components are ECG AD8232 as the ECG sensor, Arduino UNO as the microcontroller board, ESP 8266 as the Wi-Fi module and GPS NEO6 as the GPS module.

ECG AD8232 captures the electrical activity of the heart, which is represented as an electrocardiogram. In the presence of noisy conditions, it has capability to filter signals generated by remote electrode placement or motion. The exact cutoff frequency for all the filters are selected in order to align with various applications. A fast restore function is also embedded in the sensor to reduce the duration of long settling tails of the high-pass filters. To improve the common-mode rejection of the line frequencies contained in the system and other unwanted interference, it has an amplifier for driven lead applications such as Right Led Drive (RLD). After an abrupt signal change that rails the amplifier, the ECG AD8232 sensor adjusts automatically to a higher filter cutoff. This quality allows the sensor to recover on time and then take the important measurements after the electrodes have been connected to the subject [36]. The ECG AD8232 sensor has nine connections, which are called "pins". Out of the nine pins, five are connected to the Arduino UNO board. These five pins are labeled as GND, 3.3v, OUTPUT, LO- and LO+ [37].

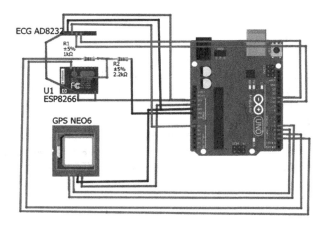

Fig. 2. The ECG Smart Jersey circuit diagram

The ESP8266 Wi-Fi module adds Wi-Fi functionality to the microcontroller to transmit the measured ECG signal to the IoT Portal/Mobile App. The hardware connections for the ESP8266 require 3.3 V power, thus the need for the R2 resistor between the Wi-Fi module and the Arduino UNO power pins in the circuit.

The Arduino Uno is a microcontroller board based on the ATmega328. It can be powered by connecting it with an AC-to-DC adapter, battery or with a USB cable connected to a computer. The Arduino Uno contains 14 digital input/output pins, 6 of these pins can be used as PWM outputs. It also contains 6 analog inputs, a 16 MHz crystal oscillator, a power jack, a USB connector, an ICSP header, and a reset button. Everything needed to support the microcontroller is contained in the Arduino Uno [39]. The ATmega328 microcontroller on the Arduino UNO board is pre-burned with a boot loader that allows the upload of new codes without the use of an external hardware programmer. The codes for capturing of the ECG signals from the sensor and transmission to the *IoT Portal/Mobile App* was written with the Arduino C language and uploaded on the microcontroller.

2.3   ECG Dataset, Features Extraction and Machine Learning Model

The raw ECG dataset as well as features extraction algorithm employed in our previous study [13] was adopted in this current work. This is because the current work is an effort to translate the study in [13] into usable product for societal impact. However, to realize the Machine Learning (ML) model shown in the architecture, we adopted the Multilayer Perceptron Artificial Neural Network (MLP-ANN) topology. Experiments were performed by varying the number of hidden layer neurons from 10 to 100, the epoch was set to 1000, with training batch size of 10 and the training dataset obtained through the application of the features extraction algorithm to the raw ECG data was portioned into 80% training and 20% testing datasets. In this study, the feature extraction algorithm was implemented using Python while Tensorflow, a state-of-the-art Python library for ML was engaged for implementing the MLP-ANN. The training accuracies and Mean Square Error (MSE) obtained from the experiments are presented in Sect. 3.

3   Results and Discussion

The designed circuitry was properly packaged and embedded on a Sporting Jersey to produce the prototype ECG Smart Jersey. The pictures in Fig. 3. show the packaged electronic module (Fig. 3a) as well as the ECG Smart Jersey (Figs. 3b and c) in which it is embedded. The ECG Smart Jersey was worn by three volunteers and the ECG data acquisition was streamed to ThingSpeak, which is the IoT Portal selected for this study. The ECG readings for each of the subjects as streamed to ThingSpeak are presented in Fig. 4. As clearly shown, the ECG reading is distinct for each of the subjects.

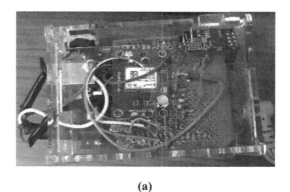

(a)

(b)

(c)

Fig. 3. (a) Packaged electronic module (b) External view of the prototype ECG Smart Jersey with embedded electronic module (on the bottom left hand corner) (c) Internal view of the prototype ECG Smart Jersey with ECG probes.

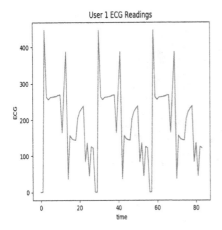
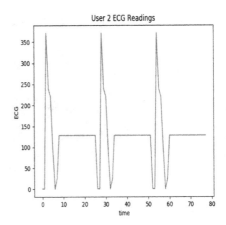

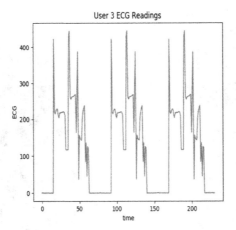

Fig. 4. ECG readings for 1st, 2nd and 3rd subjects

The result of the experiments performed to determine the optimal ML model is shown in Table 1. As shown in the Table, the MLP-ANN model with 70 neurons in the hidden layer produces the highest training accuracy of 100% with an MSE of 1.5561. This experimental result shows that the use of NGC technologies has potentials for realizing Smart Jersey towards the reduction of SDC among athletes. We are currently working on generalization testing of the ML model and the development of a Mobile App (Fig. 5) that can receive raw ECG readings from the Smart Jersey and output the heart condition of the wearer on a mobile device. This will find potential applications among Footballers, Sprinters, Mountain Climbers and host of other athletes.

Table 1. Experimental result for the ML model

No. Hidden layers	Accuracy (%)	MSE
10	75	1.5563848
20	50	1.5564306
30	55	1.5562989
40	26.25	1.5563371
50	92.5	1.5562966
60	17.5	1.5563021
70	100	1.5561352
80	70	1.5562894
90	96.25	1.5562563
100	98.75	1.5562425

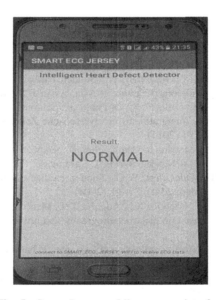

Fig. 5. Smart Jersey mobile app user interface

Similar to the efforts in this study, several works in the literature have adopted machine learning techniques both for ECG interpretation as well as in other medical domains [6, 13–17].

4   Conclusion

The ECG Smart Jersey based on NGC technologies as presented in this paper is a disruptive approach that could drastically reduce the incidence of SCD among athletes once fully implemented. The system architecture as well as the prototype of the ECG

Smart Jersey has been clearly presented and described. The prototype ECG Smart Jersey was able to stream ECG readings of subjects to the ThingSpeak online platform. Feature extraction algorithm as well as ML models were also implemented with Python and Tensorflow library respectively. Further efforts on this study will involve the use of 12-lead based ECG data acquisition from normal and Sports men with various kinds of heart defects. Various machine learning models will be explored for accurate automation of heart defects detection and monitoring. Attempts will also be made to miniaturize the electronic module in the ECG Smart Jersey to enhance its wearability by athletes during active athletic activities. This will involve experimentations with ECG metallic and graphene sensors.

Acknowledgement. This work was supported by the Covenant University Centre for Research, Innovation and Development (CUCRID), Covenant University, Canaanland, Ota, Ogun State, Nigeria

References

1. Khan, K.: Sudden cardiac death in athletes. N. Engl. J. Med. **6** (2003)
2. Halabchi, F., Seif-Barghi, T., Mazaheri, R.: Sudden cardiac death in young athletes; a literature review and special considerations in Asia. Asian J. Sports Med. **2**(1), 1 (2011)
3. Pelliccia, A., Borjesson, M., Villiger, B., Di Paolo, F., Schmied, C.: Incidence and etiology of sudden cardiac death in young athletes. Schweizerische Zeitschrift fur Sportmedizin und Sporttraumatologie **59**(2), 74 (2011)
4. Talle, M.A., et al.: Sudden cardiac death: clinical perspectives from the University of Maiduguri Teaching Hospital Nigeria. World J. Cardiovasc. Dis. **5**(05), 95 (2015)
5. Rotimi, O., Fatusi, A.O., Odesanmi, W.O.: Sudden cardiac death in Nigerians-the Ile-Ife experience. West Afr. J. Med. **23**(1), 27–31 (2004)
6. Adetiba, E., Iweanya, V.C., Popoola, S.I., Adetiba, J.N., Menon, C.: Automated detection of heart defects in athletes based on electrocardiography and artificial neural network. Cogent Eng. **4**(1), 1411220 (2017)
7. Price, D.: How to read an electrocardiogram (ECG). part 1: basic principles of the ECG. the normal ECG. S. Sudan Med. J. **3**(2), 26–31 (2010)
8. Vagott, J., Parachuru, R.: An overview of recent developments in the field of wearable smart textiles. J. Textile Sci. Eng. **8**(368), 2 (2018)
9. Mathias, D.N., Kim, S.I., Park, J.S., Joung, Y.H.: Real time ECG monitoring through a wearable smart T-shirt. Trans. Electr. Electron. Mater. **16**(1), 16–19 (2015)
10. Boehm, A., Yu, X., Neu, W., Leonhardt, S., Teichmann, D.: A novel 12-lead ECG T-shirt with active electrodes. Electronics **5**(4), 75 (2016)
11. ESP8266 WiFi Module Quick Start Guide. http://blog.electrodragon.com/esp8266-gpiotest-edited-firmware/ Pulse sensor with Arduino tutorial. http://www.instructables.com/id/Pulse-Sensor-With-Arduino-Tutorial/
12. Badamasi, Y.A.: The working principle of an Arduino. In: 2014 11th International Conference on Electronics, Computer and Computation (ICECCO), pp. 1–4. IEEE, September 2014
13. Adetiba, E., Olugbara, O.O.: Improved classification of lung cancer using radial basis function neural network with affine transforms of VOSS representation. PLoS ONE **10**(12), 1–25 (2015)

14. Adeyemo, J.O., Olugbara, O.O., Adetiba, E.: Smart city technology based architecture for refuse disposal management. In: IST-Africa Week Conference 2016, pp. 1–8. IEEE, May 2016
15. Abayomi, A., Olugbara, O.O., Adetiba, E., Heukelman, D.: Training pattern classifiers with physiological cepstral features to recognise human emotion. In: Pillay, N., Engelbrecht, A.P., Abraham, A., du Plessis, M.C., Snášel, V., Muda, A.K. (eds.) Advances in Nature and Biologically Inspired Computing. AISC, vol. 419, pp. 271–280. Springer, Cham (2016). https://doi.org/10.1007/978-3-319-27400-3_24
16. Badejo, J.A., Adetiba, E., Akinrinmade, A., Akanle, M.B.: Medical image classification with hand-designed or machine-designed texture descriptors: a performance evaluation. In: Rojas, I., Ortuño, F. (eds.) IWBBIO 2018. LNCS, vol. 10814, pp. 266–275. Springer, Cham (2018). https://doi.org/10.1007/978-3-319-78759-6_25
17. Adetiba, E., Olugbara, O.O.: Lung cancer prediction using neural network ensemble with histogram of oriented gradient genomic features. Sci. World J. **2015**, 1–17 (2015)

Biomedicine and e-Health

Analysis of Finger Thermoregulation by Using Signal Processing Techniques

María Camila Henao Higuita[1] , Macheily Hernández Fernández[2] ,
Delio Aristizabal Martínez[2] , and Hermes Fandiño Toro[1][(✉)]

[1] Grupo de Automática, Electrónica y Ciencias Computacionales, Instituto
Tecnológico Metropolitano, Carrera 31 No. 54-22, Medellín, Colombia
hermesfandino@itm.edu.co
[2] Instituto Tecnológico Metropolitano, Calle 54A No. 30-01, Medellín, Colombia

Abstract. The analysis of finger thermoregulation helps the diagnostic
of some disabling pathologies. Despite there is possible to analyze the
finger thermoregulation whether at the complete fingers geometry or at
the fingertips, there is comparatively little experimental evidence about
how different is the thermoregulation of these regions. This work presents
an analysis based on computer vision and signal processing techniques
applied to thermal infrared images, to identify regions inside the fingers
geometry that exhibit different thermoregulation patterns in response to
a controlled cold stimulation.

Keywords: Finger thermoregulation · Thermal infrared images ·
Signal processing

1 Introduction

Human beings have a thermoregulatory system that articulates physiological and
neurological processes, to maintain stable the temperatures at both the body core
and the peripheral shell [1,2]. The temperatures at acral regions such as hands,
are more directly influenced by external stimulations. Therefore, the manner
in which the acral body regions respond to external stimulations is relevant in
multiple research fields. Thus, delayed or unusual thermoregulatory responses
can be associated with the presence of some disabling pathologies such as the
Raynaud's phenomenon [3], Parkinson's disease [4], systemic sclerosis [5] and
Complex regional pain syndrome [6].

Considering the cutaneous vasculature and neural structures at acral body
regions, some researchers have evaluated the thermoregulation by focusing on
the fingers. For example, the study reported in [7] proposes a methodology to
model thermal oscillations in the hands. This study validates their results with
temperatures measured at the fingertips of five volunteer subjects. On the other
hand, the work reported in [8] analyzes the thermoregulatory response of the
sixty-six subjects: 18 healthy and 48 patients whether with primary and secondary Raynaud's phenomenon. In this last study the analysis is carried out at
the complete geometry of the fingers.

© Springer Nature Switzerland AG 2019
I. Rojas et al. (Eds.): IWBBIO 2019, LNBI 11466, pp. 537–549, 2019.
https://doi.org/10.1007/978-3-030-17935-9_48

Although these works were aimed to different purposes, they reveal there exist the possibility of analyzing finger thermoregulation whether at the fingertips or at the complete fingers. The research question of this work is if during the hand thermoregulation the fingers exhibit a uniform thermal response. From a perspective based on computer vision, there is comparatively little evidence reported to support the choice of any of the aforementioned possibilities.

2   Materials and Methods

2.1   Datasets and Image Processing

This work analyzes 30 sequences of thermal images. Each sequence contains 850 images; one acquired each second. The images show the temperatures of a hand before, during, and after its immersion in water at $15\,^{\circ}$C. Table 1 shows details of the protocol used for capturing the images [9].

Table 1. Details of the acquisition protocol. Further details can be found in [9].

Camera specifications	Model	FLIR A655SC
	Spectral range	7–$14\,\mu$m
	Image size	480×640 pixels
	Ambient temperature	26.8–$28\,^{\circ}$C
	Relative humidity	$47.18 \pm 2.16\%$
	Apparent reflected temperature = ambient temperature	
	Working distance	$0.58\,$m
	Emissivity	0.98
Acquisition duration	Basal temperature acquisition time	$120\,$s
	Cold stimulation	$70\,$s
	Recovery acquisition time	$660\,$s

We threshold and binarize the first five images of each sequence, by using the expectation-maximization method [10]. Next, we apply the AND logical operation between these binary images to get a final binary image I_b. Then, we segment the five fingers in I_b by using the following methodology: a reference point is localized inside the palm region, after applying the inverse distance function to I_b. This reference point is used to calculate their distance to the pixels in the contour of I_b. The vector with these distances represents a time-series whose peaks and valleys correspond to the fingertips and the inter-digit regions of the hand, respectively. By calculating and comparing the cumulative distances between each fingertip and the others, it is straightforward to localize the thumb fingertip and the fingertips associated with the remaining fingers. Line segments drawn between the inter-digit points are used to segment the finger regions [11].

Often, thresholding methods simpler than expectation-maximization method work well segmenting thermal infrared images. Thus, if the thermal contrast is not a problem, then the method proposed by Otsu is a proper choice [12]. In our case, the images show the hands resting on an acrylic surface. Thus, small movements of the fingers can unveil false hotspots that really are regions where the acrylic stored thermal energy. These hot spots appear as false contours of the fingers that difficult the segmentation required. For this reason, we decided to use the expectation-maximization segmentation instead of the Otsu's method.

For analyzing only the distal phalanges, we perform a manual segmentation. For each finger, we select two pixels located at both sides of their distal inter-phalangeal joint. These pixels are then connected through a black straight line. We segment the distal phalanges by eroding I_b with the superimposed black straight lines. The structural element used in this erosion is disk-shaped with a radius of three pixels.

2.2   Finger Regions with Similar Thermoregulatory Response

We analyze the temperatures in three regions-of-interest (ROI): at the complete fingers, at ROIs segmented with the k-means (see Subsect. 2.1), and at ROIs enclosing the distal phalanges. These last are segmented manually.

For identifying finger regions with similar thermoregulatory responses, we generate a matrix from pixels inside the complete finger ROI. Each row of this matrix contains the temperature values of a single pixel through the time. Figure 1 shows the structure of a row vector in such a matrix.

The row vectors in this matrix are clustered into two groups by using the k-means algorithm [13]. The clustering provides class labels to the vectors. Since each vector relates to a specific spatial location in the ROI, when each pixel is given the class label of their associated vector, a segmentation is achieved.

We measure the distance between the centroids of the two clusters and the reference point mentioned in Subsect. 2.1. We use these two distance to ensure that the regions closer to the distal phalanx have always the same class label.

We use the graphic in Fig. 2 to determine the length of the analyzed vectors. This graphic shows the average thermoregulatory response of the thirty analyzed subjects. We analyzed the signals from $t = 176$ s. At this time the signals do not exhibit temperature decreases, which indicates the hands are recovering the temperature values previous to its immersion in water. This criterion produces vectors with 675 elements.

The main result after using the methods described in Subsects. 2.1 and 2.2, is a set of ROIs for analyzing temperatures across the image sequences. We measure the average temperatures at these ROIs and use it to evaluate the convenience in modeling the thermoregulation of the finger by known functions.

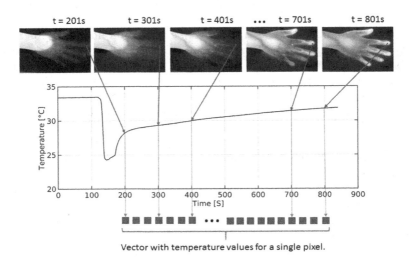

Fig. 1. The temperature values in a spatial location are used to create a vector.

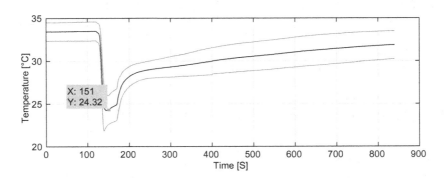

Fig. 2. In black and red, respectively, the mean and standard deviation values for the temperatures measured at the hand region in the thirty sequences analyzed. (Color figure online)

2.3   Curve Fitting Analysis

To analyze the data whose extraction is described at Subsect. 2.2, we compute the mean square error (MSE) between them and a fitting model $f(x_i)$, as stated by Eq. (1). This analysis is carried out in two cases: considering data taken from the regions enclosing the complete fingers, and considering data taken only from the manually segmented distal phalanges.

$$\text{MSE} = \sqrt{\sum_{i=1}^{n}(y_i - f(x_i))^2} \tag{1}$$

In this equation y_i, $f(x_i)$ and n are the actual temperature value and the fitted value to a function f (at time $t = i$ seconds) and the number of samples, respectively. The rewarming signals were analyzed from a time instant where the cold stimulation has been finished. Two functions are considered for this analysis. The first is a linear combination of two exponential functions expressed by Eq. (2), where A, b, C and d are the coefficients to be estimated.

$$f(t) = Ae^{bt} + Ce^{dt} \tag{2}$$

The second function considered is a polynomial function of degree 3 expressed by Eq. (3), where P_1, P_2, P_3 and P_4 are the third, second, first and zero order coefficients. We considered these two parametric functions because of their visual similarity with the rewarming signals analyzed (see Figs. 1 and 2 when $t \approx 200\,\text{s}$). This approach differs from earlier related works, where specific functions were proposed to model the thermoregulation at specific regions.

$$f(t) = P_1 t^3 + P_2 t^2 + P_3 t + P_4 \tag{3}$$

3   Results

The Figs. 3, 4 and 5 show the result of the image processing stage for one of the thirty image sequences analyzed. Figure 3 shows: (a) the first thermal image in the sequence, and (b) the achieved $\boldsymbol{I}_b$: the image resulting after applying the AND logical function to the first five thresholded images in the sequence. The thermal scale in image (a) is in Celsius degrees.

Figure 4 presents: (a) the segmentation achieved with the method proposed in [11], and (d) the distal phalanges manually segmented (in black color). The processing of the first images of a sequence is convenient because of their high thermal contrast regarding the remaining images in the sequence.

Figure 5 shows the result obtained with the methodology described in Subsect. 2.2. The figure shows the ROIs (in bright gray color) close to the distal phalanges identified by the k-means clustering. The figure also shows the ROIs close to the proximal phalanges: in black color. Each ROI enclose finger regions exhibiting similar thermoregulation.

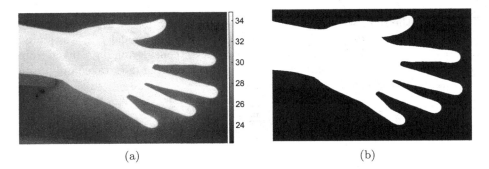

(a)   (b)

Fig. 3. (a) Example of a thermal image in one analyzed image sequence. (b) The binary image I_b obtained from the first five thresholded images in the sequence.

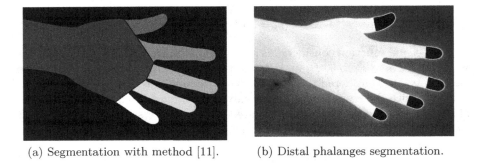

(a) Segmentation with method [11].   (b) Distal phalanges segmentation.

Fig. 4. ROIs segmented in the case of the image sequence considered in Fig. 3.

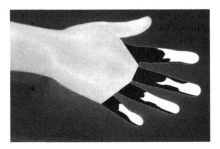

Fig. 5. ROIs segmented for an image sequence, after clustering the finger regions with a similar thermoregulatory response.

The first row in Table 2 shows the average size of the distal ROI segmented with k-means clustering, as a proportion of the total finger area segmented with method presented in [11]. For example, the values for the index finger show that in the 30 sequences analyzed, the average size of the distal ROI segmented was approximately the 45% of the total area of the complete (index) finger. The table shows that the cluster recognized two regions with different thermoregulatory patterns at each index finger analyzed.

The last four rows in the table are divided into two groups: one group contains the results of analyzing the temperatures at distal ROIs. The other group, the results of analyzing the temperatures at medial ROIs. Each group shows in turn two results. The upper shows the average temperature value along the 665 s where the thermoregulation was measured. The second shows the normality test outcome calculated for a significance level of 95%.

Table 2. Mean size of the ROIs after clustering the rewarming signals at fingers on the thirty analyzed image sequences.

		Finger			
		Index	Middle	Ring	Little
	Distal ROI size (%)	0.45±0.20	0.45±0.25	0.43±0.21	0.46±0.22
Distal ROI	T [°C]	28.29 ± 2.67	28.26 ± 2.74	28.10 ± 2.67	27.79 ± 3.27
	P-value	87.10%	31.10%	20.50%	36.90%
Medial ROI	T [°C]	27.67 ± 2.54	27.74 ± 2.59	28.11 ± 2.79	27.38 ± 2.65
	P-value	24.70%	75.90%	48.80%	54.00%

When comparing the temperatures of distal and medial ROIs, a remarkable result is that, excepting the Ring finger, temperatures at ROIs close to the distal phalanges are higher than in the ROIs closer to the medial phalanges. This result supports the hypothesis that finger thermoregulation can be properly measured at fingertips. The results also show that different thermoregulatory responses are observable in regions with surface greater than those enclosed by only the fingertips. The little fingers exhibited temperatures comparatively lower in their distal regions, but simultaneously higher temperature variations than the other fingers. In agreement with the normality test results, the thermoregulatory responses at each finger followed a normal distribution. Thus, the index, middle, ring and little fingers exhibit a characteristic thermal pattern.

The procedure described in Sect. 2.2 implies that for each segmented ROI there are thousands of vectors (pixels) with different thermoregulation responses. To reduce the data volume, we calculate for each image sequence, the mean thermoregulation signal from each ROI. This reduces the data to eight vectors per analyzed image sequence: two vectors for each ROI segmented.

To provide an additional analysis of the segmented ROIs of Fig. 5, we now consider four matrices: one per each finger. These matrices have the structure described in Subsect. 2.2. The principal component analysis (PCA) method was applied to each matrix and then, the samples were plotted in a new feature space provided by the three first principal components. Figure 6 shows these plots, where points in red color correspond to pixels belonging to the proximal phalanges. Points in blue color are the samples generated from the pixels close to the distal phalanges. The separation of the data from each finger indicates that the thermoregulatory response of pixels at each ROI is different. Moreover, the separation suggest it is possible to use the temperature evolution to characterize the thermoregulation of the fingers.

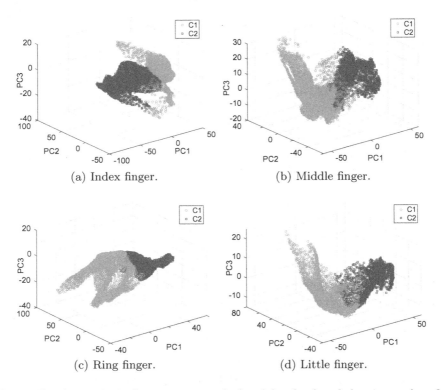

(a) Index finger.   (b) Middle finger.

(c) Ring finger.   (d) Little finger.

Fig. 6. The three principal components, calculated for the data belonging to four fingers. The image sequence is the same analyzed in Figs. 3, 4 and 5.

4   Result of the Parametric Modeling Analysis

The results in this section compare the thermoregulation of ROIs segmented by method [11] i.e. completer fingers, against the thermoregulation of the manually segmented distal phalanges. Table 3 shows the average values of the coefficients found using curve fitting. The coefficients were calculated from signals showing the average rewarming of the five fingers. For example, the first row shows the coefficients of the two fitted functions that describe the thermoregulation of the thumb fingers.

Analogously, Table 4 shows the coefficients for the distal phalanges. The first column in these two tables has the initial word of the analyzed finger. These initial words appears from top to bottom referring to the thumb, index, middle, ring and little fingers, respectively.

Table 3. Coefficients for fitting the thermal rewarming signals of the complete fingers.

	Exponential grade 2				Polynomial grade 3			
	A	b	C	d	P1	P2	P3	P4
T	-6.21×10^3	-2.14×10^{-3}	6.23×10^3	-8.34×10^{-3}	1.03×10^{-7}	-1.43×10^{-4}	6.89×10^{-2}	20.1
I	4.35×10^4	-4.96×10^{-3}	-4.35×10^4	-1.14×10^{-2}	8.74×10^{-8}	-1.22×10^{-4}	6.17×10^{-2}	19.0
M	-5.36×10^3	-4.39×10^{-3}	5.38×10^3	-8.83×10^{-3}	9.69×10^{-8}	-1.33×10^{-4}	6.59×10^{-2}	19.5
R	8.00×10^3	-1.46×10^{-3}	-7.98×10^3	-9.62×10^{-3}	9.89×10^{-8}	-1.36×10^{-4}	6.75×10^{-2}	19.2
L	2.37×10^4	-1.17×10^{-3}	-2.37×10^4	-7.10×10^{-3}	8.80×10^{-8}	-1.27×10^{-4}	6.68×10^{-2}	18.5

Table 4. Coefficients for fitting the thermal rewarming signals of the distal phalanges.

	Exponential grade 2				Polynomial grade 3			
	A	b	C	d	P1	P2	P3	P4
T	1.08×10^3	-6.56×10^{-4}	-1.06×10^3	-7.95×10^{-3}	1.13×10^{-7}	-1.57×10^{-4}	7.49×10^{-2}	19.6
I	1.83×10^4	8.94×10^{-4}	-1.83×10^4	-8.93×10^{-3}	1.19×10^{-7}	-1.65×10^{-4}	7.88×10^{-2}	18.6
M	3.10×10^3	4.17×10^{-4}	-3.09×10^3	-1.11×10^{-2}	1.22×10^{-7}	-1.66×10^{-4}	7.85×10^{-2}	18.8
R	-6.54×10^3	4.68×10^{-4}	6.56×10^3	-1.23×10^{-2}	1.25×10^{-7}	-1.69×10^{-4}	7.97×10^{-2}	18.7
L	2.31×10^5	3.22×10^{-4}	-2.31×10^5	-9.26×10^{-3}	1.16×10^{-7}	-1.61×10^{-4}	7.88×10^{-2}	17.9

Table 5 reports the errors calculated with Eq. 1 i.e. by calculating the MSE between the thermal rewarming signals and the data generated with the coefficients in Tables 3 and 4. Results in this table indicates that the exponential function describes the thermoregulation of the complete fingers better than the polynomial function. The opposite occurs with the distal phalanges, whose data fit better to the polynomial function. The overall results indicate that complete finger and distal phalanx regions can be modeled by different functions.

Table 6 shows the values of four physiological parameters of interest. These values are calculated at the complete fingers and at the distal phalanges. From left to right, the first two parameters are the mean and standard deviation of the ROI temperature along the entire image sequence. The third parameter is the percentage of total rewarming achieved by the ROI. This percentage is the

Table 5. Errors achieved by using the Eq. 1.

		Exponential grade 2	Polynomial grade 3
Complete finger	Thumb	148.33	283.13
	Index	141.12	243.88
	Middle	172.92	295.96
	Ring	169.90	276.55
	Little	168.51	233.68
Distal phalanx	Thumb	216.99	413.51
	Index	1837.08	465.72
	Middle	758.04	519.55
	Ring	279.25	505.38
	Little	295.06	426.43

ratio between the ROI temperatures measured in the last and first frames, of the image sequence. The fourth parameter is the ROI temperature measured at the 60% of the length of the total rewarming signal.

This table shows that the thumb fingers exhibited the higher temperatures and that distal phalanges of ring fingers exhibited the higher values of partial (at 60%) and total rewarming. In general, distal phalanges exhibited higher average temperature and rewarming percentages than complete fingers.

Table 6. Physiological parameters of the rewarming signals.

		$\mu \pm \sigma$	Rewarming (Rw) [%]	Temperature at 60% of Rw
Finger	Thumb	29.59 ± 3.39	94.42	20.42
	Index	28.57 ± 3.35	93.72	20.14
	Middle	28.84 ± 3.49	94.73	20.26
	Ring	28.87 ± 3.63	95.16	20.30
	Little	28.42 ± 3.81	94.54	20.15
Distal phalanx	Thumb	29.74 ± 3.63	95.67	20.47
	Index	29.20 ± 3.84	95.76	20.26
	Middle	29.32 ± 3.84	95.77	20.32
	Ring	29.40 ± 3.95	96.30	20.37
	Little	28.92 ± 4.10	95.05	20.26

Figures 7, 8 and 9 show boxplots for the data presented in Table 6. In the x-axis of these figures, the pairs 1–2, 3–4, 5–6, 7–8 and 9–10 refer to sets: distal phalanx - complete finger, of the thumb, index, middle, ring and little fingers, respectively.

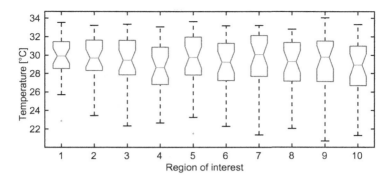

Fig. 7. Mean temperature achieved by the distal phalanges (odd numbers at x- axis) and complete fingers (even numbers at x-axis).

Figure 7 shows that, with exception of the thumb finger, the higher thermal variations occur at signals from the distal phalanges. This can be explained by the fact that in the acquired images, the thumbs does not appear perpendicular to the camera focal plane. An explanation for the higher length of the tails from signals of the distal phalanges is that such regions have higher microvascular density than regions more close to the proximal phalanx.

Figures 8 and 9 show that ring and middle fingers show the most stable thermoregulatory behavior. In agreement with the results in Fig. 7, the distal phalanges exhibit higher temperature variations. However, the Figs. 8 and 9 show that distal phalanges recover the temperature values previous to the cold stimulation, faster than in the complete fingers.

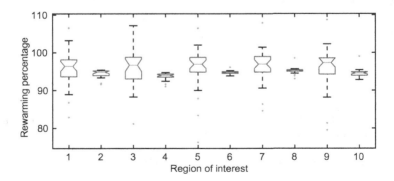

Fig. 8. Total rewarming achieved by the distal phalanges and complete fingers.

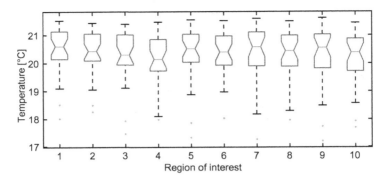

Fig. 9. Temperatures achieved by the distal phalanges and complete fingers at the 60% of the total rewarming.

5 Conclusions and Future Work

In all the image sequences analyzed it was possible to find regions inside the same finger with different thermoregulatory responses. This was verified after analyzing the thermoregulation at the complete fingers, and at distal phalanges. This conclusion indicates that the analysis of finger thermoregulation can have limited performance if considers the finger as a structure having a homogeneous thermoregulatory behavior. Thus, describing the thermoregulation at the level of the complete fingers could mask different thermoregulatory responses that occur along the finger geometry.

The analysis carried out were not focused on quantifying differences between the thermal evolution of the considered ROIs. However, the results indicated consistently that the fingertips exhibit a higher and faster thermoregulation than the remaining finger geometry. Future work should aim to characterize the finger thermoregulation using machine learning techniques, and to analyze pathological conditions that affect the finger thermoregulation.

References

1. Bouzida, N., Bendada, A., Maldague, X.P.: Visualization of body thermoregulation by infrared imaging. J. Therm. Biol. **34**, 120–126 (2009)
2. Wenger, C.B.: Human Responses to Thermal Stress. Technical report, Army Research Institute of Environmental Medicine (1996)
3. Wigley, F.M., Flavahan, N.A.: Raynaud's phenomenon. N. Engl. J. Med. **375**, 556–565 (2016)
4. Augustis, S., Saferis, V., Jost, W.: Autonomic disturbances including impaired hand thermoregulation in multiple system atrophy and Parkinson disease. J. Neural. Transm. **124**, 965–972 (2017)
5. Rosato, E., Giovanetti, A., Pisarri, S., Salsano, F.: Skin perfusion of fingers shows a negative correlation with capillaroscopic damage in patients with systemic sclerosis. J. Rheumatol. **40**, 98–99 (2013)

6. Moseley, G.L., Gallace, A., Ianetti, G.D.: Spatially defined modulation of skin temperature and hand ownership of both hands in patients with unilateral complex regional pain syndrome. Brain **135**, 3676–3686 (2012)
7. Zhang, H., He, Y., Wang, X., Shao, H., Mu, L., Zhang, J.: Dynamic infrared imaging for analysis of fingertip temperature after cold water stimulation and neurothermal modeling study. Comput. Biol. Med. **40**, 650–656 (2010)
8. Ismail, E., Orlando, G., Corradina, M.L., Amerio, P., Romani, G.L., Merla, A.: Differential diagnosis of Raynauds phenomenon based on modeling of finger thermoregulation. Physiol. Meas. **35**, 703–716 (2014)
9. Correa, A., Giraldo, J., Peña, R., Ramírez, L.M., Fandiño, H.A.: Base de datos de imágenes de termografía infrarroja para el análisis de la termorregulación de la mano. In: I Congreso Internacional de Ciencia e Ingeniería CICI (2016)
10. Lee, D.S., Yeom, S., Son, J.Y., Kim, S.H.: Automatic image segmentation for concealed object detection using the expectation-maximization algorithm. Opt. Express. **18**, 10659–10667 (2010)
11. Zapata-Osorio, N., Orrego-Serna, S., Ramirez-Arbelaez, L., Castro-Ospina, A., Fandiño-Toro, H.: Processing of thermal images oriented to the automatic analysis of hand thermoregulation. VII Latin American Congress on Biomedical Engineering CLAIB 2016, Bucaramanga, Santander, Colombia, October 26th -28th, 2016. IP, vol. 60, pp. 658–661. Springer, Singapore (2017). https://doi.org/10.1007/978-981-10-4086-3_165
12. Otsu, N.: A threshold selection method from gray-level histograms. IEEE Trans. Syst. Man Cyber. **9**, 62–66 (1979)
13. Jain, A.K.: Data clustering: 50 years beyond K-means. Pattern Recogn. Lett. **31**, 651–666 (2010)

Comparative Study of Feature Selection Methods for Medical Full Text Classification

Carlos Adriano Gonçalves[1,3], Eva Lorenzo Iglesias[1], Lourdes Borrajo[1], Rui Camacho[2,3(✉)], Adrián Seara Vieira[1], and Célia Talma Gonçalves[4,5]

[1] Computer Science Department, University of Vigo, Escola Superior de Enxeñería Informática, Ourense, Spain
[2] Faculdade de Engenharia da Universidade do Porto, Rua Dr. Roberto Frias s/n, 4200-465 Porto, Portugal
rcamacho@fe.up.pt
[3] LIAAD - INESC TEC, Campus da FEUP, Rua Dr. Roberto Frias s/n, 4200-465 Porto, Portugal
[4] CEOS.PP/ISCAP-P.PORTO, Rua Jaime Lopes Amorim s/n, 4465-004 Porto, Portugal
[5] LIACC, Rua Dr. Roberto Frias s/n, 4200-465 Porto, Portugal

Abstract. There is a lot of work in text categorization using only the title and abstract of the papers. However, in a full paper there is a much larger amount of information that could be used to improve the text classification performance. The potential benefits of using full texts come with an additional problem: the increased size of the data sets.

To overcome the increased the size of full text data sets we performed an assessment study on the use of feature selection methods for full text classification. We have compared two existing feature selection methods (Information Gain and Correlation) and a novel method called k-Best-Discriminative-Terms. The assessment was conducted using the Ohsumed corpora. We have made two sets of experiments: using title and abstract only; and full text.

The results achieved by the novel method show that the novel method does not perform well in small amounts of text like title and abstract but performs much better for the full text data sets and requires a much smaller number of attributes.

Keywords: Text classification · Feature selection · Medical texts corpus

1 Introduction

The increasing and overwhelming amount of scientific research documents available in bio-medicine scientific databases such as MEDLINE, requires the development of tools to help researchers to keep up with all the relevant work being done

© Springer Nature Switzerland AG 2019
I. Rojas et al. (Eds.): IWBBIO 2019, LNBI 11466, pp. 550–560, 2019.
https://doi.org/10.1007/978-3-030-17935-9_49

all over the world. Moreover the number of corpora of full scientific texts publicly available is also increasing rapidly. The availability of full texts may increase the chances for better analyses but also requires processing larger amounts of data. We argue that new and more powerful tools are required.

Our approach to the increase in the amount of information to process is a novel feature selection algorithm that achieves better results than existing competitors with a much smaller number of attributes. We have empirically assessed the performance of several feature selection algorithms conducting a series of experiments using: Information Gain, Correlation and the newly developed feature selection algorithm.

Text classifiers can adequately be used to extract medical/biological information from very large scientific papers repositories as shown in [20]. Techniques, like the one presented in this paper, can contribute to build better text classifiers and therefore extract better papers from those large repositories.

The rest of the paper is organized as follows. Section 2 makes an introduction to the feature selection methods focusing on the ones used in the present study. Section 3 presents the new algorithm for feature selection, k-Best-Discriminative-Terms. Section 4 will highlight the feature selection work applied to biomedical full-text document for classification. In Sect. 5 we present the results regarding our study and finally in Sect. 6 we draw the conclusions of the work described in the paper.

2   Feature Selection Methods for Text Classification

According to [2] the feature selection process or attribute reduction is the process of selecting a subset of features that best represents by itself all the data. The rational of feature selection in the context of text classification is to represent a document with a reduced number of highly representative/discriminative attributes.

The full-text document classification, specially in the biomedical domain, involves the manipulation of very large data sets. This brings several well-known problems such as the increase of the computation time. Besides that, not all the attributes are relevant and important for the classification task, which is another well-known problem that disturbs the performance of the classifiers.

We have adopted the bag-of-word approach, original documents are seen as a vector containing a huge number of words. Since we are working with a large collection of documents, the number of words increases quite dramatically, which entails memory and time restrictions to run learning algorithms. Due to the exposed situation it is seriously important to select the most important and relevant attributes for the classification process. That is the objective of Feature Selection algorithms.

According to [11] there are two main reasons for selecting some features over others. The first reason is related to the algorithm's performance, e.g., algorithms produce better results when not considering all the attributes. This is due to some attributes do not add more information instead they add noise, and removing them makes the classifier to perform better. The second reason is due to scalability, once a huge number of attributes demands computation power, memory, network bandwidth, storage, etc.) thus running a smaller subset decreases the computation time.

We have assessed the performance of three different feature selection methods:

1. Information Gain (IG)
2. Correlation (Corr)
3. k-Best-Discriminative-Terms (k-BDT)

The first two methods are now described and the k-BDT is presented in Sect. 3.

2.1   Information Gain

Information Gain (IG) is used to determine which attribute in a given set of training feature vectors is most useful for discriminating the class values to be learned [4,5].

IG is a "synonym" for *KullbackLeibler* divergence [14] and it is often used for ranking individual features [15].

In document classification, IG measures the number of bits of information gained, with respect to deciding the class to which a document belongs, by using each word frequency of occurrence in the document. However, IG only evaluates features in an individual manner.

IG is a feature selection method used prior and independent from the learning process, e.g., a filter method compares the computation score of each attribute and then selects the best attributes according to the highest scores [6].

Based on their comparative study of filter methods, [7] and [11] concluded that IG and Chi-Square (CHI) are among the most effective methods of feature selection for classification.

2.2   Correlation

According to the correlation algorithm an attribute is very relevant if it is highly correlated with the class, otherwise it is irrelevant [12].

We have used the WE*k*A *CorrelationAttributeEval* functionality, that evaluates the worth of an attribute by measuring its correlation (Pearson's Correlation) with the class. The WE*k*A *CorrelationAttributeEval* technique used requires a Ranker Search Method, that evaluates each attribute and lists the results in a ranked order.

3   k-Best-Discriminative-Terms

The rational of the k-BDT method is to find the best k terms[1] in the corpus that best discriminate the two classes of documents (assuming a binary classification

---

[1] We have used single words in our study but the k-BDT can also be used with other groupings of words like n-grams (n > 1), NERs, etc.

problem). In an informal description the documents are first separated by class value and, for each class value, the metric Tf×Df is computed for each term. This metric represents the average term frequency in the class value multiplied by the document frequency. The justification for Df is that we aim at terms that are frequent in all documents of each class value but infrequent in the "other class value". The k-BDT method is described in detail in Algorithm 1. The documents of the two class values are separate in lines 2 and 3. The documents from one of the class value (let say POS) are processed between lines 5 and 11. First the term frequency (Tf - line 7) and document frequency (Df - line 8) are computed for each term and document and then the Tf×Df is computed (line 9). Finally we compute the average Tf×Df for each term in lines 10–11. We repeat the same procedure for the other class value (NEGS) (lines 12–18). The "final values" are the difference between the Tf×Df of POS and the corresponding Tf×Df of the NEGS (lines 19–21). The final values are sorted by descending order (line 22) and the k first terms are returned in line 23. In Algorithm 1 (line 21) we have used the *abs* function but, as described in the next paragraph, we have also considered an alternative procedure.

Algorithm 1. k-Best-Discriminative-Terms algorithm

1: **procedure** K-BDTPROCEDURE(Corpus, k)
2:     Pos $\leftarrow$ relevantTexts(Corpus)
3:     Negs $\leftarrow$ irrelevantTexts(Corpus)
4:
5:     **for doc in POS do**
6:         **for term in doc do**
7:             $\text{Tf}_{term,doc} \leftarrow$ termFrequency(term, doc)
8:             $\text{Df}_{term} \leftarrow$ docFrequency(term, POS)
9:             $\text{Ti}_{value} = \text{Tf}_{term,doc} \times \text{Df}_{term}$
10:    **for term in allTerms do**
11:        $\text{T}_{posval} \leftarrow$ average(Ti_{value}, POSdocs)
12:    **for doc in NEGS do**
13:        **for term in doc do**
14:            $\text{Tf}_{term,doc} \leftarrow$ termFrequency(term, doc)
15:            $\text{Df}_{term} \leftarrow$ docFrequency(term, NEGS)
16:            $\text{Ti}_{value} = \text{Tf}_{term,doc} \times \text{Df}_{term}$
17:    **for term in allTerms do**
18:        $\text{T}_{negsval} \leftarrow$ average(Ti_{value}, NEGSdocs)
19:    $\text{T}_{values} = \emptyset$
20:    **for term in (POS $\cup$ NEGS) do**
21:        $\text{T}_{values} \leftarrow \text{T}_{values} \cup \{$ absValue(T_{posval} - $\text{T}_{negsval}$) $\}$
22:    sortInDecreasingOrder(T_{values})
23:    **Return** truncate(T_{values} , k)

Two Alternative Implementations

To choose the best k discriminative terms we have adopted and evaluated (Section 5) two alternative methods of computing the "final value" of each term (line 21). One approach, designated *abs*, sorts, in decreasing order, the absolute value of the difference between the Tf×Df value in the positives minus the value for the same term in the negatives, and chooses the best k of them. An alternative approach called *half-k* also computes, for each term, the differences of Tf×Df between the corresponding positive term and in the negative term but does not take the absolute value of their difference. It then chooses the $k/2$ terms achieving the most positive values (appear most frequently in relevant documents) and also the $k/2$ terms achieving the most negative values (appear most frequently in irrelevant documents). This late approach aims at making sure that representative terms from *positive* texts and *negative texts* are chosen.

Given a set of labeled examples, the goal of a classifier is to discriminate the elements of the different classes. K-BDT is based on a similar principle. K-BDT identifies terms that discriminate *relevant* documents from the *non-relevant* ones. It does that differently from the traditional *tf* × *idf* approach. In traditional text classification *tf* × *idf* promotes terms that are highly represented in a single document independent of the class it belongs. K-BDT promotes terms that are highly represented in a large number of documents of one of the classes[2] and at the same time rare in the documents of the other class. K-BDT promotes terms that are good at discriminating the two classes. Since K-BDT looks for terms highly represented in the whole set of documents of each class the experimental results show that we often need a small number of such terms to build a good classifier. This feature seems to be an advantage over the traditional *tf* × *idf* approach.

The k-BDT technique is suitable to be applied to a text classification problem in any domain and text corpus. There are no domain restrictions to the application of the technique.

4   Related Work on Feature Selection for Attribute Reduction in Full-Text Documents Classification

In the literature the work in feature selection is quite extensive, so we will highlight the feature selection work applied to the biomedical full-text document for classification purpose.

The recent work of [8] presents a study of the impact of feature selection on medical document classification. This study uses two data sets containing MEDLINE documents and makes a comparison between two different feature selections methods: the Gini Index and the Distinguish Feature Selector through two base classifiers: C4.5 decision tree and the Bayesian network. The authors also used documents from ten different disease categories for the experiments.

---

[2] It has been used in binary text classification but can also be adapted to non binary classification problems.

The authors concluded that the best accuracy results are a combination of the two proposed feature selection methods.

The authors in [9] present a novel method for attribute reduction using a data set of PubMed articles. The authors claim that achieved better results with their new method in terms of accuracy. The process involves a first phase of pre-processing the documents through the application of the tokenization, stemming and stop words removal. This new method is a variation of the Global Weighting Schema (GRW), that extracts unique terms from documents and these terms are weighted through the global weighting schema proposed.

The authors in [10] propose a group of scoring measures for feature selection using an SVM classifier and applied it to the OHSUMED corpus. The authors claim that the results achieved mixing their proposed scoring measures out-performed both Information Gain and Tf×IDf in some cases. According to the authors the proposed measures are more dependent of the distribution of the terms through the categories and also of the documents over the categories.

The work proposed in [16] presents a novel feature selection method to reduce the dimension of terms which takes into a new semantic space, between terms, based on the latent semantic indexing method. The idea is to appropriately capture the underlying conceptual similarity between terms and documents, which is helpful for improving the accuracy of text categorization.

Xu et al. [18] describe a work based on a very simple technique called Document Frequency thresholding (DF) that has shown to be one of the best methods in either Chinese or English text data. To improve DF Xu added the Term Frequency (TF) factor. The extended method called TFDF was tested on Reuters-21578 and OHSUMED corpora showed better results than the original DF method. Although we also use document frequency (Df), Xu approach is still quite different from the novel method reported in this paper. In Xu's work there is no concern to use directly a method that discriminates the class values by performing separate computations on each class value set of documents. Document Frequency thresholding (DF) is also a different definition than the Df used and defined in this paper.

An extensive survey on text categorization techniques can be found in [19].

5   Experimental Work and Results

Methods

The empirical evaluation was done using the OHSUMED corpus [13]. We have used five OHSUMED data sets for which we manage to collect the full texts. With that corpus we have "created" two corpus: the original corpus with the full text papers; and a corpus with the same papers but with just title and abstract. For each corpus there are five data sets (c04, c06 c14, c20 and c23) that are characterized in Table 1 for title and abstract and Table 2 for full text.

We have performed three sets of experiments. We first conducted an experiment to estimate the best values of k for the title and abstract data sets and for the full text data sets. Secondly and with the best values of k for title and

Table 1. Characterization of the data sets in the Ohsumed corpus (Title+Abstract).

Data set id	Number of relevant papers	Number of non-relevant papers
c04	2630	7755
c06	1220	8430
c14	2550	8030
c20	1220	8239
c23	3952	6778

Table 2. Characterization of the data sets in the Ohsumed corpus (full text).

Data set id	Number of relevant papers	Number of non-relevant papers
c04	5598	5598
c06	256	1582
c14	343	399
c20	239	1553
c23	683	719

abstract we have compared the performance of the three feature selection algorithms in the title and abstract corpus. Lastly and using the best values of k for full text, we have compared the algorithms in the full data set corpus.

For the experiments we have used the Support Vector Machine (SVM) algorithm from Weka [17]. A 10-fold Cross Validation procedure was used as the evaluation method. The values used for k were set to 10, 50 and 100 for the title and abstract corpus and 50, 100, 500, 1000 and 1500 for the full text corpus. Both alternative implementation (*abs* and *half-k*) were used in the assessment of the novel approach.

For the purpose of our work and concerning the Information Gain feature selection method we have used a threshold of $1.00e^{-10}$ that was a value used in a previous work [1]. In the Normalize component we have used the nominal representation.

The metric used for the evaluation of the classifiers performance was the F-measure. The F-measure value combines precision and recall, where precision is the percentage of classifications that are correct and recall is the percentage of classifications actually made by the classifier. F-measure is computed as the harmonic average of the precision and recall. The best performance of a classifier on a classification task is when the F-measure has value 1 (perfect precision and recall) and its worst performance is when the F-measure is 0.

Results

Table 3 shows the results of the experiments to assess the impact of parameter k and the two alternative methods to choose the attributes in k-BDT method. We can see from those results that the novel method does not perform well in data sets that have a small number of terms. Looking at the term-doc matrix we see a very large amount of zeros, the matrix is very sparse. There is a low probability to find a frequent term common to a large number of documents of each class value.

Table 4 shows the results of the experiments to assess the impact of parameter k and the two alternative methods to choose the attributes in k-BDT method on the full text data sets. The results are completely the opposite of the results with title and abstract. The f-measure values are well above the reference values in all data sets.

Concerning the second set of experiments we have obtained the results shown in Table 5. The results in the table show that in the case of using only title and abstract the novel method is much worse than its competitors.

Concerning the third set of experiments we have obtained the best results with the novel method in all data sets. Table 6 shows the best results of the experiments to compare the study's feature selection methods on the full text corpus. We can see that in all data sets the novel method achieves performances

Table 3. Choosing the values of k together with the best of *abs* or *half-k* alternatives. The title and abstract corpus was used. k = 100 was the best value among all alternatives tested for both *abs* and *half-k*.

Data set id	k value	method	F-measure
c04	base line value		0,899(0,009)
c04	100	abs	0,703(0,01)-
c04	100	half-k	0,703(0,01)-
c06	base line value		0,936(0,008)
c06	100	abs	0,82(0,002)-
c06	100	half-k	0,82(0,002)-
c14	base line value		0,907(0,009)
c14	100	abs	0,681(0,002)-
c14	100	half-k	0,681(0,002)-
c20	base line value		0,92(0,008)
c20	100	abs	0,823(0,005)-
c20	100	half-k	0,812(0,001)-
c23	base line value		0,715(0,012)
c23	100	abs	0,528(0,009)-
c23	100	half-k	0,528(0,009)-

Table 4. Choosing the values of k together with the best of *abs* or *half-k* alternatives. The full text corpus was used. k values are the best ones for each *abs* and *half-k* among other values tested.

Data set id	k value	method	F-measure
c04	base line value		0,888(0,01)
c04	1500	abs	0,965(0,005)+
c04	1500	half-k	0,965(0,005)+
c06	base line value		0,856(0,02)
c06	1000	abs	0,945(0,014)+
c06	100	half-k	0,951(0,014)+
c14	base line value		0,799(0,049)
c14	1000	abs	0,944(0,029)+
c14	1500	half-k	0,941(0,028)+
c20	base line value		0,873(0,021)
c20	1500	abs	0,951(0,016)+
c20	1500	half-k	0,95(0,017)+
c23	base line value		0,629(0,033)
c23	1500	abs	0,83(0,03)+
c23	1500	half-k	0,826(0,034)+

Table 5. Comparison of the feature selection methods on the corpus using only title and abstract. Cells of the table contain the average and standard deviation of F-measure of a 10-fold cross validation. IG stands for information Gain. k-BDT stands for k Best Discriminative Terms. '+' means that the value is statistically significantly better than the base line value. Base line values can be found in Table 3.

Data set id	IG	Correlation	k-BDT
c04	0,915(0,009)+	0,893(0,009)-	0,703(0,01)-
c06	0,951(0,007)+	0,936(0,008)~	0,82(0,002)-
c14	0,923(0,008)+	0,907(0,009)~	0,681(0,008)-
c20	0,941(0,007)+	0,92(0,008)	0,823(0,005)-
c23	0,752(0,01)+	0,715(0,012)~	0,528(0,009)-

well above the base line value and better than the competitors. In data set c06 and using the half-k version of the k-BDT method we need only 100 attributes to achieve a very good performance.

Table 6. Comparison of the feature selection methods on the data sets using full text. Cells of the table contain the average and standard deviation of F-measure of a 10-fold cross validation. IG stands for information Gain. k-BDT stands for k Best Discriminative Terms. '+' means that the value is statistically significantly better than the base line value. Base line values can be found in Table 4.

Data set id	IG	Correlation	k-BDT
c04	0,895(0,009)+	0,888(0,01)+	0,96(0,005)+
c06	0,913(0,018)+	0,856(0,02)~	0,951(0,014)+
c14	0,877(0,033)+	0,799(0,049)~	0,944(0,025)+
c20	0,919(0,019)+	0,873(0,021)~	0,951(0,016)+
c23	0,742(0,036)+	0,629(0,033)~	0,83(0,03)+

6 Conclusions

In this paper we have presented and empirically evaluated a novel feature selection method. The method is based on the idea of finding terms that are frequent in the documents of one of the class values and infrequent in the other class values. We have compared the novel method with too other feature selection approaches for title and abstract and for full-text document classification.

The results of the novel method are much better than its competitors in all full text data sets used. However, the novel method seems to be inadequate for data sets using title and abstract only.

The results suggest that the novel method requires a very small number of attributes to achieve good performances. In one of the data sets used in the study, the novel method just need 100 attributes to achieve the best performance among the competitors.

Acknowledgements. This work was supported by the Consellería de Educación, Universidades e Formación Profesional (Xunta de Galicia) under the scope of the strategic funding of ED431C2018/55-GRC Competitive Reference Group. This work was also partially funded by the ERDF through the COMPETE 2020 Programme within project POCI-01-0145-FEDER-006961, and by National Funds through the FCT as part of project UID/EEA/50014/2013.

References

1. Gonçalves, C.A., Iglesias, E.L., Borrajo, L., Camacho, R., Vieira, A.S., Gonçalves, C.T.: LearnSec: a framework for full text analysis. In: de Cos Juez, F., et al. (eds) Hybrid Artificial Intelligent Systems HAIS 2018, vol. 10870, pp. 502–513. Springer, Cham (2018). https://doi.org/10.1007/978-3-319-92639-1_42
2. Saeys, Y., Inza, I., Larrañaga, P.: A review of feature selection techniques in bioinformatics. Bioinformatics **23**, 2507–2517 (2007)
3. Markov, A.A., Nitussov, A.Y., Voropai, L., Link, D., Custance, G., Mahoney, M.S.: Classical Text in Translation: An Example of Statistical Investigation of the Text Eugene Onegin Concerning the Connection of Samples in Chains (2006)

4. Borasem, P.N., Kinariwala, S.A.: Image re-ranking using information gain and relative consistency through multigraph learning (2016)
5. Vieira, A.S., Iglesias, E.L., Borrajo, L.: An HMM-based text classier less sensitive to document management problems. Bioinformatics **11**, 503–515 (2016)
6. Mladenic, D., Grobelnik, M.: Feature selection for unbalanced class distribution and Naive Bayes. In: 16th International Conference on Machine Learning (ICML), pp. 258–267. Morgan Kaufmann Publishers, San Francisco (1999)
7. Yang, Y., Pedersen, J.O.: A comparative study on feature selection in text categorization. In: Fourteenth International Conference on Machine Learning, pp. 412–420. Morgan Kaufmann Publishers Inc., San Francisco (1997)
8. Parlak, B., Uysal, A.K.: The impact of feature selection on medical document classification. In: 11th Iberian Conference on Information Systems and Technologies (CISTI), pp. 1–5 (2016)
9. Imambi, S.S., Sudha, T.: Article: a novel feature selection method for classification of medical documents from Pubmed. Int. J. Comput. Appl. **26**(9), 29–33 (2011)
10. Monta, E., Ranilla, J., Fernandez, J., Combarro, E.F., Diaz, I.: Scoring and selecting terms for text categorization. IEEE Intell. Syst. **20**, 40–47 (2005)
11. Forman, G.: Feature selection for text classification. In: Liu, H., Motoda, H. (eds.) Computational Methods of Feature Selection, Data Mining and Knowledge Discoveries Series, pp. 257–276. Chapman and Hall/CRC, Boca Raton (2007)
12. Hall, M.A., Smith, L.A.: Feature selection for machine learning: comparing a correlation-based filter approach to the wrapper. In: Proceedings of the Twelfth International Florida Artificial Intelligence Research Society Conference, pp. 235–239. AAAI Press (1999)
13. Hersh, W.R., Buckley, C., Leone, T.J., Hickam, D.H.: Ohsumed: an interactive retrieval evaluation and new large test collection for research. In: 17th Annual International ACM SIGIR Conference on Research and Development in Information Retrieval. ACM Press (1994)
14. Zdravevski, E., Lameski, P., Kulakov, A., Filiposka, S., Trajanov, D., Boro, J.: Parallel computation of information gain using Hadoop and MapReduce. In: Federated Conference on Computer Science and Information Systems (2015)
15. Shang, C., Li, M., Feng, S., Jiang, Q., Fan, J.: Feature selection via maximizing global information gain for text classification. J. Know.-Based Syst. **54**, 298–309 (2013)
16. Wang, F., Li, C., Wang, J., Xu, J., Li, L.: A two-stage feature selection method for text categorization by using category correlation degree and latent semantic indexing. J. Shanghai Jiaotong Univ. (Sci.) **20**(1), 44–50 (2015)
17. Hall, M., Frank, E., Holmes, G., Pfahringer, B., Reutemann, P., Witten, I.H.: The WEKA data mining software: an update. SIGKDD Explor. Newsl. **11**(1), 10–18 (2009)
18. Xu, Y., Wang, B., Li, J.T., Jing, H.: An extended document frequency metric for feature selection in text categorization. In: Li, H., Liu, T., Ma, W.-Y., Sakai, T., Wong, K.-F., Zhou, G. (eds.) AIRS 2008. LNCS, vol. 4993, pp. 71–82. Springer, Heidelberg (2008). https://doi.org/10.1007/978-3-540-68636-1_8
19. Sebastiani, F.: Machine learning in automated text categorization. ACM Comput. Surv. **34**, 1–47 (2002)
20. Talma Gonçalves, C., Camacho, R., Oliveira, E.: BioTextRetriever: a tool to retrieve relevant papers. Int. J. Knowl. Discov. Bioinform. **2**(3), 21–36 (2011)

Estimation of Lung Properties
Using ANN-Based Inverse Modeling
of Spirometric Data

Adam G. Polak[(⊠)] [iD], Dariusz Wysoczański [iD],
and Janusz Mroczka [iD]

Wrocław University of Science and Technology, B. Prusa Str. 53/55,
50-317 Wrocław, Poland
adam.polak@pwr.edu.pl

Abstract. Spirometry is the most commonly used test of lung function because
the forced expiratory flow-volume (FV) curve is effort-independent and simultaneously sensitive to pathological processes in the lungs. Despite this, a method
for the estimation of respiratory system parameters, based on this association,
has not been yet proposed. The aim of this work was to explore a feedforward
neural network (FFNN) approximating the inverse mapping between the FV
curve and respiratory parameters. To this end, the sensitivity analysis of the
reduced model for forced expiration has been carried out, showing its local
identifiability and the importance of particular parameters. This forward model
was then applied to simulate spirometric data (8000 elements), used for training,
validating, optimizing and testing the FFNN. The suboptimal FFNN structure
had 52 input neurons (for spirometric data), two hidden nonlinear layers with 30
and 20 neurons respectively, and 10 output neurons (for parameter estimates).
The total relative error of estimation of individual parameters was between 11
and 28%. Parameter estimates yielded by this inverse FFNN will be used as
starting points for a more precise local estimation algorithm.

Keywords: Spirometry · Inverse model · Feed-forward neural network

1 Introduction

Spirometry is the most commonly used test of lung function, consisting of a maximally
strong exhalation preceded by a calm and deep inspiration. The recorded forced
expiratory flow-volume (FV) curve is effort-independent and significantly sensitive to
pathological processes in the lungs [1]. The routine approach is to calculated a set of
spirometric indexes form FV data, relate them to predicted values, and then to use such
determinants for diagnosing and monitoring respiratory disorders.

It seems that direct inferring about specific lung properties from spirometry data
would provide more benefits in clinical practice. Nevertheless, a method for the
effective estimation of respiratory system parameters, based on the aforementioned
relationship, has not been yet proposed. Certainly, this approach should be based on a
physico-mathematical model that couples respiratory mechanics with the recorded data.
Such computational models, incorporating morphological data and main physiological

© Springer Nature Switzerland AG 2019
I. Rojas et al. (Eds.): IWBBIO 2019, LNBI 11466, pp. 561–572, 2019.
https://doi.org/10.1007/978-3-030-17935-9_50

phenomena, have been fortunately published. They can be gathered into two groups: incorporating symmetrical [2, 3] or asymmetrical structure of the bronchial tree [4, 5], with specific features characterizing subsequent airway generations. Because heterogeneous and homogenous changes of the same level in small airways have an indistinguishable effect on the FV curve [6], it follows that symmetric model is sufficient to properly represent the respiratory system during forced expiration.

The early study on the mathematical model for forced expiration, applying techniques of sensitivity analysis and linear programming, exposed the main difficulties in its identification, resulting from a large number of parameters and their collinear impacts [7]. Despite this, Lambert and coworkers were partially successful when predicting individual FV curves, first manually manipulating airway areas, and then using the simulated annealing algorithm [8, 9]. It has been shown, however, that the set of key parameters determining forced expiratory flow is much bigger than the maximal airway areas adjusted in these studies [7, 10]. Since then, some additional effort to solve this inverse problem had been undertaken and finally a preliminary model for forced expiration with a set of identifiable parameters was proposed [11, 12].

Estimating model parameters from empirical data belongs to the class of inverse problems that, in case the governing relationship is ambiguous or there exists a strong correlation between the effects of parameters on the output, can be ill-posed and numerically ill-conditioned [13]. The typical approach to the identification of complex and nonlinear models consists in solving the optimization problem (i.e. in the minimization of the objective function – the distance between the model output and data), and this task should be divided into two stages because of a usually unknown and possibly complex topology of the objective function: first – global identification yielding approximate estimates of parameters near the global minimum, and then – a more accurate local estimation method. There are many popular global optimization techniques used at the first stage. The majority of them require a huge number of objective function evaluations (including model simulations), which significantly extends the time needed to obtain a result, so they are impractical in clinical applications. On the other hand, any multidimensional continuous function can be arbitrary well approximated using a feed-forward neural network (FFNN) [14, 15], including the inverse mapping [16, 17] – such as the relationship between spirometric data and lung parameters. Admittedly, FFNN training requires a huge amount of data in multidimensional problems and is time-consuming, moreover, the solution found may be a local, not the global one. On the other hand, trainings is done off-line and the global solution can be searched by multiple initializations of neural weights – both as the preprocessing steps. In return, the evaluation of the trained inverse FFNN with any new measurement data is very fast.

The aim of this work was to implement, train, optimize, and test a FFNN approximating the inverse mapping between the FV curve and respiratory parameters. In the paper, we present a revised process of deriving an identifiable forward model, its sensitivity analysis showing the importance of particular model parameters, generation of synthetic spirometric data used later for training, validating and testing the FFNN, finding the suboptimal FFNN structure, and finally, the evaluation of accuracy of this inverse model. Parameter estimates yielded by the trained inverse FFNN will be used as starting points for a more precise local method of estimation.

2 Methods

2.1 Analysis of the Forward Model for Forced Expiration

Reduced Model. The applied computational model for forced expiration was descri-bed elsewhere [3]. Breathily, the model includes the symmetrical bronchial tree con-sisting of 24 generations [18]. The mechanical properties of the airways are specified independently for each generation and include parameters describing the dependence of the airway lumen area (A) on transmural pressure (P_{tm}):

$$A(P_{tm}) = \begin{cases} A_m \alpha_0 (1 - P_{tm}/P_1)^{-n_1}, & P_{tm} \leq 0, \\ A_m[1 - (1 - \alpha_0)(1 - P_{tm}/P_2)^{-n_2}], & P_{tm} > 0, \end{cases} \tag{1}$$

where A_m is the maximal airway area, α_0 is the normalized airway area at $P_{tm} = 0$, α_0' is the slope of α_0 at $P_{tm} = 0$ (neutral compliance), $P_1 = n_1 \alpha_0 / \alpha_0'$, $P_2 = n_2(\alpha_0 - 1)/\alpha_0'$, and n_1 and n_2 are shape-adjusting exponents [2]. In addition, a nonlinear characteristics for the lung elastic recoil pressure (P_{st}) is used with its own parameters:

$$P_{st}(V_L) = \begin{cases} \frac{V_L - V_0}{C_{st}}, & V_L \leq V_{tr}, \\ \frac{V_m - V_{tr}}{C_{st}} \cdot \ln\left(\frac{V_m - V_{tr}}{V_m - V_L}\right) + \frac{V_{tr} - V_0}{C_{st}}, & V_L > V_{tr}, \end{cases} \tag{2}$$

where V_L is the lung volume, V_m, V_{tr}, and V_0 are the maximal, transition and minimal volumes, and C_{st} is the lung compliance defining the linear part of $P_{st}(V_L)$ character-istics [19]. The model has 155 parameters in total: 144 describe the properties of airway generations and 11 correspond to the lung recoil characteristics and other features of the respiratory system. Because vital capacity (VC) is assessed directly from the FV data, 154 of the parameters should be considered as unknowns.

Taking into account all essential physical phenomena and features influencing significantly the MEFV curve, the model was preliminary reduced by using sigmoidal and linear functions scaling the airway parameters along the bronchial tree relative to the generation number. Finally, the reduced model was described by 25 parameters, with 16 determining solely the descending part of the MEFV curve [11]. To further exclude the less influential and intercorrelated parameters, the estimate sensitivities to measurement disturbances were analyzed, enabling the selection of parameters based on the computed sensitivity vectors. In effect, the mathematical model for the descending part of the MEFV curve, preserving the previous complex structure of computations, was proposed, including 9 identifiable parameters that describe prop-erties of the intrapleural airways and the lung elastic recoil characteristics [12]. Because airway lengths are not correlated with a patient's height [20] (as it was assumed in the previous approach) and simultaneously the dimensional ratio of succeeding airway generations is essentially constant [21], in this work the lengths can be freely modified by an additional scalar coefficient – the 10th model parameter (see Table 1).

Scaling Functions and Ranges of Parameter Variability. The actual values of mechanical properties of a given airway generation in the complex model are calculated

Table 1. Parameters of the reduced model and the assessed ranges of their variability (VC – vital capacity, volumes in L, lung compliance in L/kPa).

Range of variability	Parameter									
	p_a	p_l	p_{z1}	p_{z2}	p_{c1}	p_{c2}	ΔV_0	ΔV_{tr}	ΔV_m	C_{st}
Lower limit	0.5	0.7	−0.26	0.0	−0.36	−6.0	−0.5	0.0	VC + 0.01	2
Upper limit	1.7	1.3	0.26	3.8	0.36	6.0	0.0	VC − 0.5	VC + 0.50	10

by multiplying their baseline values (representing the normal lung) by scaling factors. Comparing to the above preliminary study, an additional effort has been made to select more carefully the scaling functions and to assess the variability of model parameters. Now, the sigmoidal functions f_s scaling α_0 and α'_0 have the general form:

$$f_s(g) = \frac{f_u(g)}{1 + \exp(p_{s1}g + p_{s2})} + f_l, \qquad (3)$$

where f_u and f_l set the upper and lower asymptotes, g is the airway generation number, and p_{s1} and p_{s2} are the parameters of the reduced model. The nominator f_u for α_0 and α'_0 is: $(0.039\ g + 1.1)$ and 2, and the lower limit f_l: $(-0.017\ g + 0.5)$ and 0.1, respectively. These limits, used to scale parameters representing the normal lung, were deduced from published data. According to [22, 23], airway diameters of normal subjects vary in the range of ±30% around the mean, so the limits for maximal areas A_m (parameter p_a) were set approximately to 0.7^2 and 1.3^2, i.e. [0.5–1.7]. Similarly, the variability of airway lengths (parameter p_l) was evaluated as [0.7–1.3] [20]. The main impact of chronic respiratory diseases on airway mechanics is their narrowing and stiffening. This can be properly mimicked by scaling α_0 (parameters p_{z1} and p_{z2}) in the range [0.5–1.1] for the central intrapulmonary airways and in [0.1–2.0] for the small airways, together with α'_0 (parameters p_{c1} and p_{c2}) being in the range [0.1–2.0] [10, 24]. Finally, the ranges of individual parameters p_s were set in such a way that the scaling factors fall above the 5% of the lower and below the 95% of the upper asymptote. The variability of parameters representing the lung recoil characteristics: ΔV_0, ΔV_{tr} and ΔV_m (relevant volumes related to residual volume, RV), was determined as meeting the natural constraints: $\Delta V_0 < \Delta V_{tr} < VC < \Delta V_m$ and by comparing a wide sort of such real characteristics [25] with the output of this part of the model (Table 1).

Sensitivity Analysis and Estimation Accuracy. The importance of individual parameters and the approximate accuracy of their estimation can be assessed *a priori* using the technique of sensitivity analysis [26]. Because the reduced model is non-linear, the results of sensitivity analysis depend on the parameters' values. Taking the above into account, the analysis was carried out for six states of the respiratory system, characteristic for healthy and diseases lungs: (i) normal state, (ii) central intrapulmonary airways narrowed, (iii) small airways narrowed, (iv) lung recoil decreased accompanying by hyperinflation, (v) central airways narrowed and lung recoil decreased, and (vi) small airways narrowed and lung recoil decreased. First, the model output (samples of expiratory flow **Q**) sensitivity (**X**) to model parameters (**θ**) was computed numerically by increasing and decreasing succeeding parameter values by $\Delta\theta = 2.5\%$:

$$\mathbf{X} = \frac{\partial\mathbf{Q}}{\partial\boldsymbol{\theta}} = [\mathbf{x}_1\ \mathbf{x}_2 \cdots \mathbf{x}_{10}],$$

$$\mathbf{x}_k = \frac{\mathbf{Q}(\theta_k + \Delta\theta_k) - \mathbf{Q}(\theta_k - \Delta\theta_k)}{2\Delta\theta_k}. \tag{4}$$

Then $\mathbf{X}$ was recalculated into the relative sensitivity ($\mathbf{X}_d$) of parameter estimators ($\hat{\boldsymbol{\theta}}$) to random errors in measurement data ($\mathbf{e}$), assuming the minimization of the least squares criterion [27]:

$$\mathbf{X}_d = \mathbf{R}^{-1}\frac{\partial\hat{\boldsymbol{\theta}}}{\partial\mathbf{e}} = \mathbf{R}^{-1}\left(\mathbf{X}^{\mathrm{T}}\mathbf{X}\right)^{-1}\mathbf{X}^{\mathrm{T}}, \tag{5}$$

where $\mathbf{R}$ is a diagonal matrix containing the vector of parameter variability ranges $\mathbf{r}$ (see Table 1). Large numbers in $\mathbf{X}_d$ result from both the estimators sensitivity to measurement errors and correlations between the sensitivity vectors in $\mathbf{X}$. This is why the scalar measure of insignificance of individual parameters m can be defined as the norm of adequate vectors from $\mathbf{X}_d$:

$$m_k = \|\mathbf{x}_d(k)\|_2. \tag{6}$$

The larger m_k, the worse determination of parameter θ_k by the data, and the bigger total error of its estimation. Since the sensitivity vectors depend on the state of a nonlinear system, the average insignificance of parameters may be assessed as root-mean-squares (m_{rms}) of m_k representing analyzed states. In this work, the results of analysis of 6 states of the respiratory system were used to calculate the finally parameter's relative importance expressed by its weight w_k. The weights (rounded to integer values) were computed by dividing the maximal m_{rms} by individual $m_{rms}(k)$.

It may happen that the estimation of only the best determined parameters and the assignment of typical values to the most unimportant ones will reduce the overall error of estimation [27]. For a given case, this can be checked by successively eliminating the most unimportant parameters from estimation and recalculating $\mathbf{X}_d$ at each step, and simultaneously assessing the random, systematic and total error of estimation. During this procedure, $\boldsymbol{\theta}$ is divided into the sets: $\boldsymbol{\beta}$ of estimated parameters and $\boldsymbol{\lambda}$ of parameters with assigned typical values (analogously, $\mathbf{X}$ is divided into $\mathbf{X}_\beta$ and $\mathbf{X}_\lambda$). Then the covariance matrix $\mathbf{S}_\beta$ of estimated parameters is given by

$$\mathbf{S}_\beta = \sigma_e^2\left(\mathbf{X}_\beta^T\mathbf{X}_\beta\right)^{-1} \tag{7}$$

(with the variances of parameter estimator σ_β^2 on the diagonal), and their biases $\mathbf{b}_\beta$ by

$$\mathbf{b}_\beta = \left(\mathbf{X}_\beta^T\mathbf{X}_\beta\right)^{-1}\mathbf{X}_\beta^T\mathbf{X}_\lambda\Delta\boldsymbol{\lambda}, \tag{8}$$

where σ_e^2 is the variance of measurement random errors, and $\Delta\lambda$ is the vector of maximal differences between the true and assigned values of unestimated parameters [27] (assumed as the half of their ranges in this work). Finally, the relative random error d_r (from the variances of parameter estimators), relative systematic error d_s (from estimator biases) as well as total relative error of estimation d_{tot} are given as follows [12]:

$$
\begin{aligned}
d_r(s) &= \sqrt{\frac{1}{s}\sum_{k=1}^{s}\frac{\sigma_\beta^2(k)}{r_k^2}} \cdot 100\%, \\
d_s(s) &= \sqrt{\frac{1}{s}\sum_{k=1}^{s}\frac{b_\beta^2(k)}{r_k^2}} \cdot 100\%, \\
d_{tot}(s) &= \sqrt{d_r^2(s)+d_s^2(s)},
\end{aligned}
\tag{9}
$$

where s is the number of estimated parameters. The standard deviation of measurement noise of the available pneumotachometer (*Jaeger-Toennies*, model 709569), recorded with the DAQ card (*Keithley*, KPCI-3108), was computed as $\sigma_e \approx 0.01$ L/s.

2.2   Simulation of Synthetic Spirometric Data

The forward model for forced expiration is described by 10 free parameters with assed ranges of their variability. To generate synthetic spirometric data (descending parts of FV curves) appropriately representing different states of the respiratory system, necessary to train and test the inverse FFNN, the following procedure was applied. At the beginning, the sex of a virtual subject was randomly chosen as female (F) or male (M). Then her/his height and age were randomly drawn (uniform distribution) from the ranges 1.50–1.80 cm (F) or 1.60–1.90 cm (M), and 25–70 years. These data were used to calculate the predicted values of VC, as well as the basic spirometric indices for this subject, as forced expiration volume in 1 s (FEV$_1$), Tiffeneau index (FEV$_1$/FVC), forced mid-expiratory flow between 25 and 75% of FVC (FEF$_{25-75\%}$) and peak expiratory flow (PEF) [28]. Then, the whole FV curve was simulated in the range of VC (100 evenly spread samples) using the values of model parameters randomly drawn (uniform distribution) from their variability ranges (the target vector for the inverse FFNN). Finally, the first volume after PEF (V_{max}) was found and 50 samples of the descending part, evenly distributed between V_{max} and VC, were computed again. The simulated flow data were then supplemented with the values of V_{max} and VC, forming the input vector for the inverse FFNN. Because some combinations of randomly chosen parameter values may be unrealistic, returning unnatural FV curves, additionally the spirometric indices were calculated from the simulated data and compared with the upper 95 percentiles of the predicted values for the given subject. When any of these limits had been exceeded or PEF had been lesser than 2 L/s, such a FV curve was rejected. The above procedure was repeated 8000 times, resulting the sets of corresponding input (52 × 8000) and target (10 × 8000) data.

2.3   Training and Optimization of the Inverse FFNN

The FFNN with 52 input neurons, two hidden layers and 10 linear output neurons was implemented using MATLAB Neural Network Toolbox (R2017a, *MathWorks*). Hidden neurons had tan-sig transfer functions. The spirometric data generated using the forward model were divided into three sets: training, validation and test ones, in proportion 0.7:0.15:0.15 (5600:1200:1200 elements). The FFNN input data were left in their original ranges: [0.042–1.96] (L) for two volume samples and $[1.01 \times 10^{-4} - 11.2]$ (L/s) for flow samples. Since the absolute values of parameters vary by several orders, the target data (expected parameter values) were scaled by their ranges of variability (Table 1). Additionally, the error weights, equal to the previously assessed parameter weights **w**, were assigned to the corresponding FFNN outputs to enhance the accuracy of estimation of the most significant parameters. The FFNN was trained using the Levenberg–Marquardt backpropagation algorithm with the initial value of the regularization parameter $\mu = 10^{-6}$. The validation set was used to stop training before overfeeding the network (after 10 unsuccessful steps), with the network performance evaluated by the mean squared errors (MSE). The final quality of trained FFNN was assed individually for each element from the test set by the calculation of the relative error of parameter estimation:

$$\delta_k = \frac{\hat{\theta}_k - \theta_k}{r_k} \cdot 100\%, \tag{10}$$

where $\hat{\theta}_k$ is the kth parameter estimate returned by FFNN, θ_k is the parameter value used in the forward model during data simulation, and r_k is its variability range. In the end, the mean m_δ and standard deviation s_δ of relative estimation errors were computed for each parameter, as well as the total relative error of estimation:

$$\varepsilon_{tot} = \sqrt{m_\delta^2 + s_\delta^2}. \tag{11}$$

The final stage was the attempt to find the optimal FFNN structure in terms of the number of hidden neurons and to increase the chance that the inverse model was globally fitted to the data. This was done by training neural networks characterized by a combination of 10 to 40 neurons in the first and second hidden layer (with the step of 5 neurons), as well as by 30-times random initialization of network weights for each case. The structure of FFNN that yielded the least MSE for the validation set was further trained 100 times with randomly initialized network weights, and the one characterized by the smallest MSE was ultimately considered the best inverse mappings between the spirometric data and the parameters of the reduced model for forced expiration in this study. Its properties were evaluated using the test set. Because such a complex inverse model, being only an approximation of the true mapping, may occasionally return parameter estimates laying outside their variability ranges, such FFNN outputs were additionally corrected before calculating δ_k.

3   Results

The sensitivity analysis carried out for six chosen states of the respiratory system (examples of the results corresponding to two of them are shown in Fig. 1) revealed that the smallest total relative error would be reached when 10 (5 cases) or 9 (1 case) parameters were estimated. In all the analyzed cases, all inverted matrices $\mathbf{X}_\beta^T \mathbf{X}_\beta$ (see Eqs. 7 and 8) had non-zero determinants. Basing on these outcomes, the remaining work focused on estimating all of 10 parameters of the reduced model. In addition, the results also indicated that the total relative error of estimation (as defined by (9)) should be at the level of 2–10%.

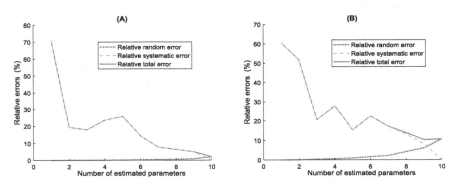

Fig. 1. Assessed relative errors of estimation for different numbers of estimated parameters: (A) healthy lungs, (B) small airways narrowed and lung recoil decreased

The simulation procedure was used to generate 8000 spirometric curves fulfilling the imposed criteria, and then to build the training, validation and test sets using these data. The adequate FV curves and inputs from the test set are illustrated in Fig. 2.

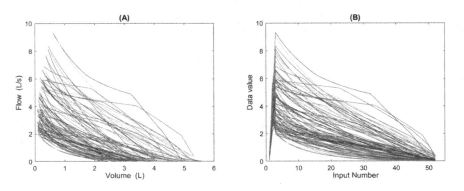

Fig. 2. (A) Descending parts of first 100 spirometric curves from the test set, (B) relevant FFNN input data

The FFNN characterized by the best performance in terms of MSE = 0.213 for the validation set was found after a few dozens of hours (using 10 threads of Intel Core i7-6950X in parallel @3.0 GHz on a PC with 32 GB RAM) and had 30 and 20 neurons in the first and second hidden layer, respectively (Fig. 3). This performance was obtained after training the network for 119 epochs.

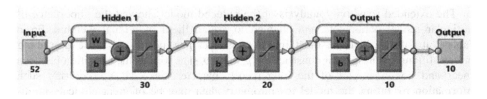

Fig. 3. Structure of the FFNN characterized by the best performance

The results of this best FFNN evaluation in terms of estimation accuracy, obtained by means of Eqs. (10) and (11), are collected in Table 2 together with the error weights applied during network training. The estimated parameter sets characterized by the smallest and largest mean square errors in relation to their target values have been used to compute the corresponding FV curves (descending parts), and these curves are compared with the ones representing the respective target parameters in Fig. 4.

Table 2. Error weights used during FFNN training (w) and statistical measures of relative errors of estimation: mean (m_δ), standard deviation (s_δ) and total error (ε_δ)

Weights and errors	Parameters									
	p_a	p_l	p_{z1}	p_{z2}	p_{c1}	p_{c2}	ΔV_0	ΔV_{tr}	ΔV_m	C_{st}
w	2	1	2	3	12	25	4	13	4	4
m_δ (%)	0.29	−1.10	−1.34	0.05	0.20	−0.59	0.59	0.55	0.50	1.60
s_δ (%)	18.6	28.4	25.9	23.7	19.9	16.0	10.8	12.2	17.2	17.3
ε_{tot} (%)	18.7	28.4	25.9	23.7	19.9	16.0	10.8	12.2	17.2	17.3

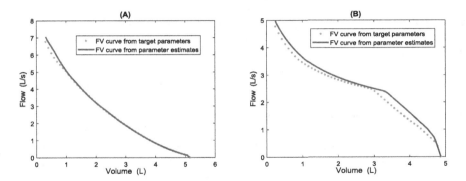

Fig. 4. Comparison of the FV curves representing target parameters from the test set with the curves computed from their estimates characterized by the smallest (A) and largest errors (B)

4 Discussion and Conclusion

The aim of this work was to analyze the performance of a FFNN used to approximate the inverse mapping between spirometric data and respiratory parameters. The role of such FFNN is to return parameter estimates enabling rough, albeit global fitting the model for forced expiration to the spirometry results. Obtained parameter estimates will then be used as starting points for a more precise local estimation method.

The extended sensitivity analysis of the reduced model showed the importance of particular model parameters and enabled to assign them individual weights. From Table 2 it follows that the most important model parameters are those describing airway stiffening during their constriction (p_{c1} and p_{c2}) as well as the transition between linear and nonlinear part of the lung recoil characteristics (ΔV_{tr}). Acquiring such information by fitting the model to spirometry data may be of great clinical significance. Moreover, it appears that all 10 parameters should be estimated to ensure the smallest estimation errors in most cases, and that the entire model is at least locally identifiable – the feature guaranteed by the non-zero determinant of the inverted information matrix [29]. The total relative errors of estimation in the range of 2–10% have been assessed assuming that effects other than measurement noise are negligible in relation to the factors included in the analyzes. In real experiments, such factors as model inadequacy (too simple description) and other disturbances (e.g. body movements during spirometry), usually make them larger.

The FFNN with two hidden layers was chosen to approximate the inverse relationship between the descending part of spirometric data and the model parameters because of its potential to capture such mapping with arbitrary accuracy, the capability of generalization and very fast computations using a fitted network. On the other hand, its optimal structure (in terms of the number of neurons in the hidden layers) should be found, as well as the global fitting to the data achieved. Searching the optimal structure by training networks with different numbers of hidden neurons and securing its global matching to data by repeatedly initializing neuronal weights for a given structure is very time-consuming. Additionally, the time of FFNN learning depends also on the number of training data. To deal with the problem, a limited number of training data was used (8000), and the optimization was done using the suboptimal enumeration method. In consequence, a suboptimal solution has been found, so the trained FFNN realizes the specified mapping with limited accuracy. This is visible in the estimation errors gathered in Table 2. The means of errors $\mathbf{m}_\delta$ are relatively small – they do not exceed 1.6% of parameter ranges – documenting that the trained FFNN has captured the main features of the inverse mapping. On the other hand, the standard deviations of errors $\mathbf{s}_\delta$ are much higher – between 11 (for ΔV_0) and 28% (for p_l), being the main components of the total errors $\mathbf{\varepsilon}_\delta$. This feature indicates that the inverse mapping performed by this FFNN is not entirely complete. Such an observation is not surprising when considering the dimensionality of the analyzed problem. To evenly probe the space spanned by 10 parameters at only 3 points in each dimension, one should take into account $3^{10} \approx 6 \times 10^4$ samples – the amount considerably greater than 5600 data used to train the FFNN in this study. Despite this, the accuracy obtained with this inverse FFNN is comparable to that yielded by a simulated annealing algorithm, where about $(7.0 + 1.5) \times 10^3$ simulations of the model output were needed for each analyzed FV curve (results not presented here). Nevertheless, some additional attempts can be undertaken in the future to possibly improve the performance of FFNN, as further reducing the problem dimensionality, trying out FFNNs with other transfer functions and other learning algorithms, using another type of artificial neural network, or using even larger sets of synthetic data.

Concluding, this study has revealed the local identifiability of the proposed reduced model for forced expiration with 10 parameters and that the FFNN with two hidden layers is able to learn the inverse mapping between spirometric data and lung properties with a relative precision of a dozen of percent. The trained FFNN can provide parameter estimates immediately, unlike typical optimization algorithms requiring a large number of model evaluations. The final decision as to the suitability of this approach will be possible after applying a local identification algorithm starting from the estimates yielded by the FFNN and analyzing its effects. Future work will concern additional reduction of free parameters (based on lung structure optimality) and using the Levenberg-Marquardt algorithm for more precise parameter estimation.

Acknowledgments. This work was supported by the grant no. 2016/21/B/ST7/02233 from the National Science Centre, Poland.

References

1. Pellegrino, R., Viegi, G., Brusasco, V., et al.: Interpretative strategies for lung function tests. Eur. Respir. J. **26**(5), 948–968 (2005)
2. Lambert, R.K., Wilson, T.A., Hyatt, R.E., Rodarte, J.R.: A computational model for expiratory flow. J. Appl. Physiol. Respir. Environ. Exerc. Physiol. **52**(1), 44–56 (1982)
3. Polak, A.G.: A forward model for maximum expiration. Comput. Biol. Med. **28**(6), 613–625 (1998)
4. Polak, A.G., Lutchen, K.R.: Computational model for forced expiration from asymmetric normal lungs. Ann. Biomed. Eng. **31**(8), 891–907 (2003)
5. Hedges, K.L., Tawhai, M.H.: Simulation of forced expiration in a biophysical model with homogeneous and clustered bronchoconstriction. J. Biomech. Eng. **138**(6), 061008 (2016)
6. Polak, A.G., Wysoczański, D., Mroczka, J.: In silico study on the impact of heterogeneous narrowing of small airways on spirometry results. Eur. Respir. J. **50**(suppl. 61), PA3008 (2017)
7. Lambert, R.K.: Sensitivity and specificity of the computational model for maximal expiratory flow. J. Appl. Physiol. Respir. Environ. Exerc. Physiol. **57**(4), 958–970 (1984)
8. Lambert, R.K., Castile, R.G., Tepper, R.S.: Model of forced expiratory flows and airway geometry in infants. J. Appl. Physiol. **96**(2), 688–692 (2004)
9. Lambert, R.K., Beck, K.C.: Airway area distribution from the forced expiration maneuver. J. Appl. Physiol. **97**(2), 570–578 (2004)
10. Morlion, B., Polak, A.G.: Simulation of lung function evolution after heart-lung transplantation using a numerical model. IEEE Trans. Biomed. Eng. **52**(7), 1180–1187 (2005)

11. Mroczka, J., Polak, A.G.: Reduced model for forced expiration and analysis of its sensitivity. In: Feng, D.D., Dubois, O., Zaytoon, J., Carson, E. (eds.) Modelling and Control in Biomedical Systems 2006 (Including Biological Systems), pp. 159–164. Elsevier, Oxford (2006)

12. Mroczka, J., Polak, A.G.: Selection of identifiable parameters from the reduced model for forced expiration. In: Magjarevic, R., Nagel, J.H. (eds.) World Congress on Medical Physics and Biomedical Engineering. IFMBE Proceedings, vol. 14, pp. 764–768. Springer, Berlin (2006). https://doi.org/10.1007/978-3-540-36841-0_180

13. Tikhonov, A.N., Goncharsky, A., Stepanov, V.V.: Numerical Methods for the Solution of Ill-Posed Problems. Kluwer, London (1995)

14. Hornik, K.: Approximation capabilities of multilayer feedforward networks. Neural Netw. 4(2), 251–257 (1991)

15. Kůrková, V.: Kolmogorov's theorem and multilayer neural networks. Neural Netw. 5(3), 501–506 (1992)

16. Ramuhalli, P., Udpa, L., Udpa, S.S.: Neural network-based inversion algorithms in magnetic flux leakage nondestructive evaluation. J. Appl. Phys. 93(10), 8274–8276 (2003)

17. Kabir, H., Wang, Y., Yu, M., Zhang, Q.J.: Neural network inverse modeling and applications to microwave filter design. IEEE Trans. Microw. Theory Tech. 56(4), 867–879 (2008)

18. Weibel, E.R.: Morphometry of the Human Lung. Springer, Berlin (1963). https://doi.org/10.1007/978-3-642-87553-3

19. Bogaard, J.M., Overbeek, S.E., Verbraak, A.F.M., et al.: Pressure-volume analysis of the lung with an exponential and linear-exponential model in asthma and COPD. Eur. Respir. J. 8(9), 1525–1531 (1995)

20. Kim, D., Son, J.S., Ko, S., Jeong, W., Lim, H.: Measurements of the length and diameter of main bronchi on three-dimensional images in Asian adult patients in comparison with the height of patients. J. Cardiothorac. Vasc. Anesth. 28(4), 890–895 (2014)

21. Canals, M., Novoa, F.F., Rosenmann, M.: A simple geometrical pattern for the branching distribution of the bronchial tree, useful to estimate optimality departures. Acta. Biotheor. 52(1), 1–16 (2004)

22. Hannallah, M.S., Benumof, J.L., Ruttimann, U.E.: The relationship between left mainstem bronchial diameter and patient size. J. Cardiothorac. Vasc. Anesth. 9(2), 119–121 (1995)

23. Majumdar, A., et al.: Relating airway diameter distributions to regular branching asymmetry in the lung. Phys. Rev. Lett. 95(16), 168101 (2005)

24. Brown, R.H., Mitzner, W.: Effect of lung inflation and airway muscle tone on airway diameter in vivo. J. Appl. Physiol. 80(5), 1581–1588 (1996)

25. Baldi, S., Miniati, M., Bellina, C.R., et al.: Relationship between extent of pulmonary emphysema by high-resolution computed tomography and lung elastic recoil in patients with chronic obstructive pulmonary disease. Am. J. Respir. Crit. Care Med. 164(4), 585–589 (2001)

26. Thomaseth, K., Cobelli, C.: Generalized sensitivity functions in physiological system identification. Ann. Biomed. Eng. 27(5), 607–616 (1999)

27. Polak, A.G.: Indirect measurements: combining parameter selection with ridge regression. Meas. Sci. Technol. 12(3), 278–287 (2001)

28. Quanjer, P.H., Tammeling, G.J., Cotes, J.E., et al.: Lung volumes and forced ventilatory flows. Eur. Respir. J. 6(suppl 16), 5–40 (1993)

29. Dötsch, H.G., Van den Hof, P.M.: Test for local structural identifiability of high-order non-linearly parametrized state space models. Automatica 32(6), 875–883 (1996)

Model of the Mouth Pressure Signal During Pauses in Total Liquid Ventilation

Jonathan Vandamme, Mathieu Nadeau, Julien Mousseau, Jean-Paul Praud, and Philippe Micheau[✉]

Inolivent, Universite de Sherbrooke, 2500 Blvd Universite,
Sherbrooke, QC J1K2R1, Canada
Philippe.Micheau@USherbrooke.ca
http://www.inolivent.ca

Abstract. Total liquid ventilation (TLV) is an innovative experimental method of mechanical ventilation in which lungs are totally filled with a breathable perfluorochemical liquid (PFC). The main objective is to develop a method to estimate the alveolar pressure from a pressure mouth measurement during pause in liquid ventilation. Experimental results show that the measured mouth pressure is disturbed by disturbances (damped oscillations due to fluid-structure tube resonances and cardiogenic oscillation). Numerical analysis of P_Y allow to obtain a fractional-order model of $\alpha = 0.7$ for the alveolar pressure.

Keywords: Liquid ventilation · Pause · Oscillation · Hankel-HSVD · Fractional order

1 Introduction

1.1 Context

In the last decade, a vast array of preclinical studies have shown the efficacy and safety of total liquid ventilation (TLV) in pediatric and adult animal models (Wolfson and Shaffer 2005). In TLV, the lungs are totally filled with a breathable perfluorocarbon (PFC) liquid and then ventilated with a tidal volume of PFC controlled by a dedicated liquid ventilator (Costantino et al. 2009). TLV offers many advantages over conventional mechanical ventilation (CMV). By eliminating the air-liquid interface, it allows the recruitment of collapsed lung regions ensuring more homogeneous alveolar ventilation (Wolfson and Shaffer 2005), an efficient lung lavaging effect (Avoine et al. 2011) and the capability to induce ultra-fast mild therapeutic hypothermia (Tissier et al. 2007). The liquid ventilator Inolivent-6 (Nadeau et al. 2014) includes a pressure mode to control the inspired and expired liquid (Micheau et al. 2011). Such pressure mode requires the measurement of the positive end-inspiratory pressure ($PEIP$) and the positive end-expiratory pressure ($PEEP$). A reference $PEEP$ value is set by the operator in order to keep the alveolar pressure stable. It is assumed that

© Springer Nature Switzerland AG 2019
I. Rojas et al. (Eds.): IWBBIO 2019, LNBI 11466, pp. 573–581, 2019.
https://doi.org/10.1007/978-3-030-17935-9_51

the PEEP control regulates the amount of liquid in the lungs at the end of expiration. The specified reference PEEP value is used by the control algorithm of the liquid ventilator to correct the inspired and expired volumes at each respiratory cycle (Micheau et al. 2011). For such pressure control mode of total liquid ventilation, accurate PEEP estimation is necessary for the control algorithm. Also, an accurate PEIP estimation is required to monitor the maximun alveolar pressure. The objective of this paper is to propose a model of the mouth pressure signal during pauses in total liquid ventilation.

2   Experimental Data

2.1   Experimental Set up

Figure 1 illustrates the PFC circuit of the liquid ventilator Inolivent 6. It comprises two independent piston pumps, respectively for inspiration and expiration. Each of these pumps is connected to the Y-piece by means of flexible Tygon tubing. They are four controlled pinch valves to direct the flow from the pumps to the lung or from the pump to the reservoir and oxygenator. During the inspiratory and expiratory pauses, the pinch valves 1 and 2 are closed.

The ventilator is connected to the endotracheal tube (ET tube) via the Y-connector. The four pinch valves are programmed to guide the liquid flow to the lung. During the inspiration phase, the valves (4b) and (4d) are open, and the valves (4a) and (4c) are closed. The inspiratory pump (Insp. pump) inserts the PFC through the endotracheal tube to the lung. During the expiration phase, the valves (4b) and (4d) are closed, and the valves (4a) and (4c) are open. The expiratory pump (exp. pump) withdraws the PFC through the endotracheal tube from the lung. During the ventilation, the pressure sensor (P sensor mouth) at the Y-connector is used to acquire the P_Y signal. This pressure sensor is positioned at the same height as the trachea on the PFC circuit.

The animal experiments were done according to a protocol approved by our institutional Ethics Committee for Animal Care and Experimentation. The lambs, weighing 3.25 ± 0.39 kg (mean $\pm$ SD), were premedicated, anesthetized, intubated, curarized and instrumented. The liquid ventilation was set with I:E ratio at 1:3 and ventilation minute at 150 ml min^{-1} kg^{-1}. Two end-inspiratory and two end-inspiratory pauses lasting at least 30 s were executed during different phases (cooling, warming, PEEP increasing and PEEP decreasing).

2.2   Typical Mouth Pressure

Figure 2 shows a typical mouth pressure signal (P_Y) recorded during inspiratory and expiratory pauses at the Y piece (Fig. 1). The pause was extended to 5 s to highlight the problems. The sudden flow changes happening when the pumps are stopped and the tubes are pinched by the valves, trigger resonances on the flow measured at the Y piece. The use of viscoelastic Tygon tubes with a distensibility greater than the compressibility of the PFC causes complex interactions between

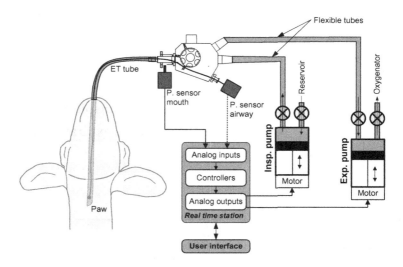

Fig. 1. PFC circuit in liquid ventilator Inolivent 6.0

the elastic tube wall and the pressure waves propagating within the fluid. At low frequencies, an acoustic plane wave, called Korteweg's wave, progagates within the PFC and is tightly coupled to the tube membrane vibrations. The dispersion and attenuation of these pressure waves strongly depend on the viscoelastic properties of the tube wall, and hence on the PFC temperature (Gautier et al. 2007).

The measured pressure, P_Y, is mostly hindered by the flexible tube resonances. It is also at the onset of respiratory pauses that P_{alv} varies the most rapidly. The second observation concerns the low-amplitude undamped oscillations present throughout the whole pause. Those perturbations seem correlated to the heart rate and are thought to be cardiogenic oscillations generated by the pulmonary blood flow (Tusman et al. 2009), or by the heart beating against the lungs.

3   Analytical Parametric Model of the P_Y Signal

3.1   Hypotheses About the Model

Two hypotheses are considered to develop a model of the measured pressure P_Y. The first hypothesis is to assume that $P_{alv}(t)$ is a slow time relaxation phenomena that can be modeled as a constant value plus a sum of exponentially damped responses. The second hypothesis is to assume the disturbance pressure signal as a sum of exponentially damped and undamped poles.

The numerical model of $P_Y(t)$ is composed of the alveolar pressure $P_{alv}(t)$ plus a disturbance model. The pressure disturbance model is represented by pairs of complex conjugated poles for the flexible tube resonance and a periodic cardiogenic signal of pulsation ω_c.

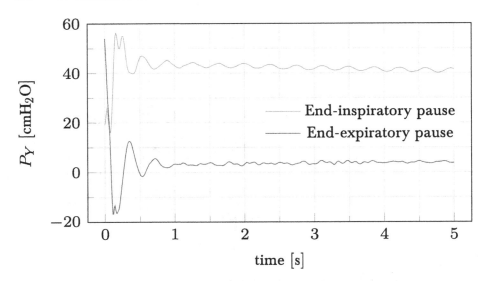

Fig. 2. Typical Y piece pressure measured during end-inspiratory and end-expiratory pauses

4   Numerical Analysis of P_Y

According to the numerical model, it is assumed that the measured signal can be approximated on a finite time interval ($t \in [0; T]$) by a linear combination of exponentials with complex-valued exponents:

$$P_Y(t) = \sum_{n=1}^{N} c_n \exp(-s_n t) \tag{1}$$

where $t = 0$ coincides with the beginning of the respiratory pause, $s_n = \sigma_n + j2\pi f_n$ is a pole (a complex frequency), and $\sigma_n = 2\pi f_n \zeta_n$.

The final value of the steady state alveolar pressure (when $t \to \infty$) is c_0 at the null pole $s_n = 0$. The relaxation phenomena of the alveolar pressure can be modeled by c_n associated with a null frequency, $f_n > 0$, and a strictly positive real part value $\sigma_n > 0$. The cardiogenic oscillations are undamped oscillations can be modeled by c_n associated with poles of null real part value, $\sigma_n = 0$. The damped oscillations due to tube resonances are represented with poles of strictly positive real part value, $\sigma_n > 0$.

The noniterative Hankel-SVD algorithm is used to estimate the values of c_n and s_n (Barkhuijsen et al. 1987) for a typical P_Y signal during a long inspiratory pause of 20 s.

In order to accurately model the signal, a series of $N = 30$ terms was used, giving a mean squared error of $0.021\,\mathrm{cmH_2O^2}$.

An arbitrary frequency of $f = 0.5\,\mathrm{Hz}$ was chosen to separate the frequencies associated to the slowly changing alveolar pressure (very low frequencies).

The 10 poles satisfying this condition are associated to the slowly varying pressure. The remaining 20 poles, from which the green curve was associated to the disturbances superimposed to the alveolar pressure.

In order to illustrate the method, the right panel of Fig. 4 shows the poles on the z-plane. The black dots located in the circumference of the unit circle are related to the damped resonance and the cardiogenic pulsation, while the red dots close to 0 Hz represent the poles related to the alveolar pressure.

Because of small heart rate variations, a large number of poles is needed to accurately model the cardiogenic pulsation throughout the whole respiratory pause. In fact, 16 poles are needed to represent the cardiogenic pulsation (with a damping ratio below 0.05). Their poles with the largest amplitudes lie around 2.4 Hz (144 beats per minute), which is close to the fundamental frequency of the resonance.

Finally, Table 1 gives a summary of the main components provided by the analysis. The first two real poles describe the alveolar pressure decay. The next two pairs of complex conjugated poles represent the first two harmonics of the flexible tube resonances. It is interesting to note that the second harmonic's natural frequency is almost three times the first mode, which is in agreement with the theoretical frequency mode of a resonating tube closed at one extremity. However, for end-expiratory pauses, only the first mode frequency is noticeable. Finally, the last complex conjugated poles represent the main component of the cardiogenic oscillations, lying around 2.4 Hz (144 beats per minute), which is close to the fundamental frequency of the resonance.

$$P_Y(t) = P_{alv}(t) + A_c \cos(\omega_c t + \phi_c) + \sum_i A_i \exp(-\sigma_i t) \cos(\omega_i t + \phi_i) \quad (2)$$

Table 1. Main components of HSVD decomposition

f_n (Hz)	σ_n (s^{-1})	ζ_n (-)	$|c_n|$ (cmH$_2$O)	$\angle c_n$ (rad)
0	0.0024	-	32.78	0
0	0.70	-	2.91	0
8.40	10.17	0.19	15.58	2.75
2.74	3.73	0.22	6.74	1.87
2.39	0.14	0.0096	1.42	1.53

4.1   Model of Alveolar Pressure Dynamics

The signal reconstructed from the estimated values, denoted by $\widehat{P}_{alv}$, is shown in red on Fig. 3 ($n \in \Omega$ such that small damped frequencies are included, $f < 0.5\,\text{Hz}$).

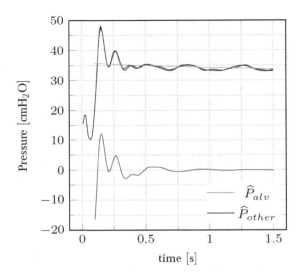

Fig. 3. Signal reconstructed from the estimated values

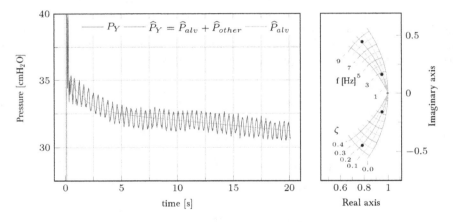

Fig. 4. Prony's analysis using HSVD of P_Y during a 20 s end-inspiratory pause. The right panel pictures the analyzed poles on the z-plane grid of constant damping factors and natural frequencies.

$$\widehat{P}_{alv}(t) = \sum_{n \in \Omega} \widehat{c}_n \exp(-\widehat{s}_n t) \tag{3}$$

The identified parameters $\widehat{c}_n$ describing the relaxation of the alveolar pressure experimentally obtained from the P_Y analysis can not be simply used in a parametric analytical model due to the lack of clear links with physiological evidences.

Hence, a dedicated alveolar pressure relaxation model in TLV is proposed to consider the respiratory system's elasticity.

Viscoelastic materials exhibiting power-law stress relaxation are ruled by fractional derivatives and possess a long memory. It is thus necessary to account for the volume history before the respiratory pause in order to provide a *representative* relaxation trace during it (Di Paola et al. 2014). On the other hand, only an acceptable behavioral model, so the volume history can be limited to the inspiration/expiration of the tidal volume V_T in τ seconds previous to the pause. The consideration of a longer volume history was deemed unjustifiable due to the required increase in model complexity. Hence, the pressure response after a volume ramp is applied is given by

$$P_{alv}(t) = B\left[(t+\tau)^\alpha - t^\alpha\right] + P_e, \tag{4}$$

where $B = QV_T/[\tau\Gamma(\alpha+1)]$ and $P_e = EV_T + P_0$ are two constants, P_0 is the alveolar pressure before the volume variation ($PEEP_s$ or $PEIP_s$).

The parameters V_T and τ are known. The parameters B, α and P_e are identified using a weighted constrained nonlinear least squares minimization routine, MATLAB's `lsqnonlin` function [The MathWorks, Natick (MA), USA].

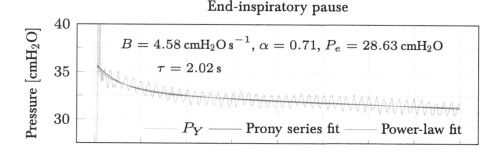

End-inspiratory pause

$B = 4.58\,\mathrm{cmH_2O\,s^{-1}}$, $\alpha = 0.71$, $P_e = 28.63\,\mathrm{cmH_2O}$

$\tau = 2.02\,\mathrm{s}$

P_Y ——— Prony series fit ——— Power-law fit

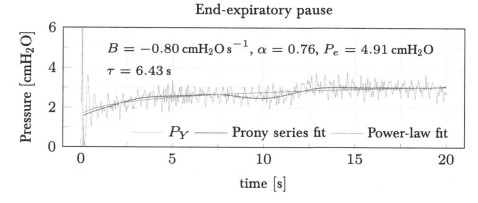

End-expiratory pause

$B = -0.80\,\mathrm{cmH_2O\,s^{-1}}$, $\alpha = 0.76$, $P_e = 4.91\,\mathrm{cmH_2O}$

$\tau = 6.43\,\mathrm{s}$

P_Y ——— Prony series fit ——— Power-law fit

Fig. 5. Alveolar pressure power-law fit versus Prony series estimation during extended end-inspiratory and end-expiratory pauses (Color figure online)

The parameter P_e was constrained to $[P_N - 5\ P_N + 5]$ and the other parameters were let free, i.e. $B \in [-\infty\ \infty]$ and $a \in [0\ 1]$. A weighing window was applied to make the fitting procedure less sensitive to the damped oscillations which are not symmetric; it was a ramp from 0 to 1 during the first 0.5 s and a constant for the rest of the pause. The values of the parameters for the inspiratory and expiratory pauses are presented on the Fig. 5.

It can be seen on Fig. 5 that both the long-time behaviour as well as the rapid pressure decrease at the beginning of the pause are well captured by the power-law fit during an extended end-inspiratory pause lasting 20 s.

The last observation relates to the slow decrease of P_Y during end-inspiratory pauses (red curve) and its slow increase during end-expiratory pauses (blue curve).

It can be seen that the pressure decreased by $\approx 4\,cmH_2O$ during the 20 s extended end-inspiratory pause. The relaxation rate is slower than the decay of an exponential, in fact, the pressure response to a step volume change decreases almost perfectly linearly with the logarithm of time. This relaxation process is typical of a fractional order dynamic with a wide and continuous distribution of relaxation time constants, such as biological tissues (Hildebrandt 1969).

This model can include the combined effect of the static lung and chest recoils (Suki et al. 1994), the viscoelasticity behavior model of fractional order α due to lung viscoelasticity properties of the lung parenchyma (Suki et al. 2005) or fluid dynamics in the fractional lung tree (Beaulieu et al. 2012).

5   Conclusion

From the typical pressure signal, it appears that the measured pressures are clearly disturbed by damped oscillations due to fluid-structure tube resonances (Beaulieu et al. 2012) and cardiogenic oscillations.

The alveolar pressure response is well described by a fractional order model of α close to 0.7, the observed phenomena could also include fluid redistribution or other yet not identified phenomena.

The numerical model can be used to develop a numerical method to estimate the alveolar pressure from a P_Y measurement during pause in spite of large dynamic disturbances. An accurate alveolar pressure appears solvable on a recorded signal of 5 s, but it becomes very challenging within the first second record of the signal due to the large dynamic of the disturbances (up to 40 cmH2O).

References

Avoine, O., et al.: Total liquid ventilation efficacy in an ovine model of severe meconium aspiration syndrome. Crit. Care Med. **39**(6), 1097–1103 (2011)

Barkhuijsen, H., de Beer, R., van Ormondt, D.: Improved algorithm for noniterative time-domain model fitting to exponentially damped magnetic resonance signals. J. Magn. Reson. (1969) **73**(3), 553–557 (1987)

Beaulieu, A., Bossé, D., Micheau, P., Avoine, O., Praud, J.-P., Walti, H.: Measurement of fractional order model parameters of respiratory mechanical impedance in total liquid ventilation. IEEE Trans. Biomed. Eng. **59**(2), 323–331 (2012)

Costantino, M.L., Micheau, P., Shaffer, T.H., Tredici, S., Wolfson, M.R.: Clinical design functions: round table discussions on the bioengineering of liquid ventilators. ASAIO J **55**(3), 206–208 (2009). 6th Internationl Symposium on Perfluorocarbon Application, and Liquid Ventilation

Di Paola, M., Fiore, V., Pinnola, F., Valenza, A.: On the influence of the initial ramp for a correct definition of the parameters of fractional viscoelastic materials. Mech. Mat. **69**, 63–70 (2014)

Gautier, F., Gilbert, J., Dalmont, J.-P., Pico Vila, R.: Wave propagation in a fluid filled rubber tube: theoretical and experimental results for Korteweg's wave. Acta Acustica United Acustica **93**(3), 333–344 (2007)

Hildebrandt, J.: Comparison of mathematical models for cat lung and viscoelastic balloon derived by laplace transform methods from pressurevolume data. Bull. Math. Biophys. **31**(4), 651–667 (1969)

Micheau, M., et al.: A Liquid ventilator prototype for total liquid ventilation preclinical studies. In: Progress in Molecular and Environmental Bioengineering - From Analysis and Modeling to Technology Applications, p. 646. Intech (2011)

Nadeau, M., et al.: Core body temperature control by total liquid ventilation using a virtual lung temperature sensor. IEEE Trans. Biomed. Eng. **61**(12), 2859–2868 (2014)

Suki, B., Barabasi, A.-L., Lutchen, K.: Lung tissue viscoelasticity: a mathematical framework and its molecular basis. J. Appl. Physiol. **76**(6), 2749–2759 (1994)

Suki, B., Ito, S., Stamenovi, D., Lutchen, K., Ingenito, E.: Invited review: biomechanics of the lung parenchyma: critical roles of collagen and mechanical forces. J. Appl. Physiol. **98**(5), 1892–1899 (2005)

Tissier, R., Hamanaka, K., Kuno, A., Parker, J.C., Cohen, M.V., Downey, J.M.: Total liquid ventilation provides ultra-fast cardioprotective cooling. J. Am. Coll. Cardiol. **49**(5), 601–605 (2007)

Tusman, G., et al.: Pulmonary blood flow generates cardiogenic oscillations. Respir. Physiol. Neurobiol. **167**(3), 247–254 (2009)

Wolfson, M.R., Shaffer, T.H.: Pulmonary applications of perfluorochemical liquids: ventilation and beyond. Paediatr. Respir. Rev. **6**(2), 117–127 (2005)

Author Index

Printed in the United States
By Bookmasters